STUDY GUIDE and
STUDENT SOLUTIONS MANUAL

College PHYSICS

Fifth Edition

BY SERWAY & FAUGHN

JOHN R. GORDON
James Madison University

CHARLES TEAGUE
Eastern Kentucky University

RAYMOND A. SERWAY
North Carolina State University

Harcourt College Publishers
Fort Worth Philadelphia San Diego New York Orlando Austin
San Antonio Toronto Montreal London Sydney Tokyo

Address for Domestic Orders
Saunders College Publishing, 6277 Sea Harbor Drive, Orlando, FL 32887-6777
800-782-4479

Address for International Orders
International Customer Service
Harcourt, Inc., 6277 Sea Harbor Drive, Orlando, FL 32887-6777
407-345-3800
(fax) 407-345-4060
(e-mail) hbintl@harcourt.com

Address for Editorial Correspondence
Saunders College Publishing, Public Ledger Building, Suite 1250,
150 S. Independence Mall West,
Philadelphia, PA 19106-3412

Web Site Address
http://www.harcourtcollege.com

Printed in the United States of America

ISBN: 0-03-022484-5

0 1 2 3 4 5 6 7 8 9 202 15 14 13 12 11 10 9 8 7 6

Preface

This Student Solutions Manual and Study Guide has been written to accompany the textbook **College Physics**, Fifth Edition, by Raymond A. Serway and Jerry S. Faughn. The purpose of this Study Guide is to provide the students with a convenient review of the basic concepts and applications presented in the textbook, together with solutions to selected end-of-chapter problems from the textbook. The Study Guide is not an attempt to rewrite the textbook in a condensed fashion. Rather, emphasis is placed upon clarifying typical troublesome points, and providing further drill in methods of problem solving.

Each chapter of the Study Guide is divided into several parts, and every textbook chapter has a matching chapter in the Study Guide. Very often, reference is made to specific equations or figures in the textbook. Every feature of the Study Guide has been included to insure that it serves as a useful supplement to the textbook. Most chapters contain the following sections:

- **Notes From Selected Chapter Sections:** This is a summary of important concepts, newly defined physical quantities, and rules governing their behavior.

- **Equations and Concepts:** This represents a review of the chapter, with emphasis on highlighting important concepts and describing important equations and formalisms.

- **Suggestions, Skills, and Strategies:** This offers hints and strategies for solving typical problems that you will often encounter in the course. In some sections, suggestions are made concerning mathematical skills that are necessary in the analysis of problems.

- **Review Checklist:** This is a list of topics and techniques that you should master after reading the chapter and working the assigned problems.

- **Solutions to Selected End-of-Chapter Problems:** Solutions are shown for approximately twenty percent of the problems from each chapter of the text. Problems were selected to illustrate important concepts in each chapter.

- **Chapter Self-Quiz:** This is a set of a twelve multiple-choice questions at the end of each chapter. They can be used to assess your understanding of the concepts presented in the chapter. Answers to the chapter self-quiz questions are given following the Self-Quiz for Chapter 30

We sincerely hope that this Study Guide will be useful to you in reviewing the material presented in the text, and in improving your ability to solve problems and score well on exams. We welcome any comments or suggestions which could help improve the content of this study guide in future editions; and we wish you success in your study.

John R. Gordon,
James Madison University
Harrisonburg, VA 22807

Charles D. Teague,
Eastern Kentucky University
Richmond, KY 40475

Raymond A. Serway,
North Carolina State University
Raleigh, NC 27650

Acknowledgments

It is a pleasure to acknowledge the excellent work of Michael Rudmin of Diversified Service Company—Publishing, whose attention to detail in the preparation of the camera-ready copy did much to enhance the quality of this fifth edition of the Student Solutions Manual and Study Guide to accompany College Physics. His graphics skills and technical expertise combined to produce illustrations for earlier editions which continue to add much to the appearance and usefulness of this volume.

Special thanks go to Senior Developmental Editor, Susan Dust Pashos and Ancillary Editor, Alexandra Buczek of Saunders College Publishing for managing all phases of this project. Finally, we express our appreciation to our families for their inspiration, patience, and encouragement.

Suggestions for Study

Very often we are asked "How should I study this subject, and prepare for examinations?" There is no simple answer to this question; however, we would like to offer some suggestions which may be useful to you.

1. It is essential that you understand the basic concepts and principles before attempting to solve assigned problems. This is best accomplished through a careful reading of the textbook before attending your lecture on that material, jotting down certain points which are not clear to you, taking careful notes in class, and asking questions. You should reduce memorization of material to a minimum. Memorizing sections of a text, equations, and derivations does not necessarily mean you understand the material. Perhaps the best test of your understanding of the material will be your ability to solve the problems in the text, or those given on exams.

2. Try to solve as many problems at the end of the chapter as possible. You will be able to check the accuracy of your calculations to the odd-numbered problems, since the answers to these are given at the back of the text. Furthermore, detailed solutions to approximately twenty percent of the problems from the text are provided in this Study Guide. Many of the worked examples in the text will serve as a basis for your study.

3. The method of solving problems should be carefully planned. First, read the problem several times until you are confident you understand what is being asked. Look for key words which will help simplify the problem, and perhaps allow you to make certain assumptions. You should also pay special attention to the information provided in the problem.

It is a good idea to write down the given information before proceeding with a solution. (For example, $a = -3.00$ m/s^2 and $v_0 = 5.00$ m/s are given. Find the velocity v, and the displacement Δx after $\Delta t = 2.00$ s.) After you have decided on the method you feel is appropriate for the problem, proceed with your solution. If you are having difficulty in working problems, we suggest that you again read the text and your lecture notes. It may take several readings before you are ready to solve certain problems. The solved problems in this Study Guide should be of value to you in this regard.

4. After reading a chapter, you should be able to define any new quantities that were introduced, and discuss the first principles that were used to derive fundamental formulas. A review is provided in each chapter of the Study Guide for this purpose, and the marginal notes in the textbook (or the index) will help you locate these topics. You should be able to correctly associate with each physical quantity the symbol used to represent that quantity (including vector notation, if appropriate) and the SI unit in which the quantity is specified. Furthermore, you should be able to express each important formula or equation in a concise and accurate prose statement.

5. We suggest that you use this Study Guide to review the material covered in the text, and as a guide in preparing for exams. You should also use the **Chapter Review, Notes From Selected Chapter Sections,** and **Equations and Concepts** to focus in on any points which require further study. Remember that the main purpose of this Study Guide is to improve upon the efficiency and effectiveness of your study hours and your overall understanding of physical concepts. However, it should not be regarded as a substitute for your textbook or individual study and practice in problem solving.

Contents

Chapter
1

INTRODUCTION

Chapter 1

INTRODUCTION

The goal of physics is to provide an understanding of nature by developing theories based on experiments. The theories are usually expressed in mathematical form. Fortunately, it is possible to explain the behavior of a variety of physical systems with a limited number of fundamental laws.

Since following chapters will be concerned with the laws of physics, we must begin by clearly defining the basic quantities involved in these laws. For example, such physical quantities as force, velocity, volume, and acceleration can be described in terms of more fundamental quantities. In the next several chapters we shall encounter three basic quantities: length (L), mass (M), and time (T). In later chapters we will need to add two other standard units to our list, for temperature (the kelvin) and for electric current (the ampere). In our study of mechanics, however, we shall be concerned only with the units of length, mass and time.

NOTES FROM SELECTED CHAPTER SECTIONS

1.1 Standards of Length, Mass, and Time

Until recently, the meter was defined as 1,650,763.73 wavelengths of orange-red light emitted from a krypton-86 lamp. However, in October 1983, the **meter** was redefined to be **the distance traveled by light in a vacuum during a time of 1/299,792,458 second.**

The SI unit of mass, the **kilogram**, is defined as **the mass of a specific platinum-iridium alloy cylinder kept at the International Bureau of Weights and Measures at Sèvres, France.**

The **second** is now defined as 9,192,631,770 **times the period of one oscillation of radiation from the cesium atom.**

Systems of units commonly used are the **SI system,** in which the units of mass, length, and time are the kilogram (kg), meter (m), and second (s), respectively; the **cgs** or **gaussian system,** in which the units of mass, length, and time are the gram (g), centimeter (cm), and second, respectively; and the **British engineering system** (sometimes called the conventional system), in which the units of mass, length, and time are the slug, foot (ft), and second, respectively.

1.2 The Building Blocks of Matter

It is useful to view the atom as a miniature Solar System with a dense, positively charged nucleus occupying the position of the Sun and negatively charged electrons orbiting like the planets. Occupying the nucleus are two basic entities, protons and neutrons. The **proton** is nature's fundamental carrier of positive charge; the **neutron** has no charge and a mass about equal to that of a proton. We shall find in Chapter 30 that even more elementary building blocks than protons and neutrons exist. Protons and neutrons are each now thought to consist of three particles called **quarks**.

1.3 Dimensional Analysis

Dimensional analysis makes use of the fact that **dimensions can be treated as algebraic quantities**. That is, quantities can be added or subtracted only if they have the same dimensions. Furthermore, the quantities on each side of an equation must have the same dimensions.

1.4 Significant Figures

When multiplying several quantities, the number of significant figures in the final answer is the same as the number of significant figures in the **least** accurate of the quantities being multiplied, where "least accurate" means "having the lowest number of significant figures." The same rule applies to division. When numbers are added (or subtracted), the number of decimal places in the result should equal the smallest number of decimal places of any term in the sum. Most of the numerical examples and end-of-chapter problems will yield answers having either two or three significant figures.

1.5 Conversion of Units

Sometimes it is necessary to convert units from one system to another. An extensive list of conversion factors can be found on the inside of the back cover of the **Student Solution Manual**.

1.6 Order-of-Magnitude Calculations

Often it is useful to estimate an answer to a problem in which little information is given. In such a case we refer to the **order of magnitude** of a quantity, by which we mean the power of ten that is closest to the actual value of the quantity. Usually, when an order-of-magnitude calculation is made, the results are reliable to within a factor of 10.

1.7 Mathematical Notation

Many mathematical symbols will be used throughout this book. Some important examples are:

$\propto$		denotes	a proportionality
$<$		means	"is less than"
$>$		means	"is greater than"
$<<$		means	"is much less than"
$>>$		means	"is much greater than"
$\approx$		indicates	approximate equality
$=$		indicates	equality
Δx	("delta x")	indicates	the change in a quantity x
$\lvert x \rvert$		means	the absolute value of x, such that the sign of $\lvert x \rvert$ is always positive, regardless of the sign of x.
Σ	(capital sigma)	represents	a sum. For example,

$$x_1 + x_2 + x_3 + x_4 + x_5 = \sum_{i=1}^{5} x_i$$

1.8 Coordinate Systems and Frames of Reference

A coordinate system used to specify locations in space consists of:

1. A fixed reference point, called the origin
2. A set of specified axes or directions
3. Instructions that tell us how to label a point in space relative to the origin and axes

1.9 Trigonometry

The portion of mathematics that is based on the special properties of a right triangle is called trigonometry. You should review the basic trigonometric functions given by Equations 1.1 and 1.2.

EQUATIONS AND CONCEPTS

The three most basic trigonometric functions of one of the acute angles of a right triangle are the sine, cosine, and tangent.

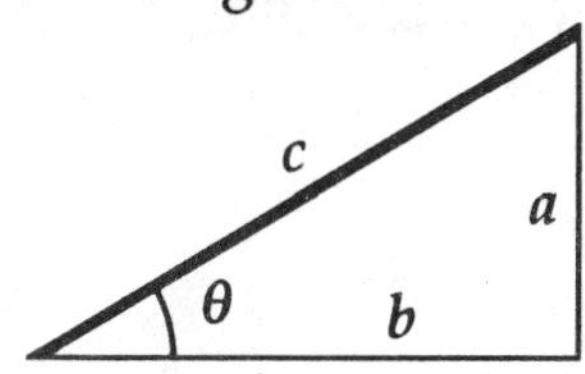

$$\sin\theta = \frac{\text{side opposite to }\theta}{\text{hypotenuse}} = \frac{a}{c}$$

$$\cos\theta = \frac{\text{side adjacent to }\theta}{\text{hypotenuse}} = \frac{b}{c} \qquad (1.1)$$

$$\tan\theta = \frac{\text{side opposite to }\theta}{\text{side adjacent to }\theta} = \frac{a}{b}$$

The Pythagorean theorem is an important relationship among the lengths of the sides of a right triangle. In this equation, c represents the hypotenuse.

$$c^2 = a^2 + b^2 \qquad (1.2)$$

Chapter 1

SUGGESTIONS, SKILLS, AND STRATEGIES

In developing problem-solving strategies, six basic steps are commonly used:

1. Read the problem carefully at least twice. Be sure you understand the nature of the problem before proceeding further.

2. Draw a suitable diagram with appropriate labels and coordinate axes, if needed.

3. As you examine what is being asked in the problem, identify the basic physical principle (or principles) that are involved, listing the knowns and unknowns.

4. Select a basic relationship or derive an equation that can be used to find the unknown, and symbolically solve the equation for the unknown.

5. Substitute the given values with the appropriate units into the equation.

6. Obtain a numerical value for the unknown. The problem is verified and receives a check mark if the following questions can be properly answered: Do the units match? Is the answer reasonable? Is the plus or minus sign proper or meaningful?

REVIEW CHECKLIST

▷ Discuss the units of length, mass and time and the standards for these quantities in SI units. (Section 1.1). Derive the quantities **force, velocity, volume, acceleration,** etc. from the three basic quantities.

▷ Perform a **dimensional analysis** of an equation containing physical quantities whose individual units are known. (Section 1.3)

▷ **Convert units** from one system to another. (Section 1.5)

▷ Carry out **order-of-magnitude calculations** or guesstimates. (Section 1.6)

▷ Describe the coordinates of a point in space using a Cartesian coordinate system. (Section 1.8)

SOLUTIONS TO SELECTED END-OF-CHAPTER PROBLEMS

3. The period of a simple pendulum, defined as the time for one complete oscillation, is measured in time units and is given by

$$T = 2\pi\sqrt{\frac{\ell}{g}}$$

where ℓ is the length of the pendulum and g is the acceleration due to gravity, in units of length divided by time squared. Show that this equation is dimensionally consistent.

Solution The length of the pendulum has units of length (L) and the acceleration due to gravity, g, is a length divided by the square of a time (L/T^2). The period, T, of the pendulum is a time (T) and $2p$ is a dimensionless constant. Substituting these dimensions into the equation for the period gives

$$(T) = \sqrt{\frac{(L)}{(L/T^2)}} = \sqrt{(L)\frac{(T^2)}{(L)}} = \sqrt{(T^2)} = (T) \qquad \lozenge$$

Thus, both sides of the equation have the same units or the equation is dimensionally consistent.

7. How many significant figures are there in (a) 78.9 ± 0.2, (b) 3.788×10^9, (c) 2.46×10^{-6}, (d) 0.0032?

Solution

(a) The notation $\pm$ 0.2 indicates an uncertainty of 2 units in the first decimal place. Thus, the number 78.9 contains 2 digits with no uncertainty and one digit with some uncertainty. Therefore, it has three significant figures. ◊

(b) In scientific notation, the first part of the number (3.788 in this case) contains all of the significant figures. The number of significant figures contained in 3.788×10^9 is therefore four. ◊

(c) As discussed in (b) above, the number of significant figures contained in 2.46×10^{-6} is three. ◊

(d) Expressing 0.0032 in scientific notation (i.e., as a number between 1 and 10 multiplied by a power of ten) gives 3.2×10^{-3}. Then, as discussed in part (b), observe that this number contains two significant figures. ◊

---

11. A farmer measures the distance around a rectangular field. The length of each long side of the rectangle is found to be 38.44 m, and the length of each short side is found to be 19.5 m. What is the total distance around the field?

Solution The total distance around the field is given by:

Perimeter = length + width + length + width

$$= 38.44 \text{ m} + 19.5 \text{ m} + 38.44 \text{ m} + 19.5 \text{ m} = 115.88 \text{ m}$$ ◊

In deciding how many significant figures to report in this answer, recall the rule for addition and subtraction:

> "When numbers are added or subtracted, the number of decimal places in the result should equal the smallest number of decimal places of any term in the sum."

The smallest number of decimal places in any term of the sum is one (in the term 19.5 m). Thus, the reported answer should only contain one decimal place. The distance around the field, including the correct number of significant figures, is therefore:

$$\text{Perimeter} = 115.9 \text{ m}$$ ◊

---

15. The speed of light is about 3.00×10^8 m/s. Convert this to miles per hour.

Solution A systematic way to convert units is to multiply the original value by one or more ratios, each ratio chosen so that its value is unity (i.e., its numerator and denominator are equal). Further, each ratio should be chosen so that multiplication by it will cancel some of the current units and move you one step closer to the desired units. Applying this technique to the stated problem gives:

$$3.00\times10^8\,\frac{\text{m}}{\text{s}} = \left(3.00\times10^8\,\frac{\text{m}}{\text{s}}\right)\left(\frac{1\text{ km}}{10^3\text{m}}\right)\left(\frac{0.621\text{ mi}}{1\text{ km}}\right)\left(\frac{3600\text{ s}}{1\text{ h}}\right) = 6.71\times10^8\,\frac{\text{mi}}{\text{h}}$$ ◊

---

23. A quart container of ice cream is to be made in the form of a cube. What should be the length of a side, in centimeters? (Use the conversion 1 gallon = 3.786 liter.)

Solution Since the container is in the form of a cube, with sides of length L, its volume may be written as $V = L \times L \times L = L^3 = 1.00$ quart. Converting to units of cubic centimeters yields:

$$L^3 = (1.00 \text{ quart})\left(\frac{1 \text{ gallon}}{4 \text{ quart}}\right)\left(\frac{3.786 \text{ liter}}{1 \text{ gallon}}\right)\left(\frac{1000 \text{ cm}^3}{1 \text{ liter}}\right) = 946 \text{ cm}^3$$

Solving for the length of one side then gives $L = \sqrt[3]{946 \text{ cm}^3} = 9.82 \text{ cm}$ ◊

27. Imagine that you are the equipment manager of a professional baseball team. One of your jobs is to keep baseballs on hand for games. Balls are sometimes lost when players hit them into the stands as either home runs or foul balls. Estimate how many baseballs you have to buy per season in order to make up for such losses. Assume your team plays an 81-game home schedule in a season.

Solution In order-of-magnitude calculations, one normally must make reasonable estimates of several quantities. In this problem, some reasonable estimates of several relevant quantities might be:

number of home games per season ≈ 81,
number of innings per game ≈ 9,
average number of hitters per inning ≈ 10,
average number of balls lost per hitter ≈ 1 ball for every 4 hitters.

With these estimates, the number of balls that you (as equipment manager) should order to replace balls lost during the season would be:

$$N = \left(\frac{\text{\# balls lost}}{\text{hitter}}\right)\left(\frac{\text{\# hitters}}{\text{inning}}\right)\left(\frac{\text{\# innings}}{\text{game}}\right)\left(\frac{\text{\# home games}}{\text{season}}\right)$$

$$N \approx \left(\frac{1 \text{ ball}}{4 \text{ hitters}}\right)\left(\frac{10 \text{ hitters}}{\text{inning}}\right)\left(\frac{9 \text{ innings}}{\text{game}}\right)\left(\frac{81 \text{ home games}}{\text{season}}\right) \approx 1800 \ \frac{\text{balls}}{\text{season}}$$ ◊

29. A point is located in a polar coordinate system by the coordinates $r = 2.5$ m and $\theta = 35°$. Find the x and y coordinates of this point, assuming the two coordinate systems have the same origin.

Solution: The point P is shown in the sketch below with both its polar and Cartesian coordinates labeled. The shaded triangle and basic trigonometry may be used to convert from the polar coordinates to the corresponding Cartesian coordinates:

$$\cos\theta = \frac{\text{adjacent side}}{\text{hypotenuse}} = \frac{x}{r} \quad \text{or} \quad x = r\cos\theta$$

$$\sin\theta = \frac{\text{opposite side}}{\text{hypotenuse}} = \frac{y}{r} \quad \text{or} \quad y = r\sin\theta$$

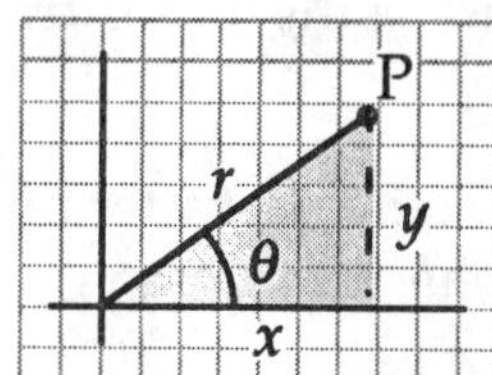

With $r = 2.5$ m and $\theta = 35°$, these yield:

$$x = 2.0 \text{ m and } y = 1.4 \text{ m}$$ ◊

33. For the triangle shown in Figure P1.33, what are (a) the length of the unknown side, (b) the tangent of θ, and (c) the sine of ϕ?

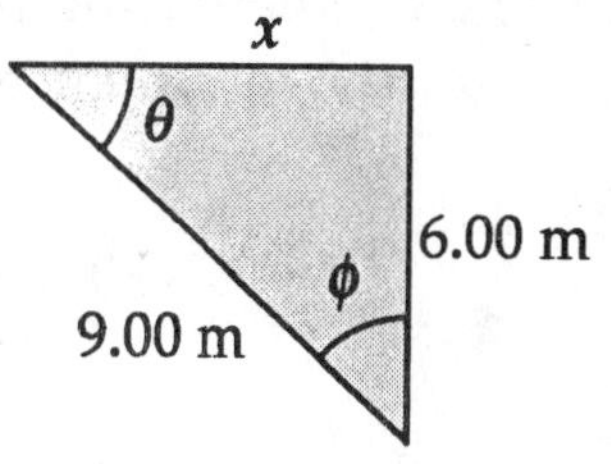

Figure P1.33

Solution (a) The unknown side, labeled x in the figure, is most easily found using the Pythagorean Theorem, $(9.00 \text{ m})^2 = x^2 + (6.00 \text{ m})^2$,

giving $\quad x^2 = 45.0 \text{ m}^2 \quad$ or $\quad x = 6.71$ m ◊

(b) $$\tan\theta = \frac{\text{side opposite } \theta}{\text{side adjacent to } \theta} = \frac{6.00 \text{ m}}{x} = \frac{6.00 \text{ m}}{6.71 \text{ m}} = 0.894$$ ◊

(c) $$\sin\phi = \frac{\text{side opposite } \phi}{\text{hypotenuse}} = \frac{x}{9.00 \text{ m}} = \frac{6.71 \text{ m}}{9.00 \text{ m}} = 0.746$$ ◊

34. A right triangle has a hypotenuse of length 3.00 m, and one of its angles is 30.0°. What are the lengths of (a) the side opposite the 30.0° angle and (b) the side adjacent to the 30.0° angle?

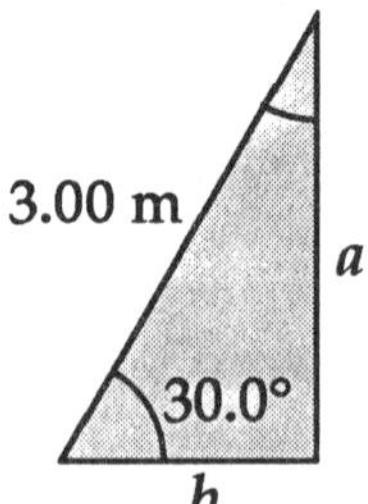

Solution: The described triangle is shown in the sketch with the unknown sides labeled a and b. These sides may be found using basic trigonometry.

(a) The sine function is chosen here since it relates the unknown side a and the known hypotenuse.

$$\sin 30.0° = \frac{\text{opposite side}}{\text{hypotenuse}} = \frac{a}{3.00 \text{ m}}$$

$$a = (3.00 \text{ m})\sin 30.0° = 1.50 \text{ m}$$ ◊

(b) Here, the cosine function is most convenient since it will relate the unknown side b to the known hypotenuse.

$$\cos 30.0° = \frac{\text{adjacent side}}{\text{hypotenuse}} = \frac{b}{3.00 \text{ m}}$$

or $$b = (3.00 \text{ m})\cos 30.0° = 2.60 \text{ m}$$ ◊

---

41. The displacement of an object moving under uniform acceleration is some function of time and the acceleration. Suppose we write this displacement as $s = k a^m t^n$, where k is a dimensionless constant. Show by dimensional analysis that this expression is satisfied if $m = 1$ and $n = 2$. Can this analysis give the value of k?

Solution For the equation to be valid, we must choose values of m and n to make it dimensionally consistent. Since s is a displacement, its dimensions are those of length (L). The acceleration, a, is a length divided by the square of a time (L/T^2). The variable t has dimensions of time (T), and the constant k has no dimensions. Substituting these dimensions into the equation yields:

$$(L) = \left(\frac{L}{T^2}\right)^m (T)^n = (L)^m (T)^{-2m} (T)^n, \quad \text{or} \quad (L)^1 (T)^0 = (L)^m (T)^{n-2m}$$

Note that the factor $(T)^0$ introduced on the left side of the second equation is equal to 1. This equation can be true only if the powers of length (L) are the same on the two sides of the equation and, simultaneously, the powers of time (T) are the same on both sides. Indeed, if another basic unit such as mass (M) were present, we would also require that its powers be identical on the two sides. Thus, we obtain a set of two simultaneous equations: $1 = m$ and $0 = n - 2m$. The solutions are therefore seen to be: $m = 1$ and $n = 2$ ◊

This technique gives no information about the possible values of the dimensionless constant k. ◊

---

45. The radius r of a circle inscribed in any triangle whose sides are $a, b,$ and c is given by $r = [(s-a)(s-b)(s-c)/s]^{1/2}$, where s is an abbreviation for $(a+b+c)/2$. Check this formula for dimensional consistency.

Solution: Each of the variables $r, a, b,$ and c are lengths (L). The variable s is a sum of lengths and is therefore a length (L). Likewise, each of the quantities $s-a$, $s-b$, and $s-c$ is the difference of two lengths and is also a length (L). Substituting these units into the formula gives:

$$(L) = \sqrt{\frac{(L)(L)(L)}{(L)}} = \sqrt{(L)^2} = (L)$$

Therefore, it is seen that the two sides of the equation have identical dimensions or the formula is dimensionally consistent. ◊

---

48. You can obtain a rough estimate of the size of a molecule by the following simple experiment. Let a droplet of oil spread out on a smooth water surface. The resulting "oil slick" will be approximately one molecule thick. Given an oil droplet of mass 9.00×10^{-7} kg and density 918 kg/m^3 that spreads out into a circle of radius 41.8 cm on the water surface, what is the diameter of an oil molecule?

Solution The density of an object is its mass divided by its volume. Recognizing this, the volume of oil in the droplet may be found from the equation (density) = (mass) / (volume):

$$\text{volume} = \frac{\text{mass}}{\text{density}} = \frac{9.00 \times 10^{-7}\ \text{kg}}{918\ \text{kg}/\text{m}^3} = 9.80 \times 10^{-10}\ \text{m}^3$$

The "oil slick" has a cylindrical shape with a height, h, equal to the diameter of an oil molecule, and a circular cross-section of radius $r = 41.8$ cm. Its volume is then given by: volume = (height)(cross-sectional area) or $\text{volume} = h(\pi r^2) = h\pi(41.8\ \text{cm})^2 = h(5.49 \times 10^3\ \text{cm}^2)$. Since the volume of oil in the "slick" is same as the volume of the droplet:

$$h(5.49 \times 10^3\ \text{cm}^2) = 9.80 \times 10^{-10}\ \text{m}^3 \quad \text{or} \quad h = \frac{9.80 \times 10^{-10}\ \text{m}^3}{5.49 \times 10^3\ \text{cm}^2}$$

Before this calculation can be completed, the length units in the numerator and denominator must be identical. Thus, we must use the conversion

$$(1\ \text{m})^2 = (10^2\ \text{cm})^2 \quad \text{or} \quad 1\ \text{m}^2 = 10^4\ \text{cm}^2$$

$$5.49 \times 10^3\ \text{cm}^2 = (5.49 \times 10^3\ \text{cm}^2)\left(\frac{1\ \text{m}^2}{10^4\ \text{cm}^2}\right) = 5.49 \times 10^{-1}\ \text{m}^2$$

and the diameter of an oil molecule is the cylinder height:

$$h = \frac{9.80 \times 10^{-10}\ \text{m}^3}{5.49 \times 10^{-1}\ \text{m}^2} = 1.78 \times 10^{-9}\ \text{m}$$ ◊

CHAPTER SELF-QUIZ

1. On planet Q, quash is a popular beverage among the natives. The average person on Q consumes 3.6 guppies of quash per month. There are 15 months per year on Q and an estimated population of 200 million. Which of the following represents the best order-of-magnitude estimate value for the total volume of quash consumed per year (in guppy units)?
   a. 10^{12}
   b. 10^{10}
   c. 10^{8}
   d. 10^{6}

2. Note the following mathematical expression: $y = x^2$. Which one of the statements below is most consistent with this expression?
   a. if y doubles, then x quadruples
   b. y is greater than x
   c. if x doubles, then y doubles
   d. if x doubles, then y quadruples

3. On planet N, the standard unit of length is the nose. If a 6.1 foot height astronaut travels to planet N and is measured to have a height of 94 noses, what would be the height **in noses** of another astronaut who measures 5.5 feet in height?
   a. 74
   b. 104
   c. 81
   d. 85

4. If the displacement of an object is related to velocity, v, according to the relation $x = Av$, the constant, A, has the dimension of which of the following?
   a. acceleration
   b. length
   c. time
   d. area

5. If a submarine on the surface dives at an angle of 15° with respect to the horizontal and follows a straight line path for a distance of 40 m, how far below the surface will it be (measured in m)?
   a. 10
   b. 39
   c. 14
   d. 600

6. The diameter of the Milky Way Galaxy is estimated at about 10^5 light years. A light year is the distance traveled by light in one year; if the speed of light is 3×10^8 m/s, about how far (in meters) is it from one side of the galaxy to the other?
   (1 year = 3.15×10^7 s)
   a. 10×10^{15}
   b. 1×10^{18}
   c. 9×10^{20}
   d. 300×10^8

7. Suppose it takes you 20 min. to walk a mile. If you trained so that you could walk steadily, day after day, for about 5 hours a day, estimate how long it would take you to walk across America (about 3000 miles).
   a. one month
   b. six months
   c. one year
   d. six years

8. Which point is nearest the x axis?
   a. (3, 4)
   b. (4, 3)
   c. (-5, 2)
   d. (-2, -5)

9. A gallon of paint (volume = $3.78 \times 10^{-3}\ m^3$) covers $25\ m^2$. What is the thickness of the paint on the wall?
   a. 0.15 mm
   b. 0.30 mm
   c. 0.75 mm
   d. 1.50 cm

10. Assume there are 50 million passenger cars in the United States, and that the average fuel consumption is 20 miles/gallon of gasoline. If the average distance traveled by each car is 10,000 miles/year, how many gallons of gasoline would be saved per year if average fuel consumption could be increased to 25 miles/gallon?
    a. 5×10^6
    b. 5×10^7
    c. 5×10^8
    d. 5×10^9

11. Calculate $(0.41 + 0.021) \times (2.2 \times 10^3)$, keeping only significant figures.
    a. 950
    b. 946
    c. 948
    d. 880

12. $S = \sum_{n=1}^{5} n^3$
    a. S = 45
    b. S = 225
    c. S = 325
    d. S = 3375

Chapter 2

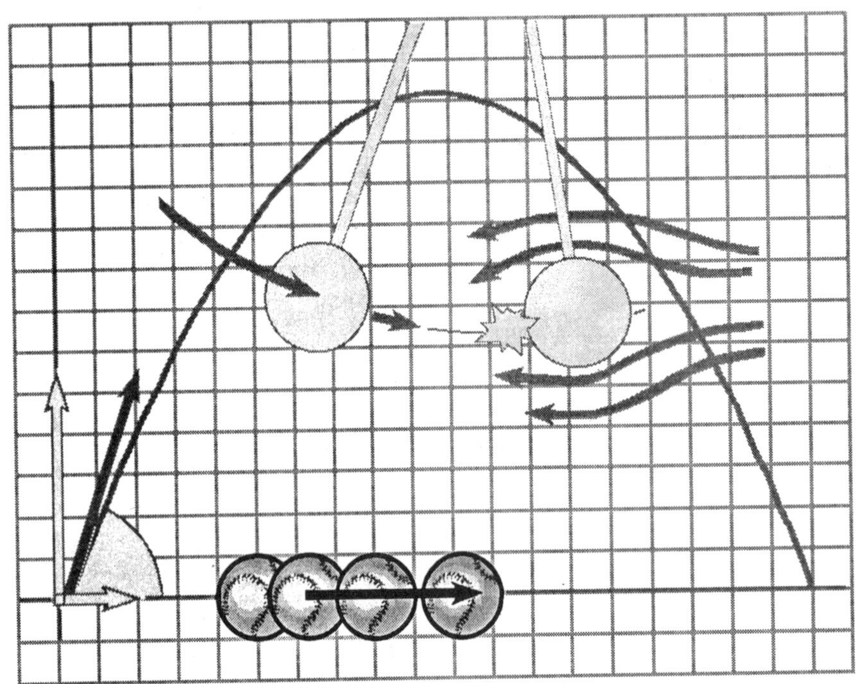

MOTION IN ONE DIMENSION

Chapter 2

MOTION IN ONE DIMENSION

The branch of physics concerned with the study of the motion of an object and the relationship of this motion to such physical concepts as force and mass is called **dynamics.** The part of dynamics that describes motion without regard to its causes is called **kinematics**. In this chapter we shall focus on kinematics and on one-dimensional motion, that is, motion along a straight line. We shall start by discussing displacement, velocity, and acceleration. These concepts will enable us to study the motion of objects undergoing constant acceleration. In Chapter 3 we shall discuss the motions of objects in two dimensions.

NOTES FROM SELECTED CHAPTER SECTIONS

2.1 Displacement

The displacement of an object, defined as its **change in position**, is given by the difference between its final and initial coordinates, or $x_f - x_i$. Displacement is an example of a vector quantity. A vector is a physical quantity that requires a specification of both direction and magnitude.

2.2 Average Velocity

The **average velocity** of an object during the time interval t_i to t_f is equal to the slope of the straight line joining the initial and final points on a graph of the position of the object plotted versus time.

2.3 Instantaneous Velocity

The **instantaneous speed** of an object, which is a scalar quantity, is defined as the magnitude of the instantaneous velocity. Hence, by definition, **speed can never be negative**.

The slope of the line tangent to the position-time curve at a point P is defined to be the instantaneous velocity at the corresponding time.

2.4 Acceleration

The **average acceleration** during a given time interval is defined as the change in velocity divided by the time interval during which this change occurs.

The **instantaneous acceleration** of an object at a certain time equals the slope of the tangent to the velocity-time graph at that instant of time.

2.5 Motion Diagrams

It is often instructive to make use of motion diagrams to describe the velocity and acceleration vectors as time progresses while an object is in motion. You should carefully study Figure 2.13 in the textbook. This figure illustrates the motion of a car in three different cases:

1. Constant positive velocity, zero acceleration.
2. Positive velocity, positive acceleration.
3. Positive velocity, negative acceleration.

2.6 One-Dimensional Motion with Constant Acceleration

This type of motion is important because it applies to many objects in nature. When an object moves with constant acceleration, the average acceleration equals the instantaneous acceleration. Equations 2.6 through 2.10 may be used to solve any problem in one-dimensional motion with constant acceleration.

2.7 Freely Falling Bodies

A freely falling body is an object moving freely under the influence of gravity only, regardless of its initial motion. Objects thrown upward or downward and those released from rest are all falling freely once they are released!

It is important to emphasize that any freely falling object experiences an **acceleration directed downward**. This is true regardless of the initial motion of the object.

An object thrown upward (or downward) will experience the same acceleration as an object released from rest. Once they are in free fall, all objects have an acceleration downward equal to the acceleration due to gravity.

EQUATIONS AND CONCEPTS

The average velocity of an object during a time interval is the ratio of the total displacement to the time interval during which the displacement occurred. Note that the instantaneous velocity of the object during the time interval might have different values from instant to instant.

$$\bar{v} = \frac{x_f - x_i}{t_f - t_i} \tag{2.2}$$

The average acceleration of an object during a time interval is the ratio of the change in velocity to the time interval during which the change in velocity occurs.

$$\bar{a} = \frac{v_f - v_i}{t_f - t_i} \tag{2.4}$$

These equations are called the equations of kinematics and can be used to describe one-dimensional motion along the x axis with constant acceleration. Note that each equation shows a different relationship among physical quantities: initial velocity, final velocity, acceleration, time, and position. Also in the form that they are shown, it is assumed that the object whose motion is described is located at the origin ($x = 0$) when $t = 0$.

$$v = v_0 + at \tag{2.6}$$

$$\bar{v} = \frac{v_0 + v}{2} \tag{2.7}$$

$$x = \frac{1}{2}(v + v_0)t \tag{2.8}$$

$$x = v_0 t + \frac{1}{2}at^2 \tag{2.9}$$

$$v^2 = v_0{}^2 + 2ax \tag{2.10}$$

The equations of kinematics for an object in free fall ($a = -g$) along the y axis. Note that since $a = -g$ in these four equations, the $+y$ axis is predefined to point upwards.

$$v = v_0 - gt$$

$$y = \frac{1}{2}(v + v_0)t$$

$$y = v_0 t - \frac{1}{2}gt^2$$

$$v^2 = v_0{}^2 - 2gy$$

SUGGESTIONS, SKILLS, AND STRATEGIES

The following procedure is recommended for solving problems involving acceleration motion:

1. Make sure all the units in the problem are consistent. That is, if distances are measured in meters, be sure that velocities have units of m/s and accelerations have units of m/s^2.

2. Choose a coordinate system.

3. Make a list of all the quantities given in the problem and a separate list of those to be determined.

4. Select from the list of kinematic equations the one or ones that will enable you to determine the unknowns.

5. Construct an appropriate motion diagram and check to see if your answers are consistent with the diagram.

REVIEW CHECKLIST

▷ Define the displacement and average velocity of a particle in motion. (Sections 2.1 and 2.2)

▷ Define the instantaneous velocity and understand how this quantity differs from average velocity. (Section 2.3)

▷ Define average acceleration and instantaneous acceleration. (Section 2.4)

▷ Construct a graph of position versus time (given a function such as $x = 5 + 3t - 2t^2$) for a particle in motion along a straight line. From this graph, you should be able to determine both average and instantaneous values of velocity by calculating the slope of the tangent to the graph. (Section 2.6)

▷ Describe what is meant by a body in **free fall** (one moving under the influence of gravity--where air resistance is neglected). Recognize that the equations of kinematics apply directly to a freely falling object and that the acceleration is then given by $a = -g$ (where $g = 9.8 \text{ m/s}^2$). (Section 2.7)

▷ Apply the equations of kinematics to any situation where the motion occurs under constant acceleration.

SOLUTIONS TO SELECTED END-OF-CHAPTER PROBLEMS

7. Two cars travel in the same direction along a straight highway, one at a constant speed of 55.0 mi/h and the other at 70.0 mi/h. (a) Assuming that they start at the same point, how much sooner does the faster car arrive at a destination 10.0 mi away? (b) How far must the faster car travel before it has a 15.0-min lead on the slower car?

Solution (a) When an object moves with constant velocity v, the distance x traveled in time t is given by $x = vt$. Since both cars travel at constant velocity, the times required for them to travel 10.0 mi are:

(Faster Car) $$t_1 = \frac{x_1}{v_1} = \frac{10.0 \text{ mi}}{70.0 \text{ mi}/\text{h}} = 0.143 \text{ h} = 1.43 \text{ x } 10^{-1} \text{ h}$$

(Slower Car) $$t_2 = \frac{x_2}{v_2} = \frac{10.0 \text{ mi}}{55.0 \text{ mi}/\text{h}} = 0.182 \text{ h} = 1.82 \text{ x } 10^{-1} \text{ h}$$

Thus, the time between the arrival of the two vehicles is:

$$t_2 - t_1 = 1.82 \text{ x } 10^{-1} \text{ h} - 1.43 \text{ x } 10^{-1} \text{ h} = \left(3.90 \text{ x } 10^{-2} \text{ h}\right)\left(\frac{60.0 \text{ min}}{1.00 \text{ h}}\right) = 2.34 \text{ min}$$ ◊

(b) When the faster car has a 15.0-min lead, it is ahead by a distance equal to that traveled by the slower car in a time of 15.0 min. This distance is given by:

$$x = vt = (55.0 \text{ mi}/\text{h})(15.0 \text{ min})\left(\frac{1.00 \text{ h}}{60.0 \text{ min}}\right) = 13.75 \text{ mi}$$

The faster car pulls ahead of the slower car at a relative velocity of $v_{rel} = 70.0 \text{ mph} - 55.0 \text{ mph} = 15.0 \text{ mph}$. Thus, the time required for it to gain a 13.8 mi lead is:

$$t = \frac{x}{v_{rel}} = \frac{13.75 \text{ mi}}{15.0 \text{ mi/h}} = 0.917 \text{ h}$$

Finally, the distance the faster car traveled while gaining a 13.8-mi lead is given by $x = vt = (70.0 \text{ mi/h})(0.917 \text{ h}) = 64.2 \text{ mi}$ ◊

11. In order to qualify for the finals in a racing event, a race car must achieve an average speed of 250 km/h on a track with a total length of 1600 m. If a particular car covers the first half of the track at an average speed of 230 km/h, what minimum average speed must it have in the second half of the event in order to qualify?

Solution

The time required to travel a distance of 1600 m at an average speed of 250 km/h is

$$t_{\text{total}} = \frac{x_{\text{total}}}{\bar{v}_{\text{total}}} = \frac{1600 \text{ m}}{250 \text{ km/h}}\left(\frac{3600 \text{ s/h}}{1000 \text{ m/km}}\right) = 23.0 \text{ s}$$

This is the maximum total elapsed time if the car is to qualify for the race. If the car travels the first 800 m at an average speed of 230 km/h, the time used for the first half of the trip is

$$t_1 = \frac{x_1}{\bar{v}_1} = \frac{800 \text{ m}}{230 \text{ km/h}}\left(\frac{3600 \text{ s/h}}{1000 \text{ m/km}}\right) = 12.5 \text{ s}$$

Thus, if the car is to qualify, the maximum time that can be used on the second half of the trip is: $t_2 = 23.0 \text{ s} - 12.5 \text{ s} = 10.5 \text{ s}$. The average speed required to cover the remaining 800 m in 10.5 s is given by

$$\bar{v}_2 = \frac{x_2}{t_2} = \frac{800 \text{ m}}{10.5 \text{ s}} = 76.2 \text{ m/s}$$

$$\bar{v}_2 = (76.2 \text{ m/s})\left(\frac{1.00 \text{ km/h}}{0.278 \text{ m/s}}\right) = 274 \text{ km/h}$$ ◊

15. Find the instantaneous velocities of the tennis player of Figure P2.6 at (a) 0.50 s, (b) 2.0 s, (c)3.0 s, (d)4.5 s.

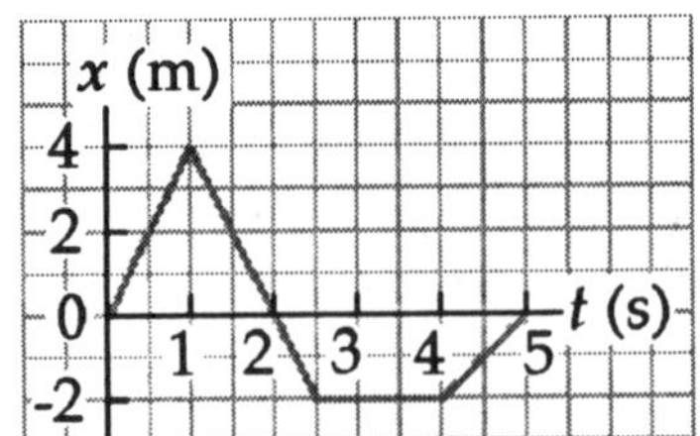

Figure P2.6

Solution

From Section 2.3 of the textbook, "**The slope of the line tangent to the position-time curve at *P* is defined to be the instantaneous velocity at that time.**"

Thus, we need to determine the slope of the tangent lines to the position-time curve shown in Figure P2.6 at each of the requested times. These slopes may be calculated as follows:

(a) $$v_{0.5\text{ s}} = \frac{x_{1.0\text{ s}} - x_{0.0\text{ s}}}{1.0\text{ s} - 0.0\text{ s}} = \frac{4.0\text{ m} - 0.0\text{ m}}{1.0\text{ s}} = 4.0\text{ m/s}$$ ◊

(b) $$v_{2.0\text{ s}} = \frac{x_{2.5\text{ s}} - x_{1.0\text{ s}}}{2.5\text{ s} - 1.0\text{ s}} = \frac{-2.0\text{ m} - 4.0\text{ m}}{1.5\text{ s}} = \frac{-6.0\text{ m}}{1.5\text{ s}} = -4.0\text{ m/s}$$ ◊

(c) $$v_{3.0\text{ s}} = \frac{x_{4.0\text{ s}} - x_{2.5\text{ s}}}{4.0\text{ s} - 2.5\text{ s}} = \frac{-2.0\text{ m} - (-2.0\text{ m})}{1.5\text{ s}} = \frac{0.0\text{ m}}{1.5\text{ s}} = 0.0\text{ m/s}$$ ◊

(d) $$v_{4.5\text{ s}} = \frac{x_{5.0\text{ s}} - x_{4.0\text{ s}}}{5.0\text{ s} - 4.0\text{ s}} = \frac{0.0\text{ m} - (-2.0\text{ m})}{1.0\text{ s}} = \frac{+2.0\text{ m}}{1.0\text{ s}} = 2.0\text{ m/s}$$ ◊

---

19. A certain car is capable of accelerating at a rate of +0.60 m/s^2. How long does it take for this car to go from a speed of 55 mi/h to a speed of 60 mi/h?

Solution The average acceleration over a time interval of duration Δt is defined as $\bar{a} = \Delta v/\Delta t$. Thus, the time required to achieve a change in velocity Δv, with an average acceleration $\bar{a}$ is $\Delta t = \Delta v / \bar{a} = (v_f - v_i)/\bar{a}$. Since the car maintains a constant acceleration during the time interval of interest, the average acceleration is the same as the constant instantaneous acceleration (i.e., $\bar{a} = a = +0.60 \text{ m/s}^2$). The required time is then:

$$\Delta t = \frac{(60 \text{ mi/h} - 55 \text{ mi/h})}{0.60 \text{ m/s}^2} = \left(\frac{5.0 \text{ mi/h}}{0.60 \text{ m/s}^2}\right)\left(\frac{1609 \text{ m/mi}}{3600 \text{ s/h}}\right) = 3.7 \text{ s} \qquad \Diamond$$

---

23. A racing car reaches a speed of 40 m/s. At this instant, it begins a uniform negative acceleration, using a parachute and a braking system, and comes to rest 5.0 s later. (a) Determine the acceleration of the car. (b) How far does the car travel after acceleration starts?

Solution (a) Since the acceleration is uniform over this 5.0 s interval, the instantaneous acceleration is the same as the average acceleration. Therefore,

$$a = \bar{a} = \frac{\Delta v}{\Delta t} = \frac{v_f - v_i}{\Delta t} = \frac{0.0 - 40 \text{ m/s}}{5.0 \text{ s}} = -8.0 \text{ m/s}^2 \qquad \Diamond$$

(b) Because the car has a uniform acceleration during the 5.0 s braking period, its average velocity during this time interval is given by

$$\bar{v} = (v_f + v_i)/2 = (0.0 + 40.0 \text{ m/s})/2 = 20.0 \text{ m/s}$$

The displacement the car undergoes during this interval is therefore

$$x = \bar{v}t = (20 \text{ m/s})(5.0 \text{ s}) = 1.0 \times 10^2 \text{ m} \qquad \Diamond$$

---

31. A record of travel along a straight path is as follows:

1. Start from rest with constant acceleration of 2.77 m/s^2 for 15.0 s.
2. Constant velocity for the next 2.05 min.
3. Constant negative acceleration –9.47 m/s^2 for 4.39 s.

(a) What was the total displacement for the complete trip? (b) What were the average speeds for legs 1, 2, and 3 of the trip as well as for the complete trip?

Solution (a) Since the object has different accelerations for each leg of the trip, you must apply the uniformly motion equations to each leg separately. Then, the displacement for the entire trip may be found by adding the individual displacements.

For first leg, $v_{1i} = 0$, $t_1 = 15.0$ s, and $a_1 = 2.77\ \text{m/s}^2$. By applying the equations $x_1 = v_{1i}t_1 + \frac{1}{2}a_1t_1^2$ and $v_{1f} = v_{1i} + a_1t_1$ with $v_{1i} = 0$, we can then find the net displacement and velocity at the end of this leg:

$$x_1 = \tfrac{1}{2}\left(2.77\ \text{m/s}^2\right)(15.0\ \text{s})^2 = 312\ \text{m} \quad \text{and} \quad v_{1f} = \left(2.77\text{m/s}^2\right)(15.0\ \text{s}) = 41.55\ \text{m/s}$$

For the second leg, the velocity is constant at $v_2 = v_{1f} = 41.55$ m/s, and the elapsed time is $t_2 = 2.05\ \text{min} = 123\ \text{s}$. Thus, the displacement during the second leg is

$$x_2 = v_2t_2 = (41.55\ \text{m/s})(123\ \text{s}) = 5.11\times10^3\ \text{m}$$

For the final leg, $v_{3i} = v_2 = 41.5$ m/s, $a_3 = -9.47\ \text{m/s}^2$, and $t_3 = 4.39$ s
Applying the equation of motion $x_3 = v_{3i}t_3 + \frac{1}{2}a_3t_3^2$, we find:

$$x_3 = (41.55\ \text{m/s})(4.39\ \text{s}) + \tfrac{1}{2}\left(-9.47\ \text{m/s}^2\right)(4.39\ \text{s})^2 = 91.2\ \text{m}$$

The total displacement is therefore $\quad x_{\text{total}} = x_1 + x_2 + x_3 = 5.51\times10^3\ \text{m}$ ◊

(b) The requested average velocities are:

$$\bar{v}_1 = x_1/t_1 = 20.8 \text{ m/s}; \quad \bar{v}_2 = x_2/t_2 = 41.6 \text{ m/s}; \quad \bar{v}_3 = x_3/t_3 = 20.8 \text{ m/s}$$

$$\bar{v}_{\text{total}} = x_{\text{total}}/t_{\text{total}} = \left(5.51\times 10^3 \text{ m}\right)/\left(15.0 \text{ s} + 123 \text{ s} + 4.39 \text{ s}\right) = 38.7 \text{ m/s} \quad \Diamond$$

---

35. A train 400 m long is moving on a straight track with a speed of 82.4 km/h. The engineer applies the brakes at a crossing, and later the last car passes the crossing with a speed of 16.4 km/h. Assuming constant acceleration, determine how long the train blocked the crossing. Disregard the width of the crossing.

Solution

If we ignore the width of the crossing, the displacement of the train from the instant the engine enters the crossing until the last car leaves the crossing is the same as the length of the train; x = 400 m.

Since the train has a constant acceleration, its average velocity while it is blocking the crossing may be found from:

$$\bar{v} = \frac{v_i + v_f}{2} = \frac{82.4 \text{ km/h} + 16.4 \text{ km/h}}{2} = 49.4 \text{ km / h}$$

The time the crossing is blocked is now found from $x = \bar{v}t$. Note that since the length of the train is given in units of meters, and the units of velocity are in km/h, a conversion of units must be performed in the calculation:

$$t = \frac{x}{\bar{v}} = \left(\frac{400 \text{ m}}{49.4 \text{ km/h}}\right)\left(\frac{3600 \text{ s/h}}{1000 \text{ m/km}}\right) = 29.1 \text{ s} \quad \Diamond$$

---

39. A small mailbag is released from a helicopter that is descending steadily at 1.50 m/s. After 2.00 s, (a) what is the speed of the mailbag, and (b) how far is it below the helicopter? (c) What are your answers to parts (a) and (b) if the helicopter is rising steadily at 1.50 m/s?

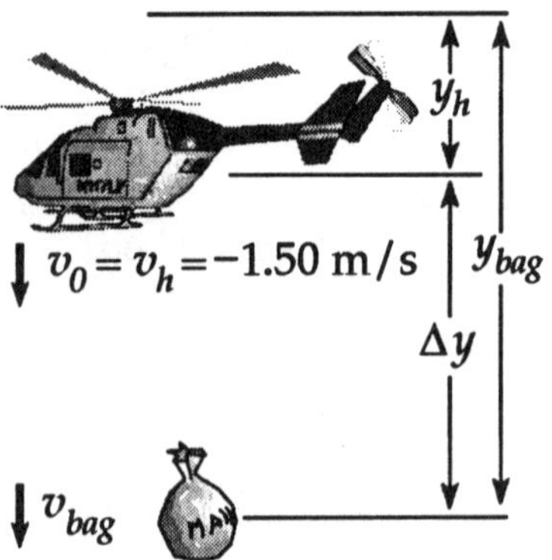

Solution (a) If we choose $+y$ to point upwards, and down as the negative direction, the initial velocity of the mailbag is $v_0 = v_h = -1.50$ m/s. As soon as the bag is released, it becomes a freely falling body with acceleration $a = -g = -9.80\ \text{m/s}^2$; the velocity of the bag 2.00 s after it is released is:

$$v_{\text{bag}} = v_0 + at = -1.50\ \text{m/s} + \left(-9.80\ \text{m/s}^2\right)(2.00\ \text{s}) = -21.1\ \text{m/s}$$

The **speed** (magnitude of velocity) is: $\text{speed} = \left|v_{\text{bag}}\right| = 21.1\ \text{m/s}$ ◊

(b) The displacement of the bag and the helicopter from the release point after 2.00 s are:

$$y_{\text{bag}} = v_{\text{bag}}t + \tfrac{1}{2}at^2 = (-1.50\ \text{m/s})(2.00\ \text{s}) + \tfrac{1}{2}\left(-9.80\ \text{m/s}^2\right)(2.00\ \text{s})^2 = -22.6\ \text{m}$$

$$y_h = v_h t = (-1.50\ \text{m/s})(2.00\ \text{s}) = -3.00\ \text{m}$$

The difference between them, $\Delta y = y_{\text{bag}} - y_h = [-22.6\ \text{m} - (-3.00\ \text{m})] = -19.6\ \text{m}$, and the bag is 19.6 m below the helicopter 2.00 s after it was released. ◊

(c) If the helicopter (and bag) was moving **upward** at the instant of release, then $v_0 = v_h = +1.50\ \text{m/s}$. Using this value for v_0 in parts (a) and (b),

$$v_{\text{bag}} = -18.1\ \text{m/s} \quad \text{and} \quad \text{speed} = \left|v_{\text{bag}}\right| = 18.1\ \text{m/s}$$ ◊

The net displacement of the helicopter and the bag are $y_{\text{bag}} = -16.6\ \text{m}$ and $y_h = +3.00\ \text{m}$, with a distance between the two of $\Delta y = y_{\text{bag}} - y_h = -19.6\ \text{m}$. Thus, the bag is again 19.6 m below the helicopter 2.00 s after its release. ◊

41. A ball thrown vertically upward is caught by the thrower after 2.00 s. Find (a) the initial velocity of the ball and (b) the maximum height it reaches.

Solution

In this problem, the ball is a freely falling body with an acceleration of $a = -g = -9.80 \text{ m/s}^2$. Choose $y = 0$ to be at the level from which the ball is thrown and $t = 0$ to coincide with the instant the ball leaves the thrower's hand. With these choices, then, we have given values of $y = 0$ at $t = 0$, and $y = 0$ at $t = 2.00$ s. We are asked to solve for v at $t = 0$ and also for y_{max}.

(a) The vertical displacement of the ball at any time during the flight is given by $y = v_0 t + \frac{1}{2}at^2$. Using the fact that $y = 0$ at $t = 2.00$ s gives:

$$0 = v_0(2.00 \text{ s}) + \tfrac{1}{2}\left(-9.80 \text{ m/s}^2\right)(2.00 \text{ s})^2$$

yielding an initial velocity of $$v_0 = \frac{\frac{1}{2}\left(9.80 \text{ m/s}^2\right)(2.00 \text{ s})^2}{2.00 \text{ s}} = 9.80 \text{ m/s} \quad \Diamond$$

(b) At the instant the ball is at its maximum height, its vertical velocity is zero. It has been moving upward ($v > 0$) prior to this instant, and will be coming downward ($v < 0$) immediately after this instant. The vertical velocity of the ball is related to its vertical displacement by the equation $v^2 = v_0^2 + 2ay$. Using the fact that $v_0 = 9.80$ m/s (from above), $a = -9.80 \text{ m/s}^2$, and $v = 0$ at $y = y_{max}$ gives:

$$0 = (9.80 \text{ m/s})^2 + 2\left(-9.80 \text{ m/s}^2\right)y_{max}, \quad \text{or} \quad y_{max} = \frac{(9.80 \text{ m/s})^2}{2\left(9.80 \text{ m/s}^2\right)} = 4.90 \text{ m} \quad \Diamond$$

---

47. A ranger in a National Park is driving at $35.0\ \mathrm{mi/h}$ when a deer jumps into the road 200 ft ahead of the vehicle. After a reaction time of t s, the ranger applies the brakes to produce an acceleration of $a=-9.00\ \mathrm{ft/s^2}$. What is the maximum reaction time allowed if she is to avoid hitting the deer?

Solution

Before the brakes are applied, the velocity of the car is:

$$v_0=(35.0\ \mathrm{mi/h})\left(\frac{5280\ \mathrm{ft/mi}}{3600\ \mathrm{s/h}}\right)=51.3\ \mathrm{ft/s}$$

Once the brakes are applied, the distance required to stop the car (i.e., reach a velocity of $v=0$) with this initial velocity and an acceleration of $a=-9.00\ \mathrm{ft/s^2}$ may be found from $v^2=v_0{}^2+2ax$:

$$0=(51.3\ \mathrm{ft/s})^2+2\left(-9.00\ \mathrm{ft/s^2}\right)x$$

so $$x=\frac{(51.3\ \mathrm{ft/s})^2}{2\left(9.00\ \mathrm{ft/s^2}\right)}=146.4\ \mathrm{ft}$$

Thus, if the deer is to be avoided, the car can travel a maximum of $(200\ \mathrm{ft}\ -\ 146.4\ \mathrm{ft})=\ 53.6\ \mathrm{ft}$ before the brakes are applied. The time required for the car to travel this distance at a speed of 35.0 mph $(51.3\ \mathrm{ft/s})$, and hence the maximum allowed reaction time, is:

$$t=\frac{x}{v}=\frac{53.6\ \mathrm{ft}}{51.3\ \mathrm{ft/s}}=1.04\ \mathrm{s}\ \Diamond$$

53. In Mostar, Bosnia, the ultimate test of a young man's courage once was to jump off a 400-year-old bridge (now destroyed) into the River Neretva, 23.0 m below the bridge. (a) How long did the jump last? (b) How fast was the diver traveling upon impact with the river? (c) If the speed of sound in air is 340 m/s, how long after the diver took off did a spectator on the bridge hear the splash?

Solution

(a) We choose the origins of position and time ($y=0$ and $t=0$) to coincide with the diver leaving the bridge. Also, we consider the velocity of the diver as he leaves the bridge to be negligible ($v_0=0$). The diver is a freely falling body ($a=-g=-9.80\text{ m/s}^2$) while in the air, and his displacement is related to the elapsed time by $y=v_0t+\frac{1}{2}at^2$. The time when he reaches the water ($y=-23.0\text{ m}$) is then given by $-23.0\text{ m}=0+\frac{1}{2}\left(-9.80\text{ m/s}^2\right)t^2$, which reduces to

$$t^2=\frac{-23.0\text{ m}}{-4.90\text{ m/s}^2}=+4.69\text{ s}^2 \quad\text{and}\quad t=2.17\text{ s} \qquad \Diamond$$

(b) The velocity of the diver as he reaches the water (at $t=2.17\text{ s}$) may be found from $v=v_0+at$ as $v=0+\left(-9.80\text{ m/s}^2\right)(2.17\text{ s})=-21.3\text{ m/s}$, so the speed (magnitude of velocity) at this time is 21.3 m/s. ◊

(c) After impact, the time for the sound of the splash to travel (at a constant velocity of 340 m/s) back to a spectator on the bridge is given by $t'=\frac{x}{v_{\text{sound}}}=\frac{23.0\text{ m}}{340\text{ m/s}}=0.068\text{ s}$. The time between start of the dive and reception of the sound is $t_{\text{total}}=t+t'=2.17\text{ s}+0.068\text{ s}=2.24\text{ s}$ ◊

---

60. A daring stunt woman sitting on a tree limb wishes to drop vertically onto a horse galloping under the tree. The speed of the horse is 10.0 m/s, and the woman is initially 3.00 m above the level of the saddle. (a) What must be the horizontal distance between the saddle and limb when the woman makes her move? (b) How long is she in the air?

Solution The most convenient way to solve this problem is to first solve part (b) and use the answer from that part in the solution to part (a) as shown below.

(b) The stunt woman starts from rest ($v_0 = 0$) and is a freely falling body $\left(a = -g = -9.80 \text{ m/s}^2\right)$ until she reaches the saddle. The time to make the 3.00 m vertical drop may be found from $y = v_0 t + \frac{1}{2}at^2$. This gives $-3.00 \text{ m} = 0 + \frac{1}{2}\left(-9.80 \text{ m/s}^2\right)t^2$ or $t = \sqrt{\frac{-3.00 \text{ m}}{-4.90 \text{ m/s}^2}} = 0.782 \text{ s}$. ◊

(a) Since the horse moves with constant velocity, the horizontal distance it travels during the 0.782 s the stunt woman is falling (and therefore the horizontal distance that should exist between her and the saddle when she makes her move) is given by:

$$x = v_{\text{horse}}t = (10.0 \text{ m/s})(0.782 \text{ s}) = 7.82 \text{ m}$$ ◊

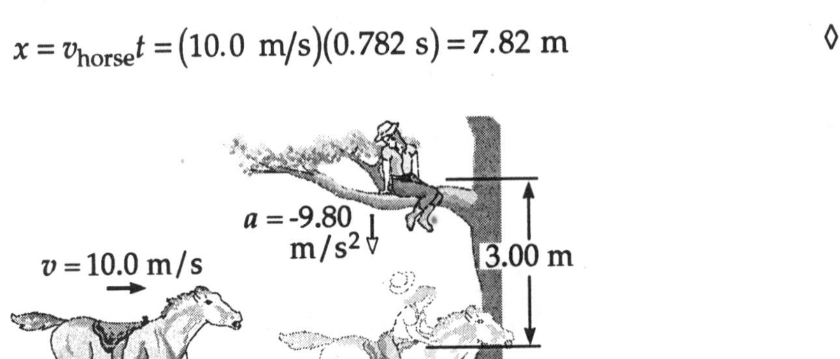

CHAPTER SELF-QUIZ

1. A European sports car dealer claims that his product will accelerate at a constant rate from rest to a speed of 90 km/hr in 8 s. If so, what is the acceleration (in m/s^2)? (First, convert the speed to m/s.)
   a. 3.1
   b. 6.2
   c. 11.3
   d. 16.9

2. A rock, released at rest from the top of a tower, hits the ground after falling for 2.00 s. What is the height of the tower? (g = 9.80 m/s^2 and air resistance is negligible)
   a. 15 m
   b. 20 m
   c. 31 m
   d. 39 m

3. A rock is thrown downward from the top of a tower with an initial speed of 12 m/s. If the rock hits the ground after 2.00 s, what is the speed of the rock as it hits the ground? (g = 9.80 m/s^2 and air resistance is negligible)
   a. 32 m/s
   b. 39 m/s
   c. 64 m/s
   d. 78 m/s

4. A rock is released at rest from the top of a 40 m tower. If g = 9.8 m/s^2 and air resistance is negligible, how long does it take the rock to reach the ground?
   a. 2.00 s
   b. 2.90 s
   c. 4.10 s
   d. 5.10 s

5. Human reaction time is usually greater than 0.10 s. If your lab partner holds a ruler between your finger and thumb and releases it without warning, how far can you expect the ruler to fall before you catch it? At least
   a. 2 cm
   b. 3 cm
   c. 4 cm
   d. 4.9 cm

6. Two stones are thrown from the top edge of a building with a speed of 10 m/s, one straight down and the other straight up. The first stone hits the street below in 4 s. How much later is it before the second stone hits the street?
   a. 5 s
   b. 4 s
   c 3 s
   d. 2 s

7. A ball is released from rest, and it rolls down a hill with constant acceleration. At the end of one second, it has traveled 3 m. How far has it traveled at the end of four seconds?
   a. 12 m
   b. 16 m
   c. 36 m
   d. 48 m

8. A car travels forward along a straight road at 40 m/s for 1000 m and then travels at 50 m/s for the next 1000 m. What is the average velocity?
   a. 45 m/s
   b. 44.4 m/s
   c. 46 m/s
   d. 0

9. A locomotive slows from 28 m/s to zero in 12 s. What distance does it travel?
   a. 168 m
   b. 196 m
   c. 336 m
   d. 392 m

10. A drag racer starts her car from rest and accelerates at 10 m/s^2 for the entire distance of 400 m (1/4 mile). How long did it take the race car to run the quarter?
   a. 8.94 s
   b. 7.74 s
   c. 6.93 s
   d. 5.99 s

11. A motorist traveling at 110 km/hr is being chased by a police car at 130 km/hr. If the police car starts from 1 kilometer back, how long does it take the police to intercept the motorist?
   a. 3 minutes
   b. 5 minutes
   c. 7.5 minutes
   d. 10 minutes

12. A toy rocket, launched from the ground, rises vertically with an acceleration of 20 m/s^2 for 2 s when it runs out of fuel. Disregarding air resistance, how high will the rocket rise?
   a. 61 m
   b. 81 m
   c. 121 m
   d. 141 m

Chapter 3

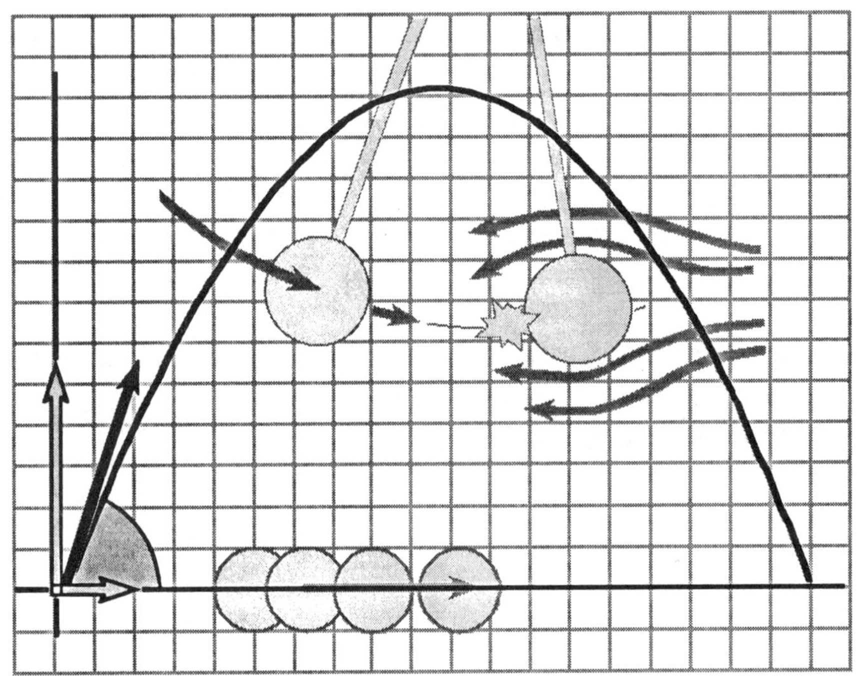

VECTORS AND TWO-DIMENSIONAL MOTION

Chapter 3

VECTORS AND TWO-DIMENSIONAL MOTION

In our discussion of one-dimensional motion in Chapter 2, we used the concept of vectors only to a very limited extent. As we progress in our study of motion, the ability to manipulate vector quantities will become increasingly important. As a result, much of this chapter will be devoted to techniques for adding vectors, subtracting them, and so forth. We will then apply our newfound skills to a special case of two-dimensional motion--projectiles. We shall also see that an understanding of vector manipulation is necessary in order to work with and understand relative motion.

NOTES FROM SELECTED CHAPTER SECTIONS

3.1 Vectors and Scalars Revisited

A **scalar** has only magnitude and no direction. On the other hand, a **vector** is a physical quantity that requires the specification of both direction and magnitude.

3.2 Some Properties of Vectors

Two vectors are **equal** if they have both the same magnitude and direction. When two or more vectors are **added**, they must have the same units. Two vectors which are the **negative** of each other have the same magnitude but opposite directions. When a vector is **multiplied** or **divided** by a positive (negative) scalar, the result is a vector in the same (opposite) direction. The magnitude of the resulting vector is equal to the product of the scalar and the magnitude of the original vector.

3.3 Components of a Vector

Any vector can be completely described by its **components**. See the recommended procedure for adding vectors algebraically in the Skills section.

3.4 Velocity and Acceleration in Two Dimensions

Consider an object moving in the plane containing the x and y axes. The displacement of the object is defined as the change in the position vector, $\Delta\mathbf{r}$. The average velocity of a particle during the time interval Δt is the ratio of the displacement to the time interval for this displacement. The average velocity is a **vector** quantity directed along $\Delta\mathbf{r}$. The instantaneous velocity, $\mathbf{v}$, is defined as the limit of the average velocity, $\Delta\mathbf{r}/\Delta t$, as Δt goes to zero. The direction of the instantaneous velocity vector is along a line that is tangent to the path of the particle and in the direction of motion. The average acceleration of an object whose velocity changes is the ratio of the net change in velocity to the time interval during which the change occurs, $\Delta\mathbf{v}/\Delta t$.

A particle can accelerate in several ways: the magnitude of the velocity vector (the speed) may change; the direction of the velocity vector may change, making a curved path, even though the speed is constant; or both the magnitude and direction of the velocity vector may change.

3.5 Projectile Motion

In the case of projectile motion, if it is assumed that air resistance is negligible and that the rotation of the Earth does not affect the motion, then:

(1) the horizontal component of velocity, v_x, remains constant because there is no horizontal component of acceleration;

(2) the vertical component of acceleration is equal to the acceleration due to gravity, $\mathbf{g}$;

(3) the vertical component of velocity, v_y, and the displacement in the y direction are identical to those of a freely falling body;

(4) projectile motion can be described as a superposition of the two motions in the x and y directions.

Review the recommended procedure in the Skills section solving projectile motion problems.

3.6 Relative Velocity

Observations made by observers in different frames of reference can be related to one another through the techniques of the transformation of relative velocities.

EQUATIONS AND CONCEPTS

Vector quantities obey the commutative law of addition. In order to add vector **A** to vector **B** using the graphical method, first construct **A**, and then draw **B** such that the tail of **B** starts at the head of **A**. The sum of **A** + **B** is the vector that completes the triangle by connecting the tail of **A** to the head of **B**.

$$\mathbf{A} + \mathbf{B} = \mathbf{B} + \mathbf{A}$$

When more than two vectors are to be added, they are all connected head-to-tail in any order and the resultant or sum is the vector which joins the tail of the first vector to the head of the last vector.

$$\mathbf{R} = \mathbf{A} + \mathbf{B} + \mathbf{C} + \mathbf{D}$$

When two or more vectors are to be added, all of them must represent the same physical quantity--that is, have the same units. In the graphical or geometric method of vector addition, the length of each vector corresponds

Comment on vector addition

to the magnitude of the vector according to a chosen scale. Also, the direction of each vector must be along a direction which makes the proper angle relative to the others.

The operation of vector subtraction utilizes the definition of the negative of a vector. The negative of vector **A** is the vector which has a magnitude equal to the magnitude of **A**, but acts or points along a direction opposite the direction of **A**.

$$\mathbf{A} - \mathbf{B} = \mathbf{A} + (-\mathbf{B}) \tag{3.1}$$

A vector **A** in a two-dimensional coordinate system can be resolved into its components along the x and y directions. The projection of **A** onto the x axis is the x component of **A**; and the projection of **A** onto the y axis is the y component of **A**.

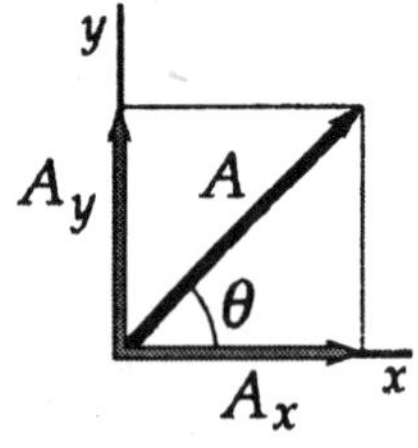

The magnitude of **A** and the angle, θ, which the vector makes with the positive x axis can be determined from the values of the x and y components of **A**.

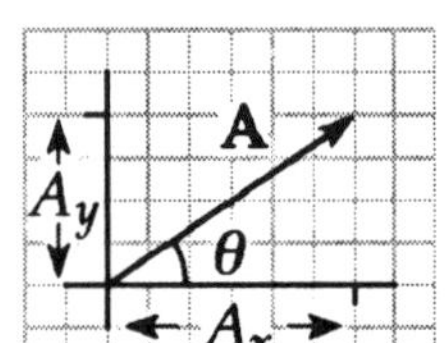

$$A_x = A\cos\theta \tag{3.2}$$

$$A_y = A\sin\theta$$

$$A = \sqrt{A_x^{\ 2} + A_y^{\ 2}} \tag{3.3}$$

$$\tan\theta = \frac{A_y}{A_x} \tag{3.4}$$

Chapter 3

The path of a projectile is curved as shown in the figure below. Such a curve is called a parabola. The velocity vector makes an angle of θ_0 with the horizontal where θ_0 is called the projectile angle. In order to analyze projectile motion, we shall separate the motion into two parts, the x (horizontal) motion and the y (vertical) motion, and solve each part separately.

The initial horizontal and vertical components of velocity of a projectile depend on the value of the initial velocity and the initial angle of launch.

$$v_{x0} = v_0 \cos\theta_0$$

$$v_{y0} = v_0 \sin\theta_0$$

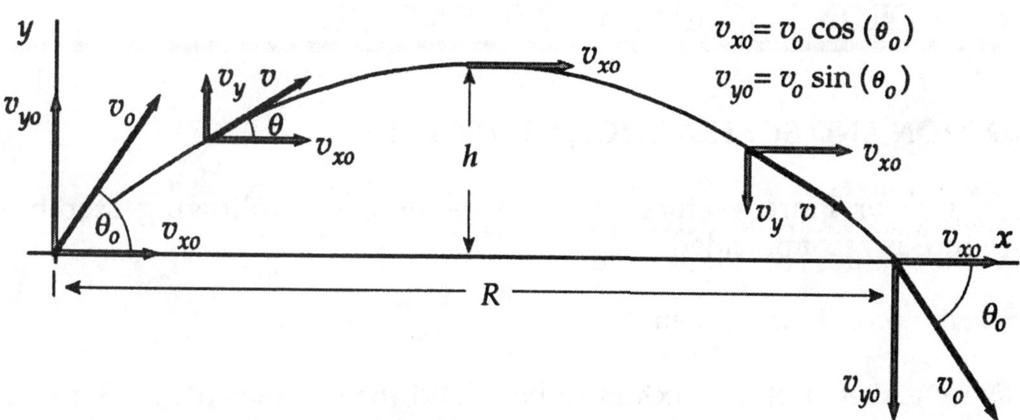

Velocity Components in projectile motion.

The horizontal component of velocity for a projectile remains constant ($a_x = 0$); while the vertical component decreases uniformly with time ($a_y = -g$).	$v_x = v_0 \cos\theta_0 = \text{constant}$	(3.9)
	$v_y = v_{y0} - gt$	(3.11)
	$v_y = v_0 \sin\theta_0 - gt$	
The x and y coordinates of the position of a projectile as functions of the elapsed time.	$x = v_{x0}t = (v_0 \cos\theta_0)t$	(3.10)
The positive direction for the vertical motion is assumed to be upward.	$y = v_{y0}t - \frac{1}{2}gt^2$	(3.12)
	$y = (v_0 \sin\theta_0)t - \frac{1}{2}gt^2$	
The initial and final velocities in the y direction as a function of vertical position.	$v_y^2 = v_{y0}^2 - 2gy$	(3.13)

SUGGESTIONS, SKILLS, AND STRATEGIES

ADDITION AND SUBTRACTION OF VECTORS

When two or more vectors are to be added, the following step-by-step procedure is recommended:

1. Select a coordinate system.

2. Draw a sketch of the vectors to be added (or subtracted), with a label on each vector.

3. Find the x and y components of all vectors.

4. Find the resultant components (the algebraic sum of the components) in both the x and y directions.

5. Use the Pythagorean theorem to find the magnitude of the resultant vector.

6. Use a suitable trigonometric function to find the angle the resultant vector makes with the x axis.

To add vector **A** to vector **B** graphically, first construct **A**, and then draw **B** such that the tail of **B** starts at the head of **A**. The sum **A** + **B** is the vector that completes the triangle as shown in the figure to the right.

The procedure for adding more than two vectors (the polygon rule) is also illustrated.

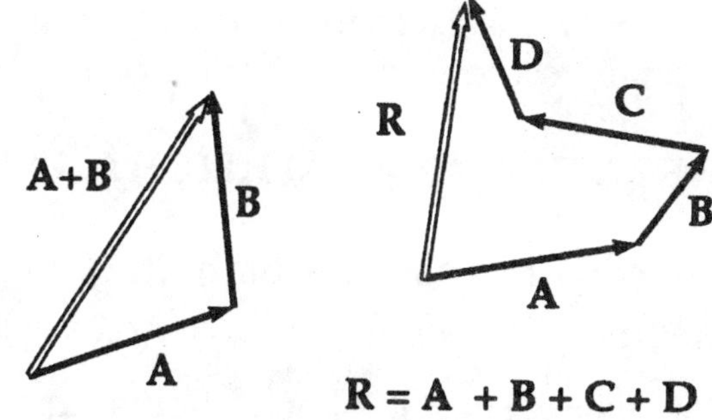

Adding vectors by (a) the triangle rule and (b) the polygon rule.

In order to subtract two vectors graphically, recognize that **A** – **B** is equivalent to the operation **A** + (–**B**). Since the vector –**B** is a vector whose magnitude is B and is opposite in direction to **B**, the construction shown in the second figure at the right is obtained.

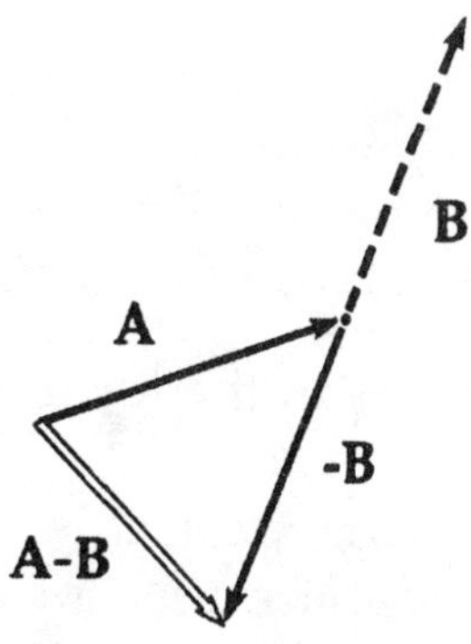

Subtracting two vectors graphically

PROJECTILE MOTION PROBLEMS

The following procedure is recommended for solving projectile motion problems:

1. Select a coordinate system.
2. Resolve the initial velocity vector into x and y components.
3. Treat the horizontal motion and the vertical motion independently.
4. Follow the techniques for solving problems with constant velocity to analyze the horizontal motion of the projectile.
5. Follow the techniques for solving problems with constant acceleration to analyze the vertical motion of the projectile.

THE PATH, MAXIMUM HEIGHT, AND RANGE OF A PROJECTILE

The equation for the path (trajectory) of a projectile can be found by combining Equation 3.10, $x = (v_0 \cos\theta_0)t$ and the equation following Equation 3.12, $y = (v_0 \sin\theta_0)t - \frac{1}{2}gt^2$. Solve equation 3.10 for t to get $t = x/(v_0 \cos\theta_0)$, and substitute this expression for t into the equation for y to find

$$y = (v_0 \sin\theta_0)\frac{x}{v_0 \cos\theta_0} - \frac{1}{2}g\left(\frac{x}{v_0 \cos\theta_0}\right)^2$$

or

$$y = x\tan\theta_0 - \frac{1}{2}\left(\frac{\sqrt{g}}{v_0 \cos\theta_0}\right)^2 x^2$$

Note that this is in the form of the equation of a parabola.

When analyzing projectile motion, there are two quantities of particular interest. These are the horizontal range, R (the maximum value of x), and the maximum height, H (the maximum value of y). These quantities can be easily determined for a projectile that impacts at the same level from which it was launched.

FINDING THE MAXIMUM HEIGHT, *H*

When the projectile reaches maximum height the y-component of the velocity will be zero; otherwise, the projectile would be continuing to rise or would be falling. Therefore, use the equation $v_y = v_0 \sin\theta_0 - gt$ with $t = t_1$ and $v_y = 0$, and solve $0 = v_0 \sin\theta_0 - gt_1$ for the time it takes the projectile to reach the peak of its curve:

$$t_1 = v_0 \sin\theta_0 / g$$

Then, noting that $y = H$ when $t = t_1$, substitute both these values into

$$y = (v_0 \sin\theta_0)t - \tfrac{1}{2}gt^2 \qquad \text{to get} \qquad H = (v_0 \sin\theta_0)\left(\frac{v_0 \sin\theta_0}{g}\right) - \frac{1}{2}g(v_0 \sin\theta_0 / g)^2$$

Simplifying,

$$H = \frac{v_0^2 \sin^2\theta_0}{2g}$$

FINDING THE RANGE, *R*

Use equation 3.10, $x = (v_0 \cos\theta_0)t$. Here t must be the **total** time of flight, t_2; and this is twice the value of t_1, the time to reach the maximum height.

So $R = v_0 \cos\theta_0 (t_2)$ or $R = v_0 \cos\theta_0 (2)(v_0 \sin\theta_0 / g) = \left(2v_0^2 \cos\theta_0 \sin\theta_0\right) / g$

Finally, use the trigonometric identity $\sin 2\theta = 2\sin\theta\cos\theta$

to simplify the expression for range to $R = \left(v_0^2 \sin 2\theta_0\right) / g$

Chapter 3

REVIEW CHECKLIST

- ▷ Recognize that two-dimensional motion in the xy plane with constant acceleration is equivalent to two independent motions: constant velocity along the x-direction and constant acceleration along the y-direction.
- ▷ Sketch a typical trajectory of a particle moving in the xy plane and draw vectors to illustrate the manner in which the displacement, velocity, and acceleration of the particle changes with time.
- ▷ Recognize the fact that if the initial speed v_0 and initial angle θ_0 of a projectile are known at a given point at $t = 0$, the velocity components and coordinates can be found at any later time t. Furthermore, one can also calculate the horizontal range R and maximum height H if v_0 and θ_0 are known.
- ▷ Understand and describe the basic properties of vectors, resolve a vector into its rectangular components, use the rules of vector addition (including graphical solutions for addition of two or more vectors), and determine the magnitude and direction of a vector from its rectangular components.
- ▷ Practice the technique demonstrated in the text to solve relative velocity problems.

SOLUTIONS TO SELECTED END-OF-CHAPTER PROBLEMS

6. A jogger runs 100 m due west, then changes direction for the second leg of the run. At the end of the run, she is 175 m away from the starting point at an angle of 15.0° north of west. What were the direction and length of her second displacement? Use graphical techniques.

Solution The vector $\mathbf{R}$ representing the jogger's net displacement from the starting point to the end of the run is

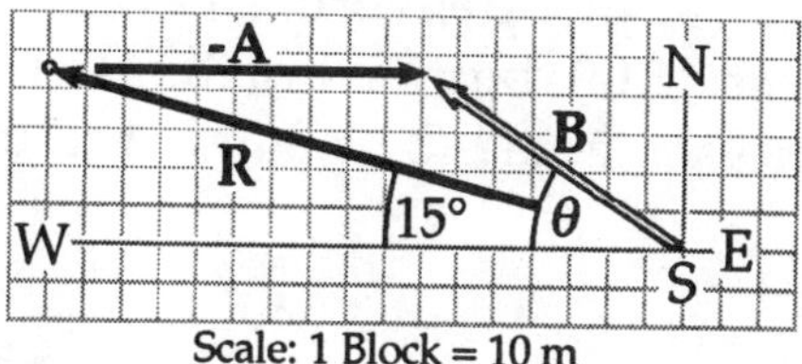

Scale: 1 Block = 10 m

$$\mathbf{R} = \mathbf{A} + \mathbf{B}$$

where **A** and **B** are the displacements that occur during each of the two legs of the run. This equation may also be written as

$$\mathbf{B} = \mathbf{R} - \mathbf{A} = \mathbf{R} + (-\mathbf{A})$$

Thus, we can solve for the second displacement **B** by using graphical techniques to add the known resultant, **R** = 175 m at 15.0° north of west, to the vector $-\mathbf{A}$.

The negative of **A** is a vector having the same magnitude as **A** but whose direction is opposite that of **A**. Therefore,

$$-\mathbf{A} = 100 \text{ m due east}$$

The sketch (above) shows a scale drawing you can make to solve for the second displacement. After the known vectors **R** and **-A** are drawn to the proper length (according to your chosen scale) and in the specified directions, the vector representing their sum is that drawn from the start of **R** to the end of $-\mathbf{A}$ as shown. The magnitude of **B** is found by measuring the length of this vector and multiplying by your scale factor. The angle θ gives the direction of **B** and may be measured on your scale drawing using a protractor.

You should find that **B** has

a magnitude of	approximately 83 m
and is oriented at	about 33° north of west. ◊

9. A girl delivering newspapers covers her route by traveling 3.00 blocks west, 4.00 blocks north, then 6.00 blocks east. (a) What is her resultant displacement? (b) What is the total distance she travels?

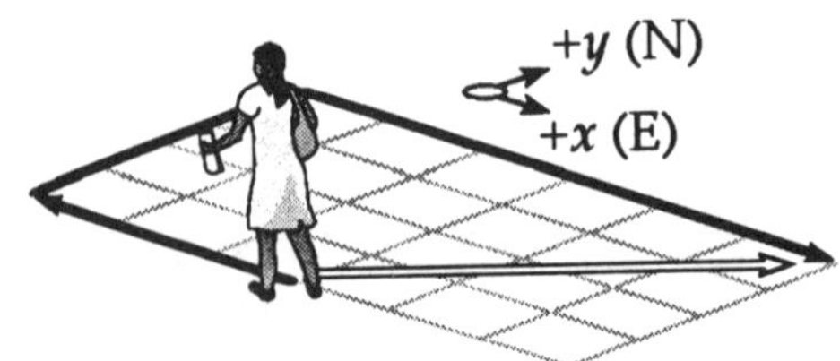

Solution

(a) If eastward and northward are chosen as the $+x$ and the $+y$ directions respectively, the components of each displacement are:

Displacement	x-component	y-component
A	–3.00 blocks	0.00
B	0.00	+4.00 blocks
C	+6.00 blocks	0.00

The components of the resultant are then:

$$R_x = A_x + B_x + C_x = +3.00 \text{ blocks}$$

$$R_y = A_y + B_y + C_y = +4.00 \text{ blocks}$$

The magnitude and direction of the resultant are then given by:

$$R = \sqrt{R_x^2 + R_y^2} = 5 \text{ blocks}$$

$$\theta = \arctan\left(\frac{R_y}{R_x}\right) = 53.1° \text{ north of east.}$$ ◊

(b) The distance traveled is (3 blocks + 4 blocks + 6 blocks) = 13 blocks. ◊

15. A man pushing a mop across a floor causes it to undergo two displacements. The first has a magnitude of 150 cm and makes an angle of 120° with the positive x axis. The resultant displacement has a magnitude of 140 cm and is directed at an angle of 35.0° to the positive x axis. Find the magnitude and direction of the second displacement.

Solution

If **A** and **B** are the first and second displacements, the resultant displacement: $\mathbf{R} = \mathbf{A} + \mathbf{B}$

If we take A_x, A_y, B_x, and B_y to be the components of the first and second displacements, the components of the resultant displacement are given by $R_x = A_x + B_x$ and $R_y = A_y + B_y$

The vectors **R** and **A** are known. Their components are as follows:

x-components:

$$R_x = R\cos\theta_R = (140\text{ cm})\cos 35.0° = +115\text{ cm}$$
$$A_x = A\cos\theta_A = (150\text{ cm})\cos 120° = -75.0\text{ cm}$$

y-components:

$$R_y = R\sin\theta_R = (140\text{ cm})\sin 35.0° = +80.3\text{ cm}$$
$$A_y = A\sin\theta_A = (150\text{ cm})\sin 120° = +130\text{ cm}$$

The components of the second displacement **B** may then be found to be:

$$B_x = R_x - A_x = 190\text{ cm} \quad \text{and} \quad B_y = R_y - A_y = -49.7\text{ cm}$$

The magnitude of **B** is $B = \sqrt{B_x^2 + B_y^2} = \sqrt{(190\text{ cm})^2 + (-49.7\text{ cm})^2} = 196\text{ cm}$ ◊

The direction of **B** is: $\theta_B = \arctan\left(\dfrac{B_y}{B_x}\right) = \arctan(-0.261) = -14.7°$ ◊

Thus, **B** = 196 cm at 14.7° below the positive x direction ◊

17. Tom the cat is chasing Jerry the mouse across a table surface 1.5 m above the floor. Jerry steps out of the way at the last second, and Tom slides off the edge of the table at a speed of 5.0 m/s. Where will Tom strike the floor, and what velocity components will he have just before he hits?

Solution Taking the positive y direction to be upward and the positive x direction to be horizontal, the components of Tom's velocity at the instant he leaves the table are:

$$v_{0x} = +5.0 \text{ m/s} \qquad \text{and} \qquad v_{0y} = 0.0$$

While Tom is in the air, the components of his acceleration are:

$$a_x = 0.0 \qquad \text{and} \qquad a_y = -g = -9.8 \text{ m/s}^2$$

The time required for Tom to reach the floor (i.e., reach $y = -1.5$ m) may be found from

$$y = v_{0y} + \tfrac{1}{2}a_y t^2 \qquad \text{or} \qquad -1.5 \text{ m} = 0 + \tfrac{1}{2}\left(-9.8 \text{ m/s}^2\right)t^2$$

which gives the time of flight as $t = 0.55$ s
Therefore, Tom will hit the floor at a horizontal displacement of:

$$x = v_{0x}t + \tfrac{1}{2}a_x t^2 = (+5.0 \text{ m/s})(0.55 \text{ s}) + 0 = +2.8 \text{ m from the table.} \quad \Diamond$$

His velocity components just before he hits are

$$v_x = v_{0x} + a_x t = 5.0 \text{ m/s} + 0 = +5.0 \text{ m/s} \quad \Diamond$$

and $$v_y = v_{0y} + a_y t = 0 + \left(-9.8 \text{ m/s}^2\right)(0.55 \text{ s}) = -5.4 \text{ m/s} \quad \Diamond$$

---

23. A projectile is launched with an initial speed of 60 m/s at an angle of 30° above the horizontal. The projectile lands on a hillside 4.0 s later. Neglect air friction. (a) What is the projectile's velocity at the highest point of its trajectory? (b) What is the straight-line distance from where the projectile was launched to where it hits?

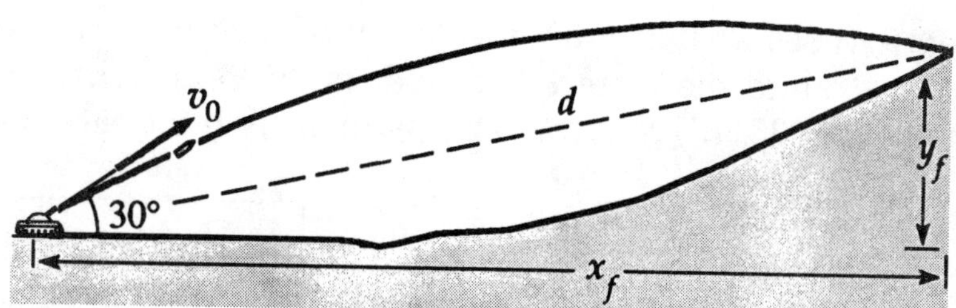

Solution: (a) Choose the origin at the point from which the projectile is launched, with $+y$ upward and $+x$ in the direction of the projectile's horizontal motion. The initial velocity of the projectile then has components of:

$$v_{0x} = v_0 \cos\theta_0 = (60 \text{ m/s})\cos 30° = +52 \text{ m/s}$$

and

$$v_{0y} = v_0 \sin\theta_0 = (60 \text{ m/s})\sin 30° = +30 \text{ m/s}$$

Since an object in free fall has acceleration components of $a_x = 0$ and $a_y = -g = -9.8 \text{ m/s}^2$, the horizontal velocity remains constant while the vertical velocity changes with time. At the highest point of the trajectory, the vertical velocity v_y is zero. The total velocity is then horizontal and has a magnitude of

$$v_x = v_{0x} = +52 \text{ m/s}$$ ◊

(b) When the projectile hits the hillside (at $t = 4.0$ s), its coordinates will be:

$$x_f = v_{0x}t + \tfrac{1}{2}a_x t^2 = (52 \text{ m/s})(4.0 \text{ s}) + 0 = 2.1\times10^2 \text{ m}$$

and

$$y_f = v_{0y}t + \tfrac{1}{2}a_y t^2 = (30 \text{ m/s})(4.0 \text{ s}) - (4.9 \text{ m/s}^2)(16 \text{ s}^2) = +42 \text{ m}$$

The straight-line distance between the launch point and impact point may then be computed using the Pythagorean Theorem as:

$$d = \sqrt{x_f^2 + y_f^2} = \sqrt{(2.1\times10^2 \text{ m})^2 + (42 \text{ m})^2} = 2.1\times10^2 \text{ m}$$ ◊

Note: This result is rounded to 2 significant figures.

27. A rowboat crosses a river with a velocity of 3.30 mi/h at an angle 62.5° north of west relative to the water. The river is 0.505 mi wide and carries an eastward current of 1.25 mi/h. How far upstream is the boat when it reaches the opposite shore?

Solution The velocity of the boat relative to the shore, $\mathbf{v}_{bs}$, may be expressed as the sum $\mathbf{v}_{bs} = \mathbf{v}_{bw} + \mathbf{v}_{ws}$, where $\mathbf{v}_{bw}$ is the velocity of the boat relative to the water and $\mathbf{v}_{ws}$ is the velocity of the water relative to the shore. Graphical addition of these vectors is illustrated in the sketch. Each component of the boat's velocity is as follows:

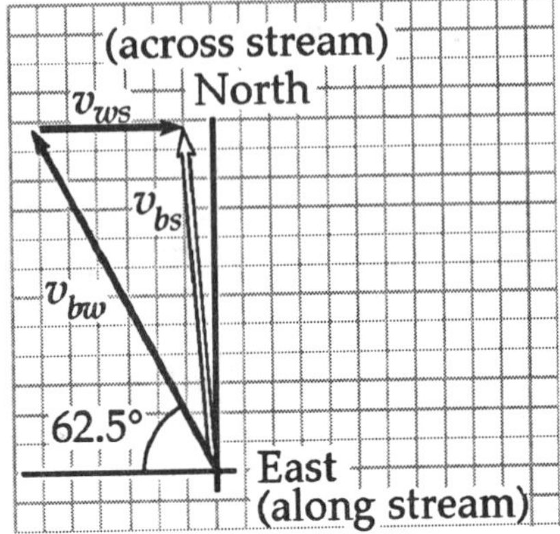

Directed across the stream (Northward),

$$(\mathbf{v}_{bs})_{\text{north}} = (\mathbf{v}_{bw})_{\text{north}} + (\mathbf{v}_{ws})_{\text{north}} = (3.30 \text{ mph})\sin 62.5° + 0 = 2.93 \text{ mph}$$

Directed parallel to the stream (Eastward),

$$(\mathbf{v}_{bs})_{\text{east}} = (\mathbf{v}_{bw})_{\text{east}} + (\mathbf{v}_{ws})_{\text{east}} = -(3.30 \text{ mph})\cos 62.5° + 1.25 \text{ mph} = -0.274 \text{ mph}$$

The time required for the boat to cross the stream (i.e., move 0.505 mi north) is therefore:

$$t = \frac{0.505 \text{ mi}}{(\mathbf{v}_{bs})_{\text{north}}} = \frac{0.505 \text{ mi}}{2.93 \text{ mi/h}} = 0.172 \text{ h}$$

The displacement of the boat parallel to the stream during this time is given by: $x = (\mathbf{v}_{bs})_{\text{east}} t = (-0.274 \text{ mi/h})(0.172 \text{ h}) = -4.71 \times 10^{-2} \text{ mi}$. Thus, as the boat crosses the river, it moves in the negative eastward (i.e., westward or upstream) direction a distance $|x| = (4.71 \times 10^{-2} \text{ mi})\left(\frac{5280 \text{ ft}}{1.00 \text{ mi}}\right) = 249 \text{ ft}$ ◊

35. A car travels due east with a speed of 50.0 km/h. Rain is falling vertically with respect to the Earth. The traces of the rain on the side windows of the car make an angle of 60.0° with the vertical. Find the velocity of the rain with respect to (a) the car and (b) the Earth.

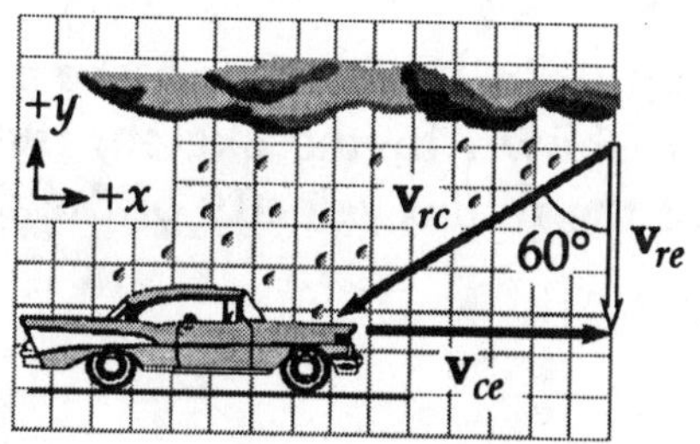

Solution (a) The velocity of the rain relative to the Earth $\mathbf{v}_{re}$ may be written as $\mathbf{v}_{re} = \mathbf{v}_{rc} + \mathbf{v}_{ce}$, where $\mathbf{v}_{rc}$ is the velocity of the rain relative to the car and $\mathbf{v}_{ce}$ is the velocity of the car relative to the Earth. From the given information, we know that $\mathbf{v}_{ce} = 50.0$ km/h directed due eastward, $\mathbf{v}_{re}$ is directed vertically downward, and $\mathbf{v}_{rc}$ is directed downward at 60.0° from the vertical. This information is summarized in the graphical addition of velocity vectors shown in the sketch. From this vector diagram, it is seen that $(\mathbf{v}_{re})_x = (\mathbf{v}_{rc})_x + (\mathbf{v}_{ce})_x$. This gives $0 = -v_{rc}\sin 60° + 50.0$ km/h, or

$$v_{rc} = \frac{50.0 \text{ km/h}}{\sin 60.0°} = 57.7 \text{ km/h}$$

The velocity of the rain relative to the car is:

$$\mathbf{v}_{rc} = 57.7 \text{ km / h at } 60.0° \text{ west of vertical.} \qquad \Diamond$$

(b) Considering the vertical components of velocity, it is seen that

$$(\mathbf{v}_{re})_y = (\mathbf{v}_{rc})_y + (\mathbf{v}_{ce})_y \qquad \text{or} \qquad -v_{re} = -v_{rc}\cos 60.0° + 0$$

and $$v_{re} = (57.7 \text{ km/h})\cos 60.0° = 28.9 \text{ km/h}$$

Thus, the velocity of the rain relative to the Earth is

$$28.9 \text{ km/h vertically downward} \qquad \Diamond$$

40. A daredevil decides to jump a canyon of width 10 m. To do so, he drives a motorcycle up an incline sloped at an angle of 15°. What minimum speed must he have in order to clear the canyon?

Solution Choosing the reference axes as shown in the sketch, the components of the initial velocity of the daredevil are

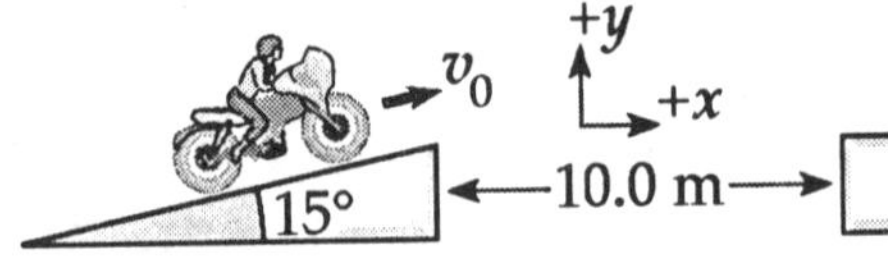

$$v_{0y} = v_0 \sin 15.0° \quad \text{and} \quad v_{0x} = v_0 \cos 15.0°$$

His acceleration components while in the air are $a_x = 0$ and $a_y = -g = -9.80 \text{ m/s}^2$. The y-coordinate of the daredevil at any time is given by $y = v_{0y}t + \frac{1}{2}a_y t^2$. When the jumper is at the level of the canyon rim ($y = 0$), this equation reduces to $0 = (v_0 \sin 15.0°)t - (4.90 \text{ m/s}^2)t^2$. This has one solution $t = 0$ which coincides with the start of the jump, and a second solution $t = (v_0 \sin 15.0°/4.90 \text{ m/s}^2)$ which gives the time when the jumper returns to the level of the canyon rim.

Since $a_x = 0$, the jumper maintains a constant horizontal velocity $v_x = v_{0x}$, his horizontal displacement at any time is given by $x = v_{0x}t$. If the jumper is to successfully cross the canyon, it is necessary to have $x \geq 10.0$ m at the time he returns to the level of the rim. Thus, the requirement for a successful jump becomes

$$x = v_{0x}t = (v_0 \cos 15.0°)\left(\frac{v_0 \sin 15.0°}{4.90 \text{ m/s}^2}\right) \geq 10.0 \text{ m}$$

with an initial speed $$v_0 \geq \sqrt{\frac{(10.0 \text{ m})(4.90 \text{ m/s}^2)}{(\sin 15.0°)(\cos 15.0°)}} = 14.0 \text{ m/s}$$ ◊

45. A quarterback throws a football toward a receiver with an initial speed of 20.0 m/s, at an angle of 30.0° above the horizontal. At that instant, the receiver is 20.0 m from the quarterback. In what direction and with what constant speed should the receiver run in order to catch the football at the level at which it was thrown?

Solution Choose reference axes with the origin at the point the ball is released, $+y$ directed upward, and $+x$ directed from the quarterback toward the receiver. The initial velocity components for the ball are

$$\mathbf{v}_{0x} = (20.0 \text{ m/s})\cos 30.0° = 17.3 \text{ m/s}$$

and

$$\mathbf{v}_{0y} = (20.0 \text{ m/s})\sin 30.0° = 10.0 \text{ m/s}$$

While the ball is in the air, its components of acceleration are

$$a_x = 0 \qquad \text{and} \qquad a_y = -g = -9.80 \text{ m/s}^2$$

The times at which the ball will be at the level from which it is thrown $(y = 0)$ may be found from $y = v_{0y}t + \frac{1}{2}a_y t^2$ as:

$$0 = (10.0 \text{ m/s})t + \tfrac{1}{2}(-9.80 \text{ m/s}^2)t^2$$

so

$$t = 0 \qquad \text{or} \qquad t = \frac{2(10.0 \text{ m/s})}{9.80 \text{ m/s}^2} = 2.04 \text{ s}$$

The x-coordinate of the ball at the time it returns to the level from which it was thrown is $x = v_{0x}t = (17.3 \text{ m/s})(2.04 \text{ s}) = 35.3 \text{ m}$. If the receiver is to be at this x-coordinate at $t = 2.04$ s, he must run a distance of

$$\Delta x = (35.3 \text{ m} - 20.0 \text{ m}) = 15.3 \text{ m} \quad \text{in a time of 2.04 s}$$

The constant velocity needed is

$$v = \frac{\Delta x}{t} = \frac{15.3 \text{ m}}{2.04 \text{ s}} = 7.50 \text{ m/s} \quad \text{directed away from the quarterback.}$$ ◊

51. A hunter wishes to cross a river that is 1.5 km wide and flows with a speed of 5.0 km/h parallel to its banks. The hunter uses a small powerboat that moves at a maximum speed of 12 km/h with respect to the water. What is the minimum time necessary for crossing?

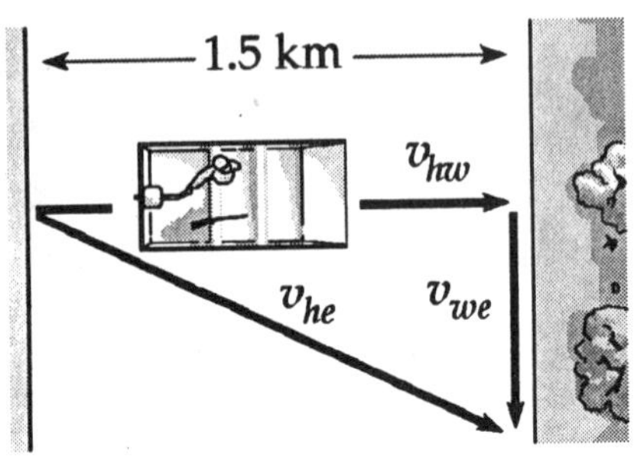

Solution The hunter will cross the river in minimum time if his velocity relative to the water carries him perpendicularly to the flow of the stream (i.e., straight across the current). The velocity of the hunter relative to the earth equals the vector sum of his velocity relative to the water and the velocity of the water relative to the earth. This is summarized in the equation

$$\mathbf{v}_{he} = \mathbf{v}_{hw} + \mathbf{v}_{we}$$

and the vector triangle shown in the sketch illustrates the conditions for minimum crossing time. Under these conditions, the time to cross the river is given by

$$t = \frac{\text{width of stream}}{v_{hw}}$$

Substituting in known values,

$$t = \frac{(1.50 \text{ km})(60 \text{ min/h})}{(12.0 \text{ km/h})} = 7.50 \text{ min}$$ ◊

Also, during this time the hunter will be carried downstream a distance of

$$d = v_{we}t = \frac{(5.00 \text{ km/h})(7.50 \text{ min})}{(60 \text{ min/h})} = 0.625 \text{ km}$$ ◊

53. A daredevil is shot out of a cannon at 45.0° to the horizontal with an initial speed of 25.0 m/s. A net is positioned a horizontal distance of 50.0 m from the cannon. At what height above the cannon should the net be placed in order to catch the daredevil?

Solution

Choose a reference frame with its origin at the point where the daredevil leaves the cannon, with the x-axis horizontal and the y-axis vertical. Then, the components of the daredevil's initial velocity and acceleration are:

$$v_{0x} = (25.0 \text{ m/s})\cos 45° = 17.7 \text{ m/s} \qquad a_x = 0$$

$$v_{0y} = (25.0 \text{ m/s})\sin 45° = 17.7 \text{ m/s} \qquad \text{and} \qquad a_y = -g = -9.80 \text{ m/s}^2$$

Thus, the horizontal velocity $v_x = v_{0x} + a_x t$ is constant and the time to travel the horizontal distance of 50.0 m to the net is

$$t = \frac{d}{v_x} = \frac{50.0 \text{ m}}{17.7 \text{ m/s}} = 2.83 \text{ s}$$

The daredevil's y-coordinate at this time is given by $y = v_{0y}t + \frac{1}{2}a_y t^2$ as:

$$y = (17.7 \text{ m/s})(2.83 \text{ s}) + \tfrac{1}{2}(-9.80 \text{ m/s}^2)(2.83 \text{ s})^2 = +10.8 \text{ m}$$

The net should be placed 10.8 m above the level of the cannon. ◊

57. Instructions for finding a buried treasure include the following: Go 75 paces at 240°, turn to 135° and walk 125 paces, then travel 100 paces at 160°. Determine the resultant displacement from the starting point.

Solution Choosing the origin of the reference frame at the starting point, the components of the individual displacements are:

$$d_{1x} = (75.0)\cos 240° = -37.5 \text{ paces} \qquad d_{1y} = (75.0)\sin 240° = -65.0 \text{ paces}$$
$$d_{2x} = (125)\cos 135° = -88.4 \text{ paces} \qquad d_{2y} = (125)\sin 135° = +88.4 \text{ paces}$$
$$d_{3x} = (100)\cos 160° = -94.0 \text{ paces} \qquad d_{3y} = (100)\sin 160° = +34.2 \text{ paces}$$

The x- and y-components of the resultant displacement are then given by:

$$R_x = \sum_{i=1}^{3} d_{ix} = (-37.5 - 88.4 - 94.0) \text{ paces} = -220 \text{ paces}$$

and $$R_y = \sum_{i=1}^{3} d_{iy} = (-65.0 + 88.4 + 34.2) \text{ paces} = +57.6 \text{ paces}$$

The magnitude of the resultant is

$$R = \sqrt{R_x^2 + R_y^2} \quad \text{or} \quad R = \sqrt{(-220 \text{ paces})^2 + (+57.6 \text{ paces})^2} = 227 \text{ paces}$$

The direction of the resultant found from

$$\theta = \arctan\left(\frac{R_y}{R_x}\right) = \arctan\left(\frac{+57.6 \text{ paces}}{-220 \text{ paces}}\right) = \arctan(-0.262)$$

This yields two possible answers: $\theta = -14.7°$, or $\theta = 165°$. When it is observed that $R_x < 0$, while $R_y > 0$, it becomes clear that the resultant vector **R** must lie in the second quadrant, not the fourth. The desired angle is then $\theta = 165°$, and the resultant is: $\mathbf{R} = 227$ paces at 165°. ◊

CHAPTER SELF-QUIZ

1. Vector **A** points north and vector **B** points east. If we subtract **A** from **B**, the direction of the vector (**B** – **A**) points
   a. north of east
   b. south of east
   c. north of west
   d. south of west

2. A baseball is thrown by the center fielder (from shoulder level) to home plate where it is caught (on the fly at shoulder level) by the catcher. At what point is the ball's speed at a minimum? (Air resistance is negligible.)
   a. just after leaving the center fielder's hand
   b. just before arriving at the catcher's mitt
   c. at the top of the trajectory
   d. speed is constant during entire trajectory

3. A baseball is thrown by the center fielder (from shoulder level) to home plate where it is caught (on the fly at shoulder level) by the catcher. At what point does the magnitude of the horizontal component of velocity have its minimum value? (Air resistance is negligible.)
   a. just after leaving the center fielder's hand
   b. just before arriving at the catcher's mitt
   c. at the top of the trajectory
   d. magnitude of the horizontal component of velocity is constant

4. A city jogger runs four blocks due east, eight blocks due south, and another two blocks due east. Assume all blocks are of equal size. What is the magnitude of the jogger's displacement, start to finish?
   a. 14.0 blocks
   b. 8.4 blocks
   c. 2.0 blocks
   d. 10.0 blocks

5. A student adds two vectors with magnitudes of 100 and 40. Of the following, which one is the only possible choice for the magnitude of the resultant?

   a. 150
   b. 120
   c. 50
   d. 25

6. A string attached to an airborne kite is maintained at an angle of 35° with the horizontal. If a total of 80 m of string is reeled in while bringing the kite back to the ground, what is the horizontal displacement of the kite in the process (assume the kite string doesn't sag)?

   a. 114 m
   b. 56 m
   c. 46 m
   d. 66 m

7. Find the resultant of the following two vectors: i) 100 units due east and ii) 200 units 30° north of west.

   a. 150 units 30° north of west
   b. 308 units 15° north of west
   c. 273 units 60° north of west
   d. 124 units 54° north of west

8. A bullet is fired from a gun at 300 m/s. It hits the ground 3 s later. At what angle above the horizon was the bullet fired? (Ignore air friction.)

   a. 3°
   b. 10°
   c. 30°
   d. 80°

9. A jet airliner moving at 600 mph due east moves into a region where the wind is blowing at 100 mph in a direction 30° north of east. What is the new velocity and direction of the aircraft?

   a. 606 mph, 7.1° N of E
   b. 629 mph, 5.6° N of E
   c. 650 mph, 4.7° N of E
   d. 688 mph, 4.2° N of E

10. A river flows due east at 2 m/s. A boat crosses the 300-m wide river by maintaining a constant velocity of 10 m/s due north relative to the water If no correction is made for the current, how far downstream does the boat move by the time it reaches the far shore?

    a. 6 m
    b. 30 m
    c. 60 m
    d. 90 m

11. A golfer wants to drive a golf ball a distance of 310 yards (283 m). If the 4-wood launches the ball at 15 ° above the horizontal, what must be the initial speed of the ball to achieve the required distance? (Ignore air friction and use $g = 9.8\ m/s^2$).

    a. 74.5 m/s
    b. 57.7 m/s
    c. 44.1 m/s
    d. 37.2 m/s

12. A baseball leaves the bat with a speed of 40 m/s and an angle of 37° above the horizontal. A very high fence is located at a horizontal distance of 128 m from the point where the ball is struck. Assuming the ball leaves the bat near ground level, how far above ground level does the ball strike the fence?

    a. 4.4 m
    b. 8.8 m
    c. 13.2 m
    d. 17.8 m

Chapter 4

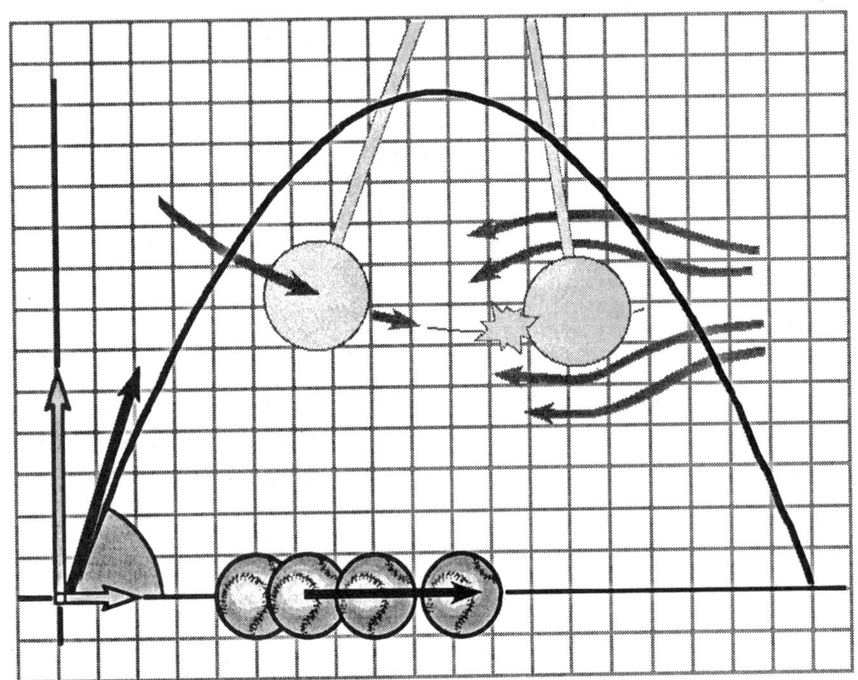

THE LAWS OF MOTION

Chapter 4

THE LAWS OF MOTION

In the foregoing chapters on kinematics, we used the definitions of displacement, velocity, and acceleration to describe motion, without concerning ourselves with the causes of that motion. In this chapter we shall investigate motion in terms of the forces that cause it. We shall then discuss the three fundamental laws of motion, which are based on experimental observations and were formulated by Sir Isaac Newton three centuries ago.

NOTES FROM SELECTED CHAPTER SECTIONS

4.1 The Concept of Force

Equilibrium is the condition under which the **net force** (vector sum of all forces) acting on an object is zero. An object in equilibrium has a zero acceleration (velocity is constant or equals zero).

Fundamental forces in nature are: (1) gravitational (attractive forces between objects due to their masses), (2) electromagnetic forces (between electric charges at rest or in motion), (3) strong nuclear forces (between subatomic particles), and (4) weak nuclear forces (accompanying the process of radioactive decay).

Classical physics is concerned with contact forces (which are the result of physical contact between two or more objects) and action-at-a-distance forces (which act through empty space and do not involve physical contact).

4.2 Newton's First Law

Newton's first law is called the **law of inertia** and states that an object at rest will remain at rest and an object in motion will remain in motion with a constant velocity unless acted on by a **net external force.**

Mass and **weight** are two different physical quantities. The weight of a body is equal to the **force of gravity** acting on the body and varies with location in the Earth's gravitational field. Mass is an **inherent property** of a body and is a measure of the body's inertia (resistance to change in its state of motion). The SI unit of mass is the **kilogram** and the unit of weight is the **newton**.

4.3 Newton's Second Law

Newton's second law, the **law of acceleration**, states that the acceleration of an object is directly proportional to the resultant force acting on it and inversely proportional to its mass. The direction of the acceleration is in the direction of the net force.

4.4 Newton's Third Law

Newton's third law, the **law of action-reaction**, states that when two bodies interact, the force which body "A" exerts on body "B" (the **action force**) is equal in magnitude and opposite in direction to the force which body "B" exerts on body "A" (the **reaction force**). A consequence of the third law is that forces occur in **pairs**. Remember that the action force and the reaction force act on **different objects**.

4.5 Some Applications of Newton's Laws

Construction of a **free-body diagram** is an important step in the application of Newton's laws of motion to the solution of problems involving bodies in equilibrium or accelerating under the action of external forces. The diagram should include an arrow labeled to identify each of the external forces acting on the body whose motion (or condition of equilibrium) is to be studied. Forces which are the **reactions** to these external forces must **not** be included. When a system consists of more than one body or mass, you must construct a free-body diagram for each mass.

4.6 Force of Friction

When a body is in motion either on a surface or through a viscous medium such as air or water, there is resistance to the motion because the body interacts with its surroundings. We call such resistance a **force of friction**. Experiments show that the frictional force arises from the nature of the two surfaces. To a good approximation, both $f_{s,\max}$ (maximum force of static friction) and f_k (force of kinetic friction) are proportional to the normal force at the interface between the two surfaces.

EQUATIONS AND CONCEPTS

A quantitative measurement of mass (the term used tó measure inertia) can be made by comparing the accelerations that a given force will produce on different bodies. If a given force acting on a body of mass m_1 produces an acceleration a_1 and the same force acting on a body of mass m_2 produces an acceleration a_2, the ratio of the two masses equals the inverse of the ratio of the two accelerations.

$$\frac{m_1}{m_2} = \frac{a_2}{a_1}$$

The acceleration of an object is proportional to the resultant force acting on it and inversely proportional to its mass. This is a statement of Newton's second law.

$$\Sigma \mathbf{F} = m\mathbf{a} \quad \text{or} \quad \mathbf{a} = \frac{\Sigma \mathbf{F}}{m} \tag{4.1}$$

When several forces act on an object, it is often convenient to write the equation expressing Newton's second law as component equations. The orientation of the coordinate system can often be chosen so that the object has a nonzero acceleration along only one direction.

$$\Sigma F_x = m a_x \tag{4.2}$$

$$\Sigma F_y = m a_y$$

$$\Sigma F_z = m a_z$$

Calculations with Equation 4.2 must be made using a consistent set of units for the quantities' force, mass, and acceleration. The SI unit of force is the newton (N), defined as the force that, when acting on a 1-kg mass, produces an acceleration of 1 m/s^2.

$$1\ \text{N} \equiv 1\ \text{kg} \cdot \text{m/s}^2 \tag{4.3}$$

$$1\ \text{dyne} \equiv 1\ \text{g} \cdot \text{cm/s}^2 \tag{4.4}$$

$$1\ \text{N} \equiv 0.225\ \text{lb} \tag{4.5}$$

Weight is not an inherent property of a body, but depends on the local value of g and varies with location.

$$w = mg \tag{4.6}$$

This is a statement of Newton's third law, which states that forces always occur in pairs; and the force exerted on body 1 by body 2 is equal in magnitude and opposite in direction to the force exerted on body 2 by body 1.

$$\mathbf{F}_{12} = \mathbf{F}_{21}$$

The force of static friction between two surfaces in contact but not in motion, relative to each other, cannot be greater than $\mu_s n$, where n is the normal (perpendicular) force between the two surfaces and μ_s (coefficient of static friction) is a dimensionless constant which depends on the nature of the pair of surfaces.

$$f_s \leq \mu_s n \quad (4.9)$$

When two surfaces are in relative motion, the force of kinetic friction on each body is directed opposite to the direction of motion of the body.

$$f_k = \mu_k n \quad (4.10)$$

SUGGESTIONS, SKILLS, AND STRATEGIES

The following procedure is recommended for problems involving objects in equilibrium.

1. Make a sketch of the object under consideration.

2. Draw a free-body diagram and label all external forces acting on the object. Try to guess the correct direction for each force. If you select a direction that leads to a negative sign in your solution for a force, do not be alarmed; this merely means that the direction of the force is the opposite of what you assumed.

3. Resolve all forces into x and y components, choosing a convenient coordinate system.

4. Use the equations $\Sigma F_x = 0$ and $\Sigma F_y = 0$. Remember to keep track of the signs of the various force components.

5. Application of Step 4 above leads to a set of equations with several unknowns. All that is left is to solve the simultaneous equations for the unknowns in terms of the known quantities.

The following procedure is recommended when dealing with problems involving the application of Newton's second law:

1. Draw a simple, neat diagram of the system.

2. Isolate the object of interest whose motion is being analyzed. Draw a free-body diagram for this object; that is, a diagram showing **all external forces acting on the object.** For systems containing more than one object, draw **separate** diagrams for each object. Do not include forces that the object exerts on its surroundings.

3. Establish convenient coordinate axes for each body and find the components of the forces along these axes.

4. Apply Newton's second law, $\Sigma \mathbf{F} = m\mathbf{a}$, in the x and y directions for each object under consideration.

5. Solve the component equations for the unknowns. Remember that you must have as many independent equations as you have unknowns in order to obtain a complete solution.

6. Often in solving such problems, one must also use the equations of kinematics (motion with constant acceleration) to find all the unknowns.

REVIEW CHECKLIST

▷ State in your own words a description of Newton's laws of motion, recall physical examples of each law, and identify the action-reaction force pairs in a multiple-body interaction problem as specified by Newton's third law.

▷ Express the normal force in terms of other forces acting on an object and write out the equation which relates the coefficient of friction, force of friction and normal force between an object and surface on which it rests or moves.

▷ Apply Newton's laws of motion to various mechanical systems using the recommended procedure discussed in Section 4.5. Most important, you should identify all external forces acting on the system, draw the **correct** free-body diagrams which apply to each body of the system, and apply Newton's second law, $\mathbf{F} = m\mathbf{a}$, in **component** form.

▷ Apply the equations of kinematics (which involve the quantities' displacement, velocity, and acceleration) as described in Chapter 2 along with those methods and equations of Chapter 4 (involving mass, force, and acceleration) to the solutions of problems where **both** the kinematic and dynamic aspects are present.

▷ Be familiar with solving several linear equations simultaneously for the unknown quantities. Recall that you must have as many **independent** equations as you have unknowns.

SOLUTIONS TO SELECTED END-OF-CHAPTER PROBLEMS

3. A bag of sugar weighs 5.00 lb on Earth. What should it weigh in newtons on the Moon, where the acceleration due to gravity is 1/6 that on Earth? Repeat for Jupiter, where g is 2.64 times Earth gravity. Find the mass in kilograms at each of the three locations.

Solution To solve this problem, one must realize that the weight, w, of an object located somewhere in space is given by $w = mg$, where m is the mass of the object and g is the acceleration due to gravity at that location. Also, the mass is a property of the object and is constant (neglecting relativistic effects). However, the acceleration due to gravity, and therefore the weight of the object, does vary as the object moves to different points in space.

If w_E is the weight of the object on Earth where the acceleration due to gravity is g_E and w_M is its weight on the moon where gravity is g_M, the ratio of these weights is

$$(w_M/w_E) = (mg_M/mg_E) = g_M/g_E$$

The weight on the moon is then given by $w_M = w_E(g_M/g_E) = w_E/6$. In an identical manner, the weight of the object on Jupiter will be: $w_J = w_E(g_J/g_E) = 2.64w_E$. Thus, a bag of sugar that has a weight on Earth of

$$w_E = 5.00 \text{ lbs} = (5.00 \text{ lbs})(4.448 \text{ N} / 1.00 \text{ lb}) = 22.24 \text{ N}$$

will have a weight on the moon or on Jupiter, of

is $$w_M = \frac{22.24 \text{ N}}{6} = 3.71 \text{ N} \quad \text{and} \quad w_J = 2.64(22.24 \text{ N}) = 58.7 \text{ N} \quad \lozenge$$

The constant mass of the object may be found as:

$$m = w_E/g_E = (22.2 \text{ N})/(9.80 \text{ m/s}^2) = 2.27 \text{ kg} \qquad \lozenge$$

5. The air exerts a forward force of 10 N on the propeller of a 0.20-kg model airplane. If the plane accelerates forward at 2.0 m/s², what is the magnitude of the resistive force exerted by the air on the airplane?

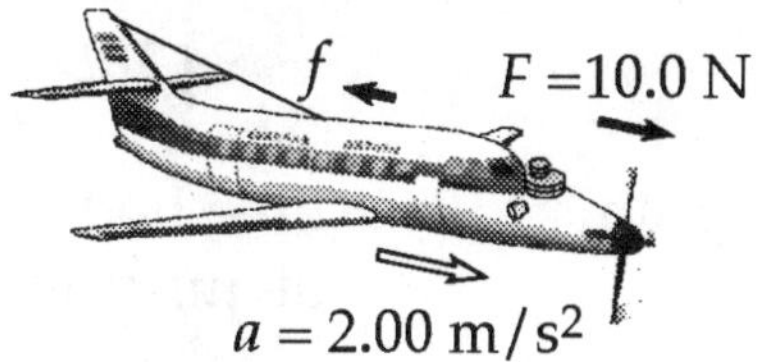

Solution

The resultant force acting on the airplane is in the direction of the acceleration and has a value of

$$F_{net} = ma = (0.20\text{ kg})(2.0\text{ m/s}^2) = 0.40\text{ N}$$

This resultant force is the **vector sum** of the forward thrust, F, due to the propeller and the rearward force, f, due to air resistance.

Thus, $$F_{\text{net}} = F - f$$

and the resistive force is $$f = F - F_{\text{net}} = 10\text{ N} - 0.40\text{ N} = 9.6\text{ N}$$ ◊

---

13. A 150-N bird feeder is supported by three cables as shown in Figure P4.13. Find the tension in each cable.

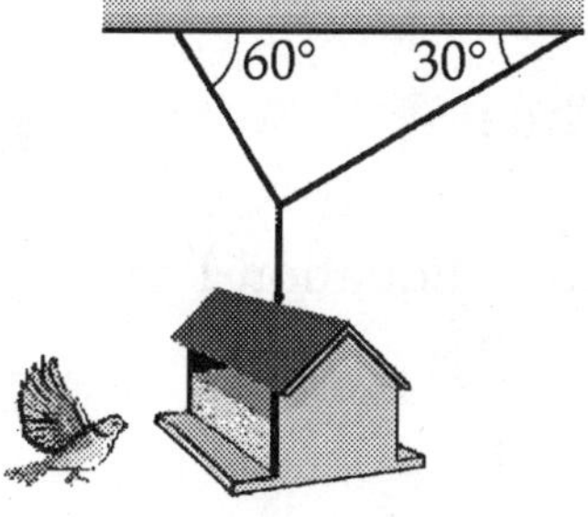

Figure P4.13

Solution The bird feeder as well as the junction in the supporting cables are held in equilibrium by the forces acting on them. Thus, the first condition of equilibrium may be applied each of these objects. Consider the free-body diagrams (A) and (B) for these objects.

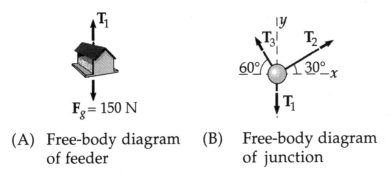

(A) Free-body diagram of feeder

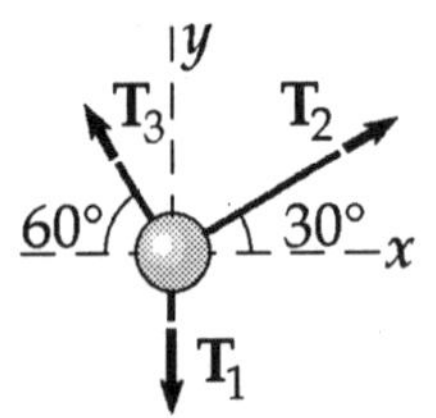

(B) Free-body diagram of junction

Note : Newton's third law has been observed in the directions of the action-reaction forces labeled T_1 in these diagrams. Considering diagram (A) gives:

$\sum F_y = +T_1 - 150 \text{ N} = 0$, or $T_1 = 150 \text{ N}$ ◊

Consideration of diagram (B) yields two equations:

$\sum F_x = +T_2 \cos 30° - T_3 \cos 60° = 0$ $\sum F_y = T_2 \sin 30° + T_3 \sin 60° - T_1 = 0$

which become, respectively

$T_3 = \left(\frac{\cos 30°}{\cos 60°}\right) T_2 = 1.73\, T_2$ [1] $(0.500)T_2 + (0.866)T_3 = 150 \text{ N}$ [2]

Substituting [1] into [2], $\left[0.500 + (0.866)(1.73)\right] T_2 = 150 \text{ N}$

Thus, $T_2 = 75 \text{ N}$ ◊

Then Equation 1 yields: $T_3 = (1.73)(75 \text{ N})$ or $T_3 = 130 \text{ N}$ ◊

---

19. A 2000-kg car is slowed down uniformly from 20.0 m/s to 5.00 m/s in 4.00 s. What average force acted on the car during this time, and how far did the car travel during the deceleration?

Solution Since the acceleration of the car is uniform, the instantaneous acceleration is the same as the average acceleration. Choosing the positive direction to be in the direction of the initial velocity, this acceleration is:

$$\mathbf{a} = \bar{\mathbf{a}} = \frac{\mathbf{v}_f - \mathbf{v}_i}{\Delta t} = \frac{5.00 \text{ m/s} - 20.0 \text{ m/s}}{4.00 \text{ s}} = -3.75 \text{ m/s}^2$$

The average resultant force acting on the car is then given by Newton's second law as:

$$\bar{\mathbf{F}} = m\bar{\mathbf{a}} = (2000 \text{ kg})(-3.75 \text{ m/s}^2) = -7.50 \times 10^3 \text{ N}$$

or 7.50×10^3 N in the direction opposite to the car's initial velocity. ◊

The displacement of the car during this period of constant acceleration is

$$\Delta\mathbf{x} = \bar{\mathbf{v}}t = \left(\frac{\mathbf{v}_f + \mathbf{v}_i}{2}\right)t = \left(\frac{5.00 \text{ m/s} + 20.0 \text{ m/s}}{2}\right)(4.00 \text{ s}) = +50.0 \text{ m}$$

or 50.0 m in the direction of the car's initial velocity. ◊

---

27. A 1000-kg car is pulling a 300-kg trailer. Together the car and trailer have an acceleration of 2.15 m/s^2 in the forward direction. Neglecting frictional forces on the trailer, determine (a) the net force on the car; (b) the net force on the trailer; (c) the force exerted on the car by the trailer; (d) the resultant force exerted on the road by the car.

Solution Choose the $+x$ direction to be horizontal and forward with the $+y$ direction to be upward. The acceleration of both the car and the trailer then has components of $a_x = +2.15 \text{ m/s}^2$ and $a_y = 0$.

(a) The net force on the car is in the direction of the car's acceleration (in the forward direction) and has the magnitude:

$$(F_{\text{car}})_{\text{net}} = m_{\text{car}}a = (1000 \text{ kg})(2.15 \text{ m/s}^2) = 2.15 \times 10^3 \text{ N}$$ ◊

(b) Likewise, the net force on the trailer is $(F_{\text{trailer}})_{\text{net}} = m_{\text{trailer}}\,a$, or

$$(F_{\text{trailer}})_{\text{net}} = (300\text{ kg})(2.15\text{ m/s}^2) = 645\text{ N} \quad \text{(also in forward direction).} \quad \Diamond$$

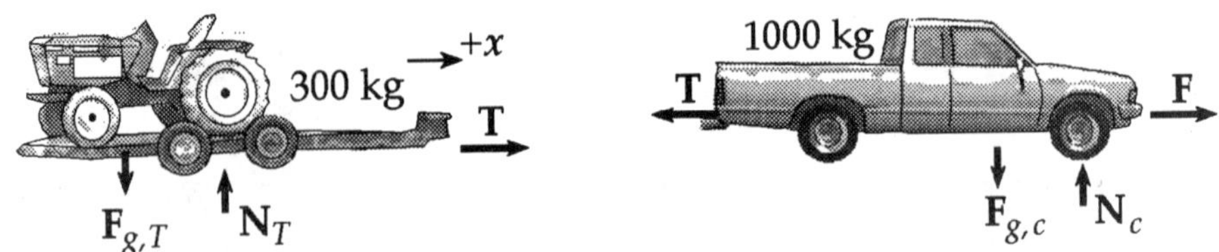

(c) Consider the free-body diagrams for the car and the trailer. The only horizontal force on the trailer is T, the tension in the link connecting the car and trailer. Thus, $T = (F_{\text{trailer}})_{\text{net}} = 645\text{ N}$ is the magnitude of the force exerted on the trailer by the car. By Newton's third law, the trailer exerts a force

$$T = 645\text{ N acting in the rearward direction on the car.} \qquad \Diamond$$

(d) The road exerts the forward force F and the normal force N_c on the car. The magnitude of these forces may be found as follows:

$$\sum F_x = ma_x: \; F - 645\text{ N} = (1000\text{ kg})(+2.15\text{ m/s}^2) \quad \text{or} \quad F = 2.80\times10^3\text{ N}$$

$$\sum F_y = ma_y: \; +N_c - F_{g,c} = 0, \quad \text{so} \quad N_c = m_{\text{car}}g = (1000\text{ kg})(9.80\text{ m/s}^2) = 9800\text{ N}$$

The resultant force exerted on the car by the road is (by the vector diagram):

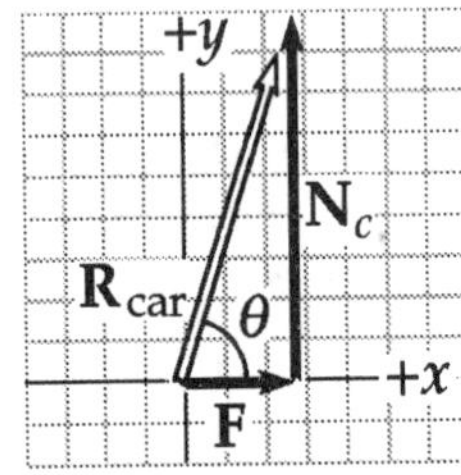

$F = 2.80 \times 10^3$ N
$N_c = 9.80 \times 10^3$ N

$$F = \sqrt{(2.80\times10^3\text{ N})^2 + (9.80\times10^3\text{ N})^2} = 1.02\times10^4\text{ N}$$

at $\quad \theta = \arctan(N_c/F) = \arctan(3.50) = 74.1°$

By Newton's third law, the resultant force exerted on the road by the car is therefore 1.02×10^4 N directed at 74.1° below the negative x direction (or equivalently at 15.9° to the rear of vertical). ◊

31. A 1000 N crate is being pushed across a level floor at a constant speed by a force **F** of 300 N at an angle of 20.0° below the horizontal as shown in Figure P4.31a. (a) What is the coefficient of kinetic friction between the crate and the floor? (b) If the 300 N force is instead pulling the block at an angle of 20.0° above the horizontal as shown in Figure 4.31b, what will be the acceleration of the crate. Assume that the coefficient of friction is the same as found in (a).

Solution

(a) Figure A at the right is a free-body diagram of the crate in Figure P4.31a of the textbook. The crate is in equilibrium since it moves at constant velocity. Looking at the vertical forces, we can find the normal force:

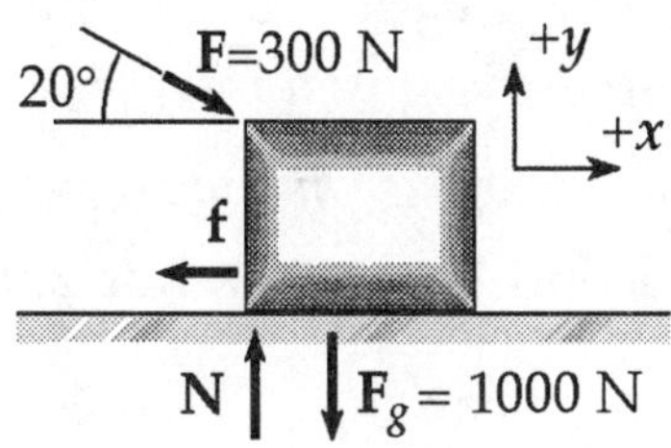

Figure A

$$\sum F_y = +N - F_g - F\sin 20.0° = 0$$

$$N = 1000\text{ N} + (300\text{ N})\sin 20.0° = 1.10\times 10^3\text{ N}$$

Then, considering the horizontal forces gives:

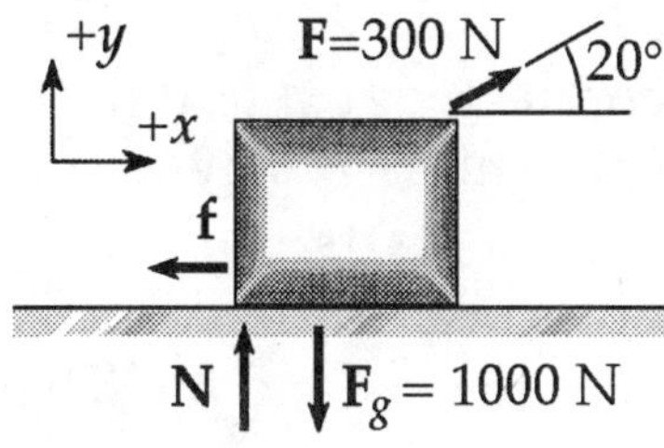

Figure B

$$\sum F_x = +F\cos 20.0° - f = 0$$

$$f = (300\text{ N})\cos 20.0° = 282\text{ N}$$

Therefore, the coefficient of kinetic friction between the crate and floor is given by:

$$\mu_k = \frac{f}{N} = \frac{282\text{ N}}{1.10\times 10^3\text{ N}}$$

or $\mu_k = 0.256$ ◊

(b) If the 300 N force pulls upward at 20.0° above the horizontal, the free-body diagram is as given in Figure B. In this case, the vertical acceleration a_y is still zero, but the crate has some unknown horizontal acceleration. Considering the vertical forces,

$$\sum F_y = ma_y = 0 \quad \text{giving} \quad +F\sin 20.0° + N - F_g = 0$$

$$\text{or} \quad N = 1000\text{ N} - (300\text{ N})\sin 20.0° = 897\text{ N}$$

Then, assuming the same coefficient of friction as found in (a), the friction force f is given by $f = \mu_k N = (0.256)(897\text{ N}) = 230\text{ N}$. Noting that the mass of the crate is $m = w/g = F_g/g = (1000\text{ N})/\left(9.80\text{ m/s}^2\right) = 102\text{ kg}$, apply Newton's second law to the horizontal motion of the crate:

$$\sum F_x = ma_x \quad \text{or} \quad a_x = \frac{F\cos 20.0° - f}{m} = \frac{(300\text{ N})\cos 20.0° - (230\text{ N})}{102\text{ kg}} = 0.509\text{ m/s}^2 \quad \Diamond$$

37. Masses m_1 = 10.0 kg and m_2 = 5.00 kg are connected by a light string that passes over a frictionless pulley as in Figure P4.25. If, when the system starts from rest, m_2 falls 1.00 m in 1.20 s, determine the coefficient of kinetic friction between m_1 and the table.

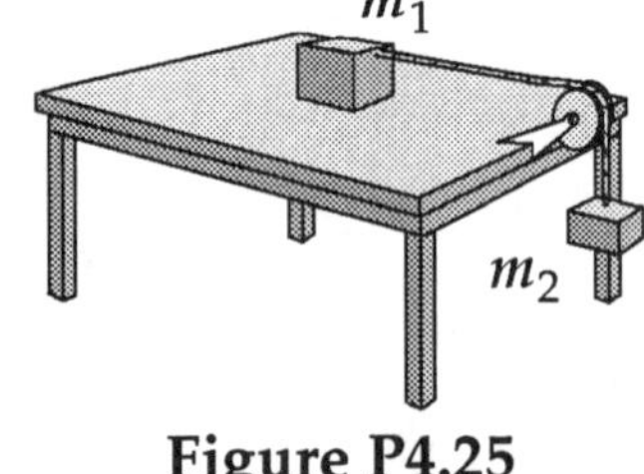

Figure P4.25

Solution The free-body diagrams of the two objects in this system are shown at the right and on the next page. Note that the accelerations of the two objects have the same magnitude, a, with the acceleration of m_1 directed horizontally to the right and the acceleration of m_2 directed vertically downward.

$a_x = +a$, $a_y = 0$; $+y$, $+x$; m_1 10.0 kg; T; f; N; $F_{g,1} = m_1 g$

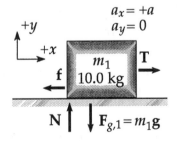

Since m_2 is observed to drop downward 1.00 m in 1.20 s when the system is released, the magnitude of the acceleration is found using $\Delta y = v_{0y}t + \frac{1}{2}a_y t^2$ as

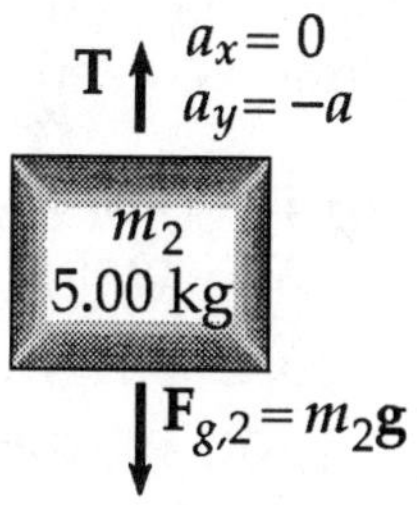

$$-1.00\text{ m} = 0 + \tfrac{1}{2}(-a)(1.20\text{ s})^2 \quad \text{so} \quad a = 1.39\text{ m/s}^2$$

$\mathbf{T}$ $a_x = 0$ $a_y = -a$

m_2 5.00 kg

$\mathbf{F}_{g,2} = m_2\mathbf{g}$

The weights of the objects are $F_{g,1} = mg = 98.0\text{ N}$ and $F_{g,2} = m_2 g = 49.0\text{ N}$. Applying Newton's second law to m_2 gives us the tension T in the cord:

$$\sum F_y = m_2 a_y \qquad +T - F_{g,2} = m_2(-a)$$

$$T = 49.0\text{ N} + (5.00\text{ kg})\left(-1.39\text{ m/s}^2\right) = 42.1\text{ N}$$

Now, consider the vertical forces acting on m_1:

$$\sum F_y = m_1 a_y \qquad +N - F_{g,1} = 0 \quad \text{so} \quad N = F_{g,1} = 98.0\text{ N}$$

Finally, considering the horizontal forces acting on m_1,

$$\sum F_x = m_1 a_x \qquad +T - f = m_1(+a)$$

$$f = T - m_1 a = 42.1\text{ N} - (10.0\text{ N})\left(1.39\text{ m/s}^2\right)$$

so the friction force is $f = 28.2\text{ N}$. The coefficient kinetic friction between m_1 and the table is therefore

$$\mu_k = \frac{f}{N} = \frac{28.2\text{ N}}{98.0\text{ N}} = 0.288 \qquad \Diamond$$

---

41. Find the acceleration experienced by each of the two masses shown in Figure P4.41 if the coefficient of kinetic friction between the 7.00-kg mass and the plane is 0.250.

7.00 kg 37.0° 12.0 kg

Figure P4.41

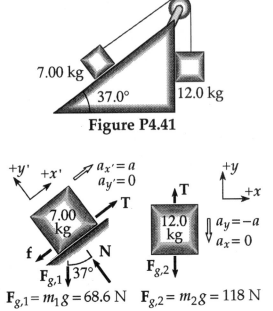

Solution Free-body diagrams of each of the two masses in Figure P4.41 are given to the right. Here, the unknown magnitude of the accelerations of the masses is labeled a. Note that it has been assumed that the 7.00-kg mass will accelerate up the incline and, consistent with that choice, it is assumed that the 12.0-kg mass will accelerate downward. First, applying Newton's second law to the 7.00-kg mass, $\sum F_{y'} = m_1 a_{y'}$ gives

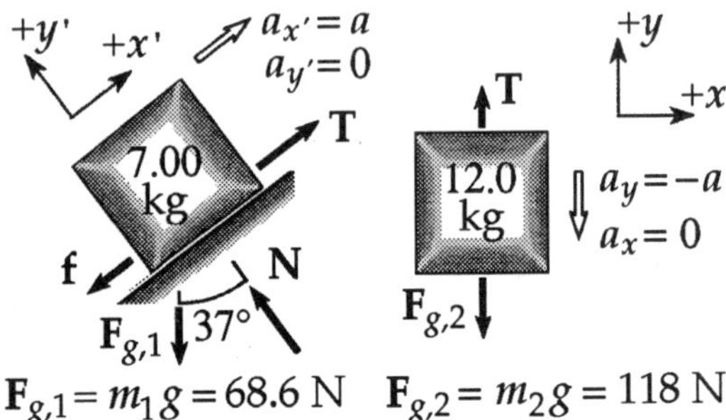

$+N - (68.6\text{ N})\cos 37.0° = 0$, or $N = 54.8$ N

Therefore, the friction force f is: $\quad f = \mu_k N = (0.250)(54.8\text{ N}) = 13.7\text{ N}$

Then, $\sum F_{x'} = m_1 a_{x'}$ gives: $\quad +T - f - (68.6\text{ N})\sin 37.0° = m_1(+a)$

or **[Equation 1]** $\quad T = 55.0\text{ N} + (7.00\text{ kg})a$

Next, applying Newton's second law to the 12.0-kg mass,

$\sum F_y = m_2 a_y$ yields $\quad +T - 118\text{ N} = (12.0\text{ kg})(-a)$

or **[Equation 2]** $\quad T + (12.0\text{ kg})a = 118\text{ N}$

Substituting T in Equation 1 into Equation 2,

$55.0\text{ N} + (7.00\text{ kg})a + (12\ .0\text{ kg})a = 118\text{ N}$ so $(19.0\text{ kg})a = 63.0\text{ N}$

Therefore, $a = 3.32\text{ m/s}^2$, so the 7.00-kg mass accelerates up the incline at 3.32 m/s^2 while the 12.0-kg mass accelerates downward at 3.32 m/s^2. ◊

49. A box rests on the back of a truck. The coefficient of static friction between box and truck bed is 0.300. (a) When the truck accelerates forward, what force accelerates the box? (b) Find the maximum acceleration the truck can have before the box slides.

Solution (a) Due to inertia, the box of mass m will tend to maintain its previous velocity (relative to the Earth) when the truck begins accelerating forward. Thus, the box will tend to slide toward the rear of the truck. The friction force exerted on the box by the truck bed will therefore be directed in the forward direction as it attempts to prevent this slippage. As seen in the free-body diagram of the box given in the sketch, this friction force f is the resultant horizontal force that will accelerate the box. ◊

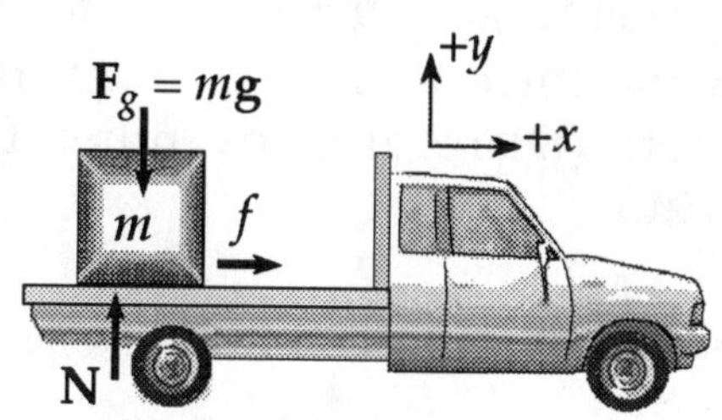

(b) The box will have zero acceleration in the vertical (y) direction. Thus,

$$\sum F_y = N - F_g = 0 \quad \text{gives} \quad N = F_g = mg$$

Therefore, the maximum magnitude a static friction force between the box and truck bed can have is $f_{max} = \mu_s N = (0.300)mg$. If the box has not started to slip, its horizontal acceleration a_x is the same as the acceleration, a, of the truck. Newton's second law, $\sum F_x = ma_x$, then gives

$$f = ma \quad \text{or} \quad a = \frac{f}{m}$$

To find the maximum acceleration of the truck before slippage will occur, use the maximum static friction force, f_{max}, to obtain:

$$a_{max} = \frac{f_{max}}{m} = \frac{(0.300)mg}{m} = (0.300)\left(9.80 \text{ m/s}^2\right) \quad \text{or} \quad a_{max} = 2.94 \text{ m/s}^2 \quad ◊$$

57. Two people pull as hard as they can on ropes attached to a 200-kg boat. If they pull in the same direction, the boat has an acceleration of 1.52 m/s^2 to the right. If they pull in opposite directions, the boat has an acceleration of 0.518 m/s^2 to the left. What is the force exerted by each person on the boat? (Disregard any other forces on the boat.)

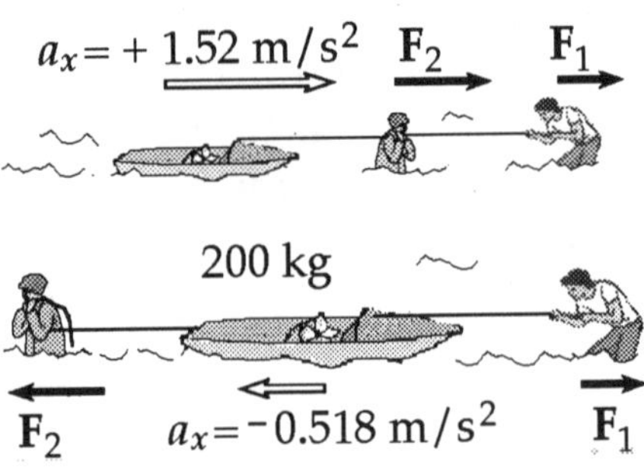

Solution

Consider the free-body diagrams above for the two situations described in this problem. Note that the $+x$ direction has been chosen toward the right, and apply Newton's second law, $\sum F_x = ma_x$, to the horizontal motion of the boat.

Pulling in the same direction,

$$F_1 + F_2 = (200\ \text{kg})(+1.52\ \text{m/s}^2)$$

$$F_1 + F_2 = 304\ \text{N} \qquad \textbf{[Equation 1]}$$

Pulling in opposite directions,

$$F_1 - F_2 = (200\ \text{kg})(-0.518\ \text{m/s}^2)$$

$$F_1 = F_2 - 104\ \text{N} \qquad \textbf{[Equation 2]}$$

Substituting Equation 2 into Equation 1 gives

$$F_2 - 104\ \text{N} + F_2 = 304\ \text{N} \qquad \text{or} \qquad F_2 = 204\ \text{N}$$

Equation 2 then yields $F_1 = 100\ \text{N}$ ◊

63. On takeoff, the combined action of the engines and wings of an airplane exerts an 8000-N force on the plane, directed upward at an angle of 65.0° above the horizontal. The plane rises with constant velocity in the vertical direction while continuing to accelerate in the horizontal direction. (a) What is the weight of the plane? (b) What is its horizontal acceleration?

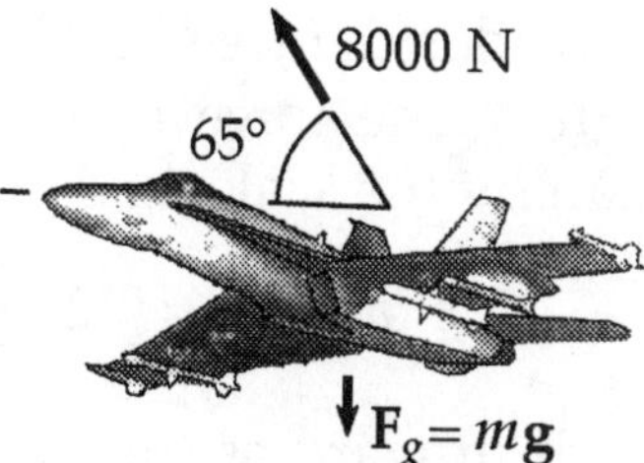

Solution The weight of the plane, $w = mg$, is the only force acting on the plane other than the 8000-N resultant force exerted by the engines and wings. Resolving the 8000-N force into its horizontal and vertical components gives:

$$F_x = (8000 \text{ N})\cos 65.0° = 3380 \text{ N}, \quad \text{and} \quad F_y = (8000 \text{ N})\sin 65.0° = 7250 \text{ N}$$

(a) Since the plane has constant velocity in the vertical direction (i.e., $a_y = 0$),

$$\sum F_y = ma_y \quad \text{yields} \quad +F_y - F_g = 0, \text{ or } F_g = F_y = 7250 \text{ N} \qquad \Diamond$$

(b) The mass of the airplane is found as $m = w/g = F_g/g = \dfrac{7250 \text{ N}}{9.80 \text{ m/s}^2} = 740 \text{ kg}$.

Then, applying Newton's second law to the horizontal motion gives:

$$a_x = \frac{\sum F_x}{m} = \frac{3380 \text{ N}}{740 \text{ kg}} = 4.57 \text{ m/s}^2 \qquad \Diamond$$

---

66. A 72-kg man stands on a spring scale in an elevator. Starting from rest, the elevator ascends, attaining its maximum speed of 1.2 m/s in 0.80 s. It travels with this constant speed for 5.0 s, undergoes a uniform **negative** acceleration for 1.5 s, and comes to rest. What does the spring scale register (a) before the elevator starts to move? (b) during the first 0.80 s? (c) while the elevator is traveling at constant speed? (d) during the negative acceleration?

Solution The spring scale will always register the magnitude of the force exerted on it by the man, which is the same as the magnitude of the force exerted on the man by the scale (recall Newton's third law).

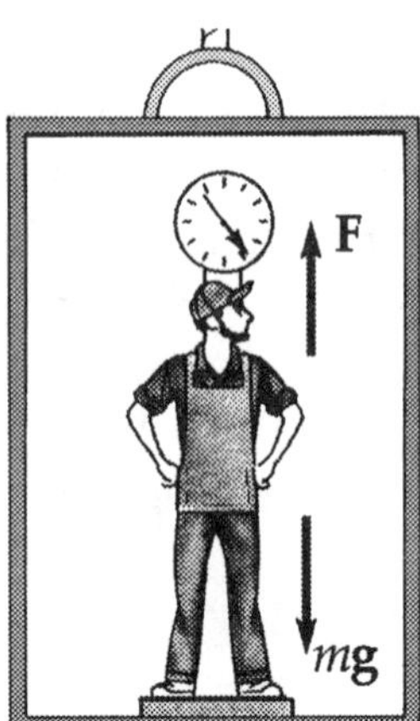

Consider the free-body diagram of the man shown at the right. The force F is the upward force exerted on the man by the scale, and his weight is

$$w = mg = (72\text{ kg})(9.8\text{ m/s}^2) = 7.1\times10^2\text{ N}$$

With upward as the positive direction, Newton's second law gives $\sum F_y = +F - w = ma$:

$$F = 7.1\times10^2\text{ N} + (72\text{ kg})a \qquad \textbf{[Equation 1]}$$

Note that a is the acceleration the man experiences.

(a) Before the elevator starts, the man is at rest and in equilibrium. Therefore, his acceleration is zero ($a = 0$) so Equation 1 gives the force exerted on him by the scale as 7.1×10^2 N (upward). Thus, the man exerts a downward force of 7.1×10^2 N on the scale. ◊

(b) During the first 0.80 s of motion, the man's acceleration is $a = \Delta v/\Delta t$, or $a = (+1.2\text{ m/s} - 0)/0.80\text{ s} = +1.5\text{ m/s}^2$. Substituting a into Equation 1 then gives: $F = 7.1\times10^2\text{ N} + (72\text{ kg})(+1.5\text{ m/s}^2) = 8.2\times10^2\text{ N}$. The man therefore exerts a force of 8.2×10^2 N directed downward on the scale. ◊

(c) While the elevator (and hence the man) is traveling upward at constant speed, his acceleration is zero and Equation 1 again gives $F = 7.1\times10^2$ N. The man exerts 7.1×10^2 N downward on the scale. ◊

(d) During the last 1.5 s, the elevator starts with an upward velocity of 1.2 m/s, and comes to rest. The man's acceleration is therefore $a = \Delta v/\Delta t$, or

$$a = \left[0.0 - (+1.2\text{ m/s})\right]/1.5\text{ s} = -0.80\text{ m/s}^2$$

By Newton's second law, the man exerts the same force downward on the scale, that the scale exerts upward on the man. The force the man exerts is thus

$$F = 7.1\times10^2\text{ N} + (72\text{ kg})(-0.80\text{ m/s}^2) = 6.5\times10^2\text{ N downwards}$$ ◊

CHAPTER SELF-QUIZ

1. There are six books in a stack, each with a weight of 3 N. The coefficient of friction between each pair of books is 0.1. With what horizontal force must I push to start sliding the top five books off the bottom one?
   a. 0.3 N
   b. 1.5 N
   c. 1.8 N
   d. 2.7 N

2. A 3.0-kg bucket is lowered into a 10-m deep well, starting from rest at the top. The tension in the rope is 9.8 N. The time to reach the bottom is
   a. 1.7 s
   b. 2.5 s
   c. 1.4 s
   d. 1.1 s

3. A woman at an airport is pulling a 15-kg suitcase with wheels at constant speed by pulling on a strap at an angle θ above the horizontal. She pulls on the strap with a 30-N force, and the frictional force is 24 N. What is θ?
   a. 30°
   b. 37°
   c. 45°
   d. 53°

4. It is 50 m between telephone poles. When a 1-kg bird lands on the telephone wire midway between the poles, the wire sags 0.2 m. How much tension in the wire does the bird produce? Ignore the weight of the wire.
   a. 1200 N
   b. 610 N
   c. 120 N
   d. 9.8 N

5. A box is dropped onto a conveyor belt moving at 2 m/s. If the coefficient of friction between the box and the belt is 0.3, how long before the box moves without slipping?
   a. 0.7 s
   b. 1.4 s
   c. 2.1 s
   d. 3.0 s

6. Two unequal masses are falling through the air with the heavier one below the lighter one. They are connected by a string. (Ignore air friction.) The tension in the string will be equal to
   a. the weight of the heavier mass
   b. the weight of the lighter mass
   c. the difference in weight of the two masses
   d. zero

7. A 15-kg block rests on a level frictionless surface and is attached by a light string to a 5.0-kg hanging mass where the string passes over a massless frictionless pulley. If $g = 9.8\ m/s^2$, what is the acceleration of the system when released?
   a. $2.45\ m/s^2$
   b. $7.35\ m/s^2$
   c. $3.27\ m/s^2$
   d. $9.8\ m/s^2$

8. A puck, upon being struck by a hockey stick, is given an initial speed of 8 m/s and continues to move in a straight path for a distance of 16 m before coming to rest. What is the coefficient of kinetic friction between puck and ice?
   a. 0.05
   b. 0.1
   c. 0.2
   d. 0.08

9. Find the tension in an elevator cable if the 1000-kg elevator is descending with an acceleration of 1.6 m/s^2, downward.
   a. 5,700 N
   b. 8,200 N
   c. 9,800 N
   d. 11,400 N

10. Two perpendicular forces, one of 30 N directed due north; and the second, 40 N directed due east; act simultaneously on an object with mass of 35 kg. What is the magnitude of the resultant acceleration of the object?
   a 1.4 m/s^2
   b. 155 m/s^2
   c. 3.5 m/s^2
   d. 0.7 m/s^2

11. Two blocks of masses 4 and 6 kg, respectively, rest on a frictionless horizontal surface and are connected by a string. A second string attached only to the 6-kg block, has a horizontal force of 20 N applied to it, causing both blocks to accelerate. What is the tension in the string between the two blocks?
   a. 12 N
   b. 28 N
   c. 8 N
   d. 10 N

12. A 150-N sled is pulled up a 28-degree slope at a constant speed by a force of 100 N parallel to the hill What force directed up the hill will allow the sled to move downhill at a constant speed?
   a. 181 N
   b. 170 N
   c. 130 N
   d. 141 N

Chapter 5

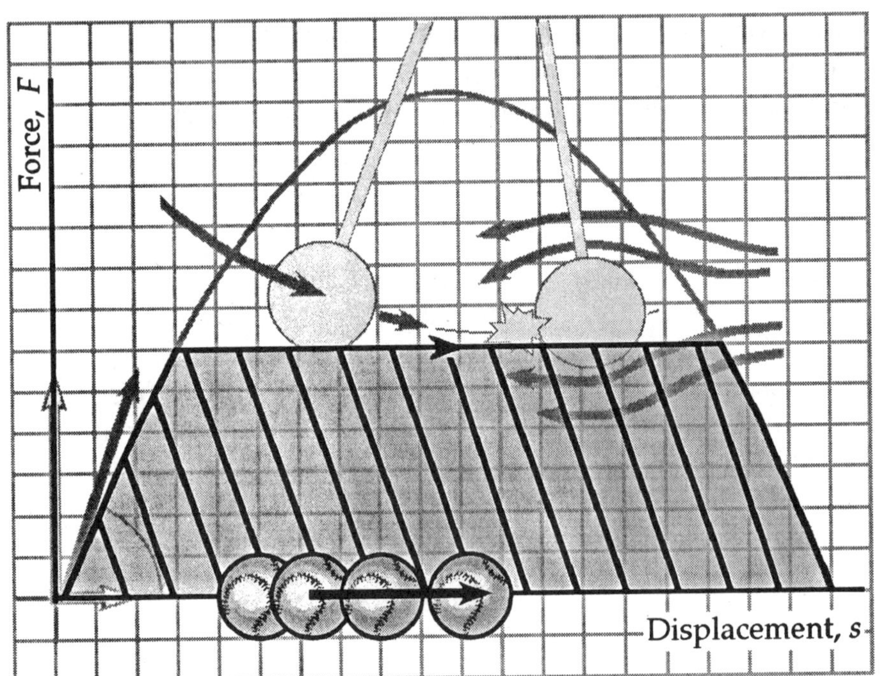

WORK AND ENERGY

Chapter 5

WORK AND ENERGY

Energy is present in the Universe in a variety of forms including mechanical energy, chemical energy, electromagnetic energy, heat energy, and nuclear energy. Although energy can be transformed from one form to another, the total amount of energy in the Universe remains the same. If an isolated system loses energy in some form, then, by the principle of conservation of energy, the system must gain an equal amount of energy in other forms.

In this chapter we are concerned only with mechanical energy. We introduce the concept of **kinetic energy**, which is defined as the energy associated with motion, and the concept of **potential energy**, the energy associated with position. We shall see that the ideas of work and energy can be used in place of Newton's laws to solve certain problems.

We begin by defining **work**, a concept that provides a link between force and energy. With this as a foundation, we can then discuss the principle of conservation of energy and apply it to problems.

NOTES FROM SELECTED CHAPTER SECTIONS

5.1 Work

In order for work to be accomplished, an object must undergo a displacement; the force associated with the work must not be perpendicular to the direction of the displacement. Work is a scalar quantity; the SI unit of work is the newton-meter (N·m) or joule (J).

5.2 Kinetic Energy and the Work-Kinetic Energy Theorem

Any object which has mass m and speed v has kinetic energy. Kinetic energy is a scalar quantity and has the same units as work and heat; therefore kinetic energy of an object will change only if work is done on the object by some external force. The relationship between work and change in kinetic energy is stated in the work-energy theorem.

5.3 Potential Energy

The work done on an object by the force of gravity is equal to the object's initial potential energy minus its final potential energy. The gravitational potential energy associated with an object depends only on the object's weight and its vertical height above the surface of the Earth. If the height above the surface increases, the potential energy will also increase; but the work done by the gravitational force will be negative. In working problems involving gravitational potential energy, it is necessary to choose an arbitrary reference level (or location) at which the potential energy is zero.

5.4 Conservative and Nonconservative Forces

A force is **conservative** if the work it does on an object moving between two points is independent of the path the object takes between the points. The work done on an object by a conservative force depends only on the initial and final positions of the object.

A force is **nonconservative** if the work it does on an object moving between two points depends on the path taken.

5.5 Conservation of Mechanical Energy

The sum of the kinetic energy plus the potential energy is called the total mechanical energy.

The law of conservation of mechanical energy states that the mechanical energy of a system remains constant if the only forces that do work on the system are **conservative forces.**

5.6 Nonconservative Forces and the Work-Kinetic Energy Theorem

In realistic situations, nonconservative forces such as friction are usually present, and the total mechanical energy of the system is not constant. In those cases, **the work done by all nonconservative forces equals the change in mechanical energy of the system**, and equation 5.14 must be used; equation 5.10 does not apply.

5.7 Conservation of Energy in General

We can generalize the energy conservation principle to include all forces, both conservative and nonconservative, acting on a system. **Energy may be transformed from one form to another, but the total energy of an isolated system is always constant.** From a universal point of view, we can say that if one part of the Universe gains energy in some form, another part must lose an equal amount of energy in some form.

EQUATIONS AND CONCEPTS

The work done on a body by a force **F**, which is constant in both magnitude and direction, is defined to be the product of the component of the force in the direction of the displacement and the magnitude of the displacement.

$$W \equiv (F\cos\theta)s \quad (5.1)$$

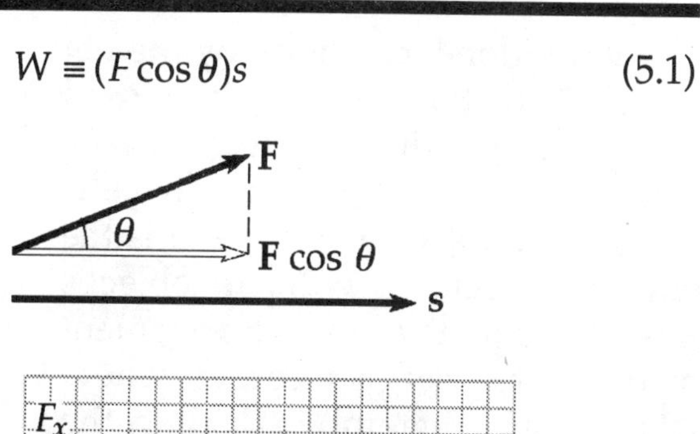

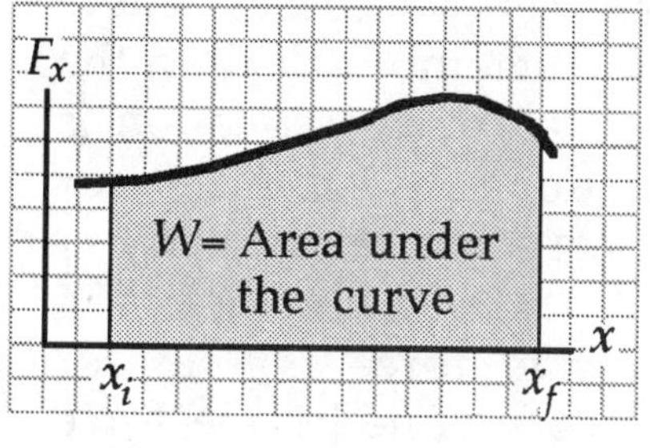

Note that the work done by a force can be positive, negative, or zero, depending on the value of θ, the angle between the direction of the force and the direction of the displacement. If $0 \le \theta < 90°$, W is positive; if $90° < \theta < 180°$, W is negative; and if $\theta = 90°$ (**F** perpendicular to **s**), then $W = 0$. In the special case where **F** and **s** are parallel, $\theta = 0$, $\cos\theta = 1$, and $W = Fs$.

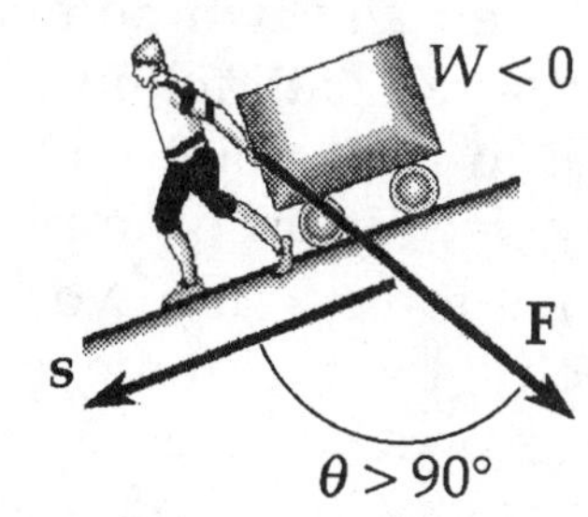

Work is a scalar quantity and the SI unit of work is the newton-meter or joule. See the summary of units of work in Table 5.1 of the textbook.

$$1\ \text{N}\cdot\text{m} = 1\ \text{J}$$

When a body of mass m experiences an acceleration due to a **net** force, the work done by the net force can be expressed in terms of the acceleration, mass, and the distance over which the acceleration is achieved.

$$W_{\text{net}} = Fs = (ma)s \qquad (5.3)$$

The work done on a body by the net or resultant force can be expressed in terms of the change in the kinetic energy of the body. Kinetic energy is a scalar quantity and is the energy associated with an object's motion. Equation 5.6 is a statement of the work-energy theorem, where it must be remembered that the work calculated by the equation is the work done by the net or resultant force acting on the body.

$$W_{\text{net}} = \tfrac{1}{2}mv^2 - \tfrac{1}{2}mv_0{}^2 \qquad (5.4)$$

$$KE \equiv \tfrac{1}{2}mv^2 \qquad (5.5)$$

$$W_{\text{net}} = KE_f - KE_i \qquad (5.6)$$

A conservative force is one for which the work done by the force in moving between any two points is independent of the path followed from the initial to final point. The force of gravity, $m\mathbf{g}$, is an example of a conservative force. The work done on a body by the force of gravity can be expressed in terms of initial and final values of the body's y-coordinates.

$$W_g = mgy_i - mgy_f$$

$$PE \equiv mgy \qquad (5.7)$$

$$W_g = PE_i - PE_f \qquad (5.8)$$

Work done by the gravitational force can also be expressed in terms of the gravitational potential energy function: the work done by the force of gravity equals the negative of the change in the gravitational potential energy function.

The units of energy (kinetic and potential) are the same as the units of work.

Comment on units.

In calculating the work done by the gravitational force, remember that the difference in potential energy between two points is independent of the location of the origin. Choose an origin which is convenient to calculate PE_i and PE_f for a particular situation.

Comment on reference level for potential energy.

When only conservative forces act on a system, the total mechanical energy ($KE + PE$) of the system remains constant; this is a statement of the law of conservation of mechanical energy.

$$KE_i + PE_i = KE_f + PE_f \qquad (5.10)$$

$$KE_i + \sum PE_i = KE_f + \sum PE_f$$

If the only conservative force is the gravitational force, the equation for conservation of mechanical energy takes a special form.

$$\tfrac{1}{2}mv_i^2 + mgy_i = \tfrac{1}{2}mv_f^2 + mgy_f \qquad (5.11)$$

The work done by a force in stretching or compressing a spring is stored in the spring as elastic potential energy. For a given displacement from the equilibrium position, the potential energy in the spring depends on the spring constant, k.

$$PE_s \equiv \frac{1}{2}kx^2 \tag{5.12}$$

If both conservative forces and nonconservative forces act on a system, the total mechanical energy will not remain constant. In this case, the work done by all nonconservative forces equals the change in the total mechanical energy of the system.

$$W_{nc} + W_c = \Delta KE$$

$$W_{nc} = (KE_f + PE_f) - (KE_i + PE_i) \tag{5.14}$$

The average power supplied by a force is the ratio of the work done by the force to the time interval over which the force acts. The average power can also be expressed in terms of the force and the average speed of the object on which the force acts.

$$\overline{P} = \frac{W}{\Delta t} \tag{5.15}$$

$$\overline{P} = F\overline{v} \tag{5.16}$$

The SI unit of power is the watt; in the British engineering system, the unit of power is the horsepower.

$$1\ \text{W} = 1\ \text{J/s} = 1\ \text{kg}\cdot\text{m}^2/\text{s}^3 \tag{5.17}$$

$$1\ \text{hp} = 550\ \frac{\text{ft}\cdot\text{lb}}{\text{s}} = 746\ \text{W} \tag{5.18}$$

SUGGESTIONS, SKILLS, AND STRATEGIES

CHOOSING A ZERO LEVEL

In working problems involving gravitational potential energy, it is always necessary to choose a location at which the gravitational potential energy is zero. This choice is completely arbitrary because the important quantity is the **difference** in potential energy, and that difference is independent of the location of zero. It is often convenient, but not essential, to choose the surface of the Earth as the reference position for zero potential energy. In most cases, the statement of the problem suggests a convenient level to use.

CONSERVATION OF ENERGY

Take the following steps in applying the principle of conservation of energy:

1. Define your system, which may consist of more than one object.
2. Select a reference position for the zero point of gravitational potential energy.
3. Determine whether or not nonconservative forces are present.
4. If mechanical energy is conserved (that is, if only conservative forces are present), you can write the total initial energy, $KE_i + PE_i$, at some point as the sum of the kinetic and potential energies at that point. Then, write an expression for the total final energy, $KE_f + PE_f$, at the final point of interest. Since mechanical energy is conserved, you can equate the two total energies and solve for the unknown.
5. If nonconservative forces such as friction are present (and thus mechanical energy is not conserved), first write expressions for the total initial and total final energies. In this case, the difference between the two total energies is equal to the work done by the nonconservative force(s). That is, you should apply Equation 5.14.

REVIEW CHECKLIST

▷ Define the work done by a constant force and work done by a force which varies with position. (Recognize that the work done by a force can be positive, negative, or zero; describe at least one example of each case.)

▷ Recognize that the gravitational potential energy function, $PE_g = mgy$, can be positive, negative, or zero, depending on the location of the reference level used to measure y. Be aware of the fact that although PE depends on the origin of the coordinate system, the **change** in potential energy, $(PE)_f - (PE)_i$, is **independent** of the coordinate system used to define PE.

▷ Understand that a force is said to be **conservative** if the work done by that force on a body moving between any two points is independent of the path taken. **Nonconservative** forces are those for which the work done on a particle moving between two points depends on the path. Account for nonconservative forces acting on a system using the work-energy theorem. In this case, the work done by all nonconservative forces equals the change in total mechanical energy of the system.

Understand the distinction between kinetic energy (energy associated with motion), potential energy (energy associated with the position or coordinates of a system), and the total mechanical energy of a system. State the law of conservation of mechanical energy, noting that mechanical energy is conserved only when conservative forces act on a system. This extremely powerful concept is very important in all areas of physics.

▷ Relate the work done by the net force on an object to the **change** in kinetic energy. The relation $W = \Delta KE = KE_f - KE_i$ is called the work-energy theorem, and is valid whether or not the (resultant) force is constant. That is, if we know the net work done on a particle as it undergoes a displacement, we also know the **change** in its kinetic energy. This is the most important concept in this chapter, so you must understand it thoroughly.

SOLUTIONS TO SELECTED END-OF-CHAPTER PROBLEMS

5. Starting from rest, a 5.00-kg block slides 2.50 m down a rough 30.0° incline. The coefficient of kinetic friction between the block and the incline is $\mu_k = 0.436$. Determine (a) the work done by the force of gravity, (b) the work done by the friction force between block and incline, and (c) the work done by the normal force.

Solution (a) The force of gravity is:

$$F_g = mg = (5.00\ \text{kg})(9.80\ \text{m/s}^2) = 49.0\ \text{N}$$

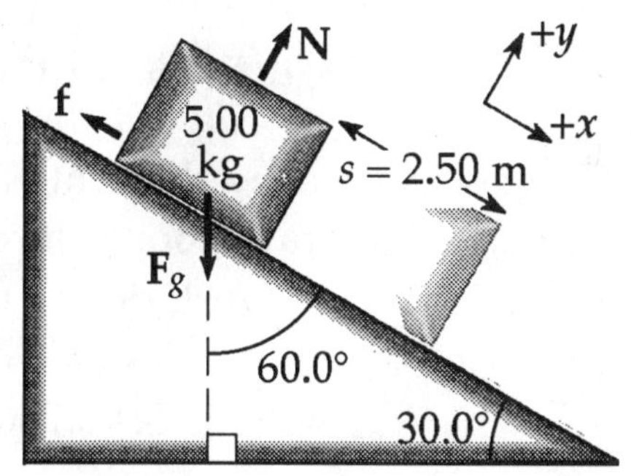

directed straight downward. The work done by this force is given by $W_g = F_g s\cos\theta$, where s is the displacement of the object and θ is the angle between the direction of the gravitational force and the direction of the displacement. Thus,

$$W_g = (49.0\ \text{N})(2.50\ \text{m})\cos 60.0° = 61.0\ \text{J} \quad \Diamond$$

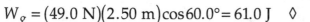

(b) To find the friction force f, it is first necessary to solve for the normal force, N. Using Newton's second law and recognizing that the block has zero acceleration directed perpendicular to the incline: $\sum F_y = N - F_g \sin 60.0° = 0$, or $N = F_g \sin 60.0° = (49.0\ \text{N})\sin 60.0° = 42.4\ \text{N}$. The friction force is then $f = \mu_k N = (0.436)(42.4\ \text{N}) = 18.5\ \text{N}$ and the work done by it is $W_f = fs\cos\theta$ where θ is now the angle between the directions of the friction force and the displacement. Therefore,

$$W_f = (18.5\ \text{N})(2.50\ \text{m})\cos 180° = -46.3\ \text{J} \quad \Diamond$$

(c) The normal force is perpendicular to the incline, and hence perpendicular to the displacement. The word done by the normal force is therefore

$$W_N = Ns\cos 90° = 0 \quad \Diamond$$

11. A person doing a chin-up weighs 700 N exclusive of the arms. During the first 25.0 cm of the lift, each arm exerts an upward force of 355 N on the torso. If the upward movement starts from rest, what is the person's speed at this point?

Solution

Three forces act on the torso of the person. These are the two 355 N forces, exerted upward by the arms, and a downward gravitational force (weight), F_g = 700 N. As the torso undergoes an upward displacement of s = 0.250 m, the **net** work done on it by these forces is

$$W_{net} = W_{arms} + W_{gravity} = 2F_{arm}\, s\cos\theta_1 + F_g\, s\cos\theta_2$$

The forces exerted by the arms are in the same direction as the displacement, so $\theta_1 = 0°$. The gravitational force is directed opposite to the displacement, or $\theta_2 = 180°$. The net work done on the torso is then:

$$W_{net} = 2(355\text{ N})(0.250\text{ m})\cos 0° + (700\text{ N})(0.250\text{ m})\cos 180°$$

$$W_{net} = +2.50\text{ J}$$

The mass of the torso is $m = F_g/g = (700\text{ N})/(9.80\text{ m/s}^2) = 71.4\text{ kg}$. Applying the work-kinetic energy theorem, $W_{net} = KE_f - KE_i$, gives:

$$W_{net} = \tfrac{1}{2}mv_f^2 - \tfrac{1}{2}mv_i^2, \quad \text{or} \quad 2.50\text{ J} = \tfrac{1}{2}(71.4\text{ kg})v_f^2 - 0$$

Therefore, $$v_f^2 = \frac{2(2.50\text{ J})}{71.4\text{ kg}} = 0.070\text{ m}^2/\text{s}^2$$

and $$v_f = 0.265\text{ m/s}$$ ◊

---

14. A 10.0-kg crate is pulled up a rough 20° incline by a 100-N force parallel to the incline. The initial speed of the crate is 1.50 m/s, the coefficient of kinetic friction is 0.40, and the crate is pulled a distance of 5.00 m. Determine how much work is done by (a) the gravitational force, and (b) the 100-N force? (c) What is the change in kinetic energy of the crate? (d) What is the speed of the crate after it is pulled 5.00 m?

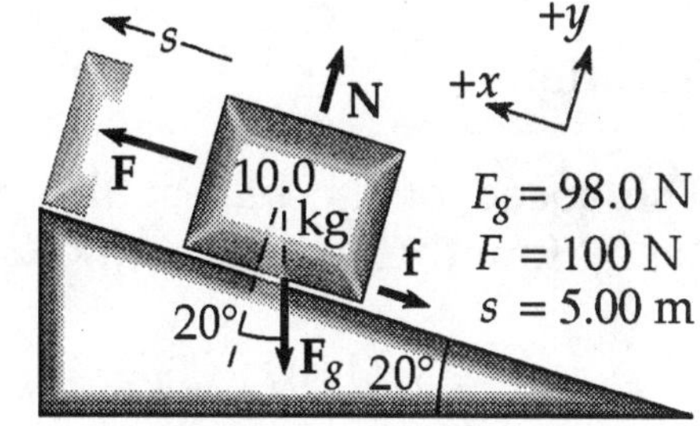

Solution (a) The work done by the gravitational force is $W_g = F_g s\cos\theta$ where $\theta = (90° + 20°) = 110°$ is the angle between the direction of $\mathbf{F}_g$ and $\mathbf{s}$. Thus,

$$W_g = (98.0\text{ N})(5.00\text{ m})\cos 110° = -168\text{ J} \quad \Diamond$$

(b) The work done by the 100-N force is $W_F = Fs\cos\theta$ with $\theta = 0°$ since **F** and **s** are parallel to each other. Thus, $W_F = (100\text{ N})(5.00\text{ m})\cos 0° = +500\text{ J}$. ◊

(c) The normal force, found from $\sum F_y = ma_y$, is

$$N - F_g\cos 20° = 0, \quad \text{or} \quad N = (98.0\text{ N})\cos 20° = 92.1\text{ N}$$

The friction force is therefore $f = \mu_k N = 0.400(92.1\text{ N}) = 36.8\text{ N}$ and the work done by friction is $W_f = fs\cos\theta = (36.8\text{ N})(5.00\text{ m})\cos 180° = -184\text{ J}$. The work-kinetic energy theorem then gives the change in the kinetic energy of the crate:

$$\Delta KE = W_{net} = W_g + W_F + W_f = -168\text{ J} + 500\text{ J} - 184\text{ J} = +148\text{ J} \quad \Diamond$$

(d) Using the above result, $\Delta KE = \frac{1}{2}mv_f^2 - \frac{1}{2}mv_i^2 = +148\text{ J}$, or

$$v_f^2 = v_i^2 + \frac{2(+148\text{ J})}{m} = (1.50\text{ m/s})^2 + \frac{2(+148\text{ J})}{10.0\text{ kg}} = 31.9\text{ m}^2/\text{s}^2$$

and $$v_f = 5.64\text{ m/s} \quad \Diamond$$

20. A softball pitcher rotates a 0.250-kg ball around a vertical circular path of radius 0.600 m before releasing it. The pitcher exerts a 30.0-N force directed parallel to the motion of the ball around the complete circular path. The speed of the ball at the top of the circle is 15.0 m/s. If the ball is released at the bottom of the circle, what is its speed upon release?

Solution

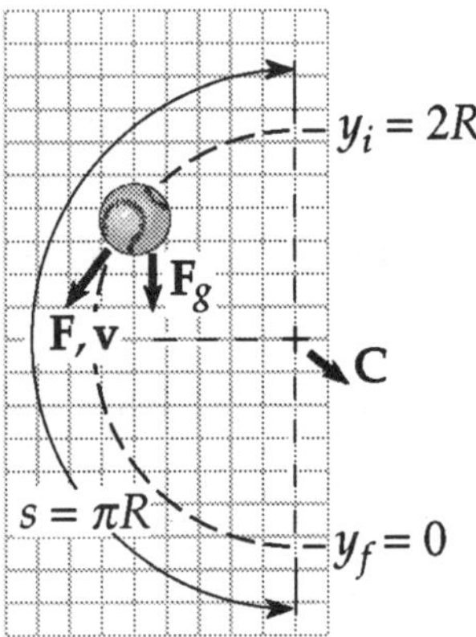

The speed of the ball at the point of release is most easily found using the work-kinetic energy theorem,

$$W_{\text{net}} = W_c + W_{nc} = KE_f - KE_i = \frac{1}{2}mv_f^2 - \frac{1}{2}mv_i^2$$

As the ball moves from the highest to the lowest point on the circular path, one conservative force (the weight of the ball, $F_g = mg = 2.45\ \text{N}$) acts on it. The work done by this force is equal to the decrease in the gravitational potential energy,

$$W_c = PE_i - PE_f = mg(y_i - y_f) = F_g(2R)$$

The radius of the circular path is $R = 0.600\ \text{m}$, so the work done by conservative forces is $W_c = (2.45\ \text{N})(1.20\ \text{m}) = 2.94\ \text{J}$. At any point along the circular path, the non-conservative force exerted by the pitcher's arm has a radial component (directed toward the center of the circular path) and a component that is tangential to the path. The radial component of this force, **C**, is always perpendicular to the motion and does no work. It is given that the tangential component has a constant magnitude, $F = 30.0\ \text{N}$. This component is always parallel to the motion and acts on the ball through a distance equal to one-half of the circumference of the circular path, $s = \pi R$. The work done by this non-conservative force is

$$W_{nc} = Fs = F(\pi R) = (30.0\ \text{N})\pi(0.600\ \text{m}) = 56.5\ \text{J}$$

The work-kinetic energy theorem then gives:

$$W_{\text{net}} = W_c + W_{nc} = \tfrac{1}{2}m\left(v_f^2 - v_i^2\right)$$

or

$$v_f^2 = v_i^2 + \frac{2(W_c + W_{nc})}{m}$$

Therefore,

$$v_f^2 = (15.0 \text{ m/s})^2 + \frac{2(2.94 \text{ J} + 56.5 \text{ J})}{0.250 \text{ kg}} = 700 \text{ m}^2/\text{s}^2$$

and

$$v_f = 26.5 \text{ m/s}$$ ◊

---

23. A child and sled with a combined mass of 50.0 kg slide down a frictionless hill. If the sled starts from rest and has a speed of 3.00 m/s at the bottom, what is the height of the hill?

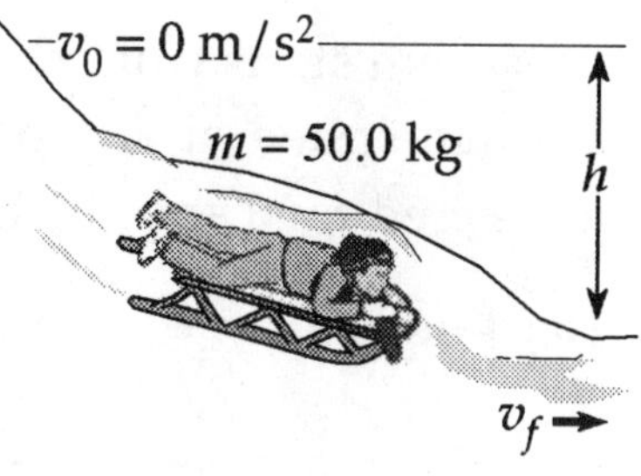

Solution As the child-sled combination slides down the frictionless hill, only one non-conservative force acts on it. This is the normal force exerted by the ground. It is always perpendicular to the displacement and hence does zero work ($W = Fs\cos\theta = 0$ if $\theta = 90°$).

Then, since no non-conservative forces do work on the system, the total mechanical energy is conserved: $KE_f + PE_f = KE_i + PE_i$. Choose the initial state to be when the system starts from rest at the top of the hill and the final state to be when the system reaches the bottom. Also, choose $y = 0$ to be at the bottom of the hill. Then $v_i = 0$, $v_f = 3.00$ m/s, $y_i = h$, and $y_f = 0$ where h is the height of the hill. The conservation of energy equation then reduces to:

$$\tfrac{1}{2}mv_f^2 + 0 = 0 + mgh \quad \text{or} \quad h = \frac{v_f^2}{2g} = \frac{(3.00 \text{ m/s})^2}{2\left(9.80 \text{ m/s}^2\right)} = 0.459 \text{ m}$$ ◊

---

27. Three masses, $m_1 = 5\text{ kg}$, $m_2 = 10\text{ kg}$, and $m_3 = 15\text{ kg}$, are attached by strings over frictionless pulleys as indicated in Figure P5.27. The horizontal surface is frictionless and the system is released from rest. Using energy concepts, find the speed of m_3 after it moves down 4.0 m.

Solution Since the pulleys and the table are all frictionless, there are no non-conservative forces doing work on the system. Thus, the total mechanical energy the system is conserved: $KE_f + PE_f = KE_i + PE_i$, or

$$\tfrac{1}{2}m_1 v_{1f}^2 + \tfrac{1}{2}m_2 v_{2f}^2 + \tfrac{1}{2}m_3 v_{3f}^2 + m_1 g y_{1f} + m_2 g y_{2f} + m_3 g y_{3f} = \tfrac{1}{2}m_1 v_{1i}^2 + \tfrac{1}{2}m_2 v_{2i}^2 + \tfrac{1}{2}m_3 v_{3i}^2 + m_1 g y_{1i} + m_2 g y_{2i} + m_3 g y_{3i}$$

The system starts from rest, so $v_{1i} = v_{2i} = v_{3i} = 0$. Also, the three masses have the same **speed** at any point in the motion, so $v_{1f} = v_{2f} = v_{3f} = v$, where v is the desired final speed. The conservation of energy equation then reduces to:

$$\tfrac{1}{2}(m_1 + m_2 + m_3)v^2 = -m_1 g\Delta y_1 - m_2 g\Delta y_2 - m_3 g\Delta y_3 \qquad \text{where} \qquad \Delta y = y_f - y_i$$

From the figure, it is observed that when m_3 drops downward 4.0 m, the vertical displacements of the three masses are:

$$\Delta y_1 = +4.0\text{ m},\ \Delta y_2 = 0 \quad \text{and} \quad \Delta y_3 = -4.0\text{ m}$$

The energy equation then gives

$$\tfrac{1}{2}(30\text{ kg})v^2 = -(49\text{ N})(+4.0\text{ m}) + 0 - (147\text{ N})(-4.0\text{ m})$$

or $$v^2 = \frac{2(-196\text{ J} + 588\text{ J})}{30\text{ kg}} = 26\text{ m}^2/\text{s}^2 \quad \text{and} \quad v = 5.1\text{ m/s}$$ ◊

33. A 70-kg diver steps off a 10-m tower and drops, from rest, straight down into the water. If he comes to rest 5.0 m beneath the surface, determine the average resistance force exerted on him by the water.

Solution

A non-conservative force acts on the diver during the 5.0-m movement through the water. This is the resistance force (directed upward) exerted by the water. The work done by this force on the diver is $W_{nc} = \bar{F}s\cos\theta = \bar{F}(5.0\text{ m})\cos 180° = -(5.0\text{ m})\bar{F}$. The only other force acting on the diver during his motion is his weight, which acts for the entire 15-m displacement. This is a conservative force, and the work done by it may be accounted for by potential energy terms in the work-energy theorem as follows:

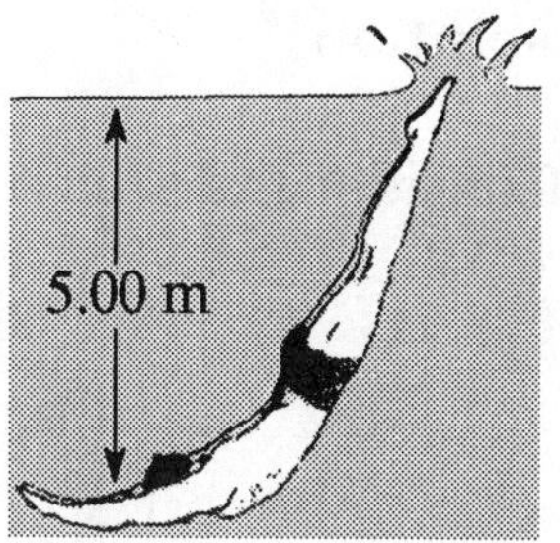

$$W_{net} = W_{nc} + W_c = W_{nc} + \left(PE_i - PE_f\right) = KE_f - KE_i$$

Since the diver starts from rest and ends up at rest, $KE_f = KE_i = 0$

Setting $y = 0$ at the bottom of the dive, $PE_f = mgy_f = 0$

and the initial potential energy is

$$PE_i = mgy_i = (70\text{ kg})\left(9.8\text{ m/s}^2\right)(5.0\text{ m} + 10\text{ m}) = (690\text{ N})(+15\text{ m})$$

By the work-energy equation, $-(5.0\text{ m})\bar{F} + (690\text{ N})(15\text{ m}) = 0$

and the average resistance force is $\bar{F} = (690\text{ N})(15\text{ m})/(5.0\text{ m}) = 2.1\times 10^3\text{ N}$ ◊

37. Starting from rest, a 10.0-kg block slides 3.00 m down a frictionless ramp (inclined at 30.0° from the floor) to the bottom. The block then slides an additional 5.00 m along the floor before coming to a stop. Determine (a) the speed of the block at the bottom of the ramp, (b) the coefficient of kinetic friction between block and floor, and (c) the mechanical energy lost due to friction.

Solution

Consider the sketch below for the two parts of this motion:

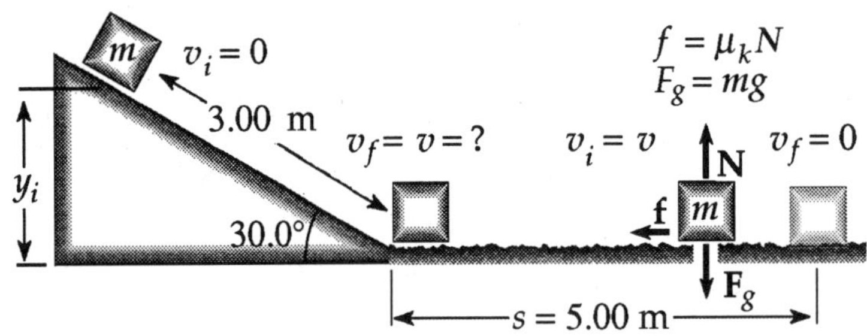

(a) Sliding down the frictionless ramp.

(b) Sliding on the rough horizontal surface

(a) As the block slides down the frictionless ramp, no non-conservative forces do work on the block. Thus, the total mechanical energy of the block is conserved as it slides down the ramp: $KE_f + PE_f = KE_i + PE_i$. Since the block starts from rest for this part of the trip, $KE_i = \frac{1}{2}mv_i^2 = 0$. Choosing $y = 0$ at the level of the bottom of the ramp, $PE_f = mgy_f = 0$, and $PE_i = mgy_i$ where $y_i = (3.00\text{ m})\sin 30.0° = 1.50\text{ m}$. Conservation of mechanical energy then gives: $\frac{1}{2}mv_f^2 + 0 = 0 + mg(1.50\text{ m})$, or

$$v_f^2 = 2\left(9.80\text{ m/s}^2\right)(1.50\text{ m}) = 29.4\text{ m}^2/\text{s}^2 \quad \text{and} \quad v = v_f = 5.42\text{ m/s} \qquad \Diamond$$

(b) As the block undergoes a displacement $s = 5.00$ m along the horizontal floor, the work done by the non-conservative friction force is

$$W_{nc} = fs\cos\theta = f(5.0\text{ m})\cos 180° = -(5.0\text{ m})f$$

This is the only work done on the block since the other two forces acting on it (a normal force and the block's weight) are perpendicular to the horizontal motion during this part of the trip. The block comes to rest so $KE_f = \frac{1}{2}mv_f^2 = 0$. For this part of the motion, the initial speed of the block is the speed at the bottom of the ramp: $v_i = v = 5.42$ m/s. The work-kinetic energy theorem, $W_{\text{net}} = KE_f - KE_i$, then gives:

$$-(5.0\text{ m})f = 0 - \tfrac{1}{2}(10.0\text{ kg})(5.42\text{ m/s})^2$$

so

$$f = \frac{147\text{ J}}{5.00\text{ m}} = 29.4\text{ N}$$

Since the block has zero acceleration in the vertical direction, $\sum F_y = +N - F_g = 0$, and the normal force is $N = F_g = mg = 98.0$ N. The coefficient of kinetic friction between the block and floor is then

$$\mu_k = \frac{f}{N} = \frac{29.4\text{ N}}{98.0\text{ N}} = 0.300 \quad \Diamond$$

(c) The loss of mechanical energy due to friction as the block slides along the horizontal floor is $E_{\text{loss}} = E_i - E_f = (KE_i + PE_i) - (KE_f + PE_f)$, or

$$E_{\text{loss}} = \left(\tfrac{1}{2}mv_i^2 + 0\right) - (0+0) = \tfrac{1}{2}(10.0\text{ kg})(5.42\text{ m/s})^2 = 147\text{ J} \quad \Diamond$$

45. A 1.50×10^3-kg car starts from rest and accelerates uniformly to 18.0 m/s in 12.0 s. Assume that air resistance remains constant at 400 N during this time. Find (a) the average power developed by the engine and (b) the instantaneous power output of the engine at t = 12.0 s.

Solution

(a) The car's acceleration is

$$a = \frac{v_f - v_i}{t} = \frac{18.0 \text{ m/s} - 0}{12.0 \text{ s}} = 1.50 \text{ m/s}^2$$

The constant forward force being exerted by the engine is found from

$$\sum F_x = F_{\text{engine}} - F_{\text{resistance}} = ma_x \qquad \text{or} \qquad F_{\text{engine}} = F_{\text{resistance}} + ma$$

Thus, $$F_{\text{engine}} = 400 \text{ N} + \left(1.50 \times 10^3 \text{ kg}\right)\left(1.50 \text{ m/s}^2\right) = 2.65 \times 10^3 \text{ N}$$

The average velocity of the car is

$$\overline{v} = \frac{\left(v_f + v_i\right)}{2} = \frac{\left(18.0 \text{ m/s} + 0\right)}{2} = 9.00 \text{ m/s}$$

The average power developed by the engine is then $\overline{P} = F_{\text{engine}}\overline{v}$

or $$\overline{P} = \left(2.65 \times 10^3 \text{ N}\right)\left(9.00 \text{ m/s}\right) = 2.39 \times 10^4 \text{ W} = 23.9 \text{ kW}$$ ◊

(b) At t = 12.0 s, the instantaneous velocity of the car is $v = 18.0$ m/s and the instantaneous power output of the engine is $P = F_{\text{engine}}v$, so

$$P = \left(2.65 \times 10^3 \text{ N}\right)\left(18.0 \text{ m/s}\right) = 4.77 \times 10^4 \text{ W} = 47.7 \text{ kW}$$ ◊

51. A toy gun uses a spring to project horizontally a 5.3-g soft rubber sphere. The spring constant is 8.0 N/m, the barrel of the gun is 15 cm long, and a constant frictional force of 0.032 N exists between barrel and projectile. With what speed does the projectile leave the barrel if the spring was compressed 5.0 cm for this launch?

Solution

In this problem, only the spring force and the frictional force do work on the ball as the ball moves along the horizontal barrel. All other forces acting on the ball are perpendicular to the motion. The spring force is conservative and the work it does may be included in elastic potential terms in the work-energy theorem:

$$W_{net} = W_{nc} + W_c = W_{nc} + \left(PE_i - PE_f\right) = KE_f - KE_i$$

The elastic potential energy is:

$$PE_s = \tfrac{1}{2}kx^2$$

where k is the force constant of the spring and x is the amount the spring is stretched or compressed.

The frictional force is non-conservative, and the work it does is $W_{nc} = fs\cos 180° = -fs$ since the friction force is directed opposite to the displacement. The ball is initially at rest, so $v_i = 0$. Hence $KE_i = \tfrac{1}{2}mv_i^2 = 0$. At the end, the spring is uncompressed so $x_f = 0$ and $(PE_s)_f = \tfrac{1}{2}kx_f^2 = 0$. Thus, the work-energy theorem gives:

$$-(0.032\text{ N})(0.15\text{ m}) + \left[\tfrac{1}{2}(8.0\text{ N/m})(0.05\text{ m})^2 - 0\right] = \tfrac{1}{2}\left(5.3\times10^{-3}\text{ kg}\right)v_f^2 - 0$$

or

$$v_f^2 = 2.0\text{ m}^2/\text{s}^2 \quad\text{and}\quad v_f = 1.4\text{ m/s}$$ ◊

59. A 5.0-kg block is pushed 3.0 m up a vertical wall with constant speed by a constant force of magnitude F applied at an angle of $\theta = 30°$ with the horizontal, as shown in Figure P5.59. If the coefficient of kinetic friction between block and wall is 0.30, determine the work done by (a) $\mathbf{F}$, (b) gravity, and (c) the normal force between block and wall. (d) By how much does the gravitational potential energy of the block increase?

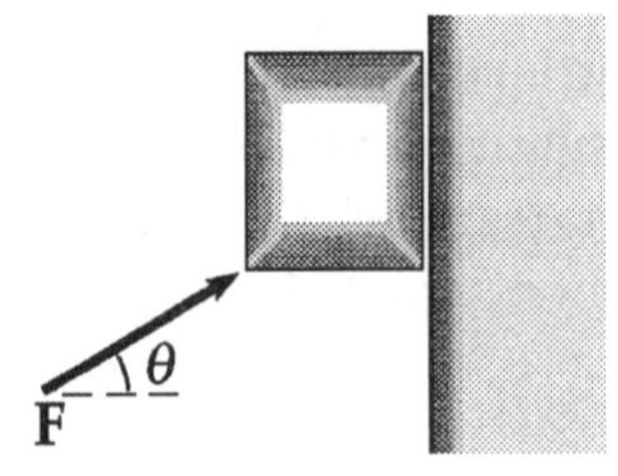

Figure P5.59

Solution (a) Since the block moves with constant velocity, it has zero acceleration. Thus, $\sum F_x = 0$, so $F\cos 30° - N = 0$, and $N = 0.87F$ **[Equation 1]**

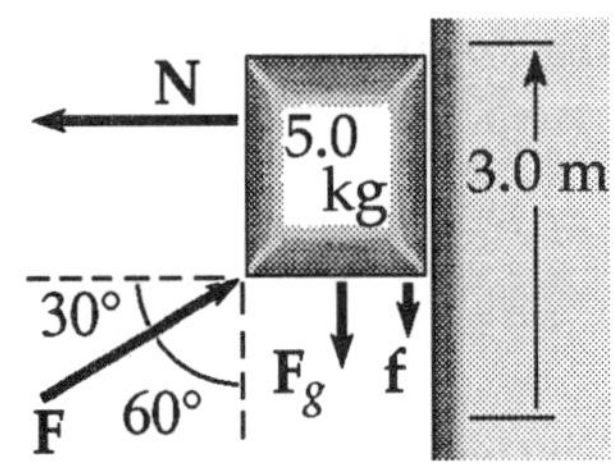

Also, $\sum F_y = 0$ giving $F\sin 30° - F_g - f = 0$

so $0.50F - f = F_g$ **[Equation 2]**

But $F_g = mg = (5.0 \text{ kg})(9.8 \text{ m/s}^2) = 49 \text{ N}$, and $f = \mu_k N$ or (using Equation 1) $f = 0.30(0.87F) = 0.26F$. Equation 2 then becomes $0.50F - 0.26F = 49 \text{ N}$ giving $F = 204 \text{ N}$. The work done by this force as the block moves 3.0 m up the wall is given by $W_F = Fs\cos\theta = (204 \text{ N})(3.0 \text{ m})\cos 60° = 3.1 \times 10^2 \text{ J}$ ◊

(b) The work done by the gravitational force (weight) as the block moves up the wall is $W_w = F_g s\cos\theta = (49 \text{ N})(3.0 \text{ m})\cos 180° = -1.5 \times 10^2 \text{ J}$ ◊

(c) Since the normal force is perpendicular to the displacement, the work it does is $W_N = Ns\cos\theta = Ns\cos 90° = 0$ ◊

(d) The change in the gravitational potential energy of the block is

$$\Delta PE = PE_f - PE_i = mgy_f - mgy_i = mg(\Delta y) = (49 \text{ N})(3.0 \text{ m}) = 1.5 \times 10^2 \text{ J}$$ ◊

Note: The increase in the gravitational potential energy is the negative of the work done by the gravitational force (see part b).

69. In the dangerous "sport" of bungee-jumping, a daring student jumps from a balloon with a specially designed elastic cord attached to his ankles, as shown in the Figure P5.69. The unstretched length of the cord is 25.0 m, the student weighs 700 N, and the balloon is 36.0 m above the surface of a river below. Calculate the required force constant of the cord if the student is to stop safely 4.00 m above the river.

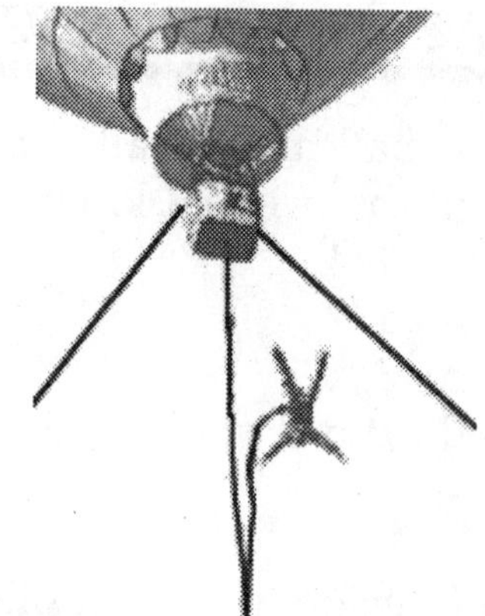

Solution Ignoring air resistance, there are no non-conservative forces acting on the student during the jump. Thus, the total mechanical energy is constant:

$$KE_f + \left(PE_g\right)_f + \left(PE_s\right)_f = KE_i + \left(PE_g\right)_i + \left(PE_s\right)_i$$

where PE_g = gravitational potential energy and PE_s = elastic potential energy

The student is at rest both before and after the jump, so $KE_f = KE_i = 0$

The cord is initially unstretched (i.e., $x_i = 0$), $\left(PE_s\right)_i = \frac{1}{2}kx_i^2 = 0$

The conservation of energy equation then reduces to:

$$\left(PE_s\right)_f = \left(PE_g\right)_i - \left(PE_g\right)_f \quad \text{or} \quad \tfrac{1}{2}kx_f^2 = mgy_i - mgy_f$$

This may be rewritten as $\frac{1}{2}kx_f^2 = mg\left(y_i - y_f\right) = F_g\left(y_i - y_f\right)$. The student is to drop from 36.0 m down to 4.00 m above the river, so $y_i - y_f = 32.0 \text{ m}$. Also, the cord is 25.0 m long. Thus, when the student has dropped 32.0 m, the cord is stretched by $x_f = (32.0 - 25.0) \text{ m} = 7.00 \text{ m}$. The energy equation then gives:

$$k = \frac{2F_g\left(y_i - y_f\right)}{x_f^2} = \frac{2(700 \text{ N})(32.0 \text{ m})}{(7.00 \text{ m})^2} = 914 \text{ N/m} \qquad \Diamond$$

CHAPTER SELF-QUIZ

1. A ball hits a wall and bounces back at half the original speed. What part of the original kinetic energy of the ball did it lose in the collision?
   a. 1/4
   b. 1/2
   c. 3/4
   d. It did not lose kinetic energy

2. I decide to make a pile out of the 100 boards that are lying out in the yard, so I stack them on top of each other. Each board is 0.05 m thick and weights 40 N. If the boards originally had zero potential energy, what is the potential energy of the new pile?
   a. 200 J
   b. 100 J
   c. 20000 J
   d. 10000 J

3. A 3-kg object starting at rest falls from a height of 10 m to the ground. In this instance, the force of the air is not negligible so that the magnitude of work done by this frictional force is 20 J. What is the object's kinetic energy prior to hitting the ground?
   a. 78 J
   b. 118 J
   c. 49 J
   d. 274 J

4. A 40-N crate is pulled 5 m up an inclined plane at a constant velocity. If the plane is inclined at an angle of 37° to the horizontal, what is the magnitude of the work done on the crate by the force of gravity?
   a. 120 J
   b. 6 J
   c. 1180 J
   d. 200 J

5. A worker pushes a 250-N weight wheelbarrow up a ramp 6.00 m in length and inclined at 20° with the horizontal. What potential energy change does the wheelbarrow experience?
   a. 513 J
   b. 1500 J
   c. 3550 J
   d. 4500 J

6. A 20-N crate starting at rest slides down a rough 3-m long ramp, inclined at 30° with the horizontal. The force of friction between crate and ramp is 6 N. What is the kinetic energy of the crate at the bottom of the ramp?
   a. zero
   b. 8 J
   c. 12 J
   d. 32 J

7. A baseball catcher puts on an exhibition by catching a 0.15-kg ball dropped from a helicopter at a height of 61 m. If the catcher "gives" with the ball for a distance of 0.75 m while catching it, what average force is exerted on the mitt by the ball?
   ($g = 9.80\ m/s^2$)
   a. 47 N
   b. 94 N
   c. 119 N
   d. 376 N

8. A simple pendulum, 2.00 m in length, is released from rest when the support string is at an angle of 25° from the vertical. What is the speed of the suspended mass at the bottom of the swing? (Ignore air resistance, $g = 9.80\ m/s^2$.)
   a. 0.60 m/s
   b. 0.80 m/s
   c. 1.30 m/s
   d. 1.90 m/s

9. A 1500-kg car travels along a highway at a speed of 20 m/s. What is its kinetic energy?
   a. 15.0×10^5 J
   b. 2.5×10^5 J
   c. 3.0×10^5 J
   d. 6.0×10^5 J

10. A pulley-cable system on a crate hoists a bucket of cement with a total weight of 20 000 N to a height of 40 m. If this is accomplished in 2 min., what is the power output by the pulley-cable system?
    a. 6.7 kW
    b. 3.3 kW
    c. 13.3 kW
    d. 400 kW

11. A horizontal force of 200 N is applied to move a 55-kg cart across a 10-m level surface. If the cart accelerates at 2.00 m/s^2, starting from rest, then what kinetic energy does it gain while moving the 10-m distance?
    a. 2200 J
    b. 1100 J
    c. 550 J
    d. 880 J

12. A 15.0-kg crate, initially at rest, slides down a ramp 2.00 m long and inclined at an angle of 20° with the horizontal. If there is a constant force of kinetic friction of 25 N between the crate and ramp, what kinetic energy would the crate have at the bottom of the ramp? (g = 9.80 m/s^2)
    a. 50 J
    b. 638 J
    c. 171 J
    d. 5.2 J

Chapter 6

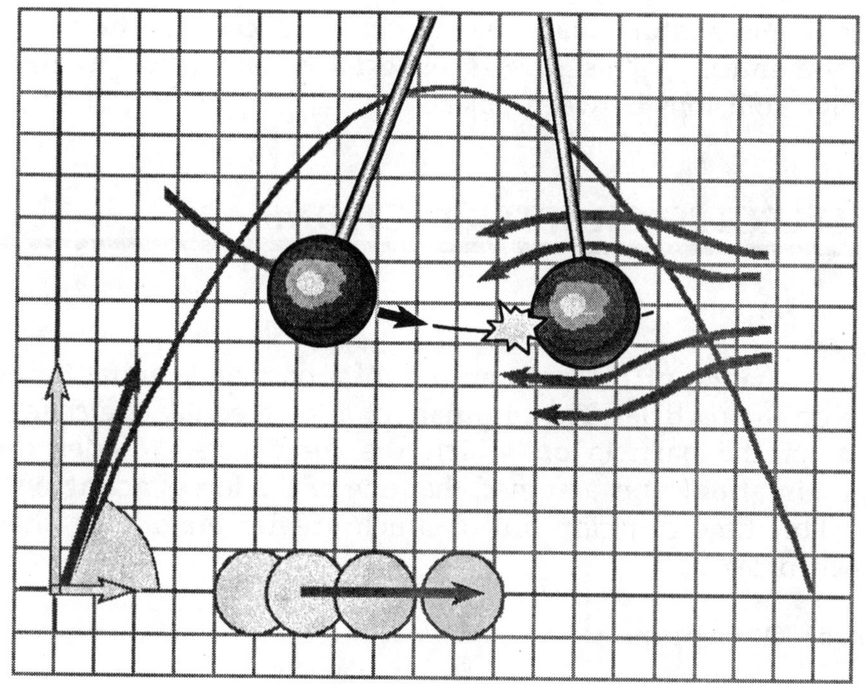

MOMENTUM AND COLLISIONS

Chapter 6

MOMENTUM AND COLLISIONS

One of the main objectives of this chapter is to help you understand how to analyze collisions and other events which involve objects which experience forces and accelerations over very short time intervals. As a first step, we shall introduce the term **momentum**. We often use the concept of momentum which is defined as the product of mass and velocity in describing objects in motion.

The concept of momentum leads us to a second conservation law: conservation of momentum. This law is especially useful for treating problems that involve collisions between objects.

NOTES FROM SELECTED CHAPTER SECTIONS

6.1 Momentum and Impulse

The **time rate of change of the momentum** of a particle is equal to the resultant force on the particle. The **impulse** of a force equals the change in momentum of the particle on which the force acts. Under the **impulse approximation,** it is assumed that one of the forces acting on a particle is of short time duration but of much greater magnitude than any of the other forces.

6.2 Conservation of Momentum

If two particles of masses m_1 and m_2 form an **isolated system,** then the total momentum of the system remains constant.

6.3 Collisions

For **any type of collision,** the total momentum before the collision equals the total momentum just after the collision.

In an **inelastic collision,** the total momentum is conserved; however, the total kinetic energy is not conserved.

In a **perfectly inelastic** collision, the two colliding objects stick together following the collision.

In an **elastic collision,** both momentum and kinetic energy are conserved.

6.4 Glancing Collisions

The law of conservation of momentum is not restricted to one-dimensional collisions. If two masses undergo a **two-dimensional** (glancing) **collision** and there are no external forces acting, the total momentum in each of the x, y, and z directions is conserved.

EQUATIONS AND CONCEPTS

An object of mass, m, and velocity, **v**, is characterized by a vector quantity called linear momentum. The SI units of linear momentum are kg·m/s. Since momentum is a vector quantity, the defining equation can be written in component form.

$$\mathbf{p} \equiv m\mathbf{v} \qquad (6.1)$$

Since momentum is a vector quantity, the components of the vector along the x direction and the y direction can be calculated.

$$p_x = mv_x$$

$$p_y = mv_y$$

The resultant force acting on an object equals the time rate of change of the object's momentum. This equation is a mathematical expression of Newton's second law.

$$\Sigma\mathbf{F} = \frac{\Delta\mathbf{p}}{\Delta t} \qquad (6.2)$$

Note that as a special case in Equation 6.2, if the resultant force $\Sigma \mathbf{F} = 0$, then the momentum does not change.

Comment.

This equation is a mathematical statement of the important impulse-momentum theorem. The product of the force acting on an object and the time interval during which it acts is called the impulse imparted to the object by the force.

$$\mathbf{F}\Delta t = m\mathbf{v}_f - m\mathbf{v}_i \tag{6.4}$$

The impulse imparted by a force during a time interval Δt is equal to the area under the force-time graph from the beginning to the end of the time interval. When the force varies in time, as illustrated in the figure, it is often convenient to define an average force, $\overline{\mathbf{F}}$, which is a constant force that imparts the same impulse in a time interval Δt as the actual time varying force.

Comment on impulse of force.

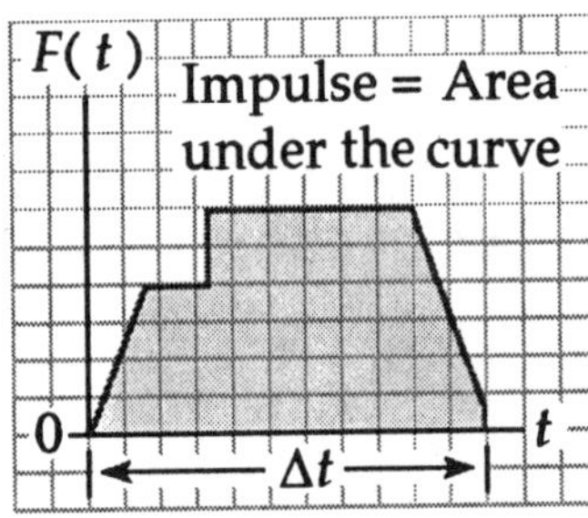

When two objects interact in a collision (and exert forces on each other) and no external forces act on the two-object system, the total momentum of the system before the collision equals the total momentum after the collision.

$$m_1\mathbf{v}_{1i} + m_2\mathbf{v}_{2i} = m_1\mathbf{v}_{1f} + m_2\mathbf{v}_{2f} \tag{6.5}$$

In general, when the external force acting on any system of objects is zero, the linear momentum of the system is conserved.

Comment on law of conservation of momentum.

An **elastic collision** is one in which both momentum and kinetic energy are conserved.

An **inelastic collision** is one in which momentum is conserved but kinetic energy is not.

A **perfectly inelastic collision** is a collision in which the colliding objects stick together so that their final velocities are equal.

Comment on types of collisions.

Conservation of momentum for a one-dimensional perfectly inelastic collision.

$$m_1 v_{1i} + m_2 v_{2i} = (m_1 + m_2) v_f \qquad (6.6)$$

Note that v_{1i}, v_{2i}, and v_f are actually components of velocity vectors and hence they may have positive or negative values — determined by the direction of motion relative to the chosen coordinate origin.

Comment on signs.

In an elastic head-on collision, both momentum and kinetic energy are conserved.

$$m_1v_{1i} + m_2v_{2i} = m_1v_{1f} + m_2v_{2f} \quad (6.7)$$

$$\tfrac{1}{2}m_1v_{1i}^2 + \tfrac{1}{2}m_2v_{2i}^2 = \tfrac{1}{2}m_1v_{1f}^2 + \tfrac{1}{2}m_2v_{2f}^2 \quad (6.8)$$

This equation states another characteristic of a perfectly elastic collision: the relative velocity of the two objects before the collision equals the negative of the relative velocity of the two objects after the collision.

$$v_{1i} - v_{2i} = -(v_{1f} - v_{2f}) \quad (6.11)$$

In a glancing collision, the colliding masses rebound along a direction which is not parallel to the direction of motion before the collision. This is an example of a two-dimensional collision. If no net external force acts on the system of colliding masses, momentum is conserved along each coordinate direction.

$$\sum p_{ix} = \sum p_{fx} \quad (6.12)$$

and

$$\sum p_{iy} = \sum p_{fy}$$

SUGGESTIONS, SKILLS, AND STRATEGIES

The following procedure is recommended when dealing with problems involving collisions between two objects:

1. Set up a coordinate system and define your velocities with respect to that system. That is, objects moving in the direction selected as the positive direction of the x axis are considered as having a positive velocity and negative if moving in the negative x direction. It is convenient to have the x axis coincide with one of the initial velocities.

2. In your sketch of the coordinate system, draw all velocity vectors with labels and include all the given information.

3. Write expressions for the momentum of each object before and after the collision. (In two-dimensional collision problems, write expressions for the x and y components of momentum before and after the collision.) Remember to include the appropriate signs for the velocity vectors.

4. Now write expressions for the **total** momentum **before** and **after** the collision and equate the two. (For two-dimensional collisions, this expression should be written for the momentum in both the x and y directions.) It is important to emphasize that it is the momentum of the **system** (the two colliding objects) that is conserved, not the momentum of the individual objects.

5. If the collision is **inelastic**, you should then proceed to solve the momentum equations for the unknown quantities.

6. If the collision is **elastic**, kinetic energy is also conserved, so you can equate the total kinetic energy before the collision to the total kinetic energy after the collision. This gives an additional relationship between the various velocities. The conservation of kinetic energy for elastic collisions leads to the expression $v_{1i} - v_{2i} = -(v_{1f} - v_{2f})$, which is often easier to use in solving elastic collision problems than is an expression for conservation of kinetic energy.

REVIEW CHECKLIST

▷ The impulse of a force acting on a particle during some time interval equals the **change** in momentum of the particle, and the impulse equals the area under the Force-Time graph.

▷ The momentum of any isolated system (one for which the net external force is zero) is conserved, regardless of the nature of the forces between the masses which comprise the system.

▷ There are two types of collisions that can occur between two particles, namely elastic and inelastic collisions. Recognize that a **perfectly** inelastic collision is an inelastic collision in which the colliding particles stick together after the collision, and hence move as a composite particle. Kinetic energy is conserved in perfectly elastic collisions, but not in the case of inelastic collisions. Linear momentum is conserved in both types of collisions, if the net external force acting on the system of colliding objects is zero.

▷ The conservation of linear momentum applies not only to head-on collisions (one-dimensional), but also to glancing collisions (two- or three-dimensional). For example, in a two-dimensional collision, the total momentum in the x direction is conserved and the total momentum in the y direction is conserved.

▷ The equations for momentum and kinetic energy can be used to calculate the final velocities in a two-body head-on elastic collision; and to calculate the final velocity and the change of kinetic energy in a two-body system for a completely inelastic collision.

SOLUTIONS TO SELECTED END-OF-CHAPTER PROBLEMS

5. A 1500-kg car moving with a speed of 15 m/s collides with a utility pole and is brought to rest in 0.30 s. Find the magnitude of the average force exerted on the car during the collision.

Solution

The initial momentum of the car is $\mathbf{P}_i = m\mathbf{v}_i$ and the final momentum is $\mathbf{P}_f = m\mathbf{v}_f$. The impulse imparted to the car during the collision is the product of the average force acting on the car and the time interval over which the force acts. The impulse is also equal to the change in the momentum of the car (impulse-momentum theorem). That is:

$$\overline{\mathbf{F}}(\Delta t) = \mathbf{P}_f - \mathbf{P}_i = m\left(\mathbf{v}_f - \mathbf{v}_i\right)$$

Taking the direction of the initial motion of the car as the positive direction and using the given data, the impulse-momentum theorem gives

$$\overline{\mathbf{F}} = \frac{m\left(\mathbf{v}_f - \mathbf{v}_i\right)}{\Delta t} = \frac{(1500\text{ kg})(0 - 15\text{ m/s})}{0.30\text{ s}} = -7.5 \times 10^4\text{ N}$$

The negative sign in the result means that $\overline{\mathbf{F}}$ is in the direction opposite to the initial velocity of the car. The magnitude of the average force is

$$\left|\overline{\mathbf{F}}\right| = 7.5 \times 10^4\text{ N} \qquad \Diamond$$

9. The force, $\mathbf{F}_x$ acting on a 2.00-kg particle varies in time as shown in Figure P6.9. Find (a) the impulse of the force, (b) the final velocity of the particle if it is initially at rest, and (c) the final velocity of the particle if it is initially moving along the x axis with a velocity of –2.00 m/s.

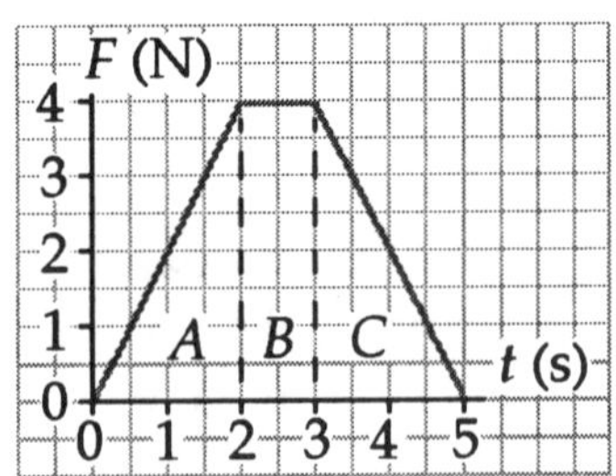

Figure P6.9

Solution (a) Graphically, the impulse of a force is equal to the area under the force-time graph. The area under the force-time graph in Figure P6.9 may be viewed as consisting of two triangles (A and C) and one rectangle (B). The impulse of this force is then:

$$\text{impulse} = (\text{area A}) + (\text{area B}) + (\text{area C})$$

or $$\text{impulse} = \left[+\tfrac{1}{2}(2.00\text{ s})(4.00\text{ N})\right] + \left[+(1.00\text{ s})(4.00\text{ N})\right] + \left[+\tfrac{1}{2}(2.00\text{ s})(4.00\text{ N})\right]$$

Which gives $$\text{impulse} = +12.0\text{ N}\cdot\text{s} = +12.0\text{ kg}\cdot\text{m/s}$$ ◊

Note: The positive sign means that the impulse is in the $+x$ direction.

(b) The impulse-momentum theorem, $\text{impulse} = \mathbf{P}_f - \mathbf{P}_i = m\mathbf{v}_f - m\mathbf{v}_i$

gives $+12.0\text{ kg}\cdot\text{m/s} = (2.00\text{ kg})\mathbf{v}_f - 0,$

solving, $\mathbf{v}_f = +6.00\text{ m/s}$ or 6.00 m/s in the $+x$ direction. ◊

(c) If the initial velocity of the car had been -2.00 m/s, the impulse-momentum equation from part (b) would have been:

$$+12.0\text{ kg}\cdot\text{m/s} = (2.00\text{ kg})\mathbf{v}_f - (2.00\text{ kg})(-2.00\text{ m/s})$$

giving $\mathbf{v}_f = +4.00\text{ m/s}$ or $\mathbf{v}_f = 4.00\text{ m/s}$ in the $+x$ direction. ◊

13. A 0.15-kg baseball is thrown with a speed of 20 m/s. It is hit straight back at the pitcher with a final speed of 22 m/s. (a) What is the impulse delivered to the ball? (b) Find the average force exerted by the bat on the ball if the two are in contact for 2.0×10^{-3} s.

Solution

(a) **Choosing toward the batter as the positive direction,** the initial and final momenta of the ball are:

$$\mathbf{P}_i = m\mathbf{v}_i = (0.15\text{ kg})(+20\text{ m/s}) = +3.0\text{ kg}\cdot\text{m/s}$$

and $$\mathbf{P}_f = m\mathbf{v}_f = (0.15\text{ kg})(-22\text{ m/s}) = -3.3\text{ kg}\cdot\text{m/s}$$

Then, the impulse delivered to the ball is

$$\text{impulse} = \mathbf{P}_f - \mathbf{P}_i = (-3.3\text{ kg}\cdot\text{m/s}) - (+3.0\text{ kg}\cdot\text{m/s}) = -6.3\text{ kg}\cdot\text{m/s}$$

or impulse $= 6.3\text{ kg}\cdot\text{m/s}$ directed **toward the pitcher.** ◊

(b) The impulse imparted to the ball may also be expressed as $\text{impulse} = \overline{\mathbf{F}}(\Delta t)$, where $\overline{\mathbf{F}}$ is the average force acting on the ball and Δt is the duration of that force. Thus, the average force exerted on the ball is

$$\overline{\mathbf{F}} = \frac{\text{impulse}}{\Delta t} = \frac{-6.3\text{ kg}\cdot\text{m/s}}{2.0\times 10^{-3}\text{ s}} = -3.2\times 10^{3}\text{ N}$$

or $\overline{\mathbf{F}} = 3.2\times 10^{3}\text{ N}$ directed **toward the pitcher.** ◊

---

19. A 65.0-kg person throws a 0.0450-kg snowball forward with a ground speed of 30.0 m/s. A second person, with a mass of 60.0 kg, catches the snowball. Both people are on skates. The first person is initially moving forward with a speed of 2.50 m/s, and the second person is initially at rest. What are the velocities of the two people after the snowball is exchanged? Disregard the friction between the skates and the ice.

Solution

Consider a system consisting of the two skaters and the snowball. The total momentum of the system **before the ball leaves the thrower's hand** (ball has same velocity as thrower, catcher still at rest) is:

$$(65.0\ \text{kg})(+2.50\ \text{m/s}) + (0.045\ \text{kg})(+2.50\ \text{m/s}) + (60.0\ \text{kg})(0) = +162.6\ \text{kg}\cdot\text{m/s}$$

Ignoring air resistance and friction, no horizontal external forces act on this system. Thus, the total horizontal momentum of the system is constant, or $\mathbf{P}_{\text{total}} = \mathbf{P}_{\text{thrower}} + \mathbf{P}_{\text{ball}} + \mathbf{P}_{\text{catcher}} = +162.6\ \text{kg}\cdot\text{m/s}$ at all times. Immediately **after the ball leaves the thrower's hand,** this gives

$$(65.0\ \text{kg})\mathbf{v}_{tf} + (0.045\ \text{kg})(+30.0\ \text{m/s}) + (60.0\ \text{kg})(0) = +162.6\ \text{kg}\cdot\text{m/s}$$

or $\mathbf{v}_{tf} = +2.48\ \text{m/s}$ as the final velocity of the thrower. ◊

The ball and catcher have the same velocity, while the thrower continues to move at $\mathbf{v}_{tf} = +2.48\ \text{m/s}$. Thus, just **after the catcher catches the ball** conservation of momentum gives:

$$(65.0\ \text{kg})(+2.48\ \text{m/s}) + (0.045\ \text{kg})\mathbf{v}_{cf} + (60.0\ \text{kg})\mathbf{v}_{cf} = +162.6\ \text{kg}\cdot\text{m/s}$$

so the final velocity of the catcher and ball is $\mathbf{v}_{cf} = +2.25\times10^{-2}\ \text{m/s}$ ◊

25. A 0.03-kg bullet is fired vertically at 200 m/s into a 0.15-kg baseball that is initially at rest. How high does the combination rise after the collision, assuming the bullet embeds in the ball?

Solution A common student error in a problem of this type is to attempt using conservation of **energy** from before collision to the end of the motion. Since this is an inelastic collision, use of energy methods over any time interval that includes the collision is extremely difficult. The best approach is to **apply conservation of momentum from just before to just after the collision.** Then energy methods can be applied **after** the collision is past, if that is desirable. Consider the sketches above:

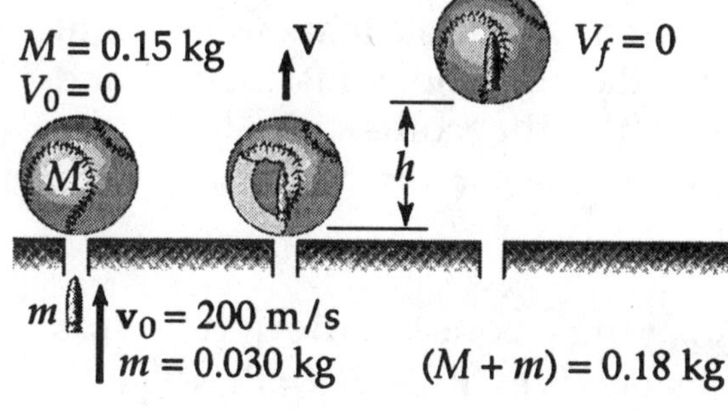

The velocity of the ball and embedded bullet, immediately after impact is V. Using momentum conservation from just before impact to just after impact, $mv_0 + MV_0 = (M+m)V$:

Substituting in known values, $$(0.030 \text{ kg})(200 \text{ m/s}) + 0 = (0.18 \text{ kg})V$$

which yields $$V = \frac{6.0 \text{ kg} \cdot \text{m/s}}{0.18 \text{ kg}} = 33.3 \text{ m/s}$$

From just **after** the collision until the ball reaches the maximum height, only conservative forces do work on the ball-bullet combination. Thus, $KE_f + PE_f = KE_i + PE_i$. If $y = 0$ at the initial level of the ball, then $y_i = 0$ and $y_f = h$; so $PE_i = 0$ and $PE_f = (M+m)gh$. Also, $V_f = 0$, so $KE_f = 0$. The energy equation then reduces to $(M+m)gh = KE_i = \frac{1}{2}(M+m)V^2$, or

$$h = \frac{V^2}{2g} = \frac{(33.3 \text{ m/s})^2}{2(9.8 \text{ m/s}^2)} = 57 \text{ m}$$ ◊

31. A 25.0-g object moving to the right at 20.0 cm/s overtakes and collides elastically with a 10.0-g object moving in the same direction at 15.0 cm/s. Find the velocity of each object after the collision.

Solution The sketches to the right show the conditions just before and just after the collision. Note the choice of the $+x$ direction.

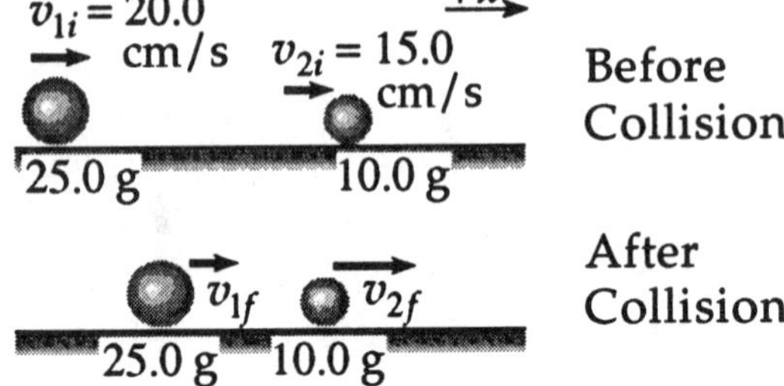

Applying conservation to this collision gives $\mathbf{P}_{1f} + \mathbf{P}_{2f} = \mathbf{P}_{1i} + \mathbf{P}_{2i}$:

$$(25.0\text{ g})\mathbf{v}_{1f} + (10.0\text{ g})\mathbf{v}_{2f} = (25.0\text{ g})(+20.0\text{ cm/s}) + (10.0\text{ g})(+15.0\text{ cm/s})$$

which reduces to: $2.50\mathbf{v}_{1f} + \mathbf{v}_{2f} = +65.0\text{ cm/s}$ **[Equation 1]**

This is a perfectly elastic, head-on collision, so Equation 6.11 in the text gives

$$\mathbf{v}_{1i} - \mathbf{v}_{2i} = -(\mathbf{v}_{1f} - \mathbf{v}_{2f}), \quad \text{or} \quad +20.0\text{ cm/s} - (+15.0\text{ cm/s}) = -\mathbf{v}_{1f} + \mathbf{v}_{2f}$$

This reduces to : $\mathbf{v}_{2f} = \mathbf{v}_{1f} + 5.00\text{ cm/s}$ **[Equation 2]**

Substitution of Equation 2 into Equation 1 yields $3.50\mathbf{v}_{1f} = +60.0\text{ cm/s}$. Thus,

$$\mathbf{v}_{1f} = +17.1\text{ cm/s} = 17.1\text{ cm/s in the } +x \text{ direction} \quad \Diamond$$

Then, Equation 2 gives:

$$\mathbf{v}_{2f} = +22.1\text{ cm/s} = 22.1\text{ cm/s in } +x \text{ direction} \quad \Diamond$$

35. A 2000-kg car moving east at 10.0 m/s collides with a 3000-kg car moving north. The cars stick together and move as a unit after the collision, at an angle of 40.0° north of east and at a speed of 5.22 m/s. Find the velocity of the 3000-kg car before the collision.

Solution

The sketch at the right shows the conditions existing just before and just after this perfectly inelastic collision. Total momentum is conserved even in perfectly inelastic collisions. Thus, $\sum \mathbf{P}_f = \sum \mathbf{P}_i$. This vector equation is equivalent to two component equations:

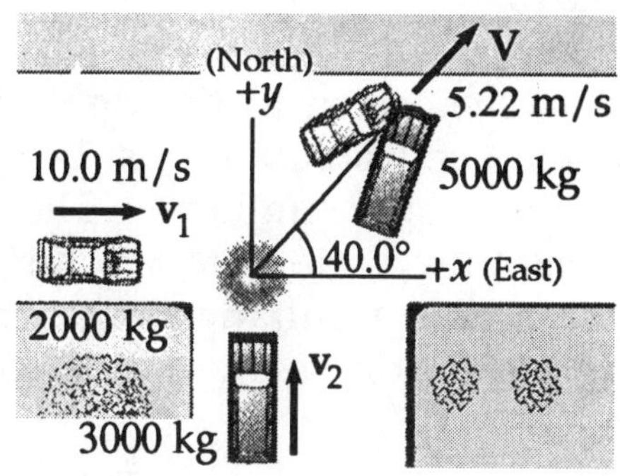

$$\left(\sum P_x\right)_f = \left(\sum P_x\right)_i, \text{ and } \left(\sum P_y\right)_f = \left(\sum P_y\right)_i$$

Since the only unknown in this problem is v_2, the initial velocity of car #2, and that is known to be in the $+y$ (or north) direction, consider the equation for the momentum in the y-direction. That gives:

$$0 + (3000 \text{ kg})v_2 = (5000 \text{ kg})(5.22 \text{ m/s})\sin 40.0°$$

or $$v_2 = \frac{(5000 \text{ kg})(5.22 \text{ m/s})\sin 40.0°}{3000 \text{ kg}}$$

Solving, $$v_2 = 5.59 \text{ m/s}$$ ◊

37. A billiard ball moving at 5.00 m/s strikes a stationary ball of the same mass. After the collision, the first ball moves at 4.33 m/s at an angle of 30.0° with respect to the original line of motion. (a) Find the velocity (magnitude and direction) of the second ball after collision. (b) Was this an inelastic collision or an elastic collision?

Solution

The sketch given to the right shows this glancing collision both just before impact and just after impact. The resultant momentum vector after collision is the same as the resultant momentum vector before collision, so

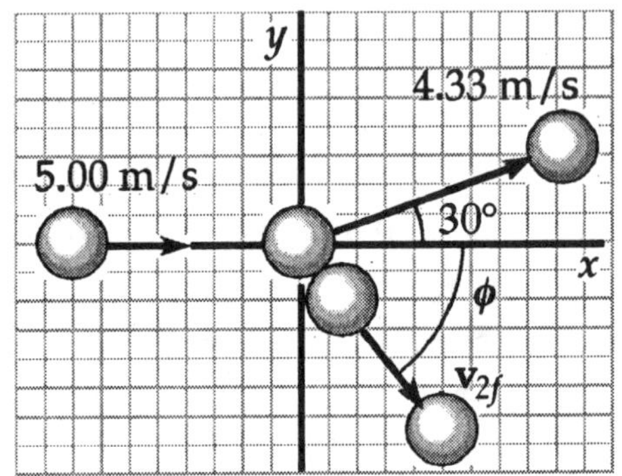

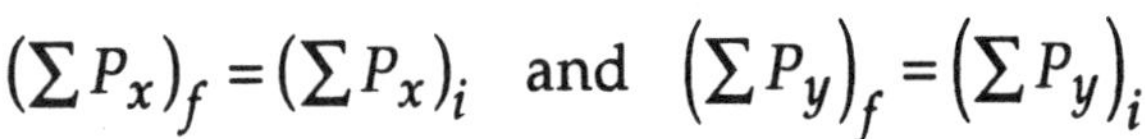

$$\left(\sum P_x\right)_f = \left(\sum P_x\right)_i \quad \text{and} \quad \left(\sum P_y\right)_f = \left(\sum P_y\right)_i$$

First, looking at the x-components of momentum, this gives

$$m(4.33 \text{ m/s})\cos 30.0° + mv_{2f}\cos\theta = m(5.00 \text{ m/s}) + 0$$

Thus $\quad 3.75 \text{ m/s} + v_{2f}\cos\theta = 5.00 \text{ m/s}$

and $\quad v_{2f}\cos\theta = 1.25 \text{ m/s}$ **[Equation 1]**

Then, considering the y-components of momentum gives

$$m(4.33 \text{ m/s})\sin 30.0° + mv_{2f}\sin\theta = 0 + 0$$

which reduces to $\quad v_{2f}\sin\theta = -2.17 \text{ m/s}$ **[Equation 2]**

Squaring Equations 1 and 2 and adding the results yields

$$v_{2f}^{\,2}\left(\cos^2\theta+\sin^2\theta\right)=(1.25\text{ m/s})^2+(-2.17\text{ m/s})^2$$

Since (by the trigonometric identity) $\cos^2\theta+\sin^2\theta=1$, the speed of the second ball after the collision is found as:

$$v_{2f}^{\,2}=6.27\text{ m}^2/\text{s}^2 \quad\text{and}\quad v_{2f}=2.50\text{ m/s}$$ ◊

Since we were asked for the velocity, we also need the angle θ of the velocity. Dividing Equation 2 by Equation 1 gives

$$\frac{\sin\theta}{\cos\theta}=\frac{-2.17\text{ m/s}}{1.25\text{ m/s}}$$

Since $\tan\theta=\dfrac{\sin\theta}{\cos\theta}$,

$$\tan\theta=\frac{-2.17\text{ m/s}}{1.25\text{ m/s}}$$

which yields two answers:

$$\theta=\arctan\left(\frac{-2.17}{1.25}\right)=-60°,\ 120°$$

Looking at the original diagram, we can see that conservation of momentum would not allow the answer 120°. Therefore, we know that only the first is correct; the velocity of the second ball is 2.50 m/s, 60° below the $+x$-axis. ◊

(b) The total kinetic energy before collision is $KE_i=KE_{1i}+KE_{2i}$, or $KE_i=\frac{1}{2}m(5.00\text{ m/s})^2+0=m\left(12.5\text{ m}^2/\text{s}^2\right)$. The total kinetic energy after collision is $KE_f=KE_{1f}+KE_{2f}=\frac{1}{2}m(4.33\text{ m/s})^2+m(2.50\text{ m/s})^2$, which yields $KE_f=m\left(12.5\text{ m}^2/\text{s}^2\right)$. Since the total kinetic energy after collision is the same as that before collision, **the collision is elastic.** ◊

39. A 0.400-kg soccer ball approaches a player horizontally with a speed of 15.0 m/s. The player illegally strikes the ball with her hand and causes it to move in the opposite direction with a speed of 22.0 m/s. What impulse was delivered to the ball by the player?

Solution

Choose the direction of the ball's initial velocity as the positive x direction. The initial momentum of the ball is then

$$P_i = mv_i = (0.400 \text{ kg})(+15.0 \text{ m/s}) = +6.00 \text{ kg} \cdot \text{m/s}$$

so the initial momentum of the ball is $6.00 \text{ kg} \cdot \text{m/s}$ in the $+x$ direction. The final momentum of the ball is:

$$P_f = mv_f = (0.400 \text{ kg})(-22.0 \text{ m/s}) = -8.80 \text{ kg} \cdot \text{m/s}$$

and the final momentum of the ball is $8.80 \text{ kg} \cdot \text{m/s}$ in the $-x$ direction. The impulse delivered to the ball is

$$\text{Impulse} = \mathbf{P}_f - \mathbf{P}_i$$

$$\text{Impulse} = -8.80 \text{ kg} \cdot \text{m/s} - (+6.00 \text{ kg} \cdot \text{m/s}) = -14.8 \text{ kg} \cdot \text{m/s} \qquad \lozenge$$

This is an impulse of magnitude $14.8 \text{ kg} \cdot \text{m/s}$ directed opposite to the direction of the **initial** velocity.

---

45. A 12.0-g bullet is fired horizontally into a 100-g wooden block initially at rest on a horizontal surface. After impact, the block slides 7.50 m before coming to rest. If the coefficient of kinetic friction between block and surface is 0.650, what was the speed of the bullet immediately before impact?

Solution Unless a collision is perfectly elastic, it is extremely difficult to apply work-energy methods to any time interval that includes that collision. The collision in this case is perfectly **inelastic** since the block and bullet move as a single unit after collision. The strategy will be to employ conservation of **momentum** from just before the collision to just after the collision. Work-energy methods may be applied after the collision is over.

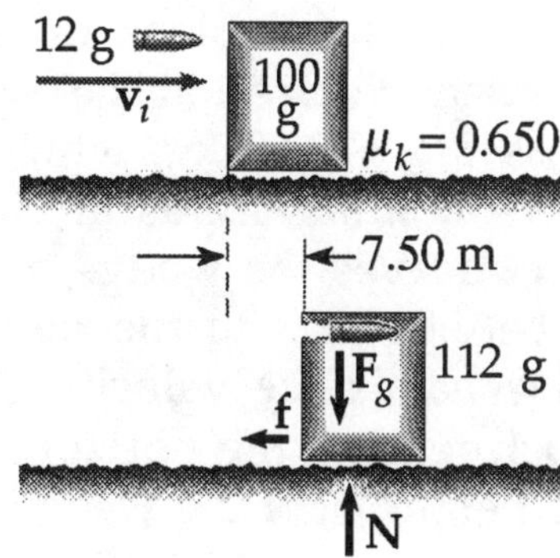

After the collision, the normal force exerted on the (block plus embedded bullet) by the surface is the weight of the block-bullet combination, or

$$N = F_g = (M+m)g = (0.100 \text{ kg} + 0.0120 \text{ kg})\left(9.80 \text{ m/s}^2\right) = 1.10 \text{ N}$$

The friction force that impedes the motion of the block-bullet combination is then $f = \mu_k N = 0.650(1.10 \text{ N}) = 0.713 \text{ N}$. This is the only force doing work on the block-bullet combination as it slides 7.50 m and comes to rest. The normal force and weight are both perpendicular to the displacement and do no work. Using the work-kinetic energy theorem, $W_{\text{net}} = KE_f - KE_i$, for the motion after collision gives $f s \cos 180° = 0 - \frac{1}{2}(M+m)V^2$:

$$-(0.713 \text{ N})(7.50 \text{ N}) = -\tfrac{1}{2}(0.112 \text{ kg})V^2$$

Thus the velocity of the block-bullet just after collision is $V = 9.77 \text{ m/s}$. Applying conservation of momentum from just before collision to just after collision gives: $P_i = P_f$. Substituting the appropriate terms,

$$m v_i + 0 = (M+m)V$$

The initial velocity of the bullet is then:

$$v_i = \frac{(M+m)V}{m} = \frac{(0.112 \text{ kg})(9.77 \text{ m/s})}{0.012 \text{ kg}} = 91.2 \text{ m/s}$$ ◊

49. A small block of mass $m_1 = 0.500$ kg is released from rest at the top of a curved wedge of mass $m_2 = 3.00$ kg, which sits on a frictionless horizontal surface as in Figure P6.49a. When the block leaves the wedge, its velocity is measured to be 4.00 m/s to the right, as in Figure P6.49b. (a) What is the velocity of the wedge after the block reaches the horizontal surface? (b) What is the height, h, of the wedge?

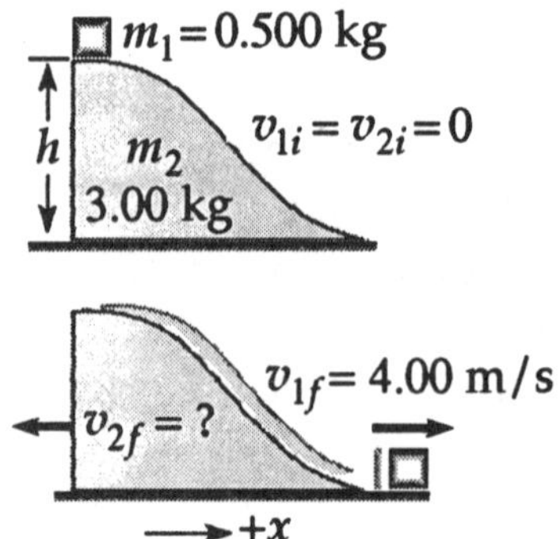

Figure P6.49 (a) and (b)

Solution (a) Consider a system that consists of the wedge and block. The wedge sits on a frictionless horizontal surface, so the only **external** forces exerted on this system are gravitation forces (weight of wedge and weight of block) and the normal force exerted on the wedge by the horizontal surface. These **external forces** are all in the vertical direction. Since no horizontal external forces act on the system, the horizontal component of the system's momentum is constant: $(P_f)_x = (P_i)_x$, or $m_1 v_{1f} + m_2 v_{2f} = m_1 v_{1i} + m_2 v_{2i}$:

This gives $$v_{2f} = \frac{m_1(v_{1i} - v_{1f}) + m_2 v_{2i}}{m_2} = \frac{0.500 \text{ kg}(0 - 4.00 \text{ m/s}) - 0}{3.00 \text{ kg}}$$

Thus, $$v_{2f} = -0.667 \text{ m/s} = 0.667 \text{ m/s in the } -x \text{ direction} \quad \Diamond$$

(b) To determine the height of the wedge, use work-energy techniques. Since no work is done on the block-wedge system by non-conservative forces, the total mechanical energy is constant. Choosing $y = 0$ at the horizontal surface,

$$(KE_{1f} + KE_{2f}) + (PE_{1f} + PE_{2f}) = (KE_{1i} + KE_{2i}) + (PE_{1i} + PE_{2i})$$

or $$\left(\tfrac{1}{2}m_1 v_{1f}^2 + \tfrac{1}{2}m_2 v_{2f}^2\right) + (0 + 0) = (0 + 0) + (m_1 g h + 0)$$

$$\tfrac{1}{2}(0.500 \text{ kg})(4.00 \text{ m/s})^2 + \tfrac{1}{2}(3.00 \text{ kg})(0.667 \text{ m/s})^2 = (0.500 \text{ kg})(9.80 \text{ m/s}^2)h$$

Hence, the height of the wedge is $$h = \frac{4.00 \text{ J} + 0.0667 \text{ J}}{4.90 \text{ N}} = 0.953 \text{ m} \quad \Diamond$$

55. As shown in Figure P6.55, a bullet of mass m and speed v passes completely through a pendulum bob of mass M. The bullet emerges with a speed of $v/2$. The pendulum bob is suspended by a stiff rod of length ℓ and negligible mass. What is the minimum value of v such that the pendulum bob will barely swing through a complete vertical circle?

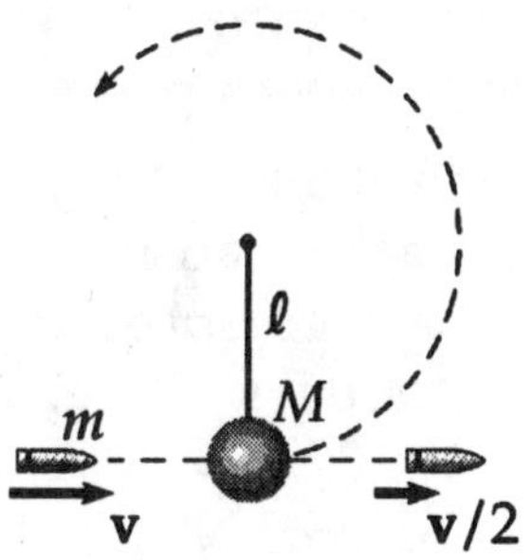

Figure P6.55

Solution If the pendulum bob has **any** speed greater than zero when it reaches the top of the circular arc, it will pass through this highest point on the arc and swing down the other side of the arc to complete the full vertical circular motion. Thus, the **minimum** speed it must have at the bottom of the circular arc is that which would allow it to reach the top with almost zero speed. To find this speed, apply conservation of energy over the interval that the bob is moving (after the bullet emerges from the bob). Conservation of energy is valid here since only a conservative force (gravity) performs work on the bob. Thus, choosing $y = 0$ at the bottom of the circular arc and V_{min} to be the bob's speed as the bullet exits the bob,

$$KE_{bottom} + PE_{bottom} = KE_{top} + PE_{top} \quad \text{or} \quad \tfrac{1}{2}MV_{min}^2 + 0 = 0 + Mg(2\ell)$$

This gives $V_{min} = \sqrt{4g\ell}$ as the minimum speed the bob must have just after the collision if it is to swing through a complete circle. To find the speed of the bullet before collision, apply the principle of conservation of **momentum**:

This gives $P_f = P_i$, or $m(v/2) + MV_{min} = mv + M(0)$

Therefore, $MV_{min} = m(v/2)$ or $v = \frac{2MV_{min}}{m}$. Using the value found for V_{min} above, this yields the required speed of the bullet:

$$v = \frac{2M}{m}\sqrt{4g\ell} \quad \text{or} \quad v = 4\left(\frac{M}{m}\right)\sqrt{g\ell}$$ ◊

CHAPTER SELF-QUIZ

1. A 0.6-kg tennis ball, initially moving at a speed of 12 m/s, is struck by a racket causing it to rebound in the opposite direction at a speed of 18 m/s. A high speed movie film determines that the racket and ball are in contact for 0.5 s. What is the average net force exerted on the ball by the racket?
   a. 32.4 N
   b. 36.0 N
   c. 1.44 N
   d. 0.72 N

2. During a snowball fight, two balls, with masses of 0.4 and 0.6 kg respectively, are thrown in such a manner that they meet head-on and combine to form a single mass of 1 kg. The magnitude of initial velocity for each is 15 m/s. The final kinetic energy of the system just after the collision is what percentage of kinetic energy just before the collision?
   a. 100%
   b. 50 %
   c. 33%
   d. 4%

3. A railroad freight car, mass 15,000 kg, is allowed to coast along a level track at a speed of 2 m/s. It collides and couples with a 50,000 kg second car, initially at rest and with brakes released. What is the speed of the two cars after coupling?
   a. 1.2 m/s
   b. 86 m/s
   c. 0.60 m/s
   d. 0.46 m/s

4. Three masses are arrayed in a two-dimensional coordinate system as follows: 4 kg at (5, −5), 6 kg at (5, 5) and 5 kg at (−4, −5). What are the coordinates of the center of mass of the system?
   a. (3, −2)
   b (4, −1)
   c. (2, 6)
   d. (2, −1)

5. A miniature, spring-loaded, radio-controlled gun is mounted on an air puck. The gun's bullet has a mass of 0.005 kg and the gun and puck have a combined mass of 0.120 kg. With the system initially at rest, the radio-controlled trigger releases the bullet causing the puck and empty gun to move with a speed of 0.5 m/s. After release, what is the speed of the center of mass of the gun-puck-bullet system?
   a. Zero
   b. 36 m/s
   c. 12 m/s
   d. 4.8 m/s

6. A 0.02-kg bullet is fired into, and becomes embedded in, a 1.7-kg wooden ballistic pendulum. If the pendulum rises a vertical distance of 0.05 m, what is the initial speed of the bullet?
   a. 167 m/s
   b. 85 m/s
   c. 67 m/s
   d. 35 m/s

7. A 0.10-kg object moving initially with a velocity of +0.20 m/s makes an elastic head-on collision with a 0.15-kg object initially at rest. What is the velocity of the 0.15-kg object after the collision?
   a. +0.16 m/s
   b. −0.16 m/s
   c. +0.04 m/s
   d. −0.045 m/s

8. A 10-gram bullet is fired into a 100-gram block of wood at rest on a horizontal surface. After impact, the block slides 8 m before coming to rest. If the coefficient of friction is $\mu = 0.6$, find the speed of the bullet before impact.
   a. 106 m/s
   b. 212 m/s
   c. 318 m/s
   d. 424 m/s

9. A firehose directs a steady stream of 15 kg/sec of water with velocity 28 m/s against a flat plate. What force is required to hold the plate in place?
   a. 110 N
   b. 420 N
   c. 1100 N
   d. 4116 N

10. Two skaters, both of mass 50 kg, are at rest on skates on a frictionless ice pond. One skater throws a 0.2-kg Frisbee at 5 m/s to her friend, who catches it and throws it back at 5 m/s. When the first skater has caught the returned Frisbee, what is the velocity of each of the two skaters?
   a. 0.02 m/s, moving apart
   b. 0.04 m/s, moving apart
   c. 0.02 m/s, moving towards each other
   d. 0.04 m/s, moving towards each other

11. A 1.00-kg duck is flying overhead at 1.50 m/s when a hunter fires straight up. The 0.010-kg bullet is moving 100 m/s when it hits the duck and stays lodged in the duck's body. What is the momentum of the duck and bullet immediately after the hit?
   a. $1.50\ kg \cdot m/s$
   b. $2.50\ kg \cdot m/s$
   c. $0.05\ kg \cdot m/s$
   d. $1.80\ kg \cdot m/s$

12. A railroad car of 1000 kg is rolling with a speed of 2.0 m/s when 2000 kg of coal is dropped from a height of 1.0 m into the car. What is the final speed of the cart after the coal is in the cart?
   a. 0.25 m/s
   b. 0.67 m/s
   c. 1.0 m/s
   d. 3.6 m/s

Chapter 7

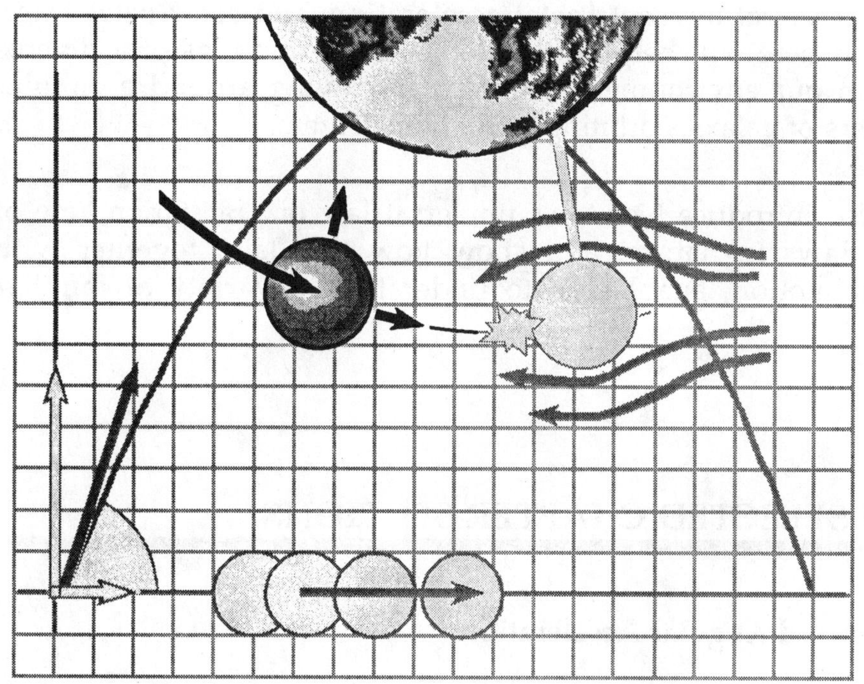

CIRCULAR MOTION AND THE LAW OF GRAVITY

Chapter 7

CIRCULAR MOTION AND THE LAW OF GRAVITY

In this chapter we shall investigate circular motion, a specific type of two-dimensional motion. We shall encounter such terms as **angular velocity, angular acceleration, centripetal acceleration, and centripetal force.** The results derived here will help you understand the motions of a diverse range of objects in our environment, from a car moving around a circular race track to clusters of galaxies orbiting a common center.

We shall also introduce Newton's universal law of gravitation, one of the fundamental laws in nature, and show how this law, together with Newton's laws of motion, enables us to understand a variety of familiar phenomena.

NOTES FROM SELECTED CHAPTER SECTIONS

7.1 Angular Speed and Angular Acceleration

Pure rotational motion refers to the motion of a rigid body about a fixed axis.

One **radian** (rad) is the angle subtended by an arc length equal to the radius of the arc.

In the case of **rotation about a fixed axis,** every particle on the rigid body has the same angular velocity and the same angular acceleration.

7.2 Rotational Motion Under Constant Angular Acceleration

The **kinematic expressions** for rotational motion under constant angular acceleration are of the **same form** as those for linear motion under constant linear acceleration with the substitutions $x \to \theta$, $v \to \omega$, and $a \to \alpha$.

7.3 Relations Between Angular and Linear Quantities

When a rigid body rotates about a fixed axis, every point in the object moves along a circular path which has its center at the axis of rotation. The instantaneous velocity of each point is directed along a tangent to the circle. Every point on the object experiences the same **angular** speed; however, points that are different distances from the axis of rotation have different tangential speeds. The value of each of the linear quantities [displacement (s), velocity (v), and acceleration (a_t)] is equal to the radial distance from the axis multiplied by the corresponding angular quantity, θ, ω, and α.

7.4 Centripetal Acceleration

In circular motion, the centripetal acceleration is directed inward toward the center of the circle and has a magnitude given either by v^2/r or $r\omega^2$.

7.5 Forces Causing Centripetal Acceleration

All centripetal forces act toward the center of the circular path along which the object moves. If the force that produces a centripetal acceleration vanishes, the object does not continue to move in its circular path; instead, it moves along a straight-line path tangent to the circle.

7.6 Describing Motion of a Rotating System

One must be very careful to distinguish real forces from fictitious ones in describing motion in an accelerating frame. An observer in a car rounding a curve is in an accelerating frame and invents a fictitious outward force to explain why he or she is thrown outward. A stationary observer outside the car, however, considers only real forces on the passenger. To this observer, the mysterious outward force **does not exist!** The only real external force on the passenger is the centripetal (inward) force due to friction or the normal force of the door.

7.7 Newton's Universal Law of Gravitation

There are several features of the universal law of gravity that deserve some attention:

1. The gravitational force is an action-at-a distance force that always exists between two particles regardless of the medium that separates them.

2. The force varies as the inverse square of the distance between the particles and therefore decreases rapidly with increasing separation.

3. The force is proportional to the product of their masses.

7.8 Kepler's Laws

Kepler's laws applied to the solar system are:

1. All planets move in elliptical orbits with the Sun at one of the focal points.

2. A line drawn from the Sun to any planet sweeps out equal areas in equal time intervals.

3. The square of the orbital period of any planet is proportional to the cube of the average distance from the planet to the Sun.

EQUATIONS AND CONCEPTS

When a particle moves along a circular path of radius r, the distance traveled by the particle is called the arc length, s. The radial line from the center of the path to the particle sweeps out an angle, θ.

$$\theta(\text{rad}) \equiv \frac{s}{r} \qquad (7.1)$$

where $\theta\,(\text{rad}) = \left(\frac{\pi}{180°}\right)\left(\theta\,(\text{deg})\right)$

The angle θ in Equation 7.1 is the ratio of two lengths (arc length to radius) and hence is a dimensionless quantity. It is common practice to refer to the angle as being in units of radians. The angle in Equation 7.1 must have units of radians.

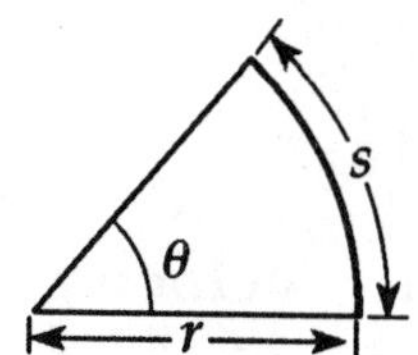

$$1\text{ rad} = \left(\frac{360°}{2\pi}\right) = 57.3°$$

The average angular velocity of a rotating object is the ratio of the angular displacement to the time interval during which the angular displacement occurs. In this equation θ must be measured in radians.

$$\overline{\omega} = \frac{\Delta\theta}{\Delta t} \qquad (7.2)$$

The average angular acceleration of a rotating object is the ratio of change in angular velocity to the time interval during which the change in velocity occurs.

$$\overline{\alpha} = \frac{\Delta\omega}{\Delta t} \qquad (7.4)$$

To the right are given the equations of rotational kinematics and the corresponding equations for linear motion with constant acceleration. Note that the rotational equations, involving the angular variables θ, ω, and α, have a one-to-one correspondence with the equations of linear motion, involving the variables x, v, and a.

$$\omega = \omega_0 + \alpha t \tag{7.5}$$

$$v = v_0 + at$$

$$\theta = \omega_0 t + \frac{1}{2}\alpha t^2 \tag{7.6}$$

$$x = v_0 t + \frac{1}{2}at^2$$

$$\omega^2 = \omega_0{}^2 + 2\alpha\theta \tag{7.7}$$

$$v^2 = v_0{}^2 + 2ax$$

The tangential velocity and tangential acceleration of a given point on a rotating object are related to the corresponding angular quantities via the radius of the circular path along which the point moves.

$$v_t = r\omega \tag{7.8}$$

$$a_t = r\alpha \tag{7.9}$$

The tangential velocity and tangential acceleration are along directions which are tangent to the circular path (and therefore perpendicular to the radius from the center of rotation). Also note that every point on a rotating object has the same value of ω and the same value of α; however, points which are at different distances from the axis of rotation have different values of v_t and a_t.

Comment on rotational velocity and acceleration.

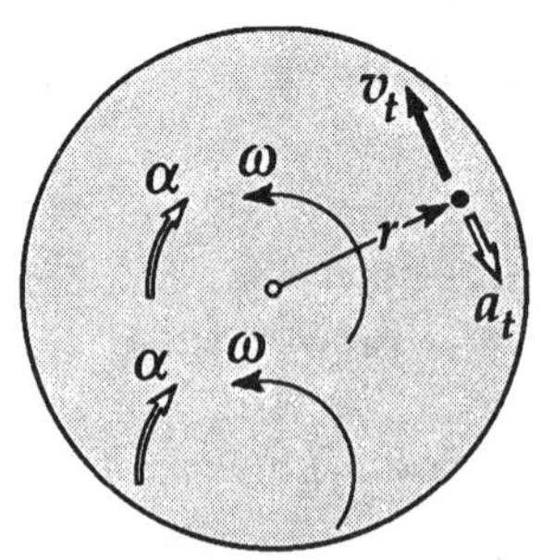

An object in circular motion has a centripetal acceleration which is directed toward the center of the circle and has a magnitude which depends on the values of the tangential velocity and the radius of the path.

$$a_c = \frac{v^2}{r} \tag{7.11}$$

or

$$a_c = r\omega^2 \tag{7.14}$$

An object which is moving along a circular path with increasing or decreasing speed has both a tangential component of acceleration (tangent to the instantaneous direction of travel) and a centripetal component of acceleration (perpendicular to the instantaneous direction of travel). The magnitude and direction of the total acceleration can be found by the usual methods of vector addition.

$$a = \sqrt{a_t^2 + a_c^2} \tag{7.15}$$

$$\theta = \tan^{-1}\left(\frac{a_t}{a_c}\right)$$

Forces which maintain motion along a circular path are directed toward the center of the path and are called centripetal forces.

$$F_c = ma_c = m\frac{v_t^2}{r} \tag{7.16}$$

The universal law of gravity states that every particle in the Universe attracts every other particle with a force that is directly proportional to the product of their masses and inversely proportional to the square of the distance between them. The constant G is called the universal gravitational constant.

$$F = G\frac{m_1 m_2}{r^2} \quad (7.17)$$

$$G = 6.673 \times 10^{-11}\,\frac{\mathrm{N \cdot m^2}}{\mathrm{kg^2}} \quad (7.18)$$

The gravitational force between two masses m_1 and m_2 is one of attraction; each mass exerts a force of attraction on the other. These two forces form an action-reaction pair. The magnitudes of the forces of gravitational attraction on the two masses are equal regardless of the relative values of m_1 and m_2.

Comments on the gravitational force.

$\mathbf{F}_1$ $\mathbf{F}_2$

m_1 m_2

Kepler's third law states that the square of the orbital period of a planet is proportional to the cube of the mean distance from the planet to the Sun. For an Earth satellite, M_s in Equation 7.21 must be replaced by M_e, the mass of the Earth.

$$T^2 = \left(\frac{4\pi^2}{GM_s}\right) r^3 = K_s r^3 \quad (7.21)$$

K_s is independent of the mass of the planet.

$$K_s = \frac{4\pi^2}{GM_s} = 2.97 \times 10^{-19}\ \frac{\mathrm{s^2}}{\mathrm{m^3}}$$

SUGGESTIONS, SKILLS, AND STRATEGIES

The following features involving centripetal forces and centripetal accelerations should be kept in mind:

1. Draw a free-body diagram of the object(s) under consideration, showing all forces that act on it.
2. Choose a coordinate system with one axis tangent to the path followed by the object and the other axis perpendicular to the plane of the circular path.
3. Find the net force toward the center of the circular path. This is the centripetal force.
4. From this point onward, the steps are virtually identical to those encountered when solving Newton's second law problems with $\Sigma \mathbf{F} = m\mathbf{a}$. Also, you should note that the magnitude of the centripetal acceleration can always be written as $a_c = v^2/r$.

REVIEW CHECKLIST

▷ Quantitatively, the angular displacement, velocity, and acceleration for a rigid body system in rotational motion are related to the distance traveled, tangential velocity, and tangential acceleration, respectively. The linear quantity is calculated by multiplying the angular quantity by the radius arm for an object or point in that system.

▷ If a body rotates about a fixed axis, every particle on the body has the same angular velocity and angular acceleration. For this reason, rotational motion can be simply described using these quantities. The formulas which describe angular motion are analogous to the corresponding set of formulas pertaining to linear motion.

▷ The nature of the acceleration of a particle moving in a circle with constant speed is such that, although $\mathbf{v}$ = constant, the **direction** of $\mathbf{v}$ varies in time, which is the origin of the radial, or centripetal acceleration.

- ▷ There are two components of acceleration for a particle moving on a curved path, when both the magnitude and direction of **v** are changing with time. In this case, the particle has a tangential component of acceleration and a radial component of acceleration.
- ▷ The forces acting on a vehicle on a banked curve can be resolved into components which are either parallel or perpendicular to the radial direction. Newton's second law can then be used to find an expression for the proper banking angle corresponding to a given speed and radius of path.
- ▷ Newton's universal law of gravitation is an example of an inverse-square law, and it describes an **attractive** force between two **particles** separated by a distance r.

SOLUTIONS TO SELECTED END-OF-CHAPTER PROBLEMS

3. A rotating body has an constant angular speed of 33 rev/min. (a) What is its angular speed in rad/s? (b) Through what angle, in radians, does it rotate in 1.5 s?

Solution (a) Convert the angular speed of the object to radians per second:

$$\omega = 33\ \frac{\text{rev}}{\text{min}} = \left(33\ \frac{\text{rev}}{\text{min}}\right)\left(\frac{2\pi\ \text{rad}}{1\ \text{rev}}\right)\left(\frac{1\ \text{min}}{60\ \text{s}}\right) = 3.5\ \frac{\text{rad}}{\text{s}} \qquad \Diamond$$

(b) The angular displacement that occurs during a time interval may be expressed as the product of the average angular velocity over that interval and the duration of the interval (i.e., $\theta = \overline{\omega}t$). In this case, the angular velocity is constant, so $\overline{\omega} = \omega = 33\ \text{rev/min} = 3.5\ \text{rad/s}$. The angular displacement that occurs in a time of 1.5 s is then:

$$\theta = \overline{\omega}t = \left(3.5\ \frac{\text{rad}}{\text{s}}\right)(1.5\ \text{s}) = 5.2\ \text{rad} \qquad \Diamond$$

If desired, this could be converted to revolutions as:

$$\theta = 5.2\ \text{rad} = (5.2\ \text{rad})\left(\frac{1.0\ \text{rev}}{2\pi\ \text{rad}}\right) = 0.84\ \text{rev}$$

9. An electric motor rotating a workshop grinding wheel at a rate of 100 rev/min is switched off. Assume constant negative angular acceleration of magnitude 2.00 rad/s^2. (a) How long does it take for the grinding wheel to stop? (b) Through how many radians has the wheel turned during the interval found in (a)?

Solution (a) In uniformly accelerated motion, the angular velocity at any time t may be expressed as $\omega = \omega_0 + \alpha t$, where ω_0 is the angular velocity at $t = 0$ and α is the constant angular acceleration. In this case,

$$\omega_0 = 100 \text{ rev/min} = (100 \text{ rev/min})\left(\frac{2\pi \text{ rad}}{1 \text{ rev}}\right)\left(\frac{1}{60 \text{ s}}\right) = 10.47 \text{ rad/s}$$

Thus, the time when the wheel comes to rest ($\omega = 0$) may be found as:

$$0 = 10.47 \text{ rad/s} + \left(-2.00 \text{ rad/s}^2\right)t \quad \text{or} \quad t = \frac{-10.47 \text{ rad/s}}{-2.00 \text{rad/s}^2} = 5.24 \text{ s} \quad \Diamond$$

(b) The average angular velocity during this 5.24 s interval is

$$\overline{\omega} = \frac{\omega + \omega_0}{2} = \frac{0 + 10.47 \text{ rad/s}}{2} = 5.24 \text{ rad/s}$$

The angular displacement the wheel undergoes as it comes to rest is therefore:

$$\theta = \overline{\omega} t = (5.24 \text{ rad/s})(5.24 \text{ s}) = 27.4 \text{ rad} \quad \Diamond$$

Note: This result could be obtained in two other ways. Using $\theta = \omega_0 t + \frac{1}{2}\alpha t^2$ gives

$$\theta = (10.47 \text{ rad/s})(5.24 \text{ s}) + \tfrac{1}{2}\left(-2.00 \text{ rad/s}^2\right)(5.24 \text{ s})^2 = 27.4 \text{ rad}$$

Or one could use $\omega^2 = \omega_0{}^2 + 2\alpha\theta$: $\theta = \dfrac{\omega^2 - \omega_0{}^2}{2\alpha} = \dfrac{0 - (10.47 \text{ rad/s})^2}{2(-2.00 \text{ rad/s})} = 27.4 \text{ rad}$

14. A car is traveling at 17.0 m/s on a straight horizontal highway. The wheels of the car have radii of 48.0 cm. If the car speeds up with an acceleration of 2.00 m/s^2 for 5.00 s, find the number of revolutions of the wheels during this period.

Solution

The linear velocity of the car at the end of the 5.00 s interval is

$$v = v_0 + at = 17.0 \text{ m/s} + \left(2.00 \text{ m/s}^2\right)(5.00 \text{ s}) = 27.0 \text{ m/s}$$

The average velocity of the car is

$$\overline{v} = \frac{v_0 + v}{2} = \frac{17.0 \text{ m/s} + 27.0 \text{ m/s}}{2} = 22.0 \text{ m/s}$$

Its linear displacement during this time is

$$s = \overline{v}t = (22.0 \text{ m/s})(5.00 \text{ s}) = 110 \text{ m}$$

If the car's wheels roll without slipping on the highway, the relation between the linear displacement of the center of a wheel (s) and the angular displacement of the wheel (θ) is $s = r\theta$ where r is the radius of the wheel.

Thus, $$\theta = \frac{s}{r} = \frac{110 \text{ m}}{0.48 \text{ m}} = 229 \text{ rad}$$

This may be converted to revolutions as:

$$\theta = (229 \text{ rad})\left(\frac{1 \text{ rev}}{2\pi \text{ rad}}\right) = 36.4 \text{ rev}$$ ◊

17. It has been suggested that rotating cylinders about 10 mi long and 5.0 mi in diameter be placed in space and used as colonies. What angular speed must such a cylinder have so that the centripetal acceleration at its surface equals Earth's gravity?

Solution If an object moves along a circular path of radius r with a tangential speed v_t, it is always accelerating toward the center of the circular path with an acceleration (called the **centripetal acceleration**) of $a_c = v_t^2/r$. It is desired to have $a_c = g_{earth} = 9.8\ \mathrm{m/s^2}$ for an object on the surface of a rotating cylinder that has a radius of

$$r = \frac{\text{diameter}}{2} = 2.5\ \mathrm{mi} = (2.5\ \mathrm{mi})\left(\frac{1.609\ \mathrm{km}}{1\ \mathrm{mi}}\right)\left(\frac{1000\ \mathrm{m}}{1\ \mathrm{km}}\right) = 4.0\times10^3\ \mathrm{m}$$

The required tangential speed is then:

$$v_t = \sqrt{ra_c} = \sqrt{\left(4.0\times10^3\ \mathrm{m}\right)\left(9.8\ \mathrm{m/s^2}\right)}$$

or

$$v_t = 2.0\times10^2\ \mathrm{m/s}$$

The angular speed, ω, of an object following a circular path of radius r is related to the tangential speed, v_t, by $v_t = r\omega$. Thus, the required angular speed is:

$$\omega = \frac{v_t}{r} = \frac{2.0\times10^2\ \mathrm{m/s}}{4.0\times10^3\ \mathrm{m}} = 5.0\times10^{-2}\ \mathrm{rad/s}$$ ◊

23. A 50.0-kg child stands at the rim of a merry-go-round of radius 2.00 m, rotating with an angular speed of 3.00 rad/s. (a) What is the child's centripetal acceleration? (b) What is the minimum force between her feet and the floor of the merry-go-round that is required to keep her in the circular path? (c) What minimum coefficient of static friction is required? Is the answer you found reasonable? In other words, is she likely to be able to stay on the merry-go-round?

Solution (a) When the child stands, without slipping, at the rim of the rotating merry-go-round, she follows a horizontal circular path of radius $r = 2.00$ m with an angular speed, $\omega = 3.00$ rad/s. Her centripetal acceleration is

$$a_c = \frac{v_t^2}{r} = \frac{(r\omega)^2}{r} = r\omega^2$$

Thus, $$a_c = (2.00 \text{ m})(3.00 \text{ rad/s})^2 = 18.0 \text{ m/s}^2$$ ◊

(b) The force, directed toward the center of the circular path, required to produce this centripetal acceleration is $F_c = ma_c$ or

$$F_c = (50.0 \text{ kg})(18.0 \text{ m/s}^2) = 900 \text{ N}$$ ◊

(c) Three forces act on the child as she stands on the rotating merry-go-round. These are: (1) a normal force, N, exerted upward on the child by the merry-go-round; (2) a downward gravitational force

$$F_g = w = mg = (50.0 \text{ kg})(9.80 \text{ m/s}^2) = 490 \text{ N}$$

and (3) a horizontal friction force between her feet and the platform. Since the child's vertical acceleration is zero, Newton's second law gives

$$\sum F_y = N - F_g = 0 \quad \text{or} \quad N = F_g = 490 \text{ N}$$

The only force that is horizontal and thus capable of producing an acceleration toward the center of the horizontal circular path is the static friction force. Therefore, it is necessary that $f_s = F_c$. Since $f_s \le \mu_s N$, this means that $F_c \le \mu_s N$, and the coefficient of static friction $\mu_s \ge \frac{F_c}{N} = \frac{900 \text{ N}}{490 \text{ N}} = 1.84$. This is much larger than 1.00. Considering common values for μ_s (see Table 4.2 in the textbook), this is clearly unrealistic. Thus, the child is unlikely to be able to stay on the merry-go-round. ◊

27. Tarzan (m = 85 kg) tries to cross a river by swinging from a 10-m-long vine. His speed at the bottom of the swing (as he just clears the water) is 8.0 m/s. Tarzan doesn't know that the vine has a breaking strength of 1000 N. Does he make it safely across the river? Justify your answer.

Solution

When Tarzan is at the bottom of the swing, the vine is vertical with its tension force, T, acting upward on Tarzan. The only other force present is his weight,

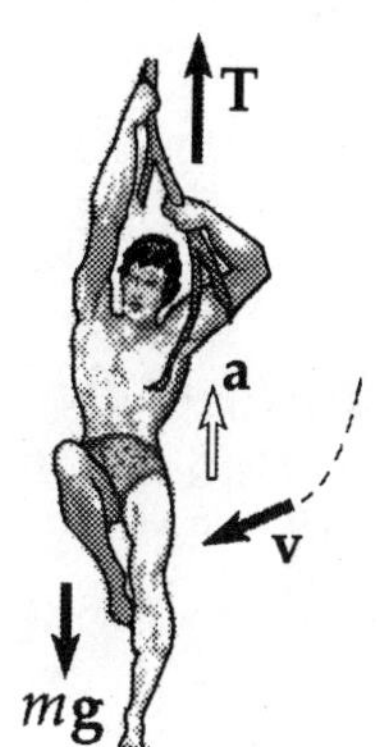

$$F_g = mg = (85\text{ kg})(9.8\text{ m/s}^2) = 8.3\times10^2\text{ N (downward)}$$

Therefore, the net force directed toward the center of Tarzan's vertical circular path is $F_{net} = T - F_g$. If Tarzan is to stay on the circular path, his needed centripetal acceleration is

$$a_c = \frac{v_t^2}{r} = \frac{(8.0\text{ m/s})^2}{10\text{ m}} = 6.4\text{ m/s}^2$$

Thus, it is necessary that $F_{net} = T - F_g = ma_c$

$$T = F_g + ma_c = 8.3\times10^2\text{ N} + (85\text{ kg})(6.4\text{ m/s}^2)$$

Solving, $T = 1.4\times10^3\text{ N}$

This exceeds the breaking strength of 1000 N. Thus, we can say that Tarzan will not cross safely. ◊

31. The average distance separating the Earth and the Moon is 384,000 km. Use the data in Table 7.3 to find the net gravitational force the Earth and the Moon exerts on a 3.00×10^4-kg spaceship located halfway between them.

Solution When a spaceship (mass m_s) is halfway between the Earth (mass m_E) and the Moon (mass m_M), it is at a distance of

$$r = 192{,}000\ \text{km}\left(\frac{1.00\times 10^3\ \text{m}}{1\ \text{km}}\right) = 1.92\times 10^8\ \text{m}$$

from each body. From Newton's Universal Law of Gravitation, the force exerted on the ship by the Earth has a magnitude of

$$F_E = Gw_E m_s / r^2 \qquad \text{(directed toward the Earth.)}$$

The force exerted on the ship by the Moon is

$$F_M = Gm_M m_s / r^2 \qquad \text{(directed toward the Moon.)}$$

From Table 7.3 in the textbook, $m_E = 5.98\times 10^{24}$ kg and $m_M = 7.36\times 10^{22}$ kg. Thus, it is observed that $F_E > F_M$ so the net gravitational force acting on the ship is directed toward the Earth and has a magnitude of:

$$F_{\text{net}} = F_E - F_M = \frac{Gm_E m_s}{r^2} - \frac{Gm_M m_s}{r^2} = \frac{Gm_s(m_E - m_M)}{r^2}$$

Substituting in the given values,

$$F_{\text{net}} = \frac{\left(6.67\times 10^{-11}\ \text{N}\cdot\text{m}^2/\text{kg}^2\right)\left(3.00\times 10^4\ \text{kg}\right)\left(5.98\times 10^{24}\ \text{kg} - 7.36\times 10^{22}\ \text{kg}\right)}{\left(1.92\times 10^8\ \text{m}\right)^2}$$

This yields $F_{\text{net}} = 321$ N directed toward the Earth. ◊

37. A 600-kg satellite is in a circular orbit about the Earth at a height above the Earth equal to the Earth's mean radius. Find (a) the satellite's orbital speed, (b) the period of its revolution, and (c) the gravitational force acting on it.

Solution The mean radius of the Earth is $R_E = 6.38 \times 10^6$ m (Table 7.3 in the text). The radius of the satellite's orbit is therefore $r = 2R_E = 1.28 \times 10^7$ m.

(a) The only force acting on the satellite, directed toward the center of its orbital path, is the gravitational force F_E exerted on it by the Earth. This force must produce the needed centripetal acceleration, $a_c = v_t^2/r$, so $F_E = m_s a_c$ where m_s is the mass of the satellite. Using Newton's Universal Law of Gravitation, this gives $Gm_E m_s/r^2 = m_s v_t^2/r$ which reduces to the orbital speed of the satellite:

$$v_t = \sqrt{\frac{Gm_E}{r}} = \sqrt{\frac{\left(6.67 \times 10^{-11} \text{ N}\cdot\text{m}^2/\text{kg}^2\right)\left(5.98 \times 10^{24} \text{ kg}\right)}{1.28 \times 10^7 \text{ m}}} = 5.58 \times 10^3 \text{ m/s} \quad \Diamond$$

(b) The period of revolution for the satellite (time for it to complete one full orbit) is $T = (\text{distance around})/(\text{orbital speed}) = 2\pi r/v_t$, or

$$T = \frac{2\pi\left(1.28 \times 10^7 \text{ m}\right)}{5.58 \times 10^3 \text{ m/s}} = 1.44 \times 10^4 \text{ s} = 240 \text{ min} = 4.00 \text{ h} \quad \Diamond$$

(c) The gravitational force acting on the satellite is directed toward the center of the Earth and has a magnitude of

$$F_E = \frac{Gm_E m_s}{r^2} = \frac{\left(6.67 \times 10^{-11} \text{ N}\cdot\text{m}^2/\text{kg}^2\right)\left(5.98 \times 10^{24} \text{ kg}\right)(600 \text{ kg})}{\left(1.28 \times 10^7 \text{ m}\right)^2} = 1.46 \times 10^3 \text{ N} \quad \Diamond$$

45. Io, a small moon of the giant planet Jupiter, has an orbital period of 1.77 days and an orbital radius equal to 4.22×10^5 km. From these data, determine the mass of Jupiter.

Solution The force available to produce the needed centripetal acceleration, a_c, and hold Io in its orbit is the gravitational force exerted on it by Jupiter. Thus, Newton's second law gives the needed magnitude of this force as

$$F_J = m_I a_c = \frac{m_I v_t^2}{r}$$

where m_I is the mass of Io, v_t is its orbital speed, and r is its orbital radius. But, Newton's Universal Law of Gravitation gives the magnitude of this gravitational force as $F_J = G m_J m_I / r^2$ where m_J is the mass of Jupiter. Therefore, it necessary that $G m_J m_I / r^2 = m_I v_t^2 / r$. Note that the mass of the satellite (Io in this case) cancels and the mass of the central body (Jupiter in this case) is given as:

$$m_J = \frac{r v_t^2}{G} \qquad \textbf{[Equation 1]}$$

The radius of Io's orbit is $r = 4.22 \times 10^5 \text{ km} = 4.22 \times 10^8 \text{ m}$, and its orbital period is $T = 1.77 \text{ days}(86400 \text{ s/day}) = 1.53 \times 10^5 \text{ s}$. The orbital speed of Io is then

$$v_t = (\text{distance around})/(\text{time for 1 complete orbit}) = 2\pi r/T$$

or $$v_t = 2\pi(4.22 \times 10^8 \text{ m})/(1.53 \times 10^5 \text{ s}) = 1.73 \times 10^4 \text{ m/s}$$

Then, Equation 1 gives the mass of Jupiter as:

$$m_J = \frac{(4.22 \times 10^8 \text{ m})(1.73 \times 10^4 \text{ m/s})^2}{6.67 \times 10^{-11} \text{ N} \cdot \text{m}^2/\text{kg}^2} = 1.90 \times 10^{27} \text{ kg} \qquad \Diamond$$

Note: This problem illustrates a very useful tool with which astronomers can determine the mass of some celestial bodies.

49. A car moves at speed v across a bridge made in the shape of a circular arc of radius r. (a) Find an expression for the normal force acting on the car when it is at the top of the arc. (b) At what minimum speed will the normal force become zero (causing occupants of the car to seem weightless) if $r = 30.0$ m?

Solution

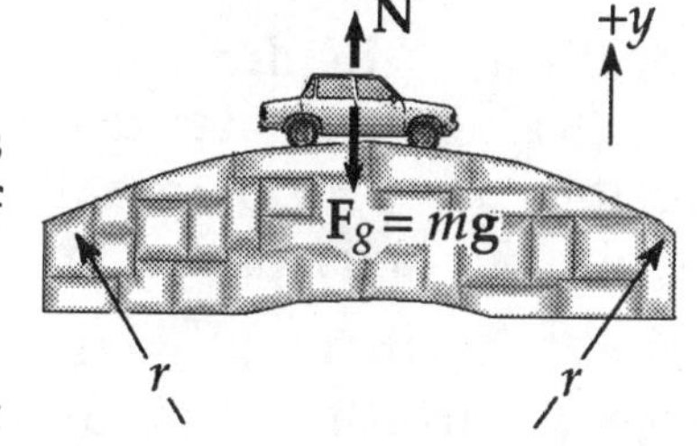

Consider the sketch at the right showing the forces acting on the car as it crosses the top of the arc of the bridge.

(a) If the car remains in contact with the bridge, it follows a circular path of radius r. Thus, it must be accelerating **toward the center** of the circular path at a rate $a_c = mv^2/r$.

When the car is at the top of the arc, the line toward the center is vertical. Taking upward as positive and applying Newton's second law to the vertical motion,

$$\Sigma F_y = ma_y \quad \text{yields} \quad N - F_g = m(-a_c) \quad \text{or} \quad N - mg = -mv^2/r$$

When the car is at the top of the arc, the normal force exerted on the car by the bridge is therefore:

$$N = mg - mv^2/r = m\left(g - v^2/r\right) \qquad \lozenge$$

(b) If the occupants of the car (and the car itself) are to seem weightless, the downward force they exert on the roadway must be zero. Then, by Newton's third law, the upward force the roadway exerts on them (i.e., the normal force N) must also be zero. Therefore, the desired speed is that for which $N = m\left(g - v^2/r\right) = 0$, or $v = \sqrt{rg}$. If the radius of the circular arc is 30.0 m, then

$$v = \sqrt{(30.0\text{ m})\left(9.80\text{ m/s}^2\right)} = 17.1\text{ m/s} \qquad \lozenge$$

57. Halley's comet approaches the Sun to within 0.570 A.U., and its orbital period is 75.6 years. (A.U. is the abbreviation for astronomical unit, where $1\text{ A.U.} = 1.50 \times 10^{11}\text{ m}$ is the mean Earth-Sun distance.) How far from the Sun will Halley's comet travel before it starts its return journey?

Solution Kepler's third law, when applied to elliptical orbits, states that the square of the orbital period is proportional to the cube of the semimajor axis of the orbit. In equation form, $T^2 = Ka^3$, where a is the semimajor axis. For objects orbiting the Sun, the proportionality constant is

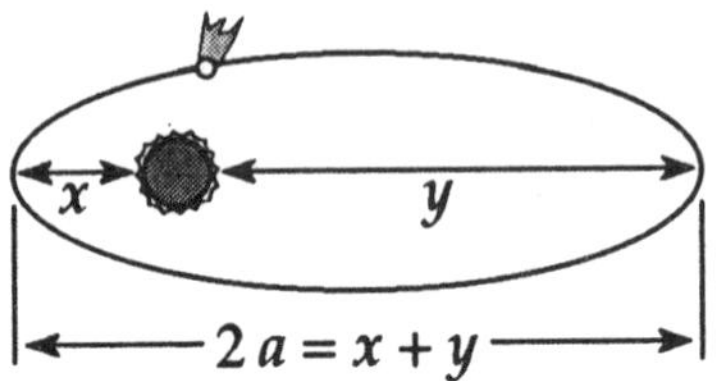

$$K = \frac{4\pi^2}{GM_{\text{sun}}} = 2.97 \times 10^{-19}\text{ s}^2/\text{m}^3$$

Halley's comet has a period of $T = (75.6\text{ years})\left(\frac{3.156 \times 10^7\text{ s}}{1\text{ year}}\right) = 2.39 \times 10^9\text{ s}$

Thus, Kepler's third law gives the semimajor axis of its orbit as

$$a = \sqrt[3]{\frac{T^2}{K}} = \sqrt[3]{\frac{\left(2.39 \times 10^9\text{ s}\right)^2}{2.97\times10^{-19}\text{ s}^2/\text{m}^3}}$$

$$a = \left(2.68 \times 10^{12}\text{ m}\right)\left(\frac{1\text{ A.U.}}{1.50 \times 10^{11}\text{ m}}\right) = 17.9\text{ A.U.}$$

From the sketch of the orbit, it is seen that the maximum distance, x, from the Sun to a point on the orbit is given by $x = 2a - y$, where y is the minimum distance between the Sun and a point on the orbit. For Halley's comet, this perihelion distance is $y = 0.570$ A.U. The maximum distance the comet ever gets from the Sun is therefore:

$$x = 2(17.9\text{ A.U.}) - 0.570\text{ A.U.} = 35.2\text{ A.U.}$$ ◊

61. Show that the escape speed from the surface of a planet of uniform density is directly proportional to the radius of the planet.

Solution The solution to this problem is from the optional **Section 7.8 - Gravitational Potential Energy Revisited**. If an object of mass m lifts off at speed v_e from the surface of a planet of mass M_p and radius R_p, its total mechanical energy is:

$$E = KE_i + PE_i = \tfrac{1}{2}mv_e^2 - \frac{GM_pm}{R_p}$$

If air resistance as the object passes through the planet's atmosphere (if any exists) may be neglected, only the gravitational force exerted by the planet (a **conservative** force) does work on the object. Thus, the total mechanical energy of the object remains constant:

$$E_f = KE_f + PE_f = KE_i + PE_i = E_i$$

If the object is to escape from the planet, it must reach a location where the gravitational force exerted on it by the planet $F = GM_p m/r^2$ is zero. This is true only when $r = \infty$. But, when $r = \infty$, the gravitational potential energy of the object is

$$PE_f = -\frac{GM_pm}{r} = -\frac{GM_pm}{\infty} = 0$$

Also, if the object leaves the planet's surface with the minimum required speed to reach this location ($r = \infty$), it will arrive with zero speed, $v_f = 0$. Thus, $KE_f = \frac{1}{2}mv_f^2 = 0$. That is, for the object to just be able to escape, its total energy must be $E_f = KE_f + PE_f = 0$. Thus, since $E_i = E_f = \text{constant}$, $E_i = \frac{1}{2}mv_e^2 - GM_pm / R_p = 0$, and the escape velocity is $v_e = \sqrt{2GM_p/R_p}$. Further, if the planet has uniform density, its mass is: $M_p = \text{density} \times \text{volume} = \rho\left(\frac{4}{3}\pi R_p^3\right)$. The escape velocity is then

$$v_e = \sqrt{\left(2G/R_p\right)\rho\left(\tfrac{4}{3}\pi R_p^3\right)} = \sqrt{\tfrac{8}{3}\pi\rho G R_p^2} = R_p\sqrt{\tfrac{8}{3}\pi\rho G} = \left(R_p\right)(\text{constant})$$

Thus, the escape velocity is directly proportional to the planet radius, R_p ◊

CHAPTER SELF-QUIZ

1. A point on a wheel rotating at 5 rev/s and located 0.2 m from the axis has what tangential velocity?
   a. 3.8 m/s
   b. 6.3 m/s
   c. 1.2 m/s
   d. 0.104 m/s

2. A point on a wheel rotating at 5 rev/s and located 0.2 m from the axis experiences what centripetal acceleration?
   a. 197 m/s^2
   b. 48 m/s^2
   c. 0.05 m/s^2
   d. 1.35 m/s^2

3. A Ferris wheel, rotating initially at an angular velocity of 0.50 rad/s, accelerates over a 5-s interval at a rate of 0.04 rad/s^2. What angular displacement does the Ferris wheel undergo in this 5-s interval?
   a. 4.1 rad
   b. 2.5 rad
   c. 3.0 rad
   d. 0.50 rad

4. A 0.3-kg mass, attached to the end of a 0.75-m string, is whirled around in a circular horizontal path. If the maximum tension that the string can withstand is 250 N, then what maximum velocity can the mass have if the string is not to break?
   a. 375 m/s
   b. 22.4 m/s
   c. 19.4 m/s
   d. 25 m/s

5. A point on the rim of a 0.20-m-radius rotating wheel has a centripetal acceleration of 4.0 m/s^2. What is the angular velocity of the wheel?
   a. 0.89 rad/s
   b. 1.6 rad/s
   c. 3.2 rad/s
   d. 4.5 rad/s

6. When a point on the rim of a 0.25-m-radius wheel experiences a centripetal acceleration of 4.0 m/s^2, what tangential acceleration does that point experience?
   a. 1.0 m/s^2
   b. 2.0 m/s^2
   c. 4.0 m/s^2
   d. Cannot determine with information given

7. A Ferris wheel, starting at rest, builds up to a final angular velocity of 0.71 rad/s while rotating through an angular displacement of 2.5 rad. What is its average angular acceleration?
   a. 0.20 rad/s^2
   b. 0.10 rad/s^2
   c. 1.8 rad/s^2
   d. 3.6 rad/s^2

8. A roller coaster, loaded with passengers, has a mass of 500 kg; the radius of curvature of the track at the bottom point of the dip is 12 m. If the vehicle has a speed of 18 m/s at this point, what force is exerted on the vehicle by the track? (g = 9.8 m/s^2)
   a. 1.6×10^4 N
   b. 1.8×10^4 N
   c. 0.5×10^4 N
   d. 1.0×10^4 N

9. Consider a point on a bicycle tire that is momentarily in contact with the ground as the bicycle rolls across the ground with constant speed. The direction for the acceleration for this point at that moment is
   a. upward
   b. down toward the ground
   c. forward
   d. at that moment the acceleration is zero

10. What angular velocity (in revolutions/second) is needed for a centrifuge to produce an acceleration of 1000*g* at a radius arm of 10 cm?
   a. 50 rev/s
   b. 75 rev/s
   c. 100 rev/s
   d. 150 rev/s

11. Geosynchronous satellites orbit the Earth at a distance of 42,000 km from the Earth's center. Their angular velocity at this height is the same as the rotation of the Earth, so they appear stationary at certain locations in the sky. What is the force acting on a 1000-kg satellite at this height?
   a. 57 N
   b. 222 N
   c. 304 N
   d. 431 N

12. What is the angular velocity about the center of the Earth for a person standing on the equator?
   a. 7.3×10^{-5} rad/s
   b. 3.6×10^{-5} rad/s
   c. 6.28×10^{-5} rad/s
   d. 3.14×10^{-5} rad/s

Chapter 8

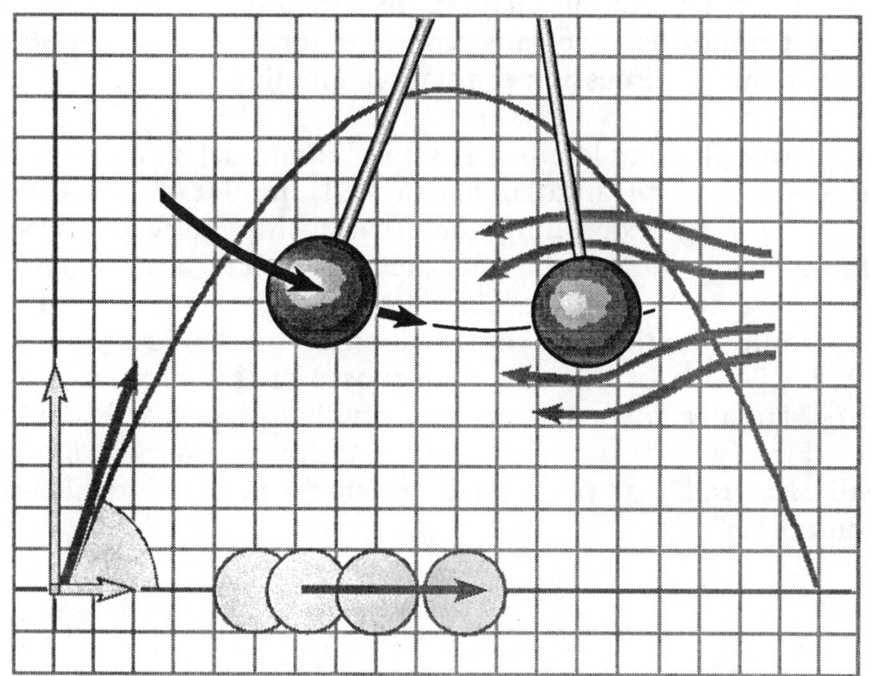

ROTATIONAL EQUILIBRIUM AND ROTATIONAL DYNAMICS

Chapter 8

ROTATIONAL EQUILIBRIUM AND ROTATIONAL DYNAMICS

We shall begin this chapter by studying objects that are at rest or moving with constant velocity; such objects are said to be in equilibrium. Knowledge of the conditions that prevail when an object is in equilibrium is important in a variety of fields. Newton's first law, as discussed in Chapter 4, is the basis for much of our work in this chapter. In addition, in order to fully understand objects in equilibrium, we must consider torque. This concept will also play a key role in our discussion of rotational motion.

In this chapter we will complete our study of rotational motion. We shall build on the definitions of angular speed and angular acceleration encountered in Chapter 7 by examining the relationship between these concepts and the forces that produce rotational motion. Specifically, we shall find the rotational analog of Newton's second law and define a term that needs to be added to our equation for conservation of mechanical energy: rotational kinetic energy. One of the central purposes of this chapter is to develop the concept of angular momentum, a quantity that plays a key role in rotational motion. Finally, just as we found that linear momentum is conserved, we shall also find that the angular momentum of any isolated system is always conserved.

NOTES FROM SELECTED CHAPTER SECTIONS

8.1 Torque

Torque is the physical quantity which is a measure of the tendency of a force to cause rotation of a body about a specified axis. It is important to remember that torque must be defined with respect to a **specific axis** of rotation. Torque which has the SI **units** of N·m must not be confused with force.

8.2 Torque and the Second Condition For Equilibrium

A body in static equilibrium must satisfy two conditions:

1. The resultant external force must be zero.
2. The resultant external torque must be zero.

You should note that it does not matter where you pick the axis of rotation for calculating the net torque if the object is in equilibrium; since the object is not rotating, the location of the axis is completely arbitrary.

8.3 The Center of Gravity

In order to calculate the torque due to the weight (gravitational force) on a rigid body, the entire weight of the object can be considered to be concentrated at a single point called the center of gravity. The center of gravity of a homogeneous, symmetric body must lie along an axis of symmetry.

8.5 Relationship Between Torque and Angular Acceleration

The angular acceleration of an object is proportional to the net torque acting on it. The moment of inertia of the object is the proportionality constant between the net torque and the angular acceleration. The force and mass in linear motion correspond to torque and moment of inertia in rotational motion. Moment of inertia of an object depends on the location of the axis of rotation and upon the manner in which the mass is distributed relative to that axis (e.g., a ring has a greater moment of inertia than a disk of the same mass and radius).

8.6 Rotational Kinetic Energy

In linear motion, the energy concept is useful in describing the motion of a system. The energy concept can be equally useful in simplifying the analysis of rotational motion. We now have expressions for three types of energy: **gravitational potential energy,** PE_g; **translational kinetic energy,** KE_t; and **rotational kinetic energy,** KE_r. We must include all these forms of energy in the equation for conservation of mechanical energy.

EQUATIONS AND CONCEPTS

This is the first condition for equilibrium. An object will remain at rest or move with uniform motion along a straight line when no net external force acts on it. This condition corresponds to translational equilibrium. The vector equation for translational equilibrium can be written in component form (in this case for a two-dimensional situation).

$$\Sigma \mathbf{F} = 0$$

$$\Sigma F_x = 0$$

$$\Sigma F_y = 0$$

A set of two or more forces are concurrent if their extended lines of action intersect at a single point. When concurrent forces act upon an object, the condition $\Sigma \mathbf{F} = 0$ is sufficient to ensure equilibrium. **Review the recommended solution procedure and example problems in the textbook.**

Comment on concurrent forces.

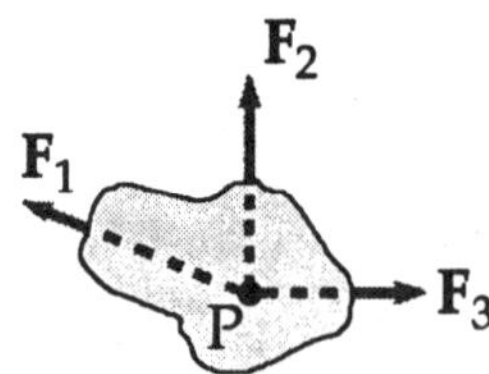

When nonconcurrent forces act on a body, the first condition ($\Sigma \mathbf{F} = 0$) is not sufficient to ensure complete equilibrium. In this case it is necessary to consider the net torque acting on the body relative to some axis. For a given force, the magnitude of the corresponding torque is the product of the magnitude of the force and the lever arm.

$$\tau = Fd \tag{8.1}$$

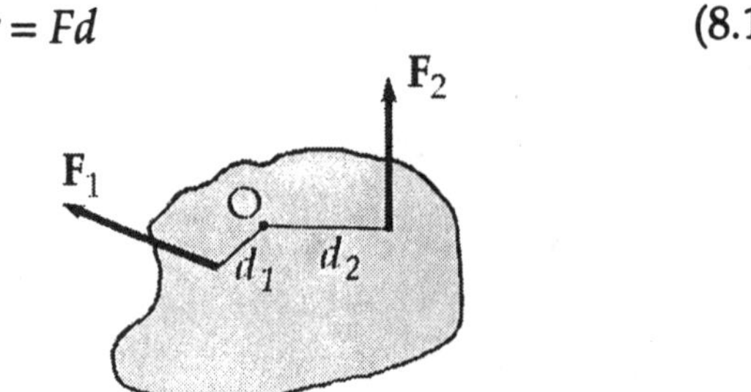

Note: For this object, $\tau_1 = -F_1 d_1$ and $\tau_2 = +F_2 d_2$

The sign of a torque due to a force is considered positive if the force has a tendency to rotate the body counterclockwise about the chosen axis, and negative if the tendency for rotation is clockwise.

Comment on sign conventions.

The second condition for equilibrium requires that the net torque acting on a body equals zero. This condition will ensure that the body is in rotational equilibrium.

$$\Sigma\tau = 0 \tag{8.2}$$

To compute the torque due to the force of gravity (an object's weight), the total weight can be considered as being concentrated at a single point called the center of gravity and having coordinates x_{cg}, y_{cg}.

$$x_{cg} = \frac{\Sigma m_i x_i}{\Sigma m_i} \tag{8.3}$$

$$y_{cg} = \frac{\Sigma m_i y_i}{\Sigma m_i} \tag{8.4}$$

The angular acceleration of a point mass, moving in a path of radius r, is proportional to the net torque acting on the mass.

$$\tau = mr^2\alpha \tag{8.6}$$

The moment of inertia of a rigid body depends on the distribution of mass relative to the axis of rotation.

$$I \equiv \Sigma mr^2 \tag{8.8}$$

The angular acceleration of an extended object is proportional to the net torque acting on the object. The proportionality constant, I, is called the moment of inertia of the object.

$$\Sigma\tau = I\alpha \tag{8.9}$$

The moment of inertia of a system depends on the mass and the manner in which the mass is distributed relative to the axis of rotation.

Comment on moment of inertia.

For a point mass, m, moving in a path of radius, r, $I = mr^2$.

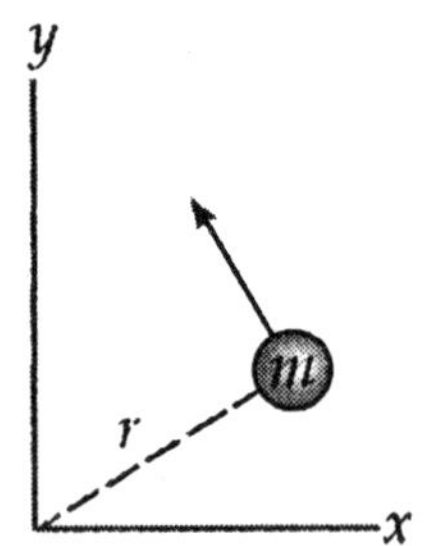

For a collection of discrete point masses, each with its own corresponding value of r, $I \equiv \Sigma mr^2$

The value of I can be calculated for an extended object with good symmetry. Expressions for the moments of inertia for a number of objects of common shape are given in Table 8.1 of your text.

The SI units of moment of inertia are kg·m^2.

Units for moment of inertia

A rigid body or mass in rotational motion has kinetic energy due to its motion. Note that this is not a new form of energy but is a convenient form for representing kinetic energy associated with rotational motion.

$$KE_r \equiv \frac{1}{2} I\omega^2 \quad (8.11)$$

When only conservative forces act on a system, the total mechanical energy of the system is conserved.

$$(KE_t + KE_r + PE_g)_i = (KE_t + KE_r + PE_g)_f \quad (8.12)$$

An object in rotational motion is characterized by a quantity, L, which is called angular momentum. Angular momentum is a vector quantity and has the direction of the angular velocity.

$$L \equiv I\omega \quad (8.13)$$

The net external torque acting on an object equals the time rate of change of its angular momentum. Note that this is the rotational analog of Newton's second law for transitional motion.

$$\tau = \frac{\Delta L}{\Delta t} \quad (8.14)$$

When the net external torque acting on a system is zero, the angular momentum of the system is conserved.

If $\Sigma\tau = 0$, then

$$I_i\omega_i = I_f\omega_f \quad (8.15)$$

SUGGESTIONS, SKILLS, AND STRATEGIES

PROBLEM-SOLVING STRATEGY FOR OBJECTS IN EQUILIBRIUM

1. Draw a simple, neat diagram of the system.

2. Isolate the object of interest being analyzed. Draw a free-body diagram for this object showing all external forces acting on the object. For systems containing more than one object, draw **separate** diagrams for each object. Do not include forces that the object exerts on its surroundings.

3. Establish convenient coordinate axes for each body and find the components of the forces along these axes. Now apply the first condition of equilibrium for each object under consideration; namely, that the net force on the object in the x and y directions must be zero.

4. Choose a convenient origin for calculating the net torque on the object. Now apply the second condition of equilibrium that says that the net torque on the object about any origin must be zero. Remember that the choice of the origin for the torque equation is arbitrary; therefore, choose an origin that will simplify your calculation as much as possible. Note that a force that acts along a line passing through the point chosen as the axis of rotation gives zero contribution to the torque.

5. The first and second conditions for equilibrium will give a set of simultaneous equations with several unknowns. To complete your solution, all that is left is to solve for the unknowns in terms of the known quantities.

Chapter 8

PROBLEM-SOLVING STRATEGY FOR ROTATIONAL MOTION

The following facts and procedures should be kept in mind when solving rotational motion problems.

1. There are actually very few new techniques that need to be learned in order to solve rotational motion problems. For example, problems involving the equation $\Sigma\tau = I\alpha$ are very similar to those encountered in Newton's second law problems, $\Sigma\mathbf{F} = m\mathbf{a}$. Note the correspondences between linear and rotational quantities in that $\mathbf{F}$ is replaced by τ, m by I, and a by α.

2. Other analogs between rotational quantities and linear quantities include the replacement of x by θ and v by ω. These are helpful as memory devices for such rotational motion quantities as rotational kinetic energy, $KE_r = \frac{1}{2}I\omega^2$, and angular momentum, $L = I\omega$.

3. With the analogs mentioned in Step 2, conservation of energy techniques remain the same as those examined in Chapter 5, except for the fact that a new kind of energy, rotational kinetic energy, must be included in the expression for the conservation of energy.

4. Likewise, the techniques for solving conservation of angular momentum problems are essentially the same as those used in solving conservation of linear momentum problems, except you are equating total angular momentum before to total angular momentum after as $I_i\omega_i = I_f\omega_f$.

Chapter 8

REVIEW CHECKLIST

▷ There are two necessary conditions for equilibrium of a rigid body: $\Sigma F = 0$ and $\Sigma\tau = 0$. Torques which cause counterclockwise rotations are positive and those causing clockwise rotations are negative.

▷ The torque associated with a force has a magnitude equal to the force times the moment arm; and the value of the torque depends on the origin about which it is evaluated. Also, the net torque on a rigid body about some axis is proportional to the angular acceleration; that is, $\tau = I\alpha$, where I is the moment of inertia about the axis about which the net torque is evaluated.

▷ The work-energy theorem can be applied to a rotating rigid body. That is, the net work done on a rigid body rotating about a fixed axis equals the change in its rotational kinetic energy; and the **law of conservation of mechanical energy** can be used in the solution of problems involving rotating rigid bodies.

▷ The time rate of change of the angular momentum of a rigid body rotating about an axis is proportional to the net torque acting about the axis of rotation. This is the rotational analog of Newton's second law.

SOLUTIONS TO SELECTED END-OF-CHAPTER PROBLEMS

3. A simple pendulum consists of a 3.0-kg point mass hanging at the end of a 2.0-m-long light string that is connected to a pivot point. Calculate the magnitude of the torque (due to the force of gravity) about this pivot point when the string makes a 5.0° angle with the vertical.

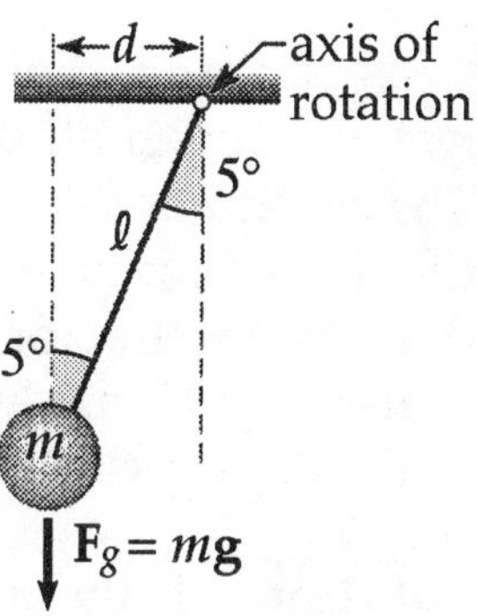

Solution The torque due to a force of magnitude F is $\tau = Fd$, where the lever arm d is the perpendicular distance from the axis of rotation to the line drawn along the direction of the force. Consider the force of gravity (i.e., the weight $F_g = mg$) acting on the bob of a simple pendulum as shown in the sketch at the right. From the sketch to the right, the lever arm, d, of the force F_g is $d = \ell \sin 5°$, where ℓ is the length of the pendulum string. Thus, the torque of this force about the indicated rotation axis is $\tau = F_g d = mg\ell \sin 5°$. Using the given data, this becomes

$$\tau = (3.0 \text{ kg})(9.8 \text{ m/s}^2)(2.0 \text{ m})\sin 5° \quad \text{or} \quad \tau = 5.1 \text{ N}\cdot\text{m}$$ ◊

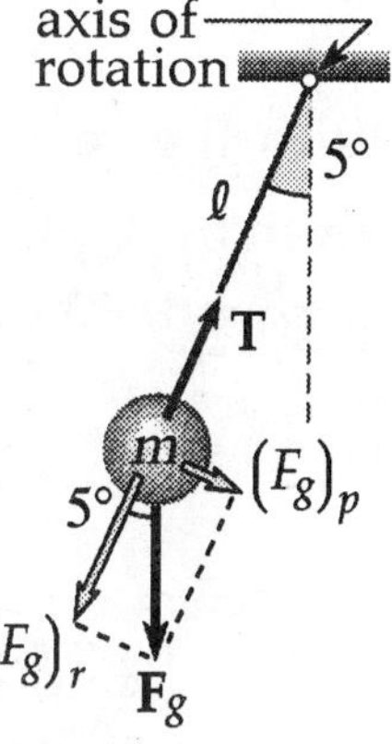

The second sketch illustrates an alternate method of computing the torque due to the force F_g about the axis of rotation. Note that F_g has **been replaced by its components,** $(F_g)_r$ and $(F_g)_p$. The component w_r lies along the line connecting the bob and the rotation axis, while the component $(F_g)_p$ is perpendicular to that line. Thus, the lever arm of $(F_g)_r$ is zero (the line along its direction passes through the axis, or there is zero distance from this line to the axis), while the lever arm of $(F_g)_p$ is the length of the string ℓ. The torque due to F_g can then be **computed as the sum of the torques due to its components** as follows:

$$\tau = \text{torque of } (F_g)_r + \text{torque of } (F_g)_p = (F_g)_r(0) + (F_g)_p(\ell) = (F_g \sin 5°)(\ell) = F_g\,\ell \sin 5°$$

This gives $\tau = 5.1 \text{ N}\cdot\text{m}$, the same as the other method. In many cases, this will be the easiest way to compute the torque due to a force. The reason for this is that the lever arms of the components are often easier to compute than the lever arm of the original force.

7. A water molecule consists of an oxygen atom with two hydrogen atoms bound to it as shown in Figure P8.7. The bonds are 0.100 nm in length and the angle between the two bonds is 106°. Use the x-y axis shown and determine the location of the center of gravity of the molecule. Consider the mass of an oxygen atom to be 16 times the mass of a hydrogen atom.

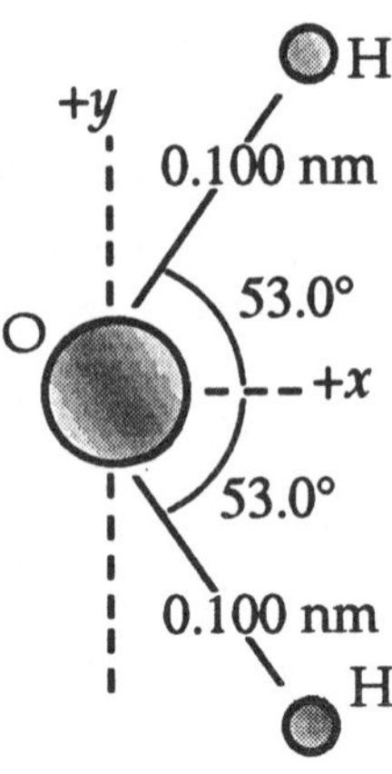

Figure P8.7

Solution If the hydrogen atoms are assigned a mass of 1 unit each (i.e., $m_H = 1.00\text{ u}$), the mass of the oxygen atom is then $m_O = 16.0\text{ u}$. The x- and y-coordinates of the center of gravity are defined respectively as:

$$x_{cg} \equiv \frac{\Sigma m_i x_i}{\Sigma m_i} \quad \text{and} \quad y_{cg} \equiv \frac{\Sigma m_i y_i}{\Sigma m_i}$$

With the reference axes shown, the x-coordinate of each hydrogen atom is given by $x_H = +(0.100\text{ nm})\cos 53° = 6.02 \times 10^{-2}\text{ nm}$ and that of the oxygen atom is $x_O = 0$. The x-coordinate of the center of mass for the water molecule is:

$$x_{cg} = \frac{(16.0\text{ u})(0) + (1.00\text{ u})\left(6.02 \times 10^{-2}\text{ nm}\right) + (1.00\text{ u})\left(6.02 \times 10^{-2}\text{ nm}\right)}{16.0\text{ u} + 1.00\text{ u} + 1.00\text{ u}}$$

$$x_{cg} = 6.69 \times 10^{-3}\text{ nm} \qquad \Diamond$$

The y-coordinate of the center of mass of the molecule is:

$$y_{cg} = \frac{(16.0\text{ u})(0) + (1.00\text{ u})[+(0.100\text{ nm})\sin 53°] + (1.00\text{ u})[-(0.100\text{ nm})\sin 53°]}{16.0\text{ u} + 1.00\text{ u} + 1.00\text{ u}} = 0 \qquad \Diamond$$

Note: The arbitrarily chosen mass unit (u) cancels out in these calculations. Also, since the molecule is symmetric about the x-axis, the result $y_{cg} = 0$ could have been obtained by observation. In general, if a body is symmetric about an axis, the center of gravity will always lie somewhere on that axis.

15. A uniform semi-circular sign 1.00 m in diameter and of weight W is supported by two wires as shown in Figure P8.15. What is the tension in each of the wires supporting the sign?

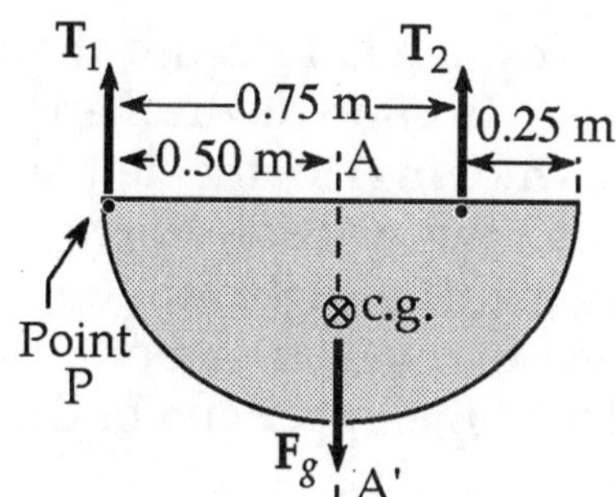

Figure P8.15

Solution Consider the free-body diagram of the sign shown at the right. Note that the center of gravity lies on the axis AA′ since the sign is symmetric about that axis. The sign is in equilibrium. Thus, the second condition of equilibrium states that $\sum \tau = 0$ for **any** rotation axis. For convenience, choose a rotation axis that passes through point P perpendicular to the page. Note that this axis was chosen because it is observed that the lever arm of the unknown force T_1 (and hence the torque due to T_1) will be zero for this axis. This will considerably simplify the torque equation. Writing out the torque equation for this axis, taking counterclockwise torques as positive, gives:

$$T_1(0) - F_g(0.5 \text{ m}) + T_2(0.75 \text{ m}) = 0 \qquad \text{or} \qquad T_2 = \frac{(0.5 \text{ m})F_g}{(0.75 \text{ m})} = \frac{2}{3}F_g \qquad \lozenge$$

Then, applying the first condition of equilibrium, $\sum F_x = 0$ and $\sum F_y = 0$, to the forces acting on the sign gives $+T_1 - F_g + T_2 = 0$.

Therefore,

$$T_1 = F_g - \tfrac{2}{3}F_g = \tfrac{1}{3}F_g \qquad \lozenge$$

---

21. An 8.0-m, 200-N uniform ladder rests against a smooth wall. The coefficient of static friction between the ladder and the ground is 0.60, and the ladder makes a 50.0° angle with the ground. How far up the ladder can an 800-N person climb before the ladder begins to slip?

SolutionThe sketch gives the free-body diagram of the ladder after the 800-N person has gone a distance ℓ up the ladder. Since the ladder is uniform, the center of gravity is at its geometric center, or 4.0 m from either end.

Gravity tends to cause the upper end of the ladder to slide down the wall with the lower end moving to the left along the ground. Thus, the friction force at the ground is toward the right to oppose this motion. Since the wall is smooth, there is no friction force at the upper end of the ladder.

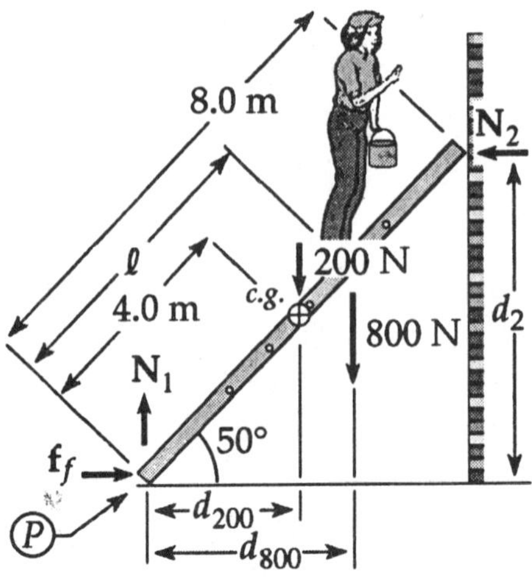

Until slippage occurs, the ladder is in equilibrium, so $\sum F_x = 0$ and $\sum F_y = 0$. Further, when slippage is just about to occur, the static friction force will be at its maximum value: $f_s = (f_s)_{max}$

$\sum F_y = 0$ gives $+N_f - 200\text{ N} - 800\text{ N} = 0$ or $N_f = 1000\text{ N}$

$f_s = (f_s)_{max} = \mu_s N_1$ or $f_s = (0.60)(1000\text{ N}) = 600\text{ N}$

$\sum F_x = 0$ gives $+f_s - N_w = 0$ or $N_w = f_s = 600\text{ N}$

The second condition of equilibrium ($\sum \tau = 0$ for any rotation axis) must also be satisfied. Choosing a rotation axis perpendicular to the page through point P, the lever arms of N_f and f_s are both zero since these forces pass through point P. The lever arms of the other forces are shown in the sketch and have values of $d_{200} = (4.0\text{ m})\cos 50°$, $d_{800} = \ell \cos 50°$, and $d_2 = (8.0\text{ m})\sin 50°$. The equation for the second condition of equilibrium (taking counterclockwise torques as positive) is then:

$$\sum \tau = 0 + 0 - (200\text{ N})(4.0\text{ m})\cos 50° - (800\text{ N})\ell \cos 50° + (600\text{ N})(8.0\text{ m})\sin 50° = 0$$

Thus, $$\ell\,(800\text{ N})\cos 50° = -(800\text{ N}\cdot\text{m})\cos 50° + (4800\text{ N}\cdot\text{m})\sin 50°$$

or $$\ell = \frac{-(800\text{ N}\cdot\text{m})\cos 50° + (4800\text{ N}\cdot\text{m})\sin 50°}{(800\text{ N})\cos 50°} = 6.2\text{ m}$$

The person can go 6.2 m up the ladder (and no farther) before the ladder will begin to slip. ◊

25. Four masses are held in position at the corners of a rectangle by light rods as shown in Figure P8.25. Find the moment of inertia of the system about (a) the x axis, (b) the y axis, and (c) an axis through O and perpendicular to the page.

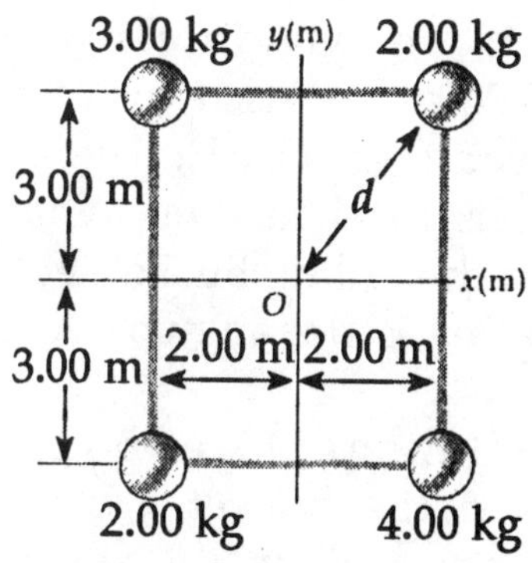

Figure P8.25

Solution The moment of inertia of a body when that body is rotating about some axis is defined as $I \equiv \sum m_i r_i^2$ where r_i is the shortest distance from the mass m_i to the rotation axis. Note that r_i is also the radius of the circular path the mass m_i follows as the body rotates about the rotation axis.

(a) If the body rotates about the x-axis, each of the four masses in Figure P8.25 will follow a circular path 3.00 m in radius. Thus, the moment of inertia of this body for rotation about the x-axis is:

$$I_x = (3.00\text{ kg})(3.00\text{ m})^2 + (2.00\text{ kg})(3.00\text{ m})^2 + (4.00\text{ kg})(3.00\text{ m})^2 + (2.00\text{ kg})(3.00\text{ m})^2 = 99.0\text{ kg}\cdot\text{m}^2 \quad \lozenge$$

(b) When this body rotates about the y-axis, each of the masses follow a circular path 2.00 m in radius. The moment of inertia of the body when rotating about the y-axis is therefore:

$$I_y = (3.00\text{ kg})(2.00\text{ m})^2 + (2.00\text{ kg})(2.00\text{ m})^2 + (4.00\text{ kg})(2.00\text{ m})^2 + (2.00\text{ kg})(2.00\text{ m})^2 = 44.0\text{ kg}\cdot\text{m}^2 \quad \lozenge$$

(c) If the body rotates about an axis perpendicular to the page and passing through point O, each of the masses will follow a circular path of radius d, where $d = \sqrt{(2.00\text{ m})^2 + (3.00\text{ m})^2} = \sqrt{13}\text{ m}$. The moment of inertia when the body rotates about this axis is then:

$$I_O = (3.00\text{ kg})\left(\sqrt{13}\text{ m}\right)^2 + (2.00\text{ kg})\left(\sqrt{13}\text{ m}\right)^2 + (4.00\text{ kg})\left(\sqrt{13}\text{ m}\right)^2 + (2.00\text{ kg})\left(\sqrt{13}\text{ m}\right)^2 = 143\text{ kg}\cdot\text{m}^2 \quad \lozenge$$

31. A cylindrical 5.00-kg pulley with a radius of 0.600 m is used to lower a 3.00-kg bucket into a well (Fig. P8.31). The bucket starts from rest and falls for 4.00 s. (a) What is the linear acceleration of the falling bucket? (b) How far does it drop? (c) What is the angular acceleration of the cylinder?

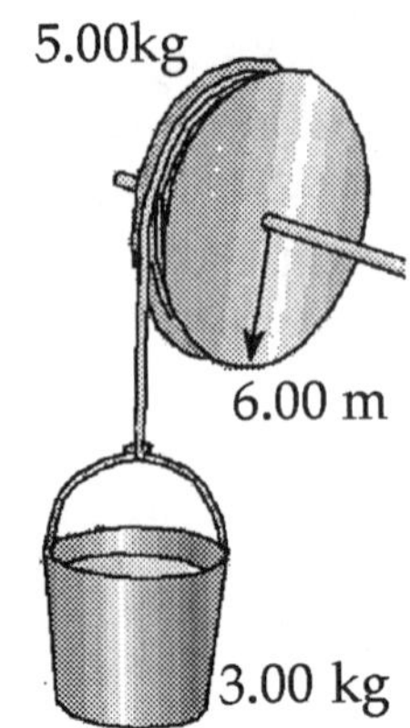

Figure P8.31

Solution Free-body diagrams of the pulley and the bucket are given in the sketch to the right. When the system is released, the bucket will accelerate downward with an acceleration of magnitude a. The pulley will rotate counter clockwise with angular acceleration α. If the cord from the bucket does not slip on the pulley, the magnitudes of these accelerations are related as $a = R\alpha$, or

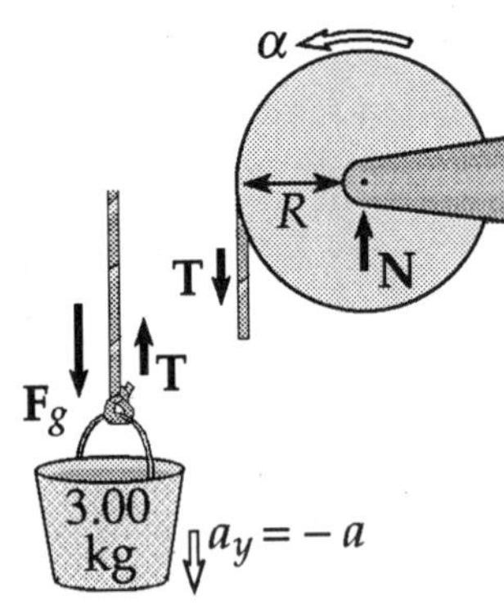

$F_g = mg = 29.4$ N

$$\alpha = a/R \qquad \textbf{[Equation 1]}$$

(a) To determine the magnitudes of these accelerations, it is necessary to apply Newton's second law to each body separately. For the rotational motion of the pulley, Newton's second law takes the form $\sum\tau = I\alpha$, where I is the moment of inertia of the pulley about a rotation axis perpendicular to the page and through its center. If the pulley is a solid cylindrical body,

$$I = \tfrac{1}{2}m_pR^2 = \tfrac{1}{2}(5.00 \text{ kg})(0.600 \text{ m})^2$$

or $$I = 0.900 \text{ kg}\cdot\text{m}^2$$

Then, taking counterclockwise rotations as positive,

$$\sum \tau = +TR = I\alpha: \qquad T = \left(\frac{I}{R}\right)\alpha = \left(\frac{0.900 \text{ kg}\cdot\text{m}^2}{0.600 \text{ m}}\right)\alpha = (1.50 \text{ kg}\cdot\text{m})\alpha$$

Using Equation 1 to substitute for α. in this result gives

$$T = (1.50 \text{ kg}\cdot\text{m})\left(\frac{a}{0.600 \text{ m}}\right) \quad \text{or} \quad T = (2.50 \text{ kg})a \qquad \textbf{[Equation 2]}$$

Taking upward as positive and applying Newton's second law to the bucket,

$\sum F_y = ma_y$ gives: $\qquad +T - 29.4 \text{ N} = (3.00 \text{ kg})(-a)$

or $\qquad T = 29.4 \text{ N} - (3.00 \text{ kg})a \qquad$ **[Equation 3]**

Combining Equations 2 and 3 gives the **downward** acceleration of the bucket as $(2.50 \text{ kg})a = 29.4 \text{ N} - (3.00 \text{ kg})a$:

or $$a = \frac{29.4 \text{ N}}{5.50 \text{ kg}} = 5.35 \text{ m/s}^2$$ ◊

(b) The vertical displacement of the bucket at time t is $y = v_{oy}t + \frac{1}{2}a_y t^2$. The bucket starts from rest, so $v_{oy} = 0$. Still taking upward as positive, the vertical acceleration of the bucket is $a_y = -a = -5.35 \text{ m/s}^2$

Thus, at $t = 4.00$ s $\qquad y = 0 + \frac{1}{2}(-5.35 \text{ m/s}^2)(4.00 \text{ s})^2 = -42.8 \text{ m}$

The displacement of the bucket after 4.00 s is: 42.8 m **downward** ◊

(c) From Equation 1, the angular acceleration of the cylindrical pulley is

$$\alpha = \frac{a}{R} = \frac{5.35 \text{ m/s}^2}{0.600 \text{ m}} \quad \text{or} \quad \alpha = 8.92 \text{ rad/s}^2$$ **(counterclockwise)** ◊

35. A 10.0-kg cylinder rolls without slipping on a rough surface. At the instant its center of mass has a speed of 10.0 m/s, determine (a) the translational kinetic energy of its center of mass, (b) the rotational kinetic energy about its center of mass, and (c) its total kinetic energy.

Solution (a) The translational kinetic energy of the cylinder is given by $KE_t = \frac{1}{2}mv^2$, where v is the translational speed of the center of mass. Thus,

$$KE_t = \tfrac{1}{2}(10.0\text{ kg})(10.0\text{ m/s})^2 \quad \text{or} \quad KE_t = 500\text{ J} \qquad \Diamond$$

(b) The rotational kinetic energy of the cylinder is $KE_r = I\omega^2/2$ where I is the moment of inertia about the axis through its center of mass, and ω is the angular speed about this axis. Assuming a uniform, solid cylinder of radius R, $I = \frac{1}{2}mR^2$, the rotational kinetic energy is $KE_r = \frac{1}{2}\left(\frac{1}{2}mR^2\right)\omega^2 = \frac{1}{4}m(R\omega)^2$. The product $R\omega$ is the same as the tangential speed of a point on the rim of the cylinder, $v_t = R\omega$, so $KE_r = \frac{1}{4}mv_t^{\,2}$. If the wheel rolls without slipping, the tangential speed of a point on the rim is the same as the translational speed of the center of mass. To understand why this is true, imagine yourself to be at the center of the wheel and moving to the left at speed v. Looking down, you would see point A at the wheel's rim moving right with the tangential speed $v_t = R\omega$. You would also see point B on the ground moving to the right at the speed v, the speed of the center of mass (and you) relative to the ground. Now, if the rim of the wheel is not slipping against the ground, the points A and B (in contact with each other) must move at the same speed. Thus, it is necessary that $v = v_t = R\omega$ if slipping does not occur. The rotational kinetic energy of the wheel is therefore,

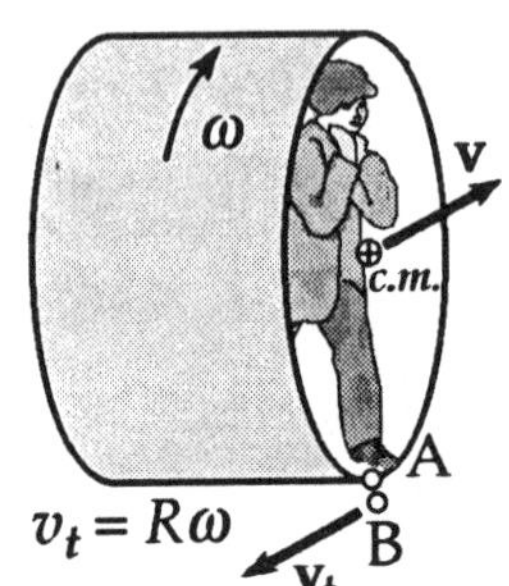

$$KE_r = \tfrac{1}{4}mv_t^{\,2} = \tfrac{1}{4}mv^2 = \tfrac{1}{4}(10.0\text{ kg})(10.0\text{ m/s})^2 = 250\text{ J} \qquad \Diamond$$

(c) The total kinetic energy of the rolling wheel is then:

$$KE_{\text{total}} = KE_t + KE_r = 500\text{ J} + 250\text{ J} = 750\text{ J} \qquad \Diamond$$

41. Use conservation of energy to determine the angular speed of the spool shown in Figure P8.33 after the mass m has fallen 4.00 m, starting from rest. The light string attached to this mass is wrapped around the spool and does not slip as it unwinds. Assume that the spool is a solid cylinder of radius 0.500 m and mass 0.500 kg, and that $m = 5.00$ kg.

Solution Consider a system consisting of the spool and the attached mass as shown in the sketch. Since gravity (a conservative force) is the only force doing work on this system as the mass falls, the total mechanical energy of the system remains constant: $KE_f + PE_f = KE_i + PE_i$. The total kinetic energy at any time is

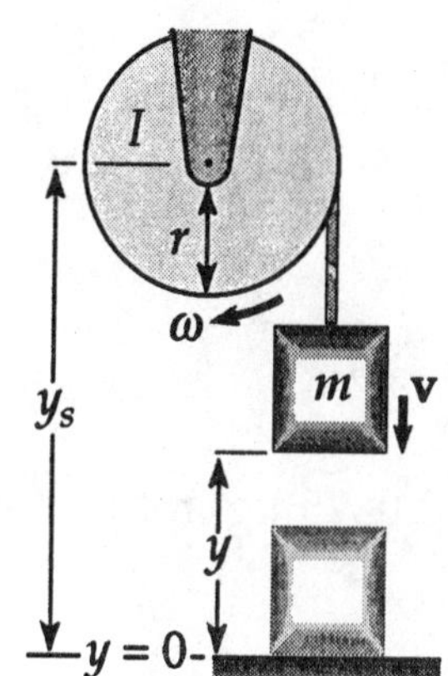

Figure P8.33 (modified)

$$KE_{\text{total}} = KE_{\text{spool}} + KE_{\text{mass}} = \tfrac{1}{2}I\omega^2 + \tfrac{1}{2}mv^2$$

where I is the moment of inertia of the spool, ω is the angular speed of the spool, and v is the linear speed of the falling mass. The total potential energy is

$$PE_{\text{total}} = PE_{\text{spool}} + PE_{\text{mass}} = Mgy_s + mgy$$

where M is the mass of the spool, y_s is its height above the reference level, and y is the height of the mass m. Since the string does not slip on the spool, the linear speed of the mass and string, v, is the same as the tangential speed of a point on the rim of the spool (i.e., $v = v_t = r\omega$). The energy equation is then:

$$\tfrac{1}{2}I\omega_f^2 + \tfrac{1}{2}m(r\omega_f)^2 + Mgy_s + mgy_f = \tfrac{1}{2}I\omega_i^2 + \tfrac{1}{2}m(r\omega_i)^2 + Mgy_s + mgy_i$$

Note that y_s is constant and the potential energy of the spool will cancel. Since the system starts from rest, $\omega_i = 0$. Choosing the reference level at the final position of the mass gives $y_f = 0$. After canceling the potential energy of the spool, the energy equation then reduces to:

$$\tfrac{1}{2} I\omega_f^2 + \tfrac{1}{2} m\left(r\omega_f\right)^2 = mgy_i$$

We substitute $\quad I = \tfrac{1}{2} Mr^2 = \tfrac{1}{2}(0.500 \text{ kg})(0.500 \text{ m})^2 = 0.0625 \text{ kg} \cdot \text{m}^2$

$m = 5.00 \text{ kg}$

$y_i = +4.00 \text{ m}$

and find:

$$\tfrac{1}{2}\left(0.0625 \text{ kg} \cdot \text{m}^2\right)\omega_f^2 + \tfrac{1}{2}(5.00 \text{ kg})(0.500 \text{ m})^2 \omega_f^2 = (5.00 \text{ kg})\left(9.80 \text{ m/s}^2\right)(4.00 \text{ m})$$

Therefore $\left(0.656 \text{ kg} \cdot \text{m}^2\right)\omega_f^2 = 196 \text{ J}$ or $\omega_f^2 = 299 \text{ s}^{-2}$ and the angular speed of the spool, after the mass has fallen 4.00 m, is $\omega = 17.3 \text{ rad/s}$ ◊

---

48. The puck in Figure P8.48 has a mass of 0.120 kg. Its original distance from the center of rotation is 40.0 cm, and the puck is moving with a speed of 80.0 cm/s. The string is pulled downward 15.0 cm through the hole in the frictionless table. Determine the work done on the puck. (**Hint:** Consider the change of kinetic energy of the puck.)

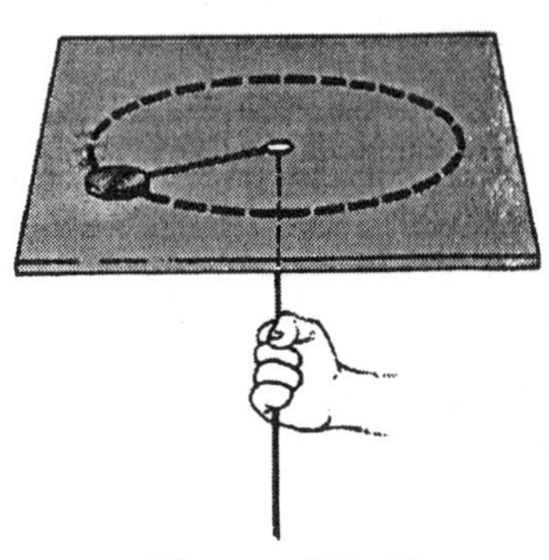

Figure P8.48

Solution The puck originally moves in a circular path with a radius of

$$r_i = 40.0 \text{ cm} = 0.400 \text{ m}$$

Pulling the string downward 15.0 cm decreases the radius of the puck's path to

$$r_f = 0.400 \text{ m} - 0.150 \text{ m} = 0.250 \text{ m}$$

The tension in the string is directed toward the center of the puck's circular path, and hence, toward the rotation axis. It exerts zero torque about the rotation axis. The weight of the puck and the normal force acting on the puck are both parallel to the rotation axis and therefore exert zero torque about that axis. The total torque acting about the rotation axis of the puck is then zero, and the angular momentum of the puck remains constant:

$$I_f\omega_f = I_i\omega_i$$

The moments of inertia about the rotation axis before and after the string is pulled are:

$$I_i = mr_i^2 = (0.120\text{ kg})(0.400\text{ m})^2 = 1.92\times10^{-2}\text{ kg}\cdot\text{m}^2$$

and

$$I_f = mr_f^2 = (0.120\text{ kg})(0.250\text{ m})^2 = 7.50\times10^{-3}\text{ kg}\cdot\text{m}^2$$

The initial angular speed of the puck is $\omega_i = \frac{v_i}{r_i} = \frac{80.0\text{ cm/s}}{40.0\text{ cm}} = 2.00\text{ rad/s}$ and the final angular speed is

$$\omega_f = \left(\frac{I_i}{I_f}\right)\omega_i = \left(\frac{1.92\times10^{-2}\text{ kg}\cdot\text{m}^2}{7.50\times10^{-3}\text{ kg}\cdot\text{m}^2}\right)(2.00\text{ rad/s}) = 5.12\text{ rad/s}$$

The kinetic energy of the puck before and after the string is pulled is then:

$$KE_i = \tfrac{1}{2}I_i\omega_i^2 = \tfrac{1}{2}\left(1.92\times10^{-2}\text{ kg}\cdot\text{m}^2\right)(2.00\text{ rad/s})^2 = 3.84\times10^{-2}\text{ J}$$

and

$$KE_f = \tfrac{1}{2}I_f\omega_f^2 = \tfrac{1}{2}\left(7.50\times10^{-3}\text{ kg}\cdot\text{m}^2\right)(5.12\text{ rad/s})^2 = 9.83\times10^{-2}\text{ J}$$

Thus, work done on the puck is given by the work-kinetic energy theorem as

$$W_{net} = KE_f - KE_i = 9.83\times10^{-2}\text{ J} - 3.84\times10^{-2}\text{ J} = 5.99\times10^{-2}\text{ J}$$ ◊

53. A person bends over and lifts a 200-N weight as in Figure P8.53a, with his back horizontal. The muscle that attaches two thirds of the way up the spine maintains the position of the back; the angle between the spine and this muscle is 12°. Using the mechanical model in Figure P8.53b and taking the weight of the upper body to be 350 N, find the tension in the back muscle and the compressional force in the spine.

Solution The free-body diagram needed to solve this equilibrium problem is shown at the right. Notice that the 350 N force acts at the center of gravity of the upper body and that the tension in the back muscle, T, has been replaced by its components:

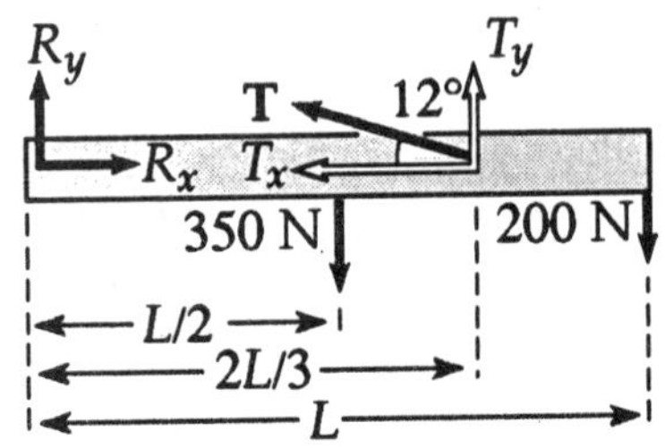

$$T_x = T\cos 12° \qquad \text{and} \qquad T_y = T\sin 12°$$

Choose a rotation axis perpendicular to the page and passing through the left end of the spine (where the forces R_x and R_y intersect). Taking counterclockwise torques as positive, the second condition of equilibrium gives:

$$\sum \tau = R_y(0) + R_x(0) - (350\text{ N})(L/2) + T_x(0) + T_y(2L/3) - (200\text{ N})L = 0$$

This reduces to $T_y = \frac{3}{2}(175\text{ N} + 200\text{ N}) = 563\text{ N}$. Hence, the tension in the back muscle is: $T = \dfrac{563\text{ N}}{\sin 12°} = 2.70 \times 10^3\text{ N}$ ◊

From $\Sigma F_x = 0$: $R_x - T_x = 0$

or $R_x = T_x = T\cos 12° = \left(2.70 \times 10^3\text{ N}\right)\cos 12° = 2.65 \times 10^3\text{ N}$

The compressional force in the spine is the magnitude of these two oppositely directed forces acting on the spine and tending to compress the spine. That is, the compressional force is $R_x = T_x = 2.65 \times 10^3\text{ N}$ ◊

59. The pulley in Figure P8.59 has a moment of inertia of 5.0 kg·m^2 and a radius of 0.50 m. The cord supporting the masses m_1 and m_2 does not slip, and the axle is frictionless. (a) Find the acceleration of each mass when $m_1 = 2.0$ kg and $m_2 = 5.0$ kg. (b) Find the tension in the cable supporting m_1 and the tension in the cable supporting m_2 (note that they are different).

Solution The needed free-body diagrams for the parts of this system are shown to the right. Note that the **magnitude** of linear acceleration is the same for the two masses. Since the cord does not slip on the pulley, this acceleration is related to the angular acceleration of the pulley by $\alpha = a/r$. Also note that, **consistent with the assumed directions of motion for the masses,** clockwise has been chosen as the positive sense of rotation of the pulley. Thus, it will be necessary to consider clockwise torques as positive in our calculation.

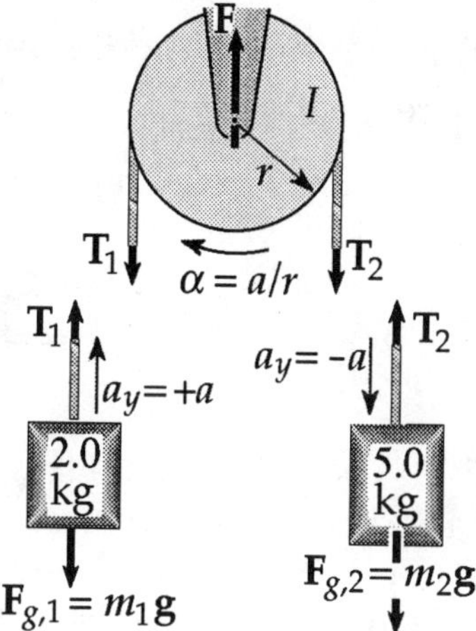

(a) Applying Newton's second law $\left(\sum F_y = ma_y\right)$ to m_1 gives:

$$T_1 - m_1 g = m_1(+a) \quad \text{or} \quad T_1 = m_1(g + a) \qquad \textbf{[Equation 1]}$$

For the mass m_2, the second law yields:

$$+T_2 - m_2 g = m_2(-a) \quad \text{or} \quad T_2 = m_2(g - a) \qquad \textbf{[Equation 2]}$$

For the pulley, the second law takes the form $\sum \tau = I\alpha = I(a/r)$. Taking clockwise torques as positive (as discussed above), this becomes:

$$T_2 r - T_1 r = I(a/r) \quad \text{or} \quad T_2 - T_1 = a\left(I/r^2\right) \qquad \textbf{[Equation 3]}$$

Substituting Equations 1 and 2 into Equation 3 will give:

$$(m_2 - m_1)g - (m_2 + m_1)a = a\left(\frac{I}{r^2}\right) \quad \text{or} \quad a = \frac{(m_2 - m_1)g}{m_2 + m_1 + I/r^2}$$

Thus, with the given data, the magnitude of the acceleration of each mass is:

$$a = \frac{(5.0\text{ kg} - 2.0\text{ kg})(9.8\text{ m/s}^2)}{5.0\text{ kg} + 2.0\text{ kg} + (5.0\text{ kg}\cdot\text{m}^2)/(0.50\text{ m})^2} = 1.1\text{ m/s}^2 \qquad \Diamond$$

(b) Equations 1 and 2 now give the tensions in the cords supporting m_1 and m_2 respectively:

$$T_1 = (2.0\text{ kg})(9.8\text{ m/s}^2 + 1.1\text{ m/s}^2) = 22\text{ N}$$

and

$$T_2 = (5.0\text{ kg})(9.8\text{ m/s}^2 - 1.1\text{ m/s}^2) = 44\text{ N} \qquad \Diamond$$

---

63. Two astronauts (Fig. P8.63), each having a mass of 75.0 kg, are connected by a 10.0-m rope of negligible mass. They are isolated in space, orbiting their center of mass at speeds of 5.00 m/s. (a) Calculate the magnitude of the angular momentum of the system by treating the astronauts as particles and (b) the rotational energy of the system. By pulling on the rope, the astronauts shorten the distance between them to 5.00 m. (c) What is the new angular momentum of the system? (d) What are their new speeds? (e) What is the new rotational energy of the system? (f) How much work is done by the astronauts in shortening the rope?

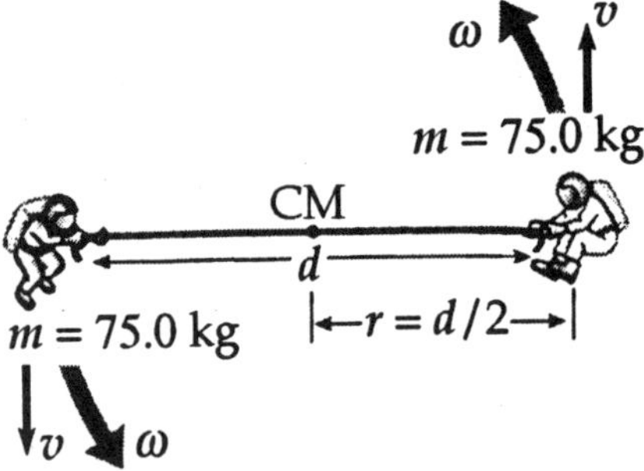

Solution (a) Initially, the astronauts are moving in a circular path of a radius $r_i = (10.0\text{ m})/2 = 5.00\text{ m}$ with linear speeds of $v_i = 5.00\text{ m/s}$ and angular speeds of

$$\omega_i = v/r = \frac{5.00\text{ m/s}}{5.00\text{ m}} = 1.00\text{ rad/s}$$

The initial moment of inertia and initial angular momentum of this system are:

$$I_i = 2I_{\substack{\text{individual}\\ \text{astronaut}}} = 2\left(mr_i^2\right) = 2(75.0\text{ kg})(5.00\text{ m})^2 = 3.75\times10^3\text{ kg}\cdot\text{m}^2$$

and $$L_i = I_i\omega_i = \left(3.75\times10^3\text{ kg}\cdot\text{m}^2\right)(1.00\text{ rad/s}) = 3.75\times10^3\text{ kg}\cdot\text{m}^2\text{/s} \qquad \Diamond$$

(b) The initial rotational kinetic energy of this system is $KE_i = \frac{1}{2}I_i\omega_i^2$, or

$$KE_i = \tfrac{1}{2}\left(3.75\times10^3\text{ kg}\cdot\text{m}^2\right)(1.00\text{ rad/s})^2 = 1.88\times10^3\text{ J} = 1.88\text{ kJ} \qquad \Diamond$$

(c) Since the astronauts are isolated in space, no external forces exert torques about the rotation axis. Thus, the angular momentum of the system is constant. That is:

$$L_f = L_i = 3.75\times10^3\text{ kg}\cdot\text{m}^2\text{/s} \qquad \Diamond$$

(d) The astronauts change the moment of inertia of the system when they change the distance they are from the center of mass. The new distance from the center of mass is $r_f = (5.00\text{ m})/2 = 2.50\text{ m}$, and the new moment of inertia is $I_f = 2(75.0\text{ kg})(2.50\text{ m})^2 = 938\text{ kg}\cdot\text{m}^2$. Then, since angular moment is conserved, $I_f\omega_f = I_i\omega_i$, and

$$\omega_f = \frac{I_i\,\omega_i}{I_f} = \left(\frac{3.75\times10^3\text{ kg}\cdot\text{m}^2\text{/s}}{938\text{ kg}\cdot\text{m}^2}\right)(1.00\text{ rad/s}) = 4.00\text{ rad/s}$$

The new linear speed is $v_f = r_f\omega_f = (2.50\text{ m})(4.00\text{ rad/s}) = 10.0\text{ m/s}$ $\Diamond$

(e) The new rotational kinetic energy of this system is $KE_f = \frac{1}{2}I_f\omega_f^2$

or $$KE_f = \tfrac{1}{2}\left(938\text{ kg}\cdot\text{m}^2\right)(4.00\text{ rad/s})^2 = 7.50\times10^3\text{ J} = 7.50\text{ kJ} \qquad \Diamond$$

(f) The work done is $W_{\text{net}} = KE_f - KE_i = 7.50\text{ kJ} - 1.88\text{ kJ} = 5.62\text{ kJ}$ $\Diamond$

CHAPTER SELF-QUIZ

1. The Earth's gravity exerts no torque on a satellite orbiting the Earth in an elliptical orbit. Compare the motion at the point nearest the Earth (perigee) to the motion at the point farthest from the Earth (apogee).
   a. At these two points, the tangential velocities are the same
   b. At these two points, the angular velocities are the same
   c. At these two points, the angular momenta are the same
   d. At these two points, the kinetic energies are the same

2. A majorette takes two batons and fastens them together in the middle at right angles to make an "x" shape. Each baton was 0.8 m long and each ball on the end is 0.30 kg. (Ignore the mass of the rods.) What is the moment of inertia if the arrangement is spun around an axis formed by one of the batons?
   a. $0.048\ \text{kg}\cdot\text{m}^2$
   b. $0.096\ \text{kg}\cdot\text{m}^2$
   c. $0.19\ \text{kg}\cdot\text{m}^2$
   d. $0.38\ \text{kg}\cdot\text{m}^2$

3. A solid cylinder $(I = MR^2/2)$, a hoop $(I = MR^2)$, and a sphere $(I = 2MR^2/5)$ are released and roll down a slope. The mass of all objects is the same, but the cylinder has twice the radius of the other two. Which reaches the bottom first?
   a. the cylinder
   b. the hoop
   c. the sphere
   d. at least two reach the bottom at the same time

4. An astronaut is on a 200-m life-line outside a spaceship, circling the ship with an angular velocity of 0.10 rad/s. How far can she be pulled in before the centripetal acceleration reaches $5g = 49\ \text{m/s}^2$?
   a. The final length of the rope will be 40 m
   b. The final length of the rope will be 69 m
   c. The final length of the rope will be 571 m
   d. The final length of the rope will be 10.1 m

5. A bus is designed to draw its power from a rotating flywheel that is brought up to its maximum speed (3000 rpm) by an electric motor. The flywheel is a solid cylinder of mass 1000 kg and diameter 1 m ($I_{cylinder} = MR^2/2$). If the bus requires an average power of 10 kilowatts, how long will the flywheel rotate?
   a. 330 s
   b. 625 s
   c. 975 s
   d. 1250 s

6. A solid cylinder ($I = 0.5MR^2$) with mass, M, and radius, R, rolls along a level surface without slipping with a linear velocity, v. What is the ratio of its rotational kinetic energy to its linear kinetic energy?
   a. 1/4
   b. 1/2
   c. 1/1
   d. 2/1

7. A phonograph turntable ($I = 2.50 \times 10^{-2}\ \text{kg} \cdot \text{m}^2$) spins freely on a frictionless bearing at 33.3 rev/min. A 0.25-kg ball of putty is dropped vertically on the turntable and sticks at a point 0.20 m from the center. What is the new rate of rotation of the system?
   a. 40.8 rev/min
   b. 23.8 rev/min
   c. 33.3 rev/min
   d. 27.2 rev/min

8. A solid sphere with a mass of 3.0 kg and radius of 0.2 m, starts from rest at the top of a ramp, inclined at 15°, and rolls to the bottom. The upper end of the ramp is 1.2 m higher than the lower end. What is the linear velocity of the sphere when it reaches the bottom of the ramp? (**Note:** $I = 0.4MR^2$ for a solid sphere and $g = 9.8\ \text{m/s}^2$.)
   a. 4.7 m/s
   b. 4.1 m/s
   c. 3.4 m/s
   d. 2.4 m/s

9. A bowling ball has a mass of 4.0 kg, a moment of inertia of 1.60×10^{-2} kg·m^2 and a radius of 0.20 m. If it rolls down the lane without slipping at a linear speed of 4.0 m/s, what is its angular velocity?
   a. 0.8 rad/s
   b. 10.0 rad/s
   c. 0.05 rad/s
   d. 20.0 rad/s

10. A bucket filled with water has a mass of 23 kg and is attached to a rope which in turn is wound around a 0.05 m radius cylinder at the top of a well. The bucket and water are first raised to the top of the well and then released. The bucket and water are moving with a speed of 7.9 m/s upon hitting the water surface in the well below. What is the angular velocity of the cylinder at this instant?
   a. 39 rad/s
   b. 79 rad/s
   c. 118 rad/s
   d. 158 rad/s

11. A phonograph record and turntable combination have a moment of inertia of 250×10^{-6} (SI units) and rotate with an angular velocity of 6 rad/sec. What net torque must be applied to bring the system to rest within 3 s?
   a. 4500×10^{-6} N·m
   b. 750×10^{-6} N·m
   c. 500×10^{-6} N·m
   d. 250×10^{-6} N·m

12. If a long rod of length L is hinged at one end, the moment of inertia as the rod rotates around that hinge is $ML^2/3$. A 4-m rod with a mass of 3 kg, hinged at one end, is held in a horizontal position, and then released as the free end is allowed to fall. What is the angular acceleration as it is released?
   a. 3.7 rad/s^2
   b. 7.35 rad/s^2
   c. 2.45 rad/s^2
   d. 4.9 rad/s^2

Chapter 9

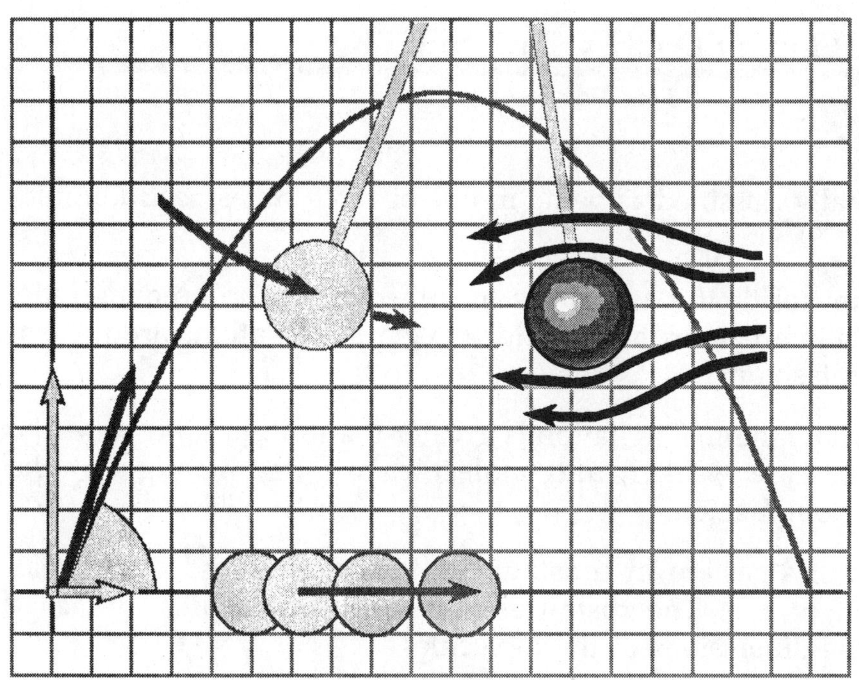

SOLIDS AND FLUIDS

Chapter 9

SOLIDS AND FLUIDS

In this chapter we consider some properties of solids and fluids (both liquids and gases). We spend some time looking at properties that are peculiar to solids, but much of our emphasis is on the properties of fluids. We open the fluids part of the chapter, with a study of fluids at rest, and finish with a discussion of the properties of fluids in motion.

NOTES FROM SELECTED CHAPTER SECTIONS

9.1 States of Matter

Matter is generally classified as being in one of three states: solid, liquid, or gas.

In a **crystalline solid,** the atoms are arranged in an ordered periodic structure; while in an **amorphous solid** (i.e. glass), the atoms are present in a disordered fashion.

In the **liquid state,** thermal agitation is greater than in the solid state, the molecular forces are weaker, and molecules wander throughout the liquid in a random fashion.

The molecules of a **gas** are in constant random motion and exert weak forces on each other. The distances separating molecules are large compared to the dimensions of the molecules.

9.2 The Deformation of Solids

The elastic properties of solids are described in terms of stress and strain. Stress is a quantity that is related to the force causing a deformation; strain is a measure of the degree of deformation. It is found that, for sufficiently small stresses, stress is proportional to strain, and the constant of proportionality depends on the material being deformed and on the nature of the deformation. We call this proportionality constant the elastic modulus.

We shall consider three types of deformation and define an elastic modulus for each:

1. **Young's modulus,** which measures the resistance of a solid to a change in its length.
2. **Shear modulus,** which measures the resistance to displacement of the planes of a solid sliding past each other.
3. **Bulk modulus,** which measures the resistance that solids or liquids offer to changes in their volume.

9.3 Density and Pressure

The **density,** ρ, of a substance of uniform composition is defined as its **mass per unit volume** and has units of kilograms per cubic meter (kg/m^3) in the SI system.

The **specific gravity** of a substance is a dimensionless quantity which is the ratio of the density of the substance to the density of water.

The **pressure,** *P*, in a fluid is the force per unit area that the fluid exerts on an object immersed in the fluid.

9.4 Variation of Pressure with Depth

In a **fluid at rest,** all points at the same depth are at the same pressure. **Pascal's law** states that a change in pressure applied to an **enclosed** fluid is transmitted undiminished to every point in the fluid and the walls of the containing vessel. **The pressure, *P*, at a depth of *h* below the surface of a liquid open to the atmosphere is greater than atmospheric pressure by an amount ρgh.**

9.5 Pressure Measurements

The **absolute pressure** of a fluid is the sum of the **gauge pressure** and atmospheric pressure. The SI unit of pressure is the Pascal (Pa). Note that $1\text{ Pa} \equiv 1\text{ N/m}^2$.

9.6 Buoyant Forces and Archimedes' Principle

Any object partially or completely submerged in a fluid experiences a buoyant force equal in magnitude to the weight of the fluid displaced by the object and acting vertically upward through the point which was the center of gravity of the displaced fluid.

9.7 Fluids in Motion

Many features of fluid motion can be understood by considering the behavior of an ideal fluid, which satisfies the following conditions:

1. **The fluid is nonviscous;** that is, there is no internal friction force between adjacent fluid layers.
2. **The fluid is incompressible,** which means that its density is constant.
3. **The fluid motion is steady,** meaning that the velocity, density, and pressure at each point in the fluid do not change in time.
4. **The fluid moves without turbulence.** This implies that each element of the fluid has zero angular velocity about its center; that is, there can be no eddy currents present in the moving fluid.

Fluids which have the "ideal" properties stated above obey two important equations:

> The **equation of continuity** states that the flow rate through a pipe is constant (i.e. the product of the cross-sectional area of the pipe and the speed of the fluid is constant).
>
> **Bernoulli's equation** states that the sum of the pressure (P), kinetic energy per unit volume $\left(\rho v^2/2\right)$, and the potential energy per unit volume (ρgh) has a constant value at all points along a streamline.

9.7 Surface Tension, Capillary Action, and Viscosity

The concept of surface tension can be thought of as the energy content of the fluid at its surface per unit surface area. In general, **any equilibrium configuration of an object is one in which the energy is minimum.** For a given volume, the spherical shape is the one that has the smallest surface area; therefore, a drop of water takes on a spherical shape. The surface tension of liquids decreases with increasing temperature.

Forces between like molecules, such as the forces between water molecules, are called **cohesive forces** and forces between unlike molecules, such as those of glass on water **adhesive forces.** If a capillary tube is inserted into a fluid for which adhesive forces dominate over cohesive forces, the liquid will rise into the tube. If a capillary tube is inserted into a liquid in which cohesive forces dominate over adhesive forces, the level of the liquid in the capillary tube will be below the surface of the surrounding fluid.

Viscosity refers to the internal friction of a fluid. At sufficiently high velocities, fluid flow changes from simple streamline flow to turbulent flow. The onset of turbulence in a tube is determined by a factor called the **Reynolds number**, which is a function of the density of the fluid, the average speed of the fluid along the direction of flow, the diameter of the tube, and the viscosity of the fluid.

9.10 Transport Phenomena

The two fundamental processes involved in fluid transport resulting from concentration differences are called **diffusion** and **osmosis.** In a **diffusion** process, molecules move from a region where their concentration is high to a region where their concentration is lower. Diffusion occurs readily in air; the process also occurs in liquids and, to a lesser extent, in solids. **Osmosis** is defined as the movement of water from a region where its concentration is high, across a selectively permeable membrane, into a region where its concentration is lower. Osmosis is often described simply as the diffusion of water across a membrane.

EQUATIONS AND CONCEPTS

A body of matter can be deformed (experience change in size or shape) by the application of external forces. Stress is a quantity which is proportional to the force which causes the deformation; strain is a measure of the degree of deformation. The elastic modulus is a general characterization of the deformation. Specific deformations are characterized by specific moduli.

$$\text{Elastic modulus} \equiv \frac{\text{stress}}{\text{strain}} \quad (9.1)$$

The SI units of pressure are newtons per square meter, or Pascal (Pa).

$$1\ \text{Pa} \equiv 1\ \text{N/m}^2 \quad (9.2)$$

Young's modulus is a measure of the resistance of a body to elongation.

$$Y = \frac{F/A}{\Delta L/L_0} \quad (9.3)$$

Within a limited range of values, the graph of stress vs. strain for a given substance will be a straight line. When the stress exceeds the elastic limit (at the yield point), the stress-strain curve will no longer be linear.

Comment on experimental observations.

The **Shear modulus** is a measure of the deformation which occurs when a force is applied along a direction parallel to one surface of a body.

$$S = \frac{F/A}{\Delta x/h} \quad (9.4)$$

The **Bulk modulus** characterizes the response of a body to uniform pressure (or squeezing) on all sides. Note that when ΔP is positive (increase in pressure), the ratio $\Delta V/V$ will be negative (decrease in volume) and vice versa. Therefore, the negative sign in the equation ensures that B will always be positive.

$$B = -\frac{\Delta P}{\Delta V / V} \tag{9.5}$$

The **density** of a homogeneous substance is defined as its ratio of mass per unit volume. The value of density is characteristic of a particular type of material and independent of the total quantity of material in the sample.

$$\rho \equiv \frac{M}{V} \tag{9.6}$$

Conversion of Units

The SI units of density are kg per cubic meter.

$$1\ \mathrm{g/cm^3} = 1000\ \mathrm{kg/m^3}$$

The (average) pressure of a fluid is defined as the normal force per unit area acting on a surface immersed in the fluid.

$$P \equiv F / A \tag{9.7}$$

Atmospheric pressure is often expressed in other units:

Conversion of Units

atmosphere: $1\ \mathrm{atm} = 1.01 \times 10^5\ \mathrm{Pa}$

mm of mercury (Torr): $1\ \mathrm{Torr} = 133.3\ \mathrm{Pa}$

pounds per sq. inch: $1\ \mathrm{lb/in^2} = 6.9 \times 10^3\ \mathrm{Pa}$

The absolute pressure, P, at a depth, h, below the surface of a liquid which is open to the atmosphere is greater than atmospheric pressure, P_0, by an amount which depends on the depth below the surface.

$$P = P_0 + \rho g h \quad (9.10)$$

$$P_0 = 1.013 \times 10^5 \text{ Pa}$$

The quantity ρgh is called the gauge pressure and P is the absolute pressure. Therefore,

absolute pressure = atmospheric pressure + gauge pressure

Pascal's law states that pressure applied to an enclosed fluid (liquid or gas) is transmitted undiminished to every point within the fluid and over the walls of the vessel which contain the fluid.

Comments on fluid pressure.

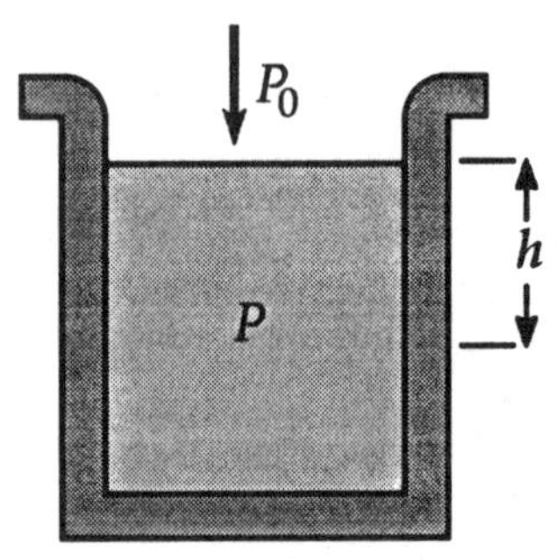

Archimedes' principle states that when an object is partially or fully submersed in a fluid, the fluid exerts an upward buoyant force on the object. The magnitude of the buoyant force depends on the density of the fluid and the volume of displaced fluid, V. In particular, note that B equals the weight of the displaced fluid.

$$B = \rho_f V g \quad (9.11)$$

Archimedes' principle

Fluid dynamics, the treatment of fluids in motion, is greatly simplified under the assumption that the fluid is ideal with the following characteristics:

Comments on ideal fluids.

(a) nonviscous - internal friction between adjacent fluid layers is negligible

(b) incompressible - the density throughout fluid is constant.

(c) steady - the velocity, density, and pressure at each point in the fluid are constant in time

(d) irrotational (without turbulence) - there are no eddy currents within the fluid (each element of the fluid has zero angular velocity about its center)

Equation 9.13 is the equation of continuity. For an **incompressible fluid** ($\rho =$ constant), the equation of continuity can be written as Eq. 9.14. The product Av is called the flow rate. Therefore, the flow rate at any point along a pipe carrying an incompressible fluid is constant.

$$\rho_1 A_1 v_1 = \rho_2 A_2 v_2 \tag{9.13}$$

$$A_1 v_1 = A_2 v_2 \tag{9.14}$$

(for incompressible fluid)

Bernoulli's equation is the most fundamental law in fluid mechanics. It is a statement of the law of conservation of mechanical energy as applied to a fluid. Bernoulli's equation states that the sum of pressure, kinetic energy per unit volume, and potential energy per unit volume remains constant along a streamline of an ideal fluid.

$$P + \frac{1}{2}\rho v^2 + \rho g y = \text{constant} \tag{9.16}$$

The **surface tension,** γ, in a film of liquid is defined as the ratio of the magnitude of the force along the surface, **F**, to the length along which the force acts.

$$\gamma \equiv \frac{F}{L} \tag{9.18}$$

In a capillary tube, the angle ϕbetween the solid surface and a line drawn tangent to the liquid at the surface is called the **angle of contact**. Note that ϕ is less than 90° for any substance in which adhesive forces are stronger than cohesive forces and greater than 90° if cohesive forces predominate.

$$h = \frac{2\gamma}{\rho g r}\cos\phi \tag{9.21}$$

The **coefficient of viscosity,** η (the lowercase Greek letter **eta**), for the fluid is defined as the ratio of the shearing stress to the rate of change of the shear strain. The SI units of viscosity are $\mathrm{N \cdot s / m^2}$.

$$\eta \equiv \frac{Fl}{Av} \tag{9.22}$$

$$1\ \text{poise} = 10^{-1}\ \mathrm{N \cdot s / m^2} \tag{9.23}$$

The onset of turbulence in a tube is determined by a factor called the **Reynolds number**, which is a function of the density of the fluid, the average speed of the fluid along the direction of flow, the diameter of the tube, and the viscosity of the fluid. In the region between 2000 and 3000, the flow is unstable, meaning that any small disturbance will cause its motion to change from streamline to turbulent flow.

$$RN = \frac{\rho v d}{\eta} \tag{9.25}$$

Streamline Flow:
$RN < \sim 2000$

Unstable Flow:
$RN = 2000 — 3000$

Turbulent Flow:
$RN > 3000$

The basic equation for diffusion is Fick's law. The left side of this equation is called the diffusion rate and is a measure of the mass being transported per unit time. The **rate of diffusion** is proportional to the cross-sectional area A and to the change in concentration per unit distance, $(C_2 - C_1)/L$, which is called the concentration gradient. The proportionality constant D is called the **diffusion coefficient** and has units of square meters per second.

$$\text{Diffusion Rate} = DA\left(\frac{C_2 - C_1}{L}\right) \tag{9.26}$$

As a sphere falls through a viscous medium, three forces act on it: the force of frictional resistance, the buoyant force of the fluid, and the weight of the sphere. When the net upward force balances the downward weight force, the sphere reaches terminal speed.

$$v_t = \frac{2r^2 g}{9\eta}\left(\rho - \rho_f\right) \tag{9.28}$$

In a centrifuge those particles having the greatest mass will have the largest terminal speed. Therefore, the most massive particles will settle out of the mixture. The factor of $k$is a coefficient of frictional resistance which must be determined experimentally.

$$v_t = \frac{m\omega^2 r}{k}\left(1 - \frac{\rho_f}{\rho}\right) \tag{9.39}$$

REVIEW CHECKLIST

- ▷ Describe the three types of deformations that can occur in a solid, and define the elastic modulus that is used to characterize each: (Young's modulus, Shear modulus, and Bulk modulus).
- ▷ Understand the concept of pressure at a point in a fluid, and the variation of pressure with depth. Understand the relationships among absolute, gauge, and atmospheric pressure values; and know the several different units commonly used to express pressure.
- ▷ Understand the origin of buoyant forces; and state and explain Archimedes' principle.
- ▷ State and understand the physical significance of the **equation of continuity** (constant flow rate) and **Bernoulli's equation** for fluid flow (relating **flow velocity, pressure,** and **pipe elevation**).

SOLUTIONS TO SELECTED END-OF-CHAPTER PROBLEMS

5. Bone has a Young's modulus of about 14.5 x 10^9 Pa. Under compression, it can withstand a stress of about 160 x 10^6 Pa before breaking. Assume that a femur (thigh bone) is 0.50 m long and calculate the amount of compression this bone can withstand before breaking.

Solution Young's Modulus for a material is defined as $Y = \text{stress}/\text{strain}$, where the tensile strain is the ratio of the change in length, ΔL to the original length, L_0. Thus, we can calculate the compression the material experiences as

$$Y = \frac{\text{stress}}{\Delta L / L_0} \qquad \text{or} \qquad \Delta L = \frac{L_0(\text{stress})}{Y}$$

The original length of the femur is $L_0 = 0.50\text{ m}$ and the femur breaks when the stress exceeds $160 \times 10^6\text{ Pa}$. Therefore, the maximum compression the femur can withstand is:

$$\Delta L = \frac{(0.50\text{ m})(160 \times 10^6\text{ Pa})}{14.5 \times 10^9\text{ Pa}} = 5.5 \times 10^{-3}\text{ m} = 5.5\text{ mm} \qquad \Diamond$$

9. A 50.0 kg ballet dancer stands on her toes during a performance with four square inches (26.0 cm^2) in contact with the floor. What is the pressure exerted by the floor over the area of contact (a) if the dancer is stationary, and (b) if the dancer is leaping upwards with an acceleration of 4.00 m/s^2?

Solution

Two forces, both vertical, act on the body of the dancer as indicated in the sketch. The force F is the normal force exerted, by the floor, on the dancer's toe. From Newton's second law, $\sum F_y = +F - F_g = ma_y$

or $$F = F_g + ma_y = m(g + a_y)$$

(a) If the dancer is stationary, $a_y = 0$ and the normal force is

$$F = m(g + a_y) = (50.0\text{ kg})(9.80\text{ m/s}^2 + 0) = 490\text{ N}$$

The pressure is then

$$P = \frac{F}{A} = \frac{490\text{ N}}{26.0\text{ cm}^2}\left[\frac{100\text{ cm}}{1.00\text{ m}}\right]^2 = 1.88 \times 10^5\text{ N/m}^2 = 1.88 \times 10^5\text{ Pa} \quad \lozenge$$

(b) When $a_y = +4.00\text{ m/s}^2$, the force exerted by the floor is

$$F = (50.0\text{ kg})(9.80\text{ m/s}^2 + 4.00\text{ m/s}^2) = 690\text{ N}$$

and the pressure is:

$$P = \frac{F}{A} = \frac{690\text{ N}}{26.0\text{ cm}^2}\left[\frac{100\text{ cm}}{1.00\text{ m}}\right]^2 = 2.65 \times 10^5\text{ N/m}^2 = 2.65 \times 10^5\text{ Pa} \quad \lozenge$$

17. A container is filled to a depth of 20.0 cm with water. On top of the water floats a 30.0-cm-thick layer of oil with specific gravity 0.700. What is the absolute pressure at the bottom of the container?

Solution If P_0 is the known absolute pressure at some chosen reference level, the absolute pressure at a depth h, in a fluid of density ρ, below that reference level is $P = P_0 + \rho g h$ where g is the acceleration due to gravity. In this problem, there are multiple fluids involved and it may appear ambiguous as to what density (i.e., density of air, oil or water) should be used in this equation for the pressure.

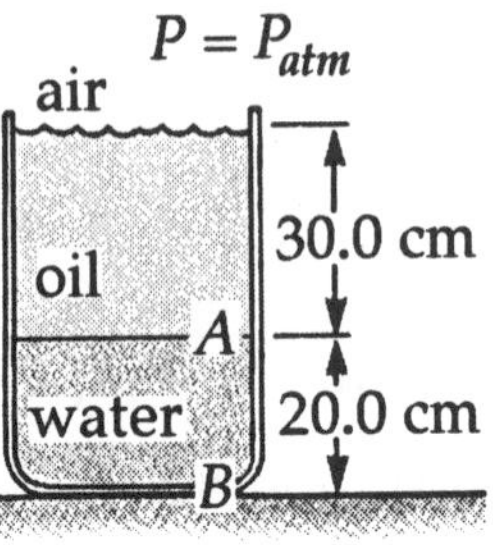

The answer to this dilemma is to deal with one fluid at a time. To find the absolute pressure at the bottom of the water layer (i.e., at the bottom of the container), first compute the absolute pressure at the boundary between the oil and water.

Atmospheric pressure exists at the upper surface of the oil (at the air-oil boundary). Choosing this boundary as the reference level, the absolute pressure at point A on the oil-water boundary is:

$$P_A = P_{\text{atm}} + \rho_{\text{oil}}\, g\, h_{\text{oil}}$$

Oil's specific gravity is $\quad (s.\ g.)_{\text{oil}} = \dfrac{\rho_{\text{oil}}}{\rho_{\text{water}}} = 0.700$

Thus, $\quad \rho_{\text{oil}} = (0.700)\rho_{\text{water}} = (0.700)\left(1000\ \text{kg/m}^3\right) = 700\ \text{kg/m}^3$

Hence, $\quad P_A = 1.013\times10^5\ \text{Pa} + \left(700\ \text{kg/m}^3\right)\left(9.80\ \text{m/s}^2\right)(0.300\ \text{m}) = 1.03\times10^5\ \text{Pa}$

Now that the pressure at point A is known, the oil-water boundary can be chosen as the new reference level (so $P_0 = P_A = 1.03\times10^5$ Pa) and the absolute pressure at point B on the bottom of the container can be computed as $P_B = P_A + \rho_{\text{water}}\, g\, h_{\text{water}}$. This yields

$$P_B = 1.03\times10^5\ \text{Pa} + \left(1000\ \text{kg/m}^3\right)\left(9.80\ \text{m/s}^2\right)(0.200\ \text{m}) = 1.05\times10^5\ \text{Pa} \quad \Diamond$$

23. An empty rubber balloon has a mass of 0.0120 kg. The balloon is filled with helium at a density of 0.181 kg/m^3. At this density the balloon is spherical with a radius of 0.500 m. If the filled balloon is fastened to a vertical line, what is the tension in the line?

Solution

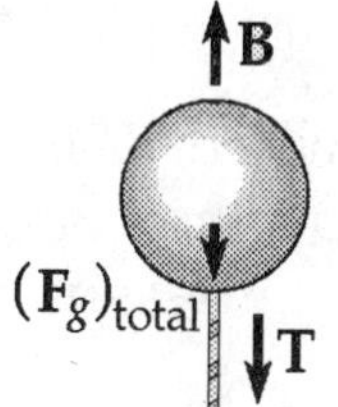

A free-body diagram of the balloon is shown in the sketch. The force T is the tension in the line, B is the buoyant force exerted on the balloon by the air surrounding it, and the total weight of the balloon and the helium filling it is

$$\left(\mathbf{F}_g\right)_{\text{total}} = \left(\mathbf{F}_g\right)_{\text{balloon}} + \left(\mathbf{F}_g\right)_{\text{helium}}$$

Since the balloon is in equilibrium, $\sum F_y = +B - \left(F_g\right)_{\text{total}} - T = 0$

or the tension in the line is: $T = B - \left(F_g\right)_{\text{total}}$ **[Equation 1]**

The weight of the balloon itself is

$$\left(F_g\right)_{\text{balloon}} = m_{\text{balloon}}\, g = (0.0120 \text{ kg})\left(9.80 \text{ m/s}^2\right) = 0.118 \text{ N}$$

The volume of helium is $V = \frac{4}{3}\pi R_{\text{balloon}}^3 = \frac{4}{3}\pi(0.500 \text{ m})^3 = 0.524 \text{ m}^3$

and $\left(F_g\right)_{\text{helium}} = \rho_{\text{helium}}\, gV = \left(0.181 \text{ kg/m}^3\right)\left(9.80 \text{ m/s}^2\right)\left(0.524 \text{ m}^3\right) = 0.929 \text{ N}$

From Archimedes' principle, the buoyant force experienced by the balloon is equal to the weight of the air it displaces. This is:

$$\mathbf{B} = m_{\text{air}} g = \left(\rho_{\text{air}} V\right) g = \left(1.29 \text{ kg /m}^3\right)\left(0.524 \text{ m}^3\right)\left(9.80 \text{ m/s}^2\right) = 6.62 \text{ N}$$

Equation 1 then gives the tension in the line holding the balloon as

$$T = 6.62 \text{ N} - (0.118 \text{ N} + 0.929 \text{ N}) = 5.57 \text{ N}$$ ◊

29. A rectangular air mattress is 2.0 m long, 0.50 m wide, and 0.08 m thick. If it has a mass of 2.0 kg, what additional mass can it support in water?

Solution The mattress, with its added mass, is in equilibrium as it floats on the water. Thus,

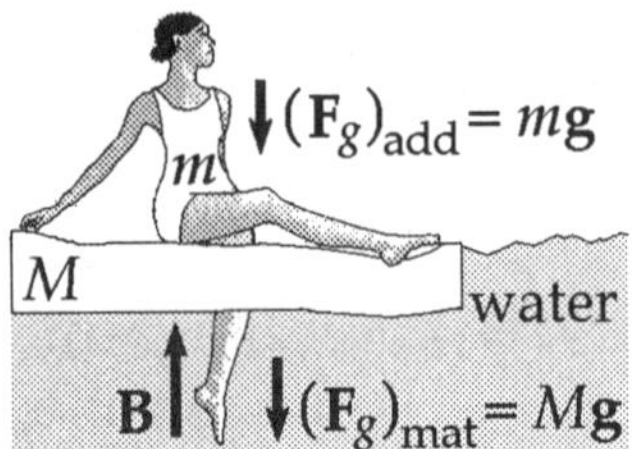

$$\sum F_y = +B - \left(F_g\right)_{mat} - \left(F_g\right)_{add} = 0$$

where B is the buoyant force exerted on the mattress by the water. The additional mass supported by the mattress is then

$$m = \frac{\left(F_g\right)_{add}}{g} = \frac{B - \left(F_g\right)_{mat}}{g} = \frac{B}{g} - M$$

This is a maximum when the buoyant force has its maximum possible value. Since the buoyant force equals the weight of the displaced water, the maximum buoyant force occurs when the mattress floats with its top even with the water surface. Then, the entire volume of the mattress is under water. The maximum buoyant is therefore is $B_{max} = \left(\rho_{water} V\right) g$, so

$$\frac{B_{max}}{g} = \rho_{water} V = \left(1000 \text{ kg/m}^3\right)\left[(2.0 \text{ m})(0.50 \text{ m})(0.08 \text{ m})\right] = 80 \text{ kg}$$

Hence, the maximum additional mass is $m_{max} = 80 \text{ kg} - 2.0 \text{ kg} = 78 \text{ kg}$ ◊

---

35. A liquid (ρ = 1.65 g/cm^3) flows through two horizontal sections of tubing joined end to end. In the first section the cross-sectional area is 10.0 cm^2, the flow speed is 275 cm/s, and the pressure is 1.20×10^5 Pa. In the second section the cross-sectional area is 2.50 cm^2. Calculate the smaller section's (a) flow speed and (b) pressure.

Solution The situation described is illustrated in the sketch at the right. The given values are:

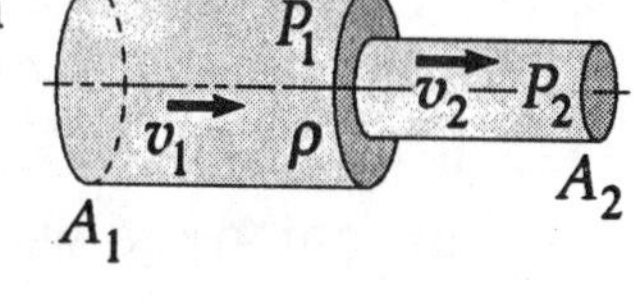

$P_1 = 1.20\times10^5$ Pa $\qquad v_1 = 275$ cm/s $= 2.75$ m/s

$A_1 = 10.0$ cm^2 $\qquad A_2 = 2.50$ cm^2

The density of the flowing liquid is

$$\rho = \left(1.65\text{ g/cm}^3\right)\left(\frac{1\text{ kg}}{1000\text{ g}}\right)\left(\frac{100\text{ cm}}{1\text{ m}}\right)^3 = 1.65\times10^3\text{ kg/m}^3$$

(a) From the equation of continuity, $A_2v_2 = A_1v_1$, the flow speed in the second section of tubing is:

$$v_2 = \frac{A_1}{A_2}v_1 = \left(\frac{10.0\text{ cm}^2}{2.50\text{ cm}^2}\right)(275\text{ cm/s}) = 1.10\times10^3\text{ cm/s} = 11.0\text{ m/s} \qquad \Diamond$$

(b) Bernoulli's equation — a consequence of the principle of conservation of energy — states that $P + \frac{1}{2}\rho v^2 + \rho g y = \text{constant}$, where P is the pressure exerted by the fluid, ρ is the fluid density, v is the flow speed, and y is the elevation above some reference level. Consider two points of the center line of the tubes, one point in the larger tube and one point in the smaller tube. Then Bernoulli's equation requires that

$$P_2 + \tfrac{1}{2}\rho v_2^2 + \rho g y_2 = P_1 + \tfrac{1}{2}\rho v_1^2 + \rho g y_1$$

Since both sections of tubing are horizontal, $y_2 = y_1$ and the potential energy terms cancel. The pressure in the smaller tube is therefore

$$P_2 = P_1 + \tfrac{1}{2}\rho\left(v_1^2 - v_2^2\right)$$

or $\qquad P_2 = 1.20\times10^5\text{ Pa} + \tfrac{1}{2}\left(1650\text{ kg/m}^3\right)\left[(2.75\text{ m/s})^2 - (11.0\text{ m/s})^2\right]$

Thus, $\qquad P_2 = 2.64\times10^4$ Pa $\qquad \Diamond$

---

39. A large storage tank, open to the atmosphere at the top and filled with water, develops a small hole in its side at a point 16.0 m below the water level. If the rate of flow from the leak is 2.50×10^{-3} m³/min, determine (a) the speed at which the water leaves the hole and (b) the diameter of the hole.

Solution (a) To determine the speed of the water as it emerges from the hole in the tank, consider two points in the water and make use of Bernoulli's equation:

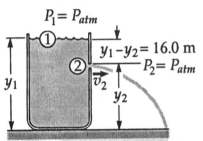

$$P_2 + \tfrac{1}{2}\rho v_2^2 + \rho g y_2 = P_1 + \tfrac{1}{2}\rho v_1^2 + \rho g y_1$$

$P_1 = P_{atm}$, $y_1 - y_2 = 16.0$ m, $P_2 = P_{atm}$, y_1, v_2, y_2

Choose the first point at the upper surface of the water and the second point in the center of the hole in the tank. Note that the water is open to the atmosphere at both of these points, so $P_2 = P_1$ and the pressure terms in Bernoulli's equation cancel. Assuming a very large tank, the speed of the water at the upper surface (i.e., at point 1) is negligible in comparison to its speed as it leaves the tank at point 2. Thus, make the approximation $v_1 \approx 0$, and reduce Bernoulli's equation to find the speed of the emerging water, v_2:

$\tfrac{1}{2}\rho v_2^2 + \rho g y_2 = \rho g y_1$:

$$v_2^2 = 2g(y_1 - y_2) = 2(9.80\text{ m/s}^2)(16.0\text{ m})$$

$$v_2 = \sqrt{314\text{ m}^2/\text{s}^2} = 17.7\text{ m/s} \qquad \Diamond$$

(b) The flow rate of a fluid through an opening is equal to the product Av, where A is the cross-sectional area of that opening and v is the flow speed as the fluid passes through the opening. Hence, if the rate of flow from the leak is $2.50\times 10^{-3}\text{ m}^3/\text{min}$, the cross-sectional area of the hole is

$$A_2 = \frac{\text{flow rate}}{v_2} = \frac{2.50\times 10^{-3}\text{ m}^3/\text{min}}{17.7\text{ m/s}}\left(\frac{1\text{ min}}{60\text{ s}}\right) = 2.35\times 10^{-6}\text{ m}^2$$

But, $A_2 = \pi r^2 = \pi d^2/4$ where d is the diameter of the hole. Hence,

$$d = \sqrt{4\frac{A_2}{\pi}} = \sqrt{(4)\frac{2.35\times 10^{-6}\text{ m}^2}{\pi}} = 1.73\times 10^{-3}\text{ m} = 1.73\text{ mm} \qquad \Diamond$$

45. Whole blood has a surface tension of 0.058 N/m and a density of 1050 kg/m^3. To what height can whole blood rise in a capillary blood vessel that has a radius of 2.0 x 10^{-6} m if the contact angle is zero?

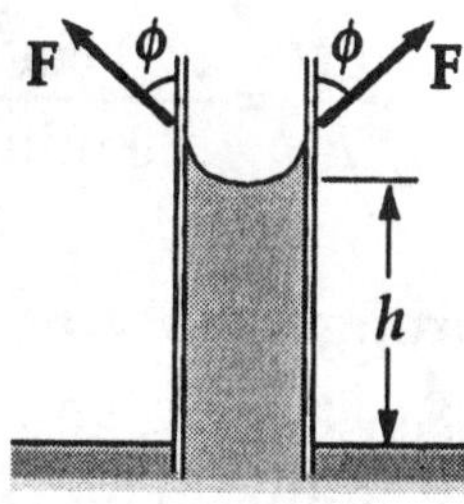

Solution The surface tension γ of a fluid is defined as the tension force per unit length (tangential to the fluid surface and tending to cause that surface to contract) along any line drawn of the surface of the fluid. Consider the line along which the upper surface of a fluid in a capillary tube meets the wall of that tube as shown in the sketch. This line has a length equal to the circumference of the tube (i.e., $L = 2\pi r$). The surface (and hence the tension force) makes an angle ϕ, known as the contact angle, with the vertical wall of the tube at points on this line. The total **upward** force the tube wall exerts on the fluid is then $F = \gamma L \cos\phi = 2\pi\gamma r \cos\phi$. The fluid then rises until the weight of the lifted fluid $w = mg = \rho V g = \rho(\pi r^2 h)g$ equals the upward force F. When equilibrium is reached, we then have

$$\rho(\pi r^2 h)g = 2\pi\gamma r \cos\phi \qquad \text{or} \qquad h = \frac{2\gamma\cos\phi}{\rho g r}$$

If the contact angle is zero when whole blood rises in a capillary blood vessel, the height to which blood will rise in a vessel with a radius of $r = 2.0\times10^{-6}$ m is:

$$h = \frac{2(0.058 \text{ N/m})\cos 0°}{(1050 \text{ kg/m}^3)(9.80 \text{ m/s}^2)(2.0\times10^{-6} \text{ m})} = 5.6 \text{ m} \qquad \Diamond$$

53. What diameter needle should be used to inject a volume of 500 cm^3 of a solution into a patient in 30 min? Assume a needle length of 2.5 cm and that the solution is elevated 1.0 m above the point of injection. Furthermore, assume the viscosity and density of the solution are those of pure water and that the pressure inside the vein is atmospheric.

Solution The sketch illustrates the intravenous injection of a fluid into a patient. In this case, it is desired to have a flow rate of

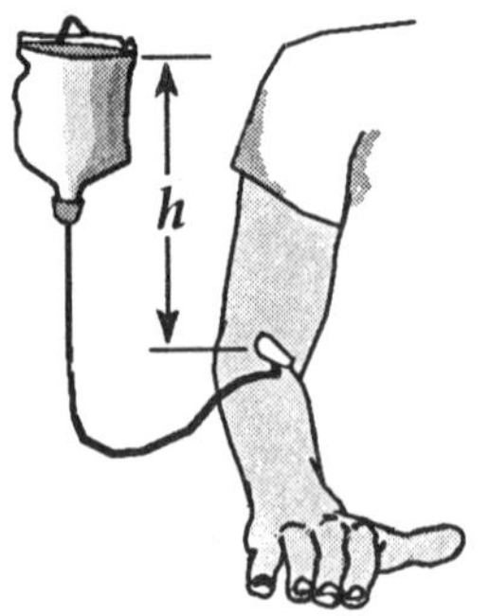

$$\frac{\Delta V}{\Delta t} = \frac{500\ \text{cm}^3}{30\ \text{min}}\left(\frac{1\ \text{min}}{60\ \text{sec}}\right)\left(\frac{1\ \text{m}}{100\ \text{cm}}\right)^3 = 2.8\times 10^{-7}\ \text{m}^3/\text{s}$$

through a needle of length $L = 2.5\times 10^{-2}$ m for a fluid of viscosity $\eta = 1.0\times 10^{-3}\ \text{N}\cdot\text{s/m}^2$ (see Table 9.4 in the text).
Poiseuille's law states that the flow rate of a fluid through a cylindrical tube is $\Delta V/\Delta t = (P_1 - P_2)\pi R^4/8L\eta$, where $(P_1 - P_2)$ is the pressure difference across the tube. In this case, the pressure in the vein is assumed to be atmospheric. Thus, the pressure difference across the needle is $(P_1 - P_2) = (P_1 - P_{atm})$, or it is the same as the **gauge** pressure at the point where the fluid enters the needle. The fluid enters the needle at a distance $h = 1.0$ m below the surface of the fluid in the IV bag. The absolute pressure at this depth below the surface is given by $P_1 = P_0 + \rho g h$ where P_0 is the pressure at the upper surface of the fluid and ρ is the fluid's density. The collapsible bag used in intravenous injections insures that the pressure P_0 is atmospheric. Thus, the pressure difference across the needle in this case becomes

$$(P_1 - P_2) = \rho g h = \left(1000\ \text{kg/m}^3\right)\left(9.8\ \text{m/s}^2\right)\left(1.0\ \text{m}\right) = 9.8\times 10^3\ \text{N/m}^2$$

The required radius of the needle is found from Poiseuille's law as

$$R^4 = \frac{8L\eta(\text{flow rate})}{\pi(P_1 - P_2)} = \frac{8\left(2.5\times 10^{-2}\ \text{m}\right)\left(1.0\times 10^{-3}\ \text{N}\cdot\text{s/m}^2\right)\left(2.8\times 10^{-7}\ \text{m}^3/\text{s}\right)}{\pi\left(9.8\times 10^3\ \text{N/m}^2\right)}$$

This gives $R^4 = 1.8\times 10^{-15}\ \text{m}^4$, and $R = 2.06\times 10^{-4}\ \text{m} = 0.21\ \text{mm}$. A needle diameter of $d = 2R = 0.41\ \text{mm}$ is needed to provide the desired flow rate. ◊

59. Small spheres of diameter 1.00 mm fall through 20° C water with a terminal speed of 1.10 cm/s. Calculate the density of the spheres.

Solution As a sphere falls through the water, three forces act on it:

(1) The weight of the sphere, $w = mg = \rho_s V g = \rho_s g\left(\frac{4}{3}\pi r^3\right)$ where ρ_s is the density of the sphere;

(2) A buoyant force, equal to the weight of the displaced water, $B = m_{\text{water}} g = \rho_w V g = \rho_w g\left(\frac{4}{3}\pi r^3\right)$ where ρ_w is the water's density;

(3) A viscous resistance force given by Stoke's law as $F = 6\pi\eta r v$. Here η is the viscosity of the water and v is the speed of the falling sphere.

The sphere reaches terminal speed (i.e., ceases to accelerate) when the net force acting on it becomes zero. Thus, at the terminal speed $(v = v_t)$, the total upward force, F and B, and the downward gravitational force, w, are equal.

That is, $F + B = w$: $\quad 6\pi\eta r v_t + \rho_w g\left(\frac{4}{3}\pi r^3\right) = \rho_s\, g\left(\frac{4}{3}\pi r^3\right)$

which reduces to $\quad \rho_s = \rho_w + \dfrac{9\eta v_t}{2r^2 g}$ **[Equation 1]**

Given values of

$$\eta = 1.00\times10^{-3}\ \mathrm{N\cdot s/m^2} \qquad \text{(Table 9.4)}$$

$$\rho_w = 1000\ \mathrm{kg/m^3} \qquad \text{(Table 9.2)}$$

$$r = \tfrac{1}{2}(\text{diameter}) = \tfrac{1}{2}\left(1.00\times10^{-3}\ \mathrm{m}\right) = 5.00\times10^{-4}\ \mathrm{m}$$

and

$$v_t = 1.10\ \mathrm{cm/s} = 1.10\times10^{-2}\ \mathrm{m/s}$$

Equation 1 becomes $\quad \rho_s = 1000\ \mathrm{kg/m^3} + \dfrac{9\left(1.00\times10^{-3}\ \mathrm{N\cdot s/m^2}\right)\left(1.10\times10^{-2}\ \mathrm{m/s}\right)}{2\left(5.00\times10^{-4}\ \mathrm{m}\right)^2\left(9.80\ \mathrm{m/s^2}\right)}$

Thus, the density of the sphere is $\rho_s = 1.02\times10^3\ \mathrm{kg/m^3}$ ◊

73. A solid copper ball with a diameter of 3.00 m at sea level is placed at the bottom of the ocean, at a depth of 10.0 km. If the density of the seawater is 1030 kg/m^3, how much does the diameter of the ball decrease when it reaches bottom?

Solution At sea level, the pressure is $P_i = P_{atm}$. At a depth of 10 000 m below the surface of the ocean, the pressure is $P_f = P_0 + \rho_w g h$ where P_0 is the pressure at the surface of the ocean (i.e., $P_0 = P_{atm}$), ρ_w is the density of the seawater, $g = 9.80\ \text{m/s}^2$, and $h = \text{depth} = 1.0 \times 10^4$ m. Thus, the change in pressure experienced by the ball as it sinks is

$$\Delta P = P_f - P_i = \rho_w g h = \left(1030\ \text{kg/m}^3\right)\left(9.80\ \text{m/s}^2\right)\left(1.0 \times 10^4\ \text{m}\right) = 1.01 \times 10^8\ \text{Pa}$$

From the definition of bulk modulus (Equation 9.5 in the text), this increase in pressure on the ball will change the volume of the ball by $\Delta V = -V(\Delta P)/B$ where V is the original volume and B is the bulk modulus of the material (copper) making up the ball. This change in volume may also be written as

$$\Delta V = V_f - V_i = \tfrac{4}{3}\pi r_f^3 - \tfrac{4}{3}\pi r_i^3 = \tfrac{4}{3}\pi\left(r_f^3 - r_i^3\right)$$

Equating the two expressions for ΔV gives $\frac{4}{3}\pi\left(r_f^3 - r_i^3\right) = -\left(\frac{4}{3}\pi r_i^3\right)\Delta P/B$, or $r_f^3 = r_i^3(1 - \Delta P/B)$. Thus, the final radius of the ball is $r_f = r_i\sqrt[3]{1 - \Delta P/B}$ and the decrease in the **diameter** of the ball radius is $\Delta D = 2\left(r_i - r_f\right)$ or $\Delta D = 2r_i\left(1 - \sqrt[3]{1 - \Delta P/B}\right) = D_i\left(1 - \sqrt[3]{1 - \Delta P/B}\right)$ where D_i is the original diameter. From Table 9.1, $B = 14 \times 10^{10}$ Pa for copper, and the decrease in diameter is

$$\Delta D = (3.00\ \text{m})\left(1 - \sqrt[3]{1 - \frac{1.01 \times 10^8\ \text{Pa}}{14 \times 10^{10}\ \text{Pa}}}\right) = 7.2 \times 10^{-4}\ \text{m} = 0.72\ \text{mm}$$ ◊

77. Oil having a density of 930 kg/m^3 floats on water. A rectangular block of wood 4.00 cm high and with a density of 960 kg/m^3 floats partly in the oil and partly in the water. The oil completely covers the block. How far below the interface between the two liquids is the bottom of the block?

Solution Assume the block floats in equilibrium with the bottom of the block a distance x below the oil-water interface. The top of the block is a distance $(4.00\text{ cm} - x)$ above the interface as shown in the sketch. At equilibrium, the total buoyant force equals the weight of the block:

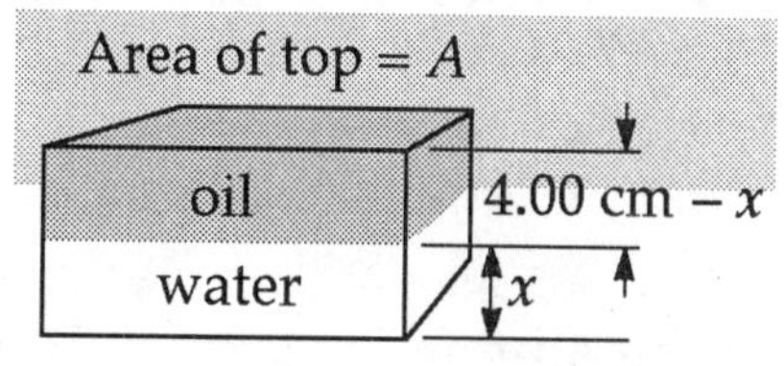

$$w = F_g = B_{\text{total}} = B_{\text{oil}} + B_{\text{water}} \qquad \textbf{[Equation 1]}$$

The block's weight is $F_g = mg = \rho_{\text{wood}}(\text{Volume of block})g = \rho_{\text{wood}}A(4.00\text{ cm})g$

and the individual buoyant forces are

$$B_{\text{oil}} = \text{weight of displaced oil} = \rho_{\text{oil}}(\text{volume of displaced oil})g$$

and $$B_{\text{water}} = \text{weight of displaced water} = \rho_{\text{water}}(\text{volume of displaced water})g$$

Substituting these values into our buoyancy equation, Equation 1,

$$\rho_{\text{wood}}A(4.00\text{ cm})g = \rho_{\text{oil}}[A(4.00\text{ cm} - x)]g + \rho_{\text{water}}[Ax]g$$

Cancelling the cross-sectional area A and the acceleration of gravity g,

$$(\rho_{\text{water}} - \rho_{\text{oil}})x = (\rho_{\text{wood}} - \rho_{\text{oil}})(4.00\text{ cm})$$

Since $\rho_{\text{water}} = 1000\text{ kg/m}^3$ $\quad\rho_{\text{oil}} = 930\text{ kg/m}^3$ $\quad\rho_{\text{wood}} = 960\text{ kg/m}^3$

we can solve for the x, the distance the block's bottom is below the oil-water interface:

$$x = \left[\frac{\rho_{\text{wood}} - \rho_{\text{oil}}}{\rho_{\text{water}} - \rho_{\text{oil}}}\right](4.00\text{ cm}) = \left[\frac{30\text{ kg/m}^3}{70\text{ kg/m}^3}\right](4.00\text{ cm}) = 1.71\text{ cm} \quad \Diamond$$

CHAPTER SELF-QUIZ

1. A liquid-filled tank has a hole on its vertical surface just above the bottom edge. If the surface of the liquid is 0.4 m above the hole, at what speed will the stream of liquid emerge from the hole?
   a. 2.8 m/s
   b. 7.84 m/s
   c. 2.0 m/s
   d. 1.7 m/s

2. An aluminum cube, 0.10 m on an edge, is subjected to a shear force of 750 N while the opposite side is clamped. What is the resultant shear strain? (For aluminum, $S = 2.5 \times 10^{10}$ N/m^2.)
   a. 3.0×10^{-6}
   b. 0.40×10^{-6}
   c. 16.0×10^{-6}
   d. 0.07×10^{-6}

3. A 4.0-kg cylinder made of solid iron is supported by a string while submerged in water. What is the tension in the string? (The density of iron and water are 7.86×10^3 kg / m^3 and 1.0×10^3 kg / m^3; $g = 9.8$ m/s^2.)
   a. 2.5 N
   b. 19.6 N
   c. 23.7 N
   d. 34.2 N

4. A solid rock, suspended in air by a spring scale, has a measured mass of 13.5 kg. When the rock is submerged in water, the scale reads 2.2 kg. What is the density of the rock? (Water density = 1.0×10^3 kg/m^3)
   a. 4.55×10^3 kg/m^3
   b. 3.5×10^3 kg/m^3
   c. 1.2×10^3 kg/m^3
   d. 2.4×10^3 kg/m^3

5. A hydraulic lift raises a 2000-kg automobile when a 500 N force is applied to the smaller piston. If the smaller piston has an area of 10 cm^2, what is the cross-sectional area of the larger piston?
   a. 40 cm^2
   b. 80 cm^2
   c. 196 cm^2
   d. 392 cm^2

6. A wire of length 4 meters, cross-sectional area 0.2×10^{-4} m^2 and Young's modulus 8×10^{10} N/m^2 has a 200-kg load hung on it. What is its increase in length? (Use $g = 9.8$ m/s^2)
   a. 0.10×10^{-3} m
   b. 0.49×10^{-3} m
   c. 4.9×10^{-3} m
   d. 9.8×10^{-3} m

7. What is the total pressure at the bottom of a 5-m deep swimming pool? (Note the pressure contribution from the atmosphere is 1.01×10^5 N/m^2, the density of water is 10^3 kg/m^3, and $g = 9.8$ m/s^2.)
   a. 1.5×10^5 N/m^2
   b. 0.72×10^5 N/m^2
   c. 0.49×10^5 N/m^2
   d. 1.01×10^5 N/m^2

8. Water (density = 1×10^3 kg/m^3) is flowing through a pipe whose radius is 0.04 m with a speed of 15 m/s. This same pipe goes up to the second floor of the building, 3 m higher, and the pressure remains unchanged. What is the radius of the pipe on the second floor?
   a. 0.046 m
   b. 0.043 m
   c. 0.037 m
   d. 0.034 m

9. A uniform pressure of 3.5×10^5 N/m^2 is applied to all six sides of an aluminum cube. What is the percentage change in volume of the cube? (For aluminum, $B = 7 \times 10^{10}$ N/m^2)
   a. $2.4 \times 10^{-2}\%$
   b. $0.4 \times 10^{-2}\%$
   c. $8.4 \times 10^{-2}\%$
   d. $0.5 \times 10^{-3}\%$

10. When a force **F** is applied to a certain rod, there is a 1 cm change in length. If that same force is applied to a rod that is three times bigger in each dimension, what will be the change in length?
    a. 9 cm
    b. 3 cm
    c. 1 cm
    d. 0.33 cm

11. A glass container is half filled with mercury and a steel ball is floating on the mercury. If water is then poured on top of the steel ball and mercury, filling the glass, what will happen to the steel ball?
    a. It will float higher in the mercury.
    b. It will float at the same height.
    c. It will float lower in the mercury.
    d. It will sink to the bottom of the mercury.

12. If the flow rate of a liquid going through a 2.00-cm radius pipe is measured at 0.8×10^{-3} m^3/s, the average fluid velocity in the pipe is which of the following?
    a. 0.64 m/s
    b. 2.0 m/s
    c. 0.04 m/s
    d. 6.3 m/s

Chapter 10

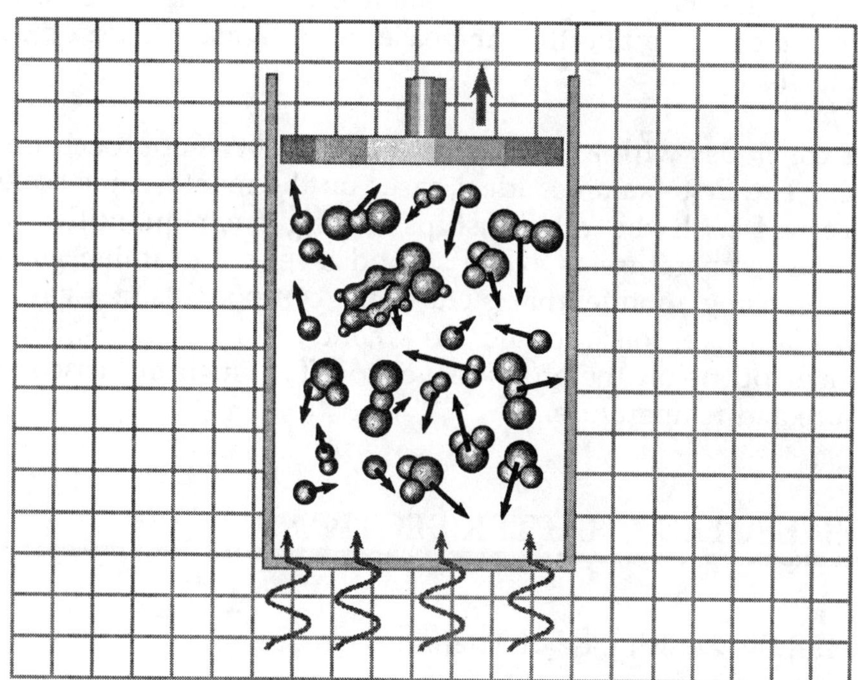

THERMAL PHYSICS

Chapter 10

THERMAL PHYSICS

Our study thus far has focused exclusively on mechanics. Such concepts as mass, force, and kinetic energy have been carefully defined in order to make the subject quantitative. We now move to a new branch of physics, thermal physics. Here we shall find that quantitative descriptions of thermal phenomena require careful definitions of the concepts of temperature, heat and internal energy. We shall examine the process of linear expansion in which a temperature increase or decrease for an object results in a change in size.

This chapter concludes with a study of ideal gases. We approach this study on two levels. The first examines ideal gases on the macroscopic scale. Here we are concerned with the relationships among such quantities as pressure, volume, and temperature. On the second level we examine gases on a microscopic scale, using a model that pictures the components of a gas as small particles. This latter approach, called the kinetic theory of gases, helps us understand what happens on the atomic level to affect such macroscopic properties as pressure and temperature.

NOTES FROM SELECTED CHAPTER SECTIONS

10.1 Temperature and the Zeroth Law of Thermodynamics

The zeroth law of thermodynamics (or the equilibrium law) can be stated as follows:

> If bodies A and B are separately in thermal equilibrium with a third body, C, then A and B will be in thermal equilibrium with each other if placed in thermal contact.

Two objects in thermal equilibrium with each other are at the same temperature.

10.2 Thermometers and Temperature Scales

The physical property used in a constant volume gas thermometer is the pressure variation with temperature of a fixed volume of gas. The temperature readings are nearly independent of the substance used in the thermometer.

The triple point of water, which is **the single temperature and pressure at which water, water vapor, and ice can coexist in equilibrium,** was chosen as a convenient and reproducible reference temperature for the Kelvin scale. It occurs at a temperature of 0.01°C and a pressure of 4.58 mm of mercury. The temperature at the triple point of water on the Kelvin scale has been assigned a value of 273.16 kelvins (K). Thus, the SI unit of temperature, the kelvin, is defined as **1/273.16 of the temperature of the triple point of water.**

The temperature 0 K is often referred to as absolute zero although this temperature has never been achieved.

10.3 Thermal Expansion of Solids and Liquids

Liquids generally increase in volume with increasing temperature and have volume expansion coefficients about ten times greater than those of solids. Water is an exception to this rule; as the temperature increases from 0 °C to 4 °C, water contracts and thus its density increases. Above 4 °C, water expands with increasing temperature. In other words, the density of water reaches its maximum value (1000 kg/m^3) 4 degrees above the freezing point.

10.5 Avogadro's Number and the Ideal Gas Law

Equal volumes of gas at the same temperature and pressure contain the same numbers of molecules.

Single moles of any gases at standard temperature and pressure contain the same numbers of molecules.

10.6 The Kinetic Theory of Gases

A microscopic **model of an ideal gas** is based on the following assumptions:

1. **The number of molecules is large, and the average separation between them is large** compared with their dimensions. Therefore, the molecules occupy a negligible volume compared with the volume of the container.

2. **The molecules obey Newton's laws of motion, but the individual molecules move in a random fashion.** By random fashion, we mean that the molecules move in all directions with equal probability and with various speeds. This distribution of velocities does not change in time, despite the collisions between molecules.

3. **The molecules undergo elastic collisions with each other.** Thus, the molecules are considered to be structureless (that is, point masses), and in the collisions both kinetic energy and momentum are conserved.

4. **The forces between molecules are negligible, except during a collision.** The forces between molecules are short-range, so that the only time the molecules interact with each other is during a collision.

5. **The gas under consideration is a pure gas.** That is, all molecules are identical.

6. **The gas is in thermal equilibrium with the walls of the container.** Hence, the wall will eject as many molecules as it absorbs, and the ejected molecules will have the same average kinetic energy as the absorbed molecules.

The pressure exerted by each component of a mixture of gases is the same as the pressure that component would exert if it were alone in the volume occupied by the mixture. This means that each component acts virtually independently of the rest of the mixture.

Vapor pressure is **the pressure exerted by a gas when it is in equilibrium with its liquid form.**

The dew point is the temperature at which the air becomes saturated with water vapor.

Boiling begins when the vapor pressure is equal to atmospheric pressure.

EQUATIONS AND CONCEPTS

T_C is the Celsius temperature and T is the Kelvin temperature (sometimes called the absolute temperature). The size of a degree on the Kelvin scale is identical to the size of a degree on the Celsius scale.

$$T_C = T - 273.15 \tag{10.1}$$

These equations are useful in converting temperature values between Fahrenheit and Celsius scales.

$$T_F = \tfrac{9}{5}T_C + 32 \tag{10.2}$$

$$T_C = \tfrac{5}{9}(T_F - 32)$$

The SI unit of temperature is the kelvin, K, which is defined as $\frac{1}{273.16}$ of the temperature of the triple point of water. The notations °C and °F refer to actual temperature values in degrees Celsius and degrees Fahrenheit.

Comment on units and notation.

These are two forms of the basic equation for the thermal expansion of a solid. The change in length is proportional to the original length and to the change in temperature. The constant α is characteristic of a particular type of material and is called the average temperature coefficient of linear expansion.

$$\Delta L = \alpha L_0 \Delta T \tag{10.4}$$

$$L - L_0 = \alpha L_0 (T - T_0)$$

If a body's temperature changes, its surface area and its volume will change by amounts which are proportional to the changes in temperature. γ (gamma) is the average temperature coefficient of area expansion; β (beta) is the average temperature coefficient of volume expansion.

$$\Delta A = \gamma A_0 \Delta T \tag{10.5}$$

$$\Delta V = \beta V_0 \Delta T \tag{10.6}$$

$$\gamma = 2\alpha$$

$$\beta = 3\alpha$$

The number of moles in a sample of an element or compound is the ratio of the mass, m, of the sample to the atomic or molecular mass, M, of the material. Also, one mole of a substance contains Avogadro's number of molecules.

$$n = \frac{m}{M} \tag{10.7}$$

$$N_A = 6.022 \times 10^{23} \ \frac{\text{molecules}}{\text{mole}}$$

Boyle's law states that when a gas is maintained at constant temperature, its pressure is inversely proportional to its volume.

$$P \propto \left(\frac{1}{V}\right); \quad T = \text{constant}$$

The law of Charles-Gay-Lussac states that when constant pressure is maintained, a gas's volume is directly proportional to its absolute temperature.

$$V \propto T; \quad P = \text{constant}$$

This is the equation of state of an ideal gas. In this equation, T must be the temperature in kelvins. R is the universal gas constant and must be expressed in units which are consistent with those used for pressure P and volume V.

$$PV = nRT \qquad (10.8)$$

$$R = 8.31 \text{ J/mol} \cdot \text{K} \qquad (10.9)$$

$$R = 0.0821 \text{ L} \cdot \text{atm/mol} \cdot \text{K}$$

One mole quantities of all gases at standard temperature and pressure contain the same numbers of all molecules. The number of molecules contained in one mole of any gas is Avogadro's number, N_A.

$$N_A = 6.02 \times 10^{23} \frac{\text{molecules}}{\text{mol}}$$

The ideal gas law can also be expressed in this alternate form where N is the total number of molecules in the sample and k is Boltzmann's constant.

$$PV = Nk_BT \qquad (10.12)$$

$$k_B = \frac{R}{N_A} = 1.38 \times 10^{-23} \text{ J/K} \qquad (10.13)$$

In an ideal gas, the pressure of the gas is proportional to the number of molecules per unit volume and proportional to the average kinetic energy of the molecules.

$$P = \frac{2}{3}\left(\frac{N}{V}\right)\left(\frac{1}{2}m\overline{v^2}\right) \qquad (10.14)$$

Equations 10.12 and 10.14 can be combined to present an important result: the average kinetic energy of gas molecules is directly proportional to the absolute temperature of the gas.

$$\frac{1}{2}m\overline{v^2} = \frac{3}{2}k_B T \qquad (10.16)$$

This expression for the root mean square (rms) speed shows that, at a given temperature, lighter molecules move faster on the average than heavier ones.

$$v_{rms} = \sqrt{\overline{v^2}} = \sqrt{\frac{3k_B T}{m}} = \sqrt{\frac{3RT}{M}} \qquad (10.19)$$

REVIEW CHECKLIST

▷ Describe the operation of the constant-volume gas thermometer and how it is used to determine the Kelvin temperature scale. Convert between the various temperature scales, especially the conversion from degrees Celsius into kelvins, degrees Fahrenheit into kelvins, and degrees Celsius into degrees Fahrenheit.

▷ Define the linear expansion coefficient and volume expansion coefficient for an isotropic solid, and understand how to use these coefficients in practical situations involving expansion or contraction.

▷ Understand the assumptions made in developing the molecular model of an ideal gas; and apply the equation of state for an ideal gas to calculate pressure, volume, temperature, or number of moles.

▷ Define each of the following terms: **molecular weight, mole, Avogadro's number, universal gas constant, and Boltzmann's constant.**

Chapter 10

SOLUTIONS TO SELECTED END-OF-CHAPTER PROBLEMS

3. Convert the following temperatures to Fahrenheit and kelvins: (a) the boiling point of liquid hydrogen –252.87°C; (b) the temperature of a room at 20°C.

Solution The Celsius and Kelvin temperature scales have the same size scale divisions, with the only difference being that the zero point on the Kelvin scale is 273.15 scale divisions below the zero point on the Celsius scale. Hence, the conversion from Celsius temperature readings to the corresponding Kelvin temperature is $T_K = T_C + 273.15$ **[Equation 1]**

The Celsius scale divisions are 9/5 (or almost double) the size of the scale divisions on the Fahrenheit thermometer. Also, the zero point on the Celsius thermometer is 32 Fahrenheit divisions below the zero point on the Fahrenheit thermometer (water freezes at 0 °C and 32 °F). Thus, the conversion from Celsius to Fahrenheit is $T_F = \frac{9}{5}T_C + 32\ °\text{F}$ **[Equation 2]**

(a) Using Equations 1 and 2, the Fahrenheit and Kelvin temperatures corresponding to a Celsius temperature of $T_C = -252.87\ °\text{C}$ are

$T_F = \frac{9}{5}(-252.87\ °\text{C}) + 32.0\ °\text{F} = -423\ °\text{F}$ and $T_K = -252.87 + 273.15 = 20.3\ \text{K}$ ◊

(b) For a Celsius temperature of $T_C = 20\ °\text{C}$, the conversions give

$$T_F = \frac{9}{5}(20\ °\text{C}) + 32.0\ °\text{F} = 68\ °\text{F} \quad \text{and} \quad T_K = 20 + 273.15 = 293\ \text{K}$$ ◊

---

5. A constant-volume gas thermometer is calibrated in dry ice (–80.0°C) and in boiling ethyl alcohol (78.0°C). The two pressures are 0.900 atm and 1.635 atm. (a) What value of absolute zero does the calibration yield? (b) What pressures would be found at the freezing and boiling points of water?

Solution When the volume occupied by a low density gas is kept constant, the graph of pressure versus temperature is a straight line. This linear relation is summarized by a equation of the form $P = A + BT$, where A and B are constants. To determine the values of these constants for this gas thermometer, use the two calibration points. The first calibration point ($P = 0.900$ atm at $T = -80.0°\text{C}$) yields the equation:

$$0.900 \text{ atm} = A - (80.0°\text{C})B \qquad \textbf{[Equation 1]}$$

The second calibration point ($P = 1.635$ atm at $T = 78.0°\text{C}$) gives the equation:

$$1.635 \text{ atm} = A + (78.0°\text{C})B \qquad \textbf{[Equation 2]}$$

Subtracting Equation 1 from Equation 2 gives

$$(158.0 \text{ °C})B = 0.735 \text{ atm} \quad \text{or} \quad B = 4.652 \times 10^{-3} \text{ atm/°C}$$

Substituting this result into either Equation 1 or Equation 2 then yields $A = 1.272$ atm. Thus, the equation of the calibration curve for this constant-volume gas thermometer is

$$P = 1.272 \text{ atm} + \left(4.652 \times 10^{-3} \text{ atm/°C}\right)T$$

(a) Gas at absolute zero would exert zero pressure. The calibration curve at $P = 0$ yields:

$$T = -1.272 \text{ atm}\Big/\left(4.652 \times 10^{-3} \text{ atm/°C}\right) = -273.5 \text{ °C} \qquad \Diamond$$

(b) At the freezing point of water, $T = 0$ °C and the calibration curve predicts a pressure of

$$P_f = 1.272 \text{ atm} \qquad \Diamond$$

At the boiling point of water ($T = 100°\text{C}$), the calibration curve yields:

$$P_b = 1.272 \text{ atm} + \left(4.652 \times 10^{-3} \text{ atm/°C}\right)(100 \text{ °C}) = 1.737 \text{ atm} \qquad \Diamond$$

15. At 20.000 °C, an aluminum ring has an inner diameter of 5.000 cm, and a brass rod has a diameter of 5.050 cm. (a) To what temperature must the aluminum ring be heated so that it will just slip over the brass rod? (b) To what temperature must **both** be heated so the aluminum ring will slip off the brass rod? Would this work?

Solution (a) The aluminum ring must be heated sufficiently to cause its diameter (a **linear** dimension of the ring) to expand from 5.000 cm to 5.050 cm. The magnitude, L, of a linear dimension at temperature T is given by $L = L_0[1+\alpha(T-T_0)]$, where L_0 is the magnitude of that dimension at temperature T_0 and α is the coefficient of linear expansion of the material making up the object. The coefficient of linear expansion for aluminum is $\alpha = 24\times10^{-6}\ (°\text{C})^{-1}$, so the temperature at which the diameter of the ring will match that of the brass rod is found from

$$5.050\text{ cm} = (5.000\text{ cm})\left[1+\left(24\times10^{-6}(°\text{C})^{-1}\right)(T-20.000°\text{C})\right]$$

Solving, $T = 437°\text{C}$ ◊

(b) When the ring is placed over the rod and allowed to cool, it attempts to contract. Since the size of the rod will not allow the ring to contract, the ring will be under tension and fit very tightly on the rod. To remove the ring from the rod would require heating **both** ring and rod to a temperature at which they have the same diameter with the ring unstretched (i.e., under zero tension). At temperature T, the diameter of the ring and of the rod are:

$$L_{\text{ring}} = L_{0\text{ring}}\left[1+\alpha_{\text{al}}(T-T_0)\right] \quad \text{and} \quad L_{\text{rod}} = L_{0\text{rod}}\left[1+\alpha_{\text{brass}}(T-T_0)\right]$$

We equate these to find the temperature at which the diameters are the same:

$$(5.00\text{ cm})\left[1+\left(24\times10^{-6}\,°\text{C}^{-1}\right)(T-20°\text{C})\right] = (5.05\text{ cm})\left[1+\left(19\times10^{-6}\,°\text{C}^{-1}\right)(T-20°\text{C})\right]$$

Solving for T, we find that the objects must be heated to $T = 2099$ °C ◊
However, aluminum melts at 660 °C, so the ring will melt before it reaches a diameter at which it could easily be slid off the rod intact. ◊

19. An underground gasoline tank at 54°F can hold 1000 gallons of gasoline. If the driver of a tanker truck fills the underground tank on a day when the temperature is 90°F, how many gallons, according to his measure on the truck, can he pour in? Assume that the temperature of the gasoline cools to 54°F upon entering the tank.

Solution

The volume V the gasoline occupies at temperature T (after entering the tank) and the volume V_0 it had at the initial temperature of T_0 are related by

$$V = V_0\left[1 + \beta(T - T_0)\right]$$

where β is the coefficient of volume expansion. For gasoline, $\beta = 9.6 \times 10^{-4}\ (°\text{C})^{-1}$ (see Table 10.1 in the text).

Upon entering the tank, the gasoline undergoes a change in temperature of

$$\Delta T = T - T_0 = 54°\text{F} - 90°\text{F} = -36°\text{F}$$

Since the scale division on a Fahrenheit thermometer is 5/9 the size of the scale divisions on a Celsius thermometer, the change in Celsius temperature is

$$\Delta T = \tfrac{5}{9}(-36) = -20°\text{C}$$

To completely fill the tank (at 54°F) the gasoline must have a final volume of V =1000 gallons. The volume V_0 the driver will measure before it enters the tank and cools is:

$$V_0 = \frac{V}{1 + \beta(\Delta T)} = \frac{1000 \text{ gal}}{1 + \left(9.6 \times 10^{-4}\ °\text{C}^{-1}\right)(-20\ °\text{C})} = 1020 \text{ gal} \qquad \Diamond$$

23. (a) An ideal gas occupies a volume of 1.0 cm^3 at 20°C and atmospheric pressure. Determine the number of molecules of gas in the container. (b) If the pressure of the 1.0 cm^3 volume is reduced to 1.0×10^{-11} Pa (an extremely good vacuum) while the temperature remains constant, how many moles of gas remain in the container?

Solution (a) The ideal gas law may be written as $PV = nRT$ where P is the absolute pressure (in Pa) of the gas, V is the volume (in m^3) the gas occupies, T is the Kelvin temperature, n is the number of moles of gas present, and $R = 8.31 \text{ J/mol} \cdot \text{K}$ is the molar gas constant. For the given gas,

$$V = 1.0 \text{ cm}^3[1 \text{ m}/100 \text{ cm}]^3 = 1.0 \times 10^{-6} \text{ m}^3$$

$$T = 20°\text{C} = 293 \text{ K}$$

and $$P = 1 \text{ atm} = 1.013 \times 10^5 \text{ Pa}$$

Therefore, $$n = \frac{PV}{RT} = \frac{\left(1.013 \times 10^5 \text{ Pa}\right)\left(1.0 \times 10^{-6} \text{ m}^3\right)}{(8.31 \text{ J/mol} \cdot \text{K})(293 \text{ K})} = 4.16 \times 10^{-5} \text{ mol}$$

This is the number of moles present; since Avogadro's number, $N_A = 6.02 \times 10^{23} \text{mol}^{-1}$, gives the number of molecules in one mole of any gas, the number of molecules in the container is:

$$N = nN_A = \left(4.16 \times 10^{-5} \text{ mol}\right)\left(6.02 \times 10^{23} \text{ mol}^{-1}\right) = 2.5 \times 10^{19}$$ ◊

(b) The equation $P_1V_1 = n_1RT_1$ refers to the initial state of the gas and the equation $P_2V_2 = n_2RT_2$ describes the final state of the gas. Dividing the second equation by the first, recognizing that the volume and temperature are both held constant in this case, gives $P_2/P_1 = n_2/n_1$. Thus, the number of moles of gas remaining in the container is

$$n_2 = \left(\frac{P_2}{P_1}\right)n_1 = \left(\frac{1.0 \times 10^{-11} \text{ Pa}}{1.013 \times 10^5 \text{ Pa}}\right)\left(4.16 \times 10^{-5} \text{ mol}\right) = 4.1 \times 10^{-21} \text{ mol}$$ ◊

29. A cylindrical diving bell, 3.00 m in diameter and 4.00 m tall with an open bottom, is submerged to a depth of 220 m in the ocean. The surface temperature is 25.0°C, and the temperature 220 m down is 5.00°C. The density of seawater is 1025 kg/m^3. How high does the seawater rise in the bell when it is submerged?

Solution At the surface of the ocean, air fills the entire volume of the diving bell. This initial volume is $V_1 = \pi r^2 h_1$, where $h_1 = 4.00\text{ m}$ is the height of the cylindrical space inside the bell. As the bell is lowered into the ocean, the trapped air is subject to the prevailing water pressure since the diving bell is open at the bottom. The pressure at a depth $H = 220\text{ m}$ in the ocean is $P_2 = P_{atm} + \rho_{seawater} g H$, or

$$P_2 = 1.013 \times 10^5\text{ Pa} + \left(1025\text{ kg/m}^3\right)\left(9.80\text{ m/s}^2\right)(220\text{ m}) = 2.31 \times 10^6\text{ Pa}$$

The initial state of the air is described by $P_1 V_1 = n_1 R T_1$ and the final state by $P_2 V_2 = n_2 R T_2$. Since no air enters or leaves the bell as it is lowered, the number of moles of air present is constant $(n_2 = n_1)$. Dividing the two state equations then gives

$$\frac{P_2 V_2}{P_1 V_1} = \frac{T_2}{T_1} \quad \text{or} \quad V_2 = \left(\frac{P_1}{P_2}\right)\left(\frac{T_2}{T_1}\right) V_1$$

Recognizing that the radius of the diving bell remains constant, the new height of the cylindrical space filled by the trapped air is

$$h_2 = \left(\frac{1}{\pi r^2}\right) V_2 = \left(\frac{1}{\pi r^2}\right)\left(\frac{P_1}{P_2}\right)\left(\frac{T_2}{T_1}\right)\pi r^2 h_1 = \left(\frac{P_1}{P_2}\right)\left(\frac{T_2}{T_1}\right) h_1$$

Thus, with initial and final temperatures of $T_1 = 25.0\text{ °C} + 273 = 298\text{ K}$ and $T_2 = 5.00\text{ °C} + 273 = 278\text{ K}$, the final height of the trapped air space is

$$h_2 = \left(\frac{1.013\times10^5\text{ Pa}}{2.31\times10^6\text{ Pa}}\right)\left(\frac{278\text{ K}}{298\text{ K}}\right)(4.00\text{ m}) = 0.164\text{ m}$$

Therefore, as the bell is submerged, the seawater rises a distance of

$$d = h_1 - h_2 = 4.00\text{ m} - 0.164\text{ m} = 3.84\text{ m}$$ ◊

34. (a) What is the total random kinetic energy of all the molecules in one mole of hydrogen at a temperature of 300 K? (b) With what speed would a mole of hydrogen have to move so that the kinetic energy of the mass as a whole would be equal to the total random kinetic energy of its molecules?

Solution Boltzmann's constant is the ratio of the molar gas constant to Avogadro's number:

$$k_B = \frac{R}{N_A} = \frac{8.31\ \text{J/gmol·K}}{6.02\times10^{-23}(\text{gmol})^{-1}} = 1.38\times10^{-23}\ \text{J/K}$$

(a) The average kinetic energy per molecule in a gas that is at an absolute temperature T is

$$\langle KE\rangle_{\text{molecule}} = \tfrac{3}{2}k_B T$$

Since one mole of hydrogen (or any other gas) contains Avogadro's number of molecules, the total kinetic energy associated with random thermal motion in one mole of any gas at 300 K is

$$KE = N_A\langle KE\rangle_{\text{molecule}} = N_A\left(\tfrac{3}{2}\right)\left(\frac{R}{N_A}\right)T = \tfrac{3}{2}RT = \tfrac{3}{2}(8.31\ \text{J/gmol}\cdot\text{K})(300\ \text{K})$$

or $\quad KE = 3.74\times10^3\ \text{J/gmol} = 3.74\ \text{kJ/gmol}$ ◊

(b) Since hydrogen forms diatomic molecules (H_2), the molecular weight of hydrogen is $2\times(\text{atomic weight in grams}) = 2\times(1.008\ \text{g}) = 2.02\ \text{g}$. The speed required for one mole of hydrogen (i.e., a mass of $m = 2.02\times10^{-3}$ kg) to possess a translational kinetic energy equal to the total random kinetic energy found in part (a), is:

$$v^2 = \frac{2KE}{m} = \frac{2\left(3.74\times10^3\ \text{J}\right)}{2.02\times10^{-3}\ \text{kg}} = 3.70\times10^6\ \text{m}^2/\text{s}^2$$

or $$v = 1.92\times10^3\ \text{m/s}$$ ◊

37. The temperature near the top of the atmosphere on Venus is 240 K. (a) Find the rms speed of hydrogen (H_2) at this point in the atmosphere. (b) Repeat for carbon dioxide (CO_2). (c) It has been found that if the rms speed exceeds one sixth of the planet's escape velocity, the gas eventually leaks out of the atmosphere and into outer space. If the escape velocity on Venus is 10.3 km/s, does hydrogen escape? Does carbon dioxide?

Solution The average kinetic energy of the molecules in a gas is $\langle KE\rangle_{\text{molecule}} = \frac{3}{2}k_B T$, where T is the absolute temperature of the gas and k_B is Boltzmann's constant. Note that this expression is independent of the type of gas involved. Thus, near the top of the atmosphere on Venus, the average kinetic energy per molecule is the same for both hydrogen and carbon dioxide.

At $T = 240$ K, $\langle KE\rangle_{\text{molecule}} = \frac{3}{2}\left(1.38\times10^{-23}\text{ J/K}\right)(240\text{ K}) = 4.97\times10^{-21}\text{ J}$

But, also, $\langle KE\rangle_{\text{molecule}} = \frac{2}{3} m_{\text{molecule}} v_{\text{rms}}^2$

so $$v_{\text{rms}} = \sqrt{\frac{2\langle KE\rangle_{\text{molecule}}}{m_{\text{molecule}}}}$$ **[Equation 1]**

(a) For hydrogen (H_2), the molecular weight is 2. Thus, the molar mass of hydrogen is $m_{\text{mole}} = 2.00\text{ g} = 2.00\times10^{-3}\text{ kg}$. Since a mole contains Avogadro's number of molecules, the mass of a single hydrogen molecule is

$$m_{\text{molecule}} = \frac{m_{\text{mole}}}{N_A} = \frac{2.00\times10^{-3}\text{ kg}}{6.02\times10^{23}} = 3.32\times10^{-27}\text{ kg}$$

Equation 1 then gives the rms speed of the hydrogen molecules as

$$v_{\text{rms}} = \sqrt{\frac{2\langle KE\rangle_{\text{molecule}}}{m_{\text{molecule}}}} = \sqrt{\frac{2\left(4.97\times10^{-21}\text{ J}\right)}{3.32\times10^{-27}\text{ kg}}} = 1.73\times10^{3}\text{ m/s} = 1.73\text{ km/s} \quad \Diamond$$

(b) The molecular weight of carbon dioxide (CO_2) is 12 + 2(16) = 44, so the molar mass and the mass per molecule are:

$$m_{\text{mole}} = 44.0\text{ g} = 4.40\times10^{-2}\text{ kg} \qquad m_{\text{molecule}} = \frac{4.40\times10^{-2}\text{ kg}}{6.02\times10^{23}} = 7.31\times10^{-26}\text{ kg}$$

From Equation 1, the rms speed of the CO_2 molecules is

$$v_{rms} = \sqrt{\frac{2\left(4.97\times10^{-21}\ \text{J}\right)}{7.31\times10^{-26}\ \text{kg}}} = 369\ \text{m/s} = 0.369\ \text{km/s} \qquad \Diamond$$

(c) One sixth of the escape velocity from Venus is

$$\frac{10.3\ \text{km/s}}{6} = 1.72\ \text{km/s}$$

Comparing this to the answers from parts (a) and (b) shows that hydrogen will escape but carbon dioxide will not. This is consistent with the finding that the atmosphere of Venus is rich in carbon dioxide but hydrogen is present only in combination with other elements. ◊

---

41. If 2.0 mol of a gas are confined to a 5.0-L vessel at a pressure of 8.0 atm, what is the average kinetic energy of a gas molecule?

Solution In standard SI units, the pressure of this gas is

$$P = (8.0\ \text{atm})\left(1.013\times10^{5}\ \text{Pa/atm}\right) = 8.1\times10^{5}\ \text{Pa}$$

and the volume it occupies is

$$V = (5.0\ \text{L})\left(1000\ \text{cm}^3/\text{L}\right)(0.01\ \text{cm/m})^3 = 5.0\times10^{-3}\ \text{m}^3$$

The absolute temperature of 2.0 mol of gas under these conditions may be found from the ideal gas law which gives

$$T = \frac{PV}{nR} = \frac{\left(8.1\times10^{5}\ \text{Pa}\right)\left(5.0\times10^{-3}\ \text{m}^3\right)}{(2.0\ \text{mol})(8.31\ \text{J/mol}\cdot\text{K})} = 244\ \text{K}$$

The average kinetic energy of a molecule in this gas is therefore

$$\langle KE\rangle_{molecule} = \tfrac{3}{2}k_B T = \tfrac{3}{2}\left(1.38\times10^{-23}\ \text{J/K}\right)(244\ \text{K}) = 5.1\times10^{-21}\ \text{J} \qquad \Diamond$$

---

45. A vertical cylinder of cross-sectional area 0.050 m^2 is fitted with a tight-fitting, frictionless piston of mass 5.0 kg (Fig. P10.45). If there are 3.0 mol of an ideal gas in the cylinder at 500 K, determine the height, h, at which the piston will be in equilibrium under its own weight?

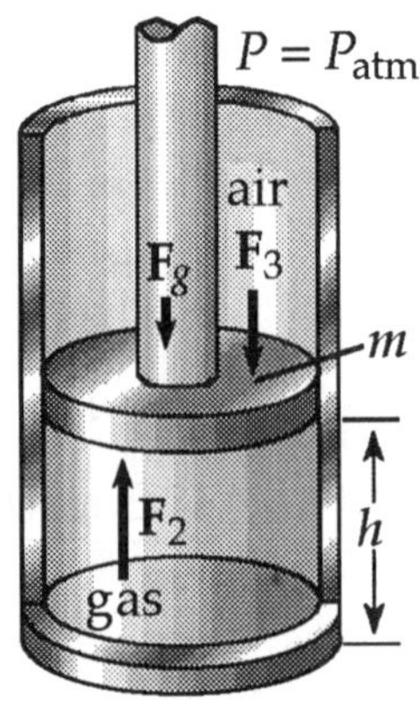

Figure P10.45 (modified)

Solution Three forces act on the piston. These are: (1) its own weight, $F_g = mg$, directed downward; (2) an upward force $F_2 = P_g A$, where P_g is the gas pressure and A is the cross-sectional area of the piston; and (3) a downward force $F_3 = P_{atm} A$ exerted on the piston by the air above it.

When the piston is in equilibrium, $\sum F_y = F_2 - F_g - F_3 = 0$

so $P_g A = mg + P_{atm} A$ **[Equation 1]**

The pressure of the gas is given by the ideal gas law as

$$P_g = \frac{nRT}{V} = \frac{nRT}{Ah}$$

where h is the height of the cylindrical volume of gas. Substituting this into Equation 1 gives

$$\left(\frac{nRT}{Ah}\right)A = mg + P_{atm}A \quad \text{or} \quad \frac{1}{h} = \frac{mg + P_{atm}A}{nRT}$$

Thus, at equilibrium, the height of the trapped column of gas is:

$$h = \frac{nRT}{mg + P_{atm}A} = \frac{(3.0\text{ mol})(8.31\text{ J/mol}\cdot\text{K})(500\text{ K})}{(5.0\text{ kg})(9.8\text{ m/s}^2) + (1.013\times10^5\text{ Pa})(0.050\text{ m}^2)} = 2.4\text{ m} \quad \Diamond$$

51. A copper rod and steel rod are heated. At 0°C the copper rod has a length of L_C, the steel one has a length L_S. When the rods are being heated or cooled, a difference of 5.00 cm is maintained between their lengths. Determine the values of L_C and L_S.

Solution If the difference in the lengths of the two rods is to remind constant as the rods expand or contract, any change in the length of one rod must be matched by an equal change in the length of the other rod. That is, it is necessary that $\Delta L_S = \Delta L_C$ for any change in temperature ΔT.

When a change in temperature ΔT occurs, the new lengths of the rods are $L_S' = L_S + \alpha_S L_S(\Delta T)$ and $L_C' = L_C + \alpha_C L_C(\Delta T)$ where α_S and α_C are the coefficients of linear expansion for steel and copper respectively. The changes that have occurred in the lengths are then: $\Delta L_S = L_S' - L_S = \alpha_S L_S(\Delta T)$, and $\Delta L_C = L_C' - L_C = \alpha_C L_C(\Delta T)$. Since these changes in length must always be equal, this requires that

$$\alpha_S L_S(\Delta T) = \alpha_C L_C(\Delta T) \quad \text{or} \quad L_S = (\alpha_C/\alpha_S)L_C \qquad \textbf{[Equation 1]}$$

A second equation relating the lengths (at 0°C) of the two rods is:

$$L_S - L_C = 5.00 \text{ cm} \qquad \textbf{[Equation 2]}$$

Solve this pair of simultaneous equations by substituting Equation 1 into Equation 2, obtaining

$$(\alpha_C/\alpha_S)L_C - L_C = 5.00 \text{ cm} \quad \text{or} \quad L_c\left((\alpha_c/\alpha_s) - 1\right) = 5.00 \text{ cm}$$

From Table 10.1 in the textbook, $\alpha_c = 17\times10^{-6}\ (°\text{C})^{-1}$ $\alpha_S = 11\times10^{-6}\ (°\text{C})^{-1}$

Therefore, $$L_C = \frac{5.00 \text{ cm}}{(17/11) - 1} = 9.17 \text{ cm} \qquad \Diamond$$

From Equation 2, $L_S = L_C + 5.00$ cm Thus, $$L_S = 14.17 \text{ cm} \qquad \Diamond$$

57. A mercury thermometer is constructed as in Figure P10.46. The capillary tube has a diameter of 0.0050 cm, and the bulb has a diameter of 0.30 cm. Neglecting the expansion of the glass, find the change in height of the mercury column for a temperature change of 25°C.

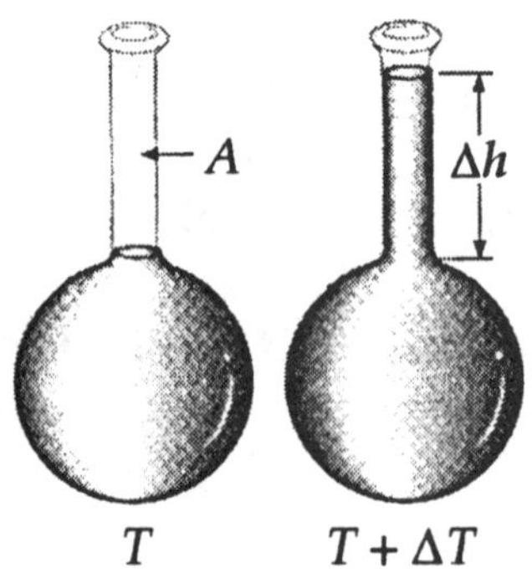

Figure P10.46

Solution Note that at temperature T the mercury just fills the spherical bulb of the thermometer. At temperature $T + \Delta T$, the mercury has expanded into the cylindrical stem of the thermometer. If the expansion of the glass is neglected, the increase in the volume of the mercury equals the volume of mercury now in the stem. This is a cylindrical volume of cross-sectional area A and height Δh, so $\Delta V = A(\Delta h)$. The original volume of mercury equals the volume of the spherical bulb, or $V_0 = 4\pi R^3 / 3$ where R is the radius of the bulb.

After the change in temperature, the new volume of the mercury is given by $V = V_0[1 + \beta(\Delta T)]$, where $\beta = 1.82 \times 10^{-4}\ °\text{C}^{-1}$ is the coefficient of volume expansion for mercury (see Table 10.1 in textbook). The change in volume is thus $\Delta V = V - V_0 = \beta V_0(\Delta T)$. Therefore, the height the mercury has risen in the stem is

$$\Delta h = \Delta V / A = \beta V_0(\Delta T)/A$$

The diameter of the stem is $d = 5.0 \times 10^{-3}$ cm and the radius of the bulb is $R = (0.30\text{ cm})/2 = 0.15\text{ cm}$. Hence, when the temperature increases 25°C, the mercury rises:

$$\Delta h = \frac{\Delta V}{A} = \frac{\beta\left(4\pi R^3/3\right)(\Delta T)}{\pi(d/2)^2} = \frac{16\beta R^3(\Delta T)}{3d^2} = \frac{16\left(1.82 \times 10^{-4}\ °\text{C}^{-1}\right)(0.15\text{ cm})^3(25\ °\text{C})}{3\left(5.0 \times 10^{-3}\text{ cm}\right)^2}$$

Solving, we find the distance that the mercury rises to be $\Delta h = 3.3\text{ cm}$ ◊

CHAPTER SELF-QUIZ

1. Consider two containers that have the same volume and temperature. The first contains a certain number of oxygen molecules while the second contains the same number of hydrogen molecules.
   a. The heavier oxygen molecules will exert greater pressure on the walls of the first bottle.
   b. The faster hydrogen molecules will exert greater pressure on the walls of the second bottle.
   c. The pressure on the walls of both bottles will be equal.
   d. Not enough information is given to allow an accurate calculation of relative pressure.

2. Suppose for a brief moment that the gas molecules hitting a wall stuck to the wall instead of bouncing off the wall. How would the pressure on the wall be affected during that brief time?
   a. The pressure would be zero.
   b. The pressure would be half as big.
   c. The pressure would remain unchanged.
   d. The pressure would be twice as big.

3. Three moles of an ideal gas are confined to a 30-liter container at a pressure of 2 atmospheres. What is the gas temperature? (R = 0.0821 L·atm/mol·K)
   a. 975 K
   b. 487 K
   c. 365 K
   d. 244 K

4. Suppose one sphere is made of a metal that has twice the coefficient of linear expansion of a second sphere. The coefficient of volume expansion for the first sphere will be bigger than the second by a factor of
   a. 2
   b. 3
   c. 4
   d. 8

5. Suppose the ends of a 30-m-long steel rail are rigidly clamped at 0°C to prevent expansion. The rail has a cross-sectional area of 30 cm^2. What force does the rail exert when it is heated to 40°C? ($\alpha_{steel} = 1.1 \times 10^{-5}/°C$, $Y_{steel} = 2 \times 10^{11}$ N/m^2)
   a. 2.6×10^5 N
   b. 5.6×10^4 N
   c. 1.3×10^3 N
   d. 650 N

6. Oxygen condenses into a liquid at approximately 90° Kelvin. What temperature, in degrees Fahrenheit does this correspond to?
   a. −118°F
   b. −193°F
   c. −265°F
   d. −297°F

7. An auditorium has dimensions 10 m × 20 m × 30 m. How many molecules of air are needed to fill the auditorium at standard temperature and pressure?
   a. 1.6×10^{29}
   b. 1.6×10^{27}
   c. 1.6×10^{25}
   d. 1.6×10^{23}

8. A helium-filled balloon has a volume of 1 m^3. As it rises in the Earth's atmosphere, its volume expands. What will be its new volume (in cubic meters) if its original temperature and pressure are 20°C and 1 atm, and it final temperature and pressure are −40°C and 0.1 atm?
   a. 4 m^3
   b. 6 m^3
   c. 8 m^3
   d. 10 m^3

9. A spherical air bubble originating from a scuba diver has a radius of 5 mm at some depth, *h*. When the bubble reaches the surface, it has a radius of 7 mm. Assuming constant temperature, what is the depth of the diver?.
   a. 8 m
   b. 13 m
   c. 18 m
   d. 23 m

10. If the rms velocity of a helium atom at room temperature is 1350 m/s, what is the rms velocity of an oxygen (O_2) molecule?
   a. 675 m/s
   b. 477.3 m/s
   c. 337.5 m/s
   d. 168.7 m/s

11. The density of gasoline is 730 kg/m^3 at 0 °C. Its volume expansion coefficient is 9.6×10^{-4} /°C. If 1 gallon occupies 0.0038 m^3, how many extra kilograms of gasoline do you get when you buy 10 gallons of gasoline at 0 °C rather than at 20 °C?
   a. 0.78 kg
   b. 0.52 kg
   c. 0.26 kg
   d. Zero

12. An automobile tire is filled with air to a gauge pressure of 30 PSI at 10 °C. After driving into desert country, the temperature of the tire is 29 °C. What pressure would be measured (in PSI) at that temperature? (Assume tire volume doesn't change.)
   a. 31 PSI
   b. 32 PSI
   c. 33 PSI
   d. 34 PSI

Chapter 11

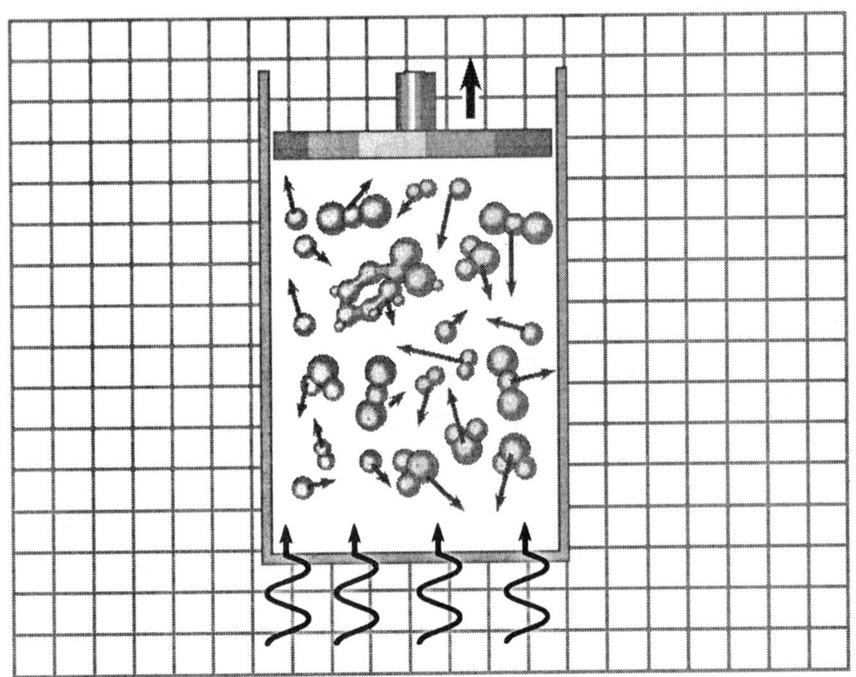

HEAT

Chapter 11

HEAT

It is an experimentally established fact that when two objects at different temperatures are placed in thermal contact with each other, the temperature of the warmer object decreases and the temperature of the cooler object increases. If the two are left in contact for some time, they eventually reach a common equilibrium temperature that is intermediate between the two initial temperatures. When such a process occurs, we say that heat is transferred from the object at the higher temperature to the one at the lower temperature.

The principle of conservation of energy is a universal principle of nature in which heat is treated as another form of energy that can be transformed into mechanical energy. It has been demonstrated that whenever heat is gained or lost by a system during some process, the gain or loss can be accounted for by an equivalent quantity of mechanical work done on the system.

The focus of this chapter is to introduce the concept of heat and some of the processes that enable heat to be transferred between a system and its surroundings.

NOTES FROM SELECTED CHAPTER SECTIONS

11.1 The Mechanical Equivalent of Heat

When two systems at different temperatures are in contact with each other, energy will transfer between them until they reach the same temperature (that is, when they are in thermal equilibrium with each other). This energy is called heat, or thermal energy, and the term "heat flow" refers to an energy transfer as a consequence of a temperature difference.

The unit of heat is the **calorie** (cal), defined as the amount of heat necessary to increase the temperature of 1 g of water from 14.5 °C to 15.5 °C. The **mechanical equivalent of heat,** first measured by Joule, is given by 1 cal = 4.186 J.

11.2 Specific Heat

The specific heat, *c*, of any substance is defined as the amount of heat required to increase the temperature of that substance by one Celsius degree. Its units are cal/°C.

11.3 Conservation of Energy: Calorimetry

Calorimetry is a procedure in which only the transfer of thermal energy between the system and its surroundings is considered: a negligible amount of mechanical work is done in the process. The law of conservation of energy in a calorimeter requires that the heat that leaves the warmer substance (of unknown specific heat) equals the heat that enters the water.

11.4 Latent Heat and Phase Changes

A substance usually undergoes a change in temperature when heat is transferred between it and its surroundings. There are situations, however, in which the flow of heat does not result in a change in temperature. This is the case whenever the substance undergoes a physical alteration from one form to another, referred to as a **phase change**. Some common phase changes are solid to liquid, liquid to gas, and a change in crystalline structure of a solid. Every phase change involves a change in internal energy.

The **latent heat of fusion** is a parameter used to characterize a solid-to-liquid phase change; the **latent heat of vaporization** characterizes the liquid-to-gas phase change.

11.5 Heat Transfer by Conduction
11.6 Convection
11.7 Radiation

There are three basic processes of heat transfer. These are (1) conduction, (2) convection, and (3) radiation.

Conduction is a heat transfer process which occurs when there is a **temperature gradient** across the body. That is, conduction of heat occurs only when the body's temperature is **not** uniform. For example, if you heat a metal rod at one end with a flame, heat will flow from the hot end to the colder end. The rate of flow of heat along the rod, sometimes called the **heat current,** is proportional to the cross-sectional area of the rod, the temperature gradient, and k, the thermal conductivity of the material of which the rod is made.

When heat transfer occurs as the result of the motion of material, such as the mixing of hot and cold fluids, the process is referred to as **convection.** Convection heating is used in conventional hot-air and hot-water heating systems. Convection currents produce changes in weather conditions when warm and cold air masses mix in the atmosphere.

Heat transfer by **radiation** is the result of the continuous emission of electromagnetic radiation by all bodies.

EQUATIONS AND CONCEPTS

When two systems initially at different temperatures are placed in contact with each other, energy will be transferred from the system at higher temperature to the system at lower temperature until the two systems reach a common temperature (thermal equilibrium).

Comment on heat energy.

The energy transferred from one system to the other is called heat or thermal energy; and the term "heat flow" refers to the process of energy transfer due to a temperature difference.

The unit of heat energy is the calorie (cal), defined as the quantity of heat energy required to increase the temperature of 1 g of water from 14.5°C to 15.5°C.

Comment on units.

The **mechanical equivalent of heat** was first measured by Joule.

$$1 \text{ cal} = 4.186 \text{ J} \quad (11.1)$$

The quantity of heat energy required to increase the temperature of a given mass by a specified amount varies from one substance to another. Every substance is characterized by a unique value of specific heat, c.

$$Q = mc\,\Delta T \quad (11.3)$$

The heat energy required to cause a quantity of substance of mass, m, to undergo a phase change (solid to liquid or liquid to gas) depends on the value of the latent heat of the substance. The latent heat of fusion, L_f, is used when the phase change is from solid to liquid and the latent heat of vaporization, L_v, when the phase change is from liquid to gas.

$$Q = mL \quad (11.4)$$

There are three basic processes of heat transfer: (1) conduction, (2) convection, and (3) radiation.

Comment on heat transfer processes.

This is the basic **law of heat conduction.** The constant k is called the thermal conductivity and is characteristic of a particular material. The rate of flow of heat energy along the conducting material is the heat current, H. Conduction of heat occurs only when there is a temperature gradient (or temperature difference between two points in the conducting material).

$$H = \frac{Q}{\Delta t} = kA\left(\frac{T_2 - T_1}{L}\right) \quad (11.6)$$

(Conduction process)

To calculate the rate of heat transfer through a compound slab a summation is made over all portions of the slab. For this calculation, T_1 and T_2 are the temperatures of the **outer extremities** of the slab. In engineering practice, the term L/k for a particular substance is referred to as the **R value** of the material.

$$\frac{Q}{\Delta t} = \frac{A(T_2 - T_1)}{\sum_i L_i / k_i} \quad (11.7)$$

The heat current, H, can be expressed in cal/s, Btu/h, or watts (where 1 W = 1 J/s).

Comment on units.

The rate at which heat is transferred by a fluid to a surface of area A depends on the temperature difference and the convection coefficient h.

$$\frac{\Delta Q}{\Delta t} = hA\,\Delta T$$

(Convection process)

When heat transfer occurs as the result of the movement of a heated substance (usually a fluid) between points at different temperatures, the process is called convection.

Comment on convection.

The rate of emission of heat energy by radiation (the power radiated) is given by Stefan's law. The radiated power is proportional to the fourth power of the absolute temperature.

$$P = \sigma A e T^4 \quad (11.9)$$

(Radiation process)

In Equation 11.12, σ is a universal constant, $\sigma = 5.6696 \times 10^{-8}\ \mathrm{W/m^2}$, T is the absolute temperature in K, and e (the emissivity of the radiating body) can have a value between 0 and 1 depending on the nature of the surface.

Comment on Stefan's law.

An object at temperature T in surroundings at temperature T_0 will experience a net radiated ($T > T_0$) or absorbed ($T < T_0$) power. At thermal equilibrium ($T = T_0$), an object radiates and absorbs energy at the same rate and the temperature of the object remains constant.

$$P_{\text{net}} = \sigma A e (T^4 - T_0^{\,4}) \quad (11.10)$$

SUGGESTIONS, SKILLS, AND STRATEGIES

If you are having difficulty with calorimetry problems, one or more of the following factors should be considered:

1. Be sure your units are consistent throughout. That is, if you are using specific heats measured in cal/g·°C, be sure that masses are in grams and temperatures are in Celsius units throughout.

2. Losses and gains in heat are found by using $Q = mc\Delta T$ only for those intervals in which no phase changes occur. Likewise, the equations $Q = mL_f$ and $Q = mL_v$ are to be used only when phase changes **are** taking place.

3. Often sign errors occur in heat loss = heat gain equations. One way to determine whether your equation is correct is to examine the sign of all ΔT's that appear in your equation. Every ΔT that appears should be a positive number.

REVIEW CHECKLIST

▷ Understand the concepts of heat, internal energy, and thermodynamic processes.

▷ Define and discuss the calorie, Btu, specific heat, and latent heat. Convert between calories, Btu's, and joules.

▷ Use equations for specific heat, latent heat, temperature change and energy gain (loss) to solve calorimetry problems.

▷ Discuss the possible mechanisms which can give rise to heat transfer between a system and its surroundings; that is, heat conduction, convection and radiation; and give a realistic example of each heat transfer mechanism.

▷ Apply the basic law of heat conduction, and Stefan's law for heat transfer by radiation.

SOLUTIONS TO SELECTED END-OF-CHAPTER PROBLEMS

1. As part of an exercise routine a 50.0-kg person climbs 10.0 meters up a vertical rope. How many (food) Calories are expended in a single climb up the rope? [1 (food) Calorie = 10^3 calories]

Solution The climber begins and ends at rest ($v_f = v_i = 0$) and climbs a distance of $y_f - y_i = 10.0\text{ m}$. The mechanical energy of the climber changes by

$$\Delta E = \Delta KE + \Delta PE = \frac{1}{2}m\left(v_f^2 - v_i^2\right) + mg\left(y_f - y_i\right)$$

Thus, $$\Delta E = 0 + (50.0\text{ kg})\left(9.80\text{ m/s}^2\right)(10.0\text{ m}) = 4900\text{ J}$$

Converting units to those of heat energy,

$$\Delta E = (4900\text{ J})(1\text{ cal} / 4.186\text{ J})\left(10^{-3}\text{ Cal/cal}\right) = 1.17\text{ Cal}$$ ◊

9. A 200-g aluminum cup contains 800 g of water in thermal equilibrium at 80 °C. The combination of cup and water is cooled uniformly so that the temperature decreases by 1.5 °C per minute. At what rate is heat energy being removed? Express your answer in watts.

Solution Consider a 1.0-minute time interval. During this interval, the temperature of the system (consisting of the aluminum cup and the water in the cup) decreases by 1.5 °C. The heat energy removed from the system is

$$Q_{\text{total}} = Q_{\text{cup}} + Q_{\text{water}} = m_{\text{cup}}c_{\text{Al}}\Delta T_{\text{cup}} + m_{\text{water}}c_{\text{water}}\Delta T_{\text{water}}$$

From Table 11.1, the specific heats $c_{\text{Al}} = 900\text{ J/kg·°C}$ and $c_{\text{water}} = 4186\text{ J/kg·°C}$ The heat energy removed is thus:

$$Q_{\text{total}} = (0.20\text{ kg})(900\text{ J/kg·°C})(1.5\text{ °C}) + (0.80\text{ kg})(4186\text{ J/kg·°C})(1.5\text{ °C}) = 5300\text{ J}$$

This happens over 60 seconds. Thus, energy is removed at a rate of

$$P = Q_{\text{total}}/t = (5300\text{ J}/60\text{ s}) = 88\text{ J/s} = 88\text{ W}$$ ◊

15. If 200 g of water is contained in a 300-g aluminum vessel at 10°C and an additional 100 g of water at 100°C is poured into the container, what is the final equilibrium temperature of the mixture?

Solution Assuming this system (initially consisting of the cool water, aluminum cup, and some hot water) is isolated from its surroundings, its total energy content must remain constant. Thus, any heat energy lost by the hot water must be absorbed by the cooler water and cup. If T is the final equilibrium temperature, the heat **lost** by the hot water is $Q_{\text{lost}} = m_{hw}c_{\text{water}}(100\ °\text{C} - T)$ and the heat **gained** by the cooler water and cup is

$$Q_{\text{gain}} = Q_{\text{cool water}} + Q_{\text{cup}} = m_{cw}\,c_{\text{water}}(T - 10\ °\text{C}) + m_{\text{cup}}c_{\text{Al}}(T - 10\ °\text{C})$$

Note that the change in temperature has been written as $\Delta T = T_{\text{initial}} - T_{\text{final}}$ for parts that are losing heat, but as $\Delta T = T_{\text{final}} - T_{\text{initial}}$ for a part that gains heat. That is, we write only the magnitude or absolute value of the change in temperature for all parts of the system. The fact that the temperature of some parts of the system is actually decreasing will be taken into account when Q for that part of the system is included on the **loss** side rather than the **gain** side of the energy balance equation. The energy balance equation (a restatement of the principle of conservation of energy) is $Q_{\text{gain}} = Q_{\text{lost}}$:

$$m_{cw}\,c_{\text{water}}(T - 10\ °\text{C}) + m_{\text{cup}}c_{\text{Al}}(T - 10\ °\text{C}) = m_{hw}\,c_{\text{water}}(100\ °\text{C} - T)$$

or
$$\frac{\left(m_{cw}\,c_{\text{water}} + m_{\text{cup}}c_{\text{Al}}\right)}{m_{hw}\,c_{\text{water}}}(T - 10\ °\text{C}) = (100\ °\text{C} - T)$$

Thus, obtaining the specific heats from Table 11.1 in the text,

$$\frac{(0.20\text{ kg})(4186\text{ J/kg}\cdot°\text{C}) + (0.30\text{ kg})(900\text{ J/kg}\cdot°\text{C})}{(0.10\text{ kg})(4186\text{ J/kg}\cdot°\text{C})}(T - 10\ °\text{C}) = (100\ °\text{C} - T)$$

Isolating the temperature on the left gives $\left(1.5\times10^{3}\text{ J/°C}\right)T = 5.3\times10^{4}\text{ J}$

Thus the equilibrium temperature of the mixture is $T = 35\ °\text{C}$ ◊

19. A student drops two metallic objects into a 120-g steel container holding 150 g of water at 25 °C. One object is a 200-g cube of copper that is initially at 85 °C, and the other is a chunk of aluminum that is initially at 5 °C. To the surprise of the student, the water reaches a final temperature of 25 °C, exactly where it started. What is the mass of the aluminum chunk?

Solution Notice that the initial temperature of the copper cube is higher than the final temperature. Thus, the copper cube loses heat energy in this process. The initial temperature of the aluminum chunk is less than the final temperature, so it gains heat energy. The temperature of the water and the steel container is unchanged so these parts of the system have zero net change in heat energy. Thus, assuming the system is isolated from its surroundings, the energy balance equation is $Q_{\text{gain}} = Q_{\text{lost}}$, or $m_{\text{Al}}c_{\text{Al}}(\Delta T)_{\text{Al}} = m_{\text{Cu}}c_{\text{Cu}}(\Delta T)_{\text{Cu}}$. Using Table 11.1 in the textbook, this becomes (see the note in the solution of Problem 15 above):

$$m_{\text{Al}}(900\ \text{J/kg}\cdot{}^\circ\text{C})(25\ {}^\circ\text{C} - 5\ {}^\circ\text{C}) = (0.20\ \text{kg})(387\ \text{J/kg}\cdot{}^\circ\text{C})(85\ {}^\circ\text{C} - 25\ {}^\circ\text{C})$$

Solving for the mass of the aluminum chunk then gives

$$m_{\text{Al}} = \frac{(0.20\ \text{kg})(387\ \text{J/kg}\cdot{}^\circ\text{C})(60\ {}^\circ\text{C})}{(900\ \text{J/kg}\cdot{}^\circ\text{C})(20\ {}^\circ\text{C})} = 0.26\ \text{kg} = 260\ \text{g}$$ ◊

---

23. What mass of steam that is initially at 120 °C is needed to warm 350 g of water and its 300-g aluminum container from 20 °C to 50 °C?

Solution To answer this problem, we must recognize that energy must be conserved: heat lost by the cooling steam must equal the heat gained by the cup and the cold water ($Q_{\text{lost}} = Q_{\text{gain}}$). In this isolated system, the steam loses heat, the 350 g of water and the aluminum cup gain heat, and the final equilibrium temperature of the system is 50 °C. Since this final temperature is below the boiling point of water (100 °C), the steam must undergo a phase change, and the heat energy lost by the steam must be computed in three steps. The heat lost as the steam cools to the boiling point is:

$$Q_{\substack{\text{cool to}\\ 100\ °\text{C}}} = m_{\text{steam}} c_{\text{steam}} (120\ °\text{C} - 100\ °\text{C})$$

By Table 11.1,
$$Q_{\substack{\text{cool to}\\ 100\ °\text{C}}} = m_{\text{steam}} (2010\ \text{J/kg}\cdot°\text{C})(20\ °\text{C}) = \left(4.0\times10^{4}\ \text{J/kg}\right) m_{\text{steam}}$$

Taking L_v to be the heat of vaporization for water, the heat loss as the steam condenses is

$$Q_{\substack{\text{condense}\\ \text{steam}}} = m_{\text{steam}} L_v$$

By Table 11.2,
$$Q_{\substack{\text{condense}\\ \text{steam}}} = \left(2.26\times10^{6}\ \text{J/kg}\right) m_{\text{steam}}$$

For the next stage of cooling, the condensate is liquid water, and the specific heat of water is used. Therefore, the heat lost as the condensed steam cools to 50°C is

$$Q_{\substack{\text{cool}\\ \text{condensate}}} = m_{\text{steam}} c_{\text{water}} (100\ °\text{C} - 50\ °\text{C})$$

$$Q_{\substack{\text{cool}\\ \text{condensate}}} = m_{\text{steam}} (4186\ \text{J/kg}\cdot°\text{C})(50\ °\text{C}) = \left(2.1\times10^{5}\ \text{J/kg}\right) m_{\text{steam}}$$

Summing,
$$Q_{\text{lost}} = Q_{\substack{\text{cool to}\\ 100\ °\text{C}}} + Q_{\substack{\text{condense}\\ \text{the steam}}} + Q_{\substack{\text{cool}\\ \text{condensate}}} = \left(2.5\times10^{6}\ \text{J/kg}\right) m_{\text{steam}}$$

Note that $(\Delta T)_{\text{cup}} = (\Delta T)_{\text{water}}$. Thus, heat gain for the cup and the cool water are:

$$Q_{\text{gain}} = Q_{\text{cup}} + Q_{\text{water}} = m_{\text{cup}} c_{\text{Al}} (\Delta T)_{\text{cup}} + m_{\text{water}} c_{\text{water}} (\Delta T)_{\text{water}}$$

$$Q_{\text{gain}} = \left[(0.30\ \text{kg})(900\ \text{J/kg}\cdot°\text{C}) + (0.35\ \text{kg})(4186\ \text{J/kg}\cdot°\text{C})\right](50\ °\text{C} - 20\ °\text{C})$$

$$Q_{\text{gain}} = 5.2\times10^{4}\ \text{J}$$

$\left(Q_{\text{lost}} = Q_{\text{gain}}\right)$ therefore gives: $\left(2.5\times10^{6}\ \text{J/kg}\right) m_{\text{steam}} = 5.2\times10^{4}\ \text{J}$

and the mass of steam is $m_{\text{steam}} = 0.021\ \text{kg} = 21\ \text{g}$ ◊

27. A 40-g block of ice is cooled to –78 °C. It is added to 560 g of water in an 80-g copper calorimeter at a temperature of 25 °C. Determine the final temperature. (If not all the ice melts, determine how much ice is left.) Remember that the ice must first warm to 0 °C, melt, and then continue warming as water. The specific heat of ice is 0.500 cal/g · °C = 2090 J/kg · °C.

Solution

There are 3 possible outcomes of the process that must be considered. These are:

(1) The water and cup could reach 0 °C and all the water could freeze before the block of ice warms to 0 °C. In this case, the equilibrium temperature T_f would be below 0 °C.

(2) The water and cup may reach 0 °C after the block of ice has reached 0 °C but before all the ice has melted. Then, the equilibrium temperature will be $T_f = 0$ °C.

(3) The block of ice could reach 0 °C and completely melt before the water and cup have cooled to 0 °C. In this case, $T_f > 0$ °C.

To distinguish between these possible scenarios, consider the quantities of heat energy that would be transferred in various steps. The heat energy the block of ice will need to absorb before its temperature will reach 0 °C is

$$Q_{\text{warm ice to 0 °C}} = Q_1 = m_{\text{ice}}\, c_{\text{ice}}\left(0\ \text{°C} - T_{\text{ice, i}}\right)$$

or $$Q_1 = (0.040\ \text{kg})(2090\ \text{J/kg·°C})(78\ \text{°C}) = 6.5 \times 10^3\ \text{J}$$

After its temperature has reached 0 °C, the additional heat energy needed to completely melt the ice (with L_f being the heat of fusion for water) is:

$$Q_{\text{melt ice}} = Q_2 = m_{\text{ice}} L_f$$

By Table 11.2, $$Q_2 = (0.040\ \text{kg})(3.33 \times 10^5\ \text{J/kg}) = 1.3 \times 10^4\ \text{J}$$

The quantity of heat energy the water and cup could give up before their temperature reaches 0 °C is

$$Q_3 = \left(m_{\text{water}}c_{\text{water}} + m_{\text{cup}}c_{\text{cup}}\right)\left(T_{\text{water+cup, i}} - 0\ °\text{C}\right)$$

$$Q_3 = \left[(0.56\ \text{kg})(4186\ \text{J/kg}\cdot°\text{C}) + (0.080\ \text{kg})(387\ \text{J/kg}\cdot°\text{C})\right](25\ °\text{C}) = 5.9\times10^4\ \text{J}$$

Observe that $Q_3 > Q_1 + Q_2$. Thus, the water and cup are capable of providing more than enough heat energy to warm the ice to 0 °C and completely melt the ice. Therefore, the equilibrium temperature is greater than 0 °C. This temperature can be determined from an energy balance equation for the isolated system: $Q_{\text{gain}} = Q_{\text{loss}}$.

Specifically, that is

$$Q_{\substack{\text{warm ice}\\ \text{to 0 °C}}} + Q_{\substack{\text{melt}\\ \text{ice}}} + Q_{\substack{\text{warm}\\ \text{melted ice}}} = Q_{\substack{\text{cool warm}\\ \text{water + cup}}}$$

Recognize that after the ice melts, its specific heat is simply that of liquid water. Then, the energy balance equation becomes

$$Q_1 + Q_2 + m_{\text{ice}}c_{\text{water}}\left(T_f - 0\ °\text{C}\right) = \left(m_{\text{water}}c_{\text{water}} + m_{\text{cup}}c_{\text{cup}}\right)\left(25\ °\text{C} - T_f\right)$$

$$6.5\times10^3\ \text{J} + 1.3\times10^4\ \text{J} + (0.040\ \text{kg})(4186\ \text{J/kg}\cdot°\text{C})T_f = \left[(0.56\ \text{kg})(4186\ \text{J/kg}\cdot°\text{C}) + (0.080\ \text{kg})(387\ \text{J/kg}\cdot°\text{C})\right]\left(25°\text{C} - T_f\right)$$

This yields $$2.0\times10^4\ \text{J} + \left(1.7\times10^2\ \text{J/°C}\right)T_f = 6.0\times10^4\ \text{J} - \left(2.4\times10^3\ \text{J/°C}\right)T_f$$

and $$T_f = \frac{6.0\times10^4\ \text{J} - 2.0\times10^4\ \text{J}}{1.7\times10^2\ \text{J/°C} + 2.4\times10^3\ \text{J/°C}} = 16\ °\text{C}$$ ◊

33. A Thermopane window consists of two glass panes, each 0.50 cm thick, with a 1.0-cm-thick sealed layer of air between. If the inside temperature is 23.0°C and the outside temperature is 0.0°C, determine the rate of heat transfer through 1.0 m^2 of the window. Compare this with the rate of heat transfer through 1.0 m^2 of a single 1.0-cm-thick pane of glass.

Solution The rate of heat transfer through a slab which is made of multiple layers of materials may be written as $Q/\Delta t = A(T_2 - T_1)/\sum R_i$, where A is the surface area of one side of the slab, $(T_2 - T_1)$ is the total change in temperature going from one side of the slab to the other, and R_i is the R value for the ith layer making up the slab. The R value for any given layer is the ratio of its thickness L to the thermal conductivity, k, of the material making up the layer. Table 11.3 in the textbook gives the thermal conductivities of glass and air as $0.84\ \text{J/s}\cdot\text{m}\cdot°\text{C}$ and $0.0234\ \text{J/s}\cdot\text{m}\cdot°\text{C}$, respectively. For the Thermopane, the R values for the various layers are

$$R_{\substack{\text{glass}\\\text{pane}}} = (5.0\times10^{-3}\ \text{m})/(0.84\ \text{J/s}\cdot\text{m}\cdot°\text{C}) = 0.0060\ \text{s}\cdot\text{m}^2\cdot°\text{C/J}$$

and

$$R_{\substack{\text{air}\\\text{layer}}} = (1.0\times10^{-2}\ \text{m})/(0.0234\ \text{J/s}\cdot\text{m}\cdot°\text{C}) = 0.43\ \text{s}\cdot\text{m}^2\cdot°\text{C/J}$$

Thus, with a temperature change of $(T_2 - T_1) = 23\ °\text{C} - 0\ °\text{C}$ across it and a surface area $A = 1.0\ \text{m}^2$, the rate of heat transfer through the Thermopane is

$$(Q/\Delta t)_{\text{Thermopane}} = \frac{(1.0\ \text{m}^2)(23\ °\text{C} - 0\ °\text{C})}{(6.0\times10^{-3} + 0.43 + 6.0\times10^{-3})\ \text{s}\cdot\text{m}^2\cdot°\text{C/J}} = 52\ \text{J/s} \quad \lozenge$$

For the single pane window, the R value of the thick glass pane is

$$R_{\text{thick pane}} = 1.0\times10^{-2}\ \text{m}/(0.84\ \text{J/s}\cdot\text{m}\cdot°\text{C}) = 1.2\times10^{-2}\ \text{s}\cdot\text{m}^2\cdot°\text{C/J}$$

Then, for the same area and temperature difference as the Thermopane, the rate of heat transfer for the single pane is

$$(Q/\Delta t)_{\substack{\text{single}\\\text{pane}}} = \frac{(1.0\ \text{m}^2)(23\ °\text{C} - 0\ °\text{C})}{1.2\times10^{-2}\ \text{J/s}\cdot\text{m}\cdot°\text{C}} = 1.9\times10^3\ \text{J/s} \quad \lozenge$$

37. A sphere that is to be considered as a perfect black-body radiator has a radius of 0.060 m and is at 200°C in a room where the temperature is 22 °C. Calculate the net rate at which the sphere radiates energy.

Solution

The rate at which an body of surface area A and at **absolute temperature** T radiates energy is given by Stefan's law as $P_{\text{radiation}} = \sigma A e T^4$, where e is the emissivity of that body and $\sigma = 5.6696 \times 10^{-8}\ \text{W/m}^2\cdot\text{K}^4$ is a constant. If the temperature of the surroundings of that body is T_0 the body also **absorbs** energy (i.e., has a negative flow of energy from the body) at a rate $P_{\text{absorption}} = \sigma A e T_0{}^4$. The **net** rate of energy transfer **away** from the body is then given by

$$P_{\text{net}} = P_{\text{radiation}} - P_{\text{absorption}} = \sigma A e\left(T^4 - T_0{}^4\right)$$

The surface area of a sphere of radius R is $A = 4\pi R^2$ and the emissivity of a perfect black-body is $e = 1.0$.

Thus, if a spherical black-body radiator with a radius of 0.060 m is at a temperature $T = 200\ °\text{C} = 473\ \text{K}$ and the temperature of the surroundings is $T_0 = 22\ °\text{C} = 295\ \text{K}$, the net rate of radiation from the sphere is

$$P_{\text{net}} = \left(5.6696 \times 10^{-8}\ \text{W/m}^2\cdot\text{K}^4\right)\left[4\pi(0.060\ \text{m})^2\right](1.0)\left[(473\ \text{K})^4 - (295\ \text{K})^4\right]$$

or $\quad P_{\text{net}} = 1.1 \times 10^2\ \text{W}$ ◊

45. A 40-g ice cube floats in 200 g of water in a 100-g copper cup; all are at a temperature of 0 °C. A piece of lead at 98 °C is dropped into the cup, and the final equilibrium temperature is 12 °C. What is the mass of the lead?

Solution This problem is easily solved by the methods of calorimetry (i.e., conservation of energy). However, students must be careful to remember that the ice must completely melt before its temperature (and hence that of the water and cup) will rise above 0 °C.

Also, after the ice melts, the cup will contain a total of 240 g liquid water. The energy balance equation $Q_{\text{loss}} = Q_{\text{gain}}$ for this isolated system becomes:

$$Q_{\text{cool lead}} = Q_{\text{melt ice}} + Q_{\text{warm water}} + Q_{\text{warm cup}}$$

$$m_{\text{lead}} c_{\text{lead}} \left(T_{\text{lead, i}} - T_f\right) = m_{\text{ice}} L_f + m_{\text{water + ice}} c_{\text{water}} \left(T_f - 0\ ^\circ\text{C}\right) + m_{\text{cup}} c_{cu} \left(T_f - 0\ ^\circ\text{C}\right)$$

Thus, referring to Tables 11.1 and 11.2 in the textbook,

$$m_{\text{lead}}(128\ \text{J/kg}\cdot{}^\circ\text{C})(98\ ^\circ\text{C} - 12\ ^\circ\text{C}) = (0.040\ \text{kg})\left(3.33 \times 10^5\ \text{J/kg}\right) + (0.240\ \text{kg})(4186\ \text{J/kg}\cdot{}^\circ\text{C})(12\ ^\circ\text{C}) + (0.100\ \text{kg})(387\ \text{J/kg}\cdot{}^\circ\text{C})(12\ ^\circ\text{C})$$

or

$$m_{\text{lead}} = \frac{2.6 \times 10^4\ \text{J}}{1.1 \times 10^4\ \text{J/kg}} = 2.4\ \text{kg} \qquad \Diamond$$

---

51. An aluminum rod and an iron rod are joined end to end in a good thermal contact. The two rods have equal lengths and radii. The free end of the aluminum rod is maintained at a temperature of 100 °C, and the free end of the iron rod is maintained at 0 °C. (a) Determine the temperature of the interface between the two rods. (b) If each rod is 15 cm long and each has a cross-sectional area of 5.0 cm^2, what quantity of heat energy is conducted across the combination in 30 min?

Solution (a) At equilibrium, the rate of transfer of heat energy to the interface through the aluminum rod equals the rate of transfer of heat energy away from the interface through the iron rod. That is:

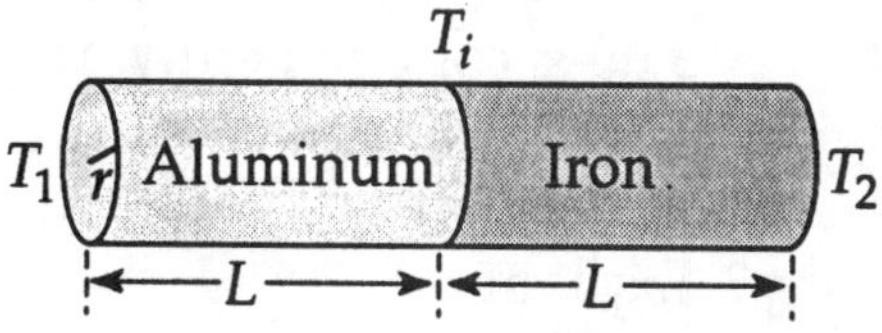

$$\frac{Q}{\Delta t} = \frac{k_{Al}A_{Al}(\Delta T)_{Al}}{L_{Al}} = \frac{k_{Fe}A_{Fe}(\Delta T)_{Fe}}{L_{Fe}}$$

where k_{Al} and k_{Fe} are the thermal conductivities of aluminum and iron respectively. The rods have equal lengths $(L_{Al} = L_{Fe})$, and since they have the same radii, their cross-sectional areas are also equal $(i.e., A_{Al} = A_{Fe})$. The energy transfer equation then reduces to $k_{Al}(\Delta T)_{Al} = k_{Fe}(\Delta T)_{Fe}$, or $k_{Al}(T_1 - T_i) = k_{Fe}(T_i - T_2)$. Since $T_1 = 100\ °C$ and $T_2 = 0\ °C$, this gives

$$(238\ J/s\cdot m\cdot °C)(100\ °C - T_i) = (79.5\ J/s\cdot m\cdot °C)(T_i - 0\ °C)$$

where the thermal conductivity values are from Table 11.3 in the textbook. Thus, we can solve for the temperature at the interface between the two rods as

$$2.38\times 10^4\ J/s\cdot m = (238\ J/s\cdot m\cdot °C + 79.5\ J/s\cdot m\cdot °C)\, T_i \quad \text{and} \quad T_i = 75\ °C \quad \lozenge$$

(b) If the aluminum rod has a length of $L_{Al} = 15\ cm = 0.15\ m$ and a cross-sectional area of $A_{Al} = 5.0\ cm^2 = 5.0\times 10^{-4}\ m^2$, the rate of transfer through that rod is

$$\frac{Q}{\Delta t} = \frac{k_{Al}A_{Al}(T_1 - T_i)}{L_{Al}} = \frac{(238\ J/s\cdot m\cdot °C)(5.0\times 10^{-4}\ m^2)(25\ °C)}{0.15\ m} = 20\ J/s$$

The energy transferred in 30 min is

$$Q = (20\ J/s)\Delta t = (20\ J/s)(30\ min)(60\ s/min) = 3.6\times 10^4\ J \quad \lozenge$$

57. A 3.00-g copper penny at 25.0°C drops 50.0 m to the ground. (a) If 60.0% of its initial potential energy goes into increasing the internal energy, determine its final temperature. (b) Does the result depend on the mass of the penny? Explain.

Solution

(a) The initial potential energy of the penny is

$$PE_i = mgy_i = (0.003 \text{ kg})(9.80 \text{ m/s}^2)(50.0 \text{ m}) = 1.47 \text{ J}$$

where the ground has been chosen as the zero potential energy level. If 60.0% of this initial potential energy is converted into heat energy when the penny collides with the ground, the heat energy produced is

$$Q = (0.600)PE_i = (0.600)(1.47 \text{ J}) = 0.882 \text{ J}$$

From $Q = mc(\Delta T)$, with $c = 387$ J/kg·°C from Table 11.1 in the text, the expected rise in temperature is

$$\Delta T = \frac{Q}{mc} = \frac{0.882 \text{ J}}{(0.003 \text{ kg})(387 \text{ J/kg·°C})} = 0.760 \text{ °C}$$

The final temperature is $T_f = T_i + \Delta T = 25.0 \text{ °C} + 0.760 \text{ °C} = 25.8 \text{ °C}$ ◊

(b) Looking at the steps of the previous calculation, one observes that

$$T_f = T_i + \Delta T = T_i + \frac{Q}{mc} = T_i + \frac{(0.600)PE_i}{mc} = T_i + \frac{(0.600)\cancel{m}gy_i}{\cancel{m}c}$$

The mass cancels out; thus, the final temperature independent of the mass. ◊

61. A 60-kg runner dissipates 300 W of power while running a marathon. Assuming that 10% of the runner's energy is dissipated in the muscle tissue and that the excess heat is removed from the body primarily by sweating, determine the volume of bodily fluid (assume it is water) lost per hour. (At 37 °C the latent heat of vaporization of water is 575 kcal/kg.)

Solution

Ninety percent of the dissipated power is loss by sweating. Thus, the energy dissipated by sweating in one hour is

$$Q = 0.90P(\Delta t) = 0.90(300\text{ J/s})(1.0\text{ h})(3600\text{ s/h}) = 9.7\times 10^5\text{ J}$$

This heat energy evaporates surface water which, at body temperature of 37°C, has a heat of vaporization of

$$L_v = (575\text{ kcal/kg})(1000\text{ cal/kcal})(4.186\text{ J/cal}) = 2.41\times 10^6\text{ J/kg}$$

The mass of water evaporated in one hour is then

$$m = \frac{Q}{L_v} = \frac{9.7\times 10^5\text{ J}}{2.41\times 10^6\text{ J/kg}} = 0.40\text{ kg} = 4.0\times 10^2\text{ g}$$

The density of water is $\rho = 1000\text{ kg/m}^3 = 1.00\text{ g/cm}^3$ (see Table 9.2 in the text). Thus, the volume of liquid water evaporated in one hour is

$$V = \frac{m}{\rho} = \frac{4.0\times 10^2\text{ g}}{1.00\text{ g/cm}^3} = 4.0\times 10^2\text{ cm}^3 \qquad \Diamond$$

CHAPTER SELF-QUIZ

1. In winter, light-colored clothes will keep you warmer than dark-colored clothes if you are
   a. warmer than your surroundings
   b. at the same temperature as your surroundings
   c. cooler than your surroundings
   d. standing in sunlight

2. A 2-kg block is made of a metal that has higher specific heat than the metal out of which a 1-kg block is made. The 2-kg block was originally at 50 °C and the 1-kg block at 20 °C. When the two blocks are placed in contact, what will the final equilibrium temperature be?
   a. below 35 °C
   b. 35 °C
   c. between 35 and 40 °C
   d. above 40 °C

3. There are four identical blocks, all at 10 °C. The first block is heated to 70 °C while the others are left at 10 °C. The first block is then placed in contact with the second block until an equilibrium temperature is reached. Then the first block is placed in contact with the third block until an equilibrium temperature is reached. Then the first block is placed in contact with the fourth block until an equilibrium temperature is reached. What is the final temperature of the first block?
   a. 25 °C
   b. 20 °C
   c. 17.5 °C
   d. 15 °C

4. The primary reason why a sandy beach gets so hot on a sunny day is because sand
   a. absorbs sunlight so well
   b reflects sunlight so well
   c. has a large specific heat
   d. has a small specific heat

5. A silver bar of length 30 cm and cross-sectional area 1 cm^2 is used to transfer heat from a 100°C reservoir to a 0°C reservoir. How much heat is transferred per second?
   $(k_{silver} = 427\ W/m{\cdot}°C)$
   a. 7.1 W
   b. 9.2 W
   c. 11.7 W
   d. 14.2 W

6. A 3-gram copper penny at 20 °C is dropped from a height of 300 m and strikes the ground. If 60% of the energy goes into increasing the internal energy of the coin, determine its final temperature. (The specific heat of copper is 387 J/kg ·°C).
   a. 22.22 °C
   b. 24.55 °C
   c. 26.78 °C
   d. 29.03 °C

7. A solar heating system has a 25% conversion efficiency; the solar radiation incident on the panels is 500 W/m^2. What is the increase in temperature of 30 kg of water in a 1.0 h period by a 4.0 m^2 area collector?
   a. 14 °C
   b. 22 °C
   c. 29 °C
   d. 44 °C

8. If the absolute temperature of a spherical object were tripled, by what factor would the rate of radiated energy emitted from its surface be changed?
   a. 3.0
   b 9.0
   c. 27.0
   d. 81.0

9. On a cold day, a piece of metal feels much colder to the touch than a piece of wood. This is attributed to a difference in which property with respect to the two objects?
   a. mass density
   b. specific heat
   c. temperature
   d. thermal conductivity

10. A glass pane 0.4 cm thick has an area of 2×10^4 cm^2. On a winter day the temperature difference between the inside and outside surfaces of the pane is 25 °C. What is the rate of heat flow through this window? (Thermal conductivity for glass is $0.837\ \text{J/s}\cdot\text{m}\cdot{}^\circ\text{C}$)
   a. $837\ \text{J/s}$
   b. $10.5\ \text{kJ/s}$
   c. $16.7\ \text{kJ/s}$
   d. $34.3\ \text{kJ/s}$

11. A 1-kg block of copper at 20°C is dropped into a large vessel of liquid nitrogen at 77 K. How many kilograms of nitrogen boil away by the time the copper reaches 77 K? (The specific heat of copper is $385\ \text{J/kg}\cdot{}^\circ\text{C}$. The heat of vaporization of nitrogen is $2\times10^5\ \text{J/kg}$.)
   a. 0.212 kg
   b. 0.416 kg
   c. 0.636 kg
   d. 0.898 kg

12. How much heat energy is required to melt a 20-gram block of silver at 20 °C? The melting point of silver is 960 °C, its specific heat is $2.34\times10^2\ \text{J/kg}\cdot{}^\circ\text{C}$, and its heat of fusion is $8.83\times10^4\ \text{J/kg}$.
   a 4395 J
   b. 6165 J
   c. 7518 J
   d. 8795 J

Chapter 12

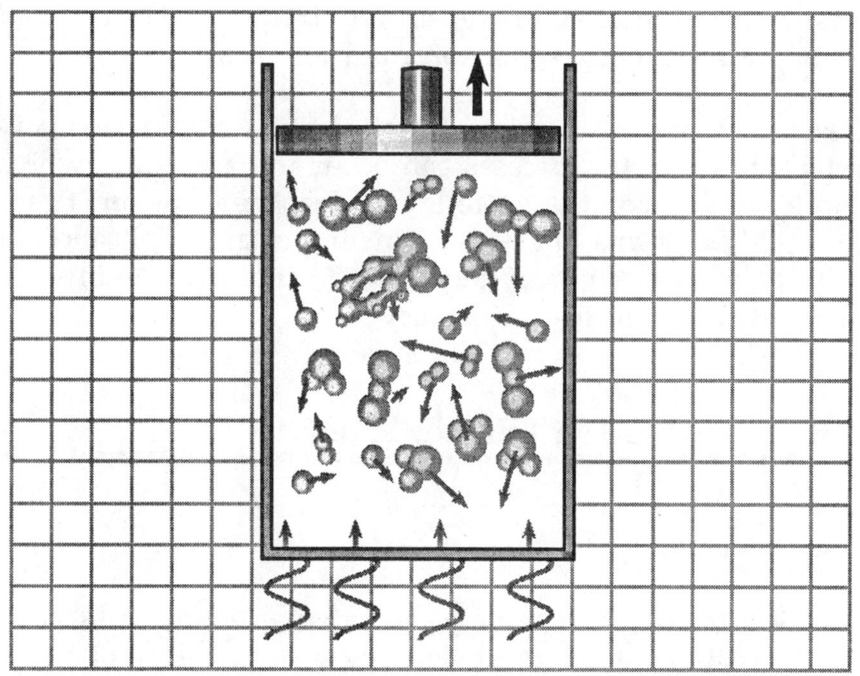

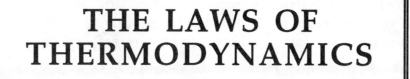

THE LAWS OF THERMODYNAMICS

Chapter 12

THE LAWS OF THERMODYNAMICS

The first law of thermodynamics is essentially the principle of conservation of energy generalized to include heat as a form of energy. It tells us that an increase in one form of energy must be accompanied by a decrease in some other form of energy. The law considers both heat and work and places no restrictions on the types of energy conversions that occur. According to the first law, the internal energy of an object can be increased either by heat added to the object or by work done on it.

The second law of thermodynamics, which can be stated in many equivalent ways, establishes which processes can occur in nature and which cannot. For example, the second law tells us that heat never flows spontaneously from a cold body to a hot body. One important application of this law is in the study of heat engines, such as the internal combustion engine, and the principles that limit their efficiency.

NOTES FROM SELECTED CHAPTER SECTIONS

12.2 Heat and Internal Energy

A major distinction must be made between internal energy and heat. **Internal energy** is all of the energy belonging to a system while it is stationary (neither translating nor rotating), including heat as well as nuclear, chemical, and elastic energy. **Thermal (or heat) energy** is the portion of internal energy that changes when the temperature of the system changes. The work done on (or by) a system is a measure of the energy transferred between the system and its surroundings. It is important to realize that energy can be transferred between two systems even when no **thermal** energy transfer occurs. When this happens, the change in internal energy is equal to the change in thermal energy and is measured by a corresponding change in temperature.

12.2 Work and Heat

The work done in the expansion from the initial state to the final state is the area under the curve in a PV diagram, as shown in the figure.

If the gas is compressed, $V_f < V_i$, and the work is negative. That is, work is done **on** the gas. If the gas expands, $V_f > V_i$, the work is positive, and the gas does work on the piston. If the gas expands at **constant pressure,** called an **isobaric process,** then $W = P(V_f - V_i)$.

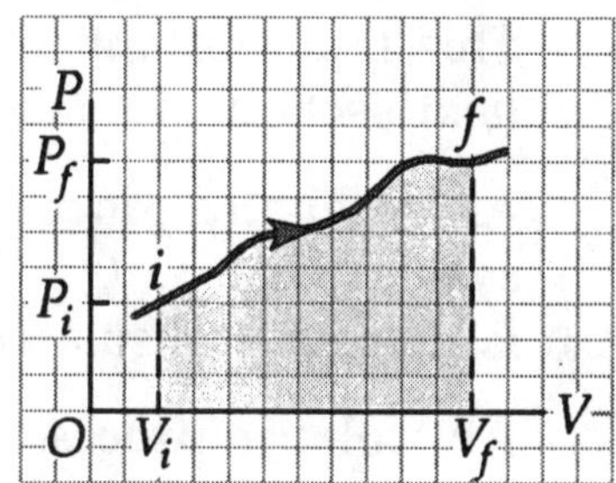

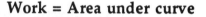

Work = Area under curve

The work done by a system depends on the process by which the system goes from the initial to the final state. In other words, the work done depends on the initial, final, and intermediate states of the system.

The amount of heat gained or lost by a system depends on the initial, final, and intermediate states of the system.

12.3 The First Law of Thermodynamics

In the first law of thermodynamics, $\Delta U = Q - W$, Q is the heat added to the system and W is the work done by the system. Note that by convention, Q is **positive** when heat enters the system and **negative** when heat is removed from the system. Likewise, W can be positive or negative as mentioned earlier. The initial and final states must be **equilibrium** states; however, the intermediate states are, in general, nonequilibrium states since the thermodynamic coordinates undergo finite changes during the thermodynamic process.

A process that occurs at constant temperature is called an **isothermal process,** and a plot of P versus V at constant temperature for an ideal gas yields a hyperbolic curve called an **isotherm.** The internal energy of an ideal gas is a function of temperature only.

An **isolated system** is one which does not interact with its surroundings. In such a system, $Q = W = 0$. That is, the internal energy of an isolated system cannot change.

A **cyclic process** is one that originates and ends up at the same state. The work done per cycle equals the heat added to the system per cycle. This is important to remember when dealing with heat engines in the next section.

An **adiabatic process** is a process in which no heat enters or leaves the system; that is, $Q = 0$. A system may undergo an adiabatic process if it is thermally insulated from its surroundings.

An **isobaric process** is a process which occurs at constant pressure. For such a process, the heat transferred and the work done are nonzero.

An **isovolumetric process** is one which occurs at constant volume. By definition, $W = 0$ for such a process (since the volume does not change). All of the heat added to the system kept at constant volume goes into increasing the internal energy of the system.

12.4 Heat Engines and the Second Law of Thermodynamics

A heat engine is a device that converts thermal energy to other useful forms, such as electrical and mechanical energy. A heat engine carries some working substance through a cyclic process during which (1) heat is absorbed from a source at a high temperature, (2) work is done by the engine, and (3) heat is expelled by the engine to a source at a lower temperature.

The engine absorbs a quantity of heat, Q_h, from a hot reservoir, does work W, and then gives up heat Q_c to a cold reservoir. Because the working substance goes through a cycle, its initial and final internal energies are equal, so $\Delta U = 0$. Hence, from the first equation we see that **the net work, W, done by a heat engine equals the net heat flowing into it.**

If the working substance is a gas, **the net work done for a cyclic process is the area enclosed by the curve representing the process on a PV diagram.**

The **thermal efficiency,** e, of a heat engine is the ratio of the net work done to the heat absorbed at the higher temperature during one cycle.

The second law of thermodynamics can be stated as follows: **It is impossible to construct a heat engine that, operating in a cycle, produces no other effect than the absorption of heat from a reservoir and the performance of an equal amount of work.**

12.5 Reversible and Irreversible Processes

A process is **irreversible** if the system and its surroundings cannot be returned to their initial states. A process is **reversible** if the system passes from the initial to the final state through a succession of equilibrium states.

12.6 The Carnot Engine

The most efficient cyclic process is called the **Carnot cycle,** described in the PV diagram shown in the figure. The Carnot cycle consists of two adiabatic and two isothermal processes, all being reversible.

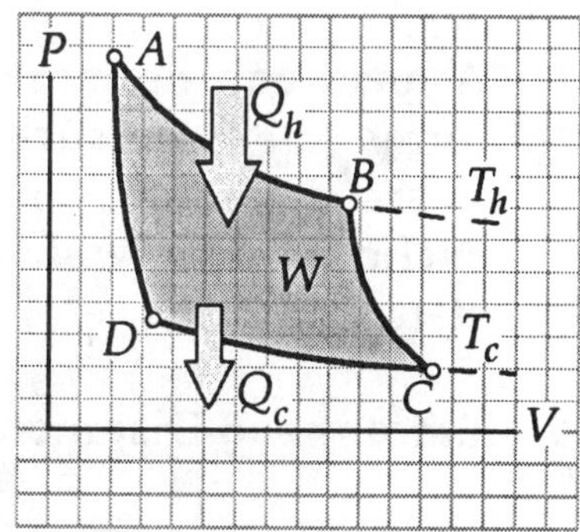

The Carnot Cycle

1. The process $A \rightarrow B$ is an isotherm (constant T), during which time the gas expands at constant temperature T_h, and absorbs heat Q_h from the hot reservoir.

2. The process $B \rightarrow C$ is an adiabatic expansion ($Q = 0$), during which time the gas expands and cools to a temperature T_c.

3. The process $C \rightarrow D$ is a second isotherm, during which time the gas is compressed at constant temperature T_c, and gives up heat Q_c to the cold reservoir.

4. The final process $D \rightarrow A$ is an adiabatic compression in which the gas temperature increases to a final temperature of T_h.

In practice, no working engine is 100% efficient, even when losses such as friction are neglected. One can obtain some theoretical limits on the efficiency of a real engine by comparison with the ideal Carnot engine. A **reversible engine** is one which will operate with the same efficiency in the forward and reverse directions. The Carnot engine is one example of a reversible engine.

> All Carnot engines operating reversibly between T_h and T_c have the **same** efficiency given by Equation 12.5.
>
> No real (irreversible) engine can have an efficiency greater than that of a reversible engine operating between the same two temperatures.

A **heat engine** is a device that converts thermal energy into other forms of energy such as mechanical and electrical energy. During its operation, a heat engine carries some working substance through a **cyclic process,** which is a process which begins and ends at the same state.

12.9 Entropy and Disorder

Entropy is a quantity used to measure the degree of **disorder** in a system. For example, the molecules of a gas in a container at a high temperature are in a more disordered state (higher entropy) than the same molecules at a lower temperature.

When heat is added to a system, the entropy **increases.** When heat is removed, the entropy **decreases.** Note that only **changes** in entropy are defined by Equation 12.8; therefore, the concept of entropy is most useful when a system undergoes a **change in its state**.

The second law of thermodynamics can be stated in terms of entropy as follows: **The total entropy of an isolated system always increases in time if the system undergoes an irreversible process.** If an isolated system undergoes a **reversible** process, the total entropy **remains constant.**

EQUATIONS AND CONCEPTS

This equation can be used to calculate the work done on or by a gas sample, if the pressure of the gas remains constant during a compression or an expansion. When ΔV is positive, the work done is positive (work is done by the gas); when ΔV is negative, the work done is negative (work is done on the gas). The work done is equal to the area under the pressure-volume curve.

$$W = P\Delta V \tag{12.1}$$

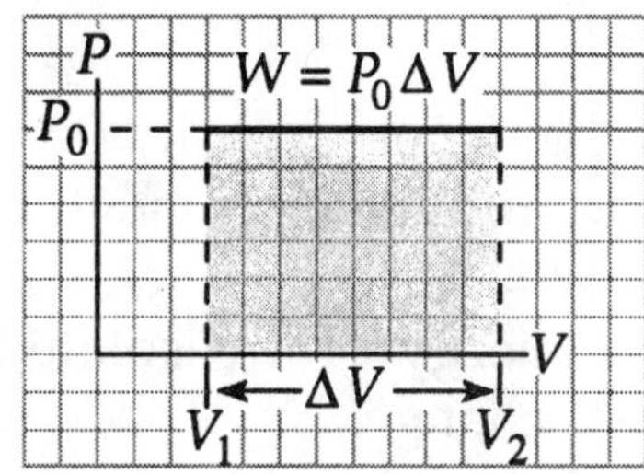

This is a statement of the first law of thermodynamics in equation form. This law is a generalization of the law of conservation of energy.

$$\Delta U = Q - W \tag{12.2}$$

Q = heat added to system

W = work done by system

ΔU = change in the internal energy of the system

Q is positive when heat energy is added to the system. W is positive when work is done by the system on its surroundings.

Comment on the first law of thermodynamics.

The values of both Q and W depend on the path or sequence of processes by which a system changes from an initial to a final state. The internal energy U depends only on the initial and final states of the system.

The following definitions will be important in describing the remaining equations in this chapter:

Isolated system

A system which does not interact with its surroundings. In such a system $Q = W = 0$, so that $\Delta U = 0$. The internal energy of an isolated system cannot change.

Cyclic process

A process for which the initial and final states are the same. For a cyclic process, $\Delta U = 0$ and $Q = W$. The net work done per cycle equals the heat energy added to the system per cycle.

Adiabatic process

A process in which no heat enters or leaves the system. $Q = 0$ and, therefore, $\Delta U = -W$. A system can undergo an adiabatic process if it is thermally insulated from its surroundings.

Isobaric process

A process which occurs at constant pressure. In an isobaric process, the work done and the heat transferred are both nonzero.

Isovolumetric process

A process which occurs at constant volume. Since at constant volume, $\Delta V = 0$, $W = 0$; and therefore, $\Delta U = Q$. All of the heat added at constant volume goes to increasing the internal energy.

Chapter 12

A process which occurs at constant temperature. The change in internal energy during an isothermal process results both from heat added and work done.	Isothermal process	
A device that converts thermal energy into other forms of energy by carrying a substance through a cycle: (1) heat is absorbed from a source at a high temperature, (2) work is done by the engine, and (3) heat is expelled by the engine to a source at a low temperature.	Heat engine	
A measure of the degree of disorder in a system.	Entropy	
The net work done by a heat engine equals the net heat flowing into it. Q_h is the quantity of heat absorbed from the high temperature reservoir and Q_c is the quantity of heat expelled to the low (cold) temperature reservoir. Both Q_h and Q_c are taken to be positive quantities.	$W = Q_h - Q_c$	(12.3)
The thermal efficiency of a heat engine is the ratio of the net work done to the heat absorbed during one cycle of the process.	$e \equiv \frac{Q_h - Q_c}{Q_h} = 1 - \frac{Q_c}{Q_h}$	(12.4)

It is impossible to construct a heat engine that, operating in a cycle, produces no other effect than the absorption of thermal energy from a reservoir and the performance of an equal amount of work.

Statement on the second law of thermodynamics.

The Carnot cycle is the most efficient cyclic process and the thermal efficiency of an ideal Carnot engine depends on the temperatures of the hot and cold reservoirs.

$$e_c = \frac{T_h - T_c}{T_h} = 1 - \frac{T_c}{T_h} \qquad (12.5)$$

Entropy, S, is a thermodynamic variable which characterizes the degree of disorder in a system. All physical processes tend toward a state of increasing entropy; and in going from an initial to a final state, the change in entropy ΔS is the ratio of the heat energy added to the system to the absolute temperature of the system.

$$\Delta S = \frac{\Delta Q_r}{T} \qquad (12.6)$$

The entropy of the Universe increases in all natural processes. This is an alternate way of expressing the second law of thermodynamics.

Chapter 12

REVIEW CHECKLIST

- ▷ Understand how work is defined when a system undergoes a change in state, and the fact that work (like heat) depends on the path taken by the system. You should also know how to sketch processes on a *PV* diagram, and calculate work using these diagrams.

- ▷ State the first law of thermodynamics ($\Delta U = Q - W$), and explain the meaning of the three forms of energy contained in this statement. Discuss the implications of the first law of thermodynamics as applied to (i) an isolated system, (ii) a cyclic process, (iii) an adiabatic process, and (iv) an isothermal process.

- ▷ Describe the processes via which an ideal heat engine goes through a **Carnot cycle.** Express the efficiency of an ideal heat engine (Carnot engine) as a function of work and heat exchange with its environment. Express the maximum efficiency of an ideal heat engine as a function of its input and output temperatures.

- ▷ Understand the concept of entropy. Define **entropy** for a system in terms of its heat energy gain or loss, and its temperature. State the **second law of thermodynamics** as it applies to entropy changes in a thermodynamic system.

SOLUTIONS TO SELECTED END-OF-CHAPTER PROBLEMS

1. The only form of energy possessed by molecules of a monatomic ideal gas is translational kinetic energy. Using the results from the discussion of kinetic theory in Section 10.6, show that the internal energy of a monatomic ideal gas at pressure P and occupying volume V may be written as $U = \frac{3}{2}PV$.

Solution

The average kinetic energy per molecule in a monatomic ideal gas is $KE_{\text{molecule}} = \frac{3}{2}k_B T$, where T is the absolute temperature of the gas and the Boltzmann constant is

$$k_B = \frac{\text{molar gas constant}}{\text{Avogadro's number}} = \frac{R}{N_A}$$

If the gas contains N molecules, the total kinetic energy associated with random thermal motions is

$$KE = N\left(KE_{\text{molecule}}\right) = \frac{3}{2}N\left(\frac{R}{N_A}\right)T = \frac{3}{2}\left(\frac{N}{N_A}\right)RT$$

The total number of molecules, N, divided by the number of molecules in a mole, N_A, gives the number of moles of gas present, n. Thus, the total kinetic energy becomes $KE = \frac{3}{2}nRT$. In an ideal gas, there are no intermolecular forces, so there are no potential energies contributing to the internal energy of the gas. The internal energy is therefore the same as the total kinetic energy, or $U = \frac{3}{2}nRT$. Making use of the ideal gas law, $PV = nRT$, this internal energy may be written as $U = \frac{3}{2}PV$. ◊

5. A gas expands from I to F along the three paths indicated in Figure P12.5. Calculate the work done by the gas along paths (a) IAF, (b) IF, and (c) IBF.

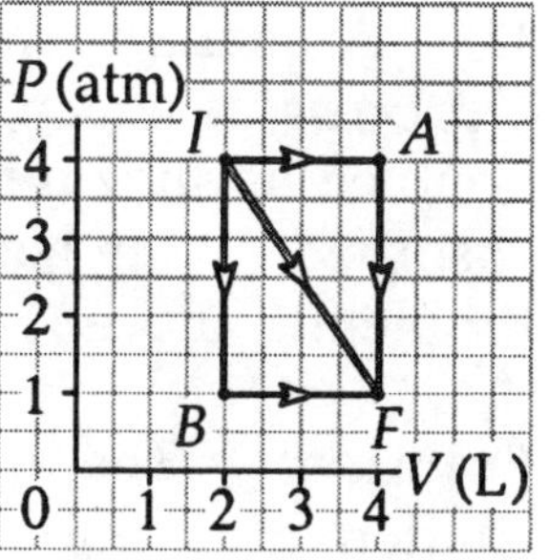

Figure P12.15

Solution In any process, the work done by the gas is equal to the area under the curve representing that process on the PV diagram.

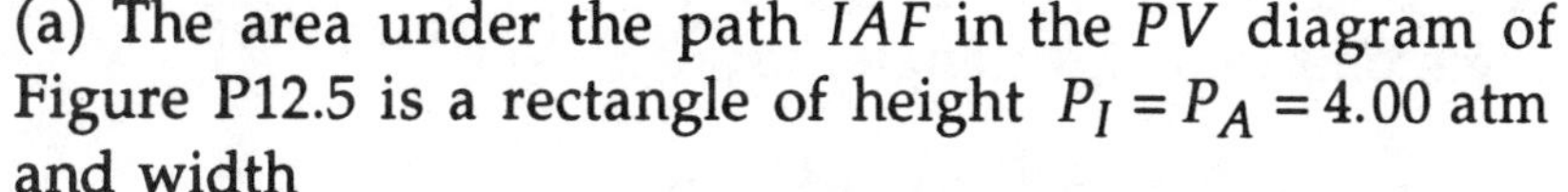

(a) The area under the path IAF in the PV diagram of Figure P12.5 is a rectangle of height $P_I = P_A = 4.00$ atm and width

$$V_A - V_I = (4.00 - 2.00) \text{ liters} = 2.00 \text{ liters}$$

The work done is then $W_{IAF} = (4.00 \text{ atm})(2.00 \text{ liters}) = 8.00 \text{ atm} \cdot \text{liter}$, which can be converted to standard SI units by the following conversion factor:

$$1 \text{ atm} \cdot \text{liter} = (1 \text{ atm} \cdot \text{liter})\left(\frac{1.013 \times 10^5 \text{ N/m}^2}{1 \text{ atm}}\right)\left(\frac{10^{-3} \text{ m}^3}{1 \text{ liter}}\right) = 101.3 \text{ J}$$

Thus, $$W_{IAF} = 8.00 \text{ atm} \cdot \text{liter}(101.3 \text{ J/atm} \cdot \text{liter}) = 810 \text{ J}$$ ◊

(b) The area under the path IF on the PV diagram consists of a triangle on top of a rectangle. The triangle has a height $P_I - P_B = 3.00$ atm and the height of the rectangle is $P_B = 1.00$ atm. The base of the triangle and the width of the rectangle both equal $V_F - V_B = 2.00$ liters. Thus,

$$W_{IF} = \frac{1}{2}(2.00 \text{ liters})(3.00 \text{ atm}) + (1.00 \text{ atm})(2.00 \text{ liters}) = 5.00 \text{ atm} \cdot \text{liter}$$

or $$W_{IF} = 5.00 \text{ atm} \cdot \text{liter}\left(\frac{101.3 \text{ J}}{1 \text{ atm} \cdot \text{liter}}\right) = 507 \text{ J}$$ ◊

(c) The area under the path IBF is a rectangle of height $P_B = 1.00$ atm and width $V_F - V_B = 2.00$ liters. The work done by the gas along this path is

$$W_{IbF} = (1.00 \text{ atm})(2.00 \text{ liters}) = 2.00 \text{ atm} \cdot \text{liter}\left(\frac{101.3 \text{ J}}{1 \text{ atm} \cdot \text{liter}}\right) = 203 \text{ J}$$ ◊

11. A quantity of a monatomic ideal gas undergoes a process in which both its pressure and volume are doubled as shown in Figure P12.11. What is the heat absorbed by the gas during this process? (**Hint:** See Problem 1 above.)

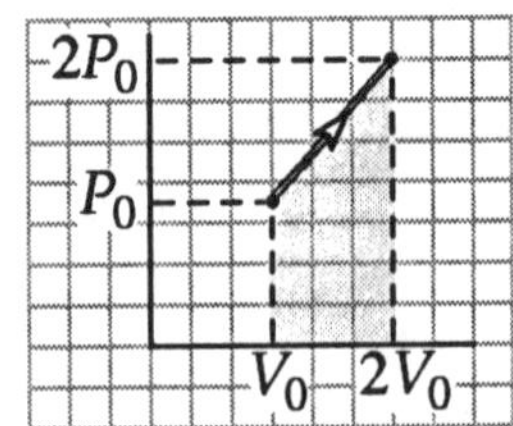

Figure P12.11 (modified)

Solution The work done by the gas is equal to the shaded area under the process curve on the PV diagram. This is a triangle on top of a rectangle, so

$$W = \frac{1}{2}(2P_0 - P_0)(2V_0 - V_0) + P_0(2V_0 - V_0) = 1.5P_0V_0$$

Using Problem 12.1, the change in the internal energy of the gas is seen to be

$$\Delta U = U_f - U_0 = \frac{3}{2}P_fV_f - \frac{3}{2}P_0V_0 = \frac{3}{2}\left[(2P_0)(2V_0) - P_0V_0\right] = 4.5P_0V_0$$

The first law of thermodynamics, $\Delta U = Q - W$, then gives the heat absorbed by the gas as

$$Q = \Delta U + W = 4.5P_0V_0 + 1.5P_0V_0 = 6P_0V_0$$ ◊

---

15. A gas is taken through the cyclic process described by Figure P12.15. (a) Find the net heat transferred to the system during one complete cycle. (b) If the cycle is reversed—that is, the process follows the path $ACBA$—what is the net heat transferred per cycle?

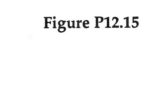

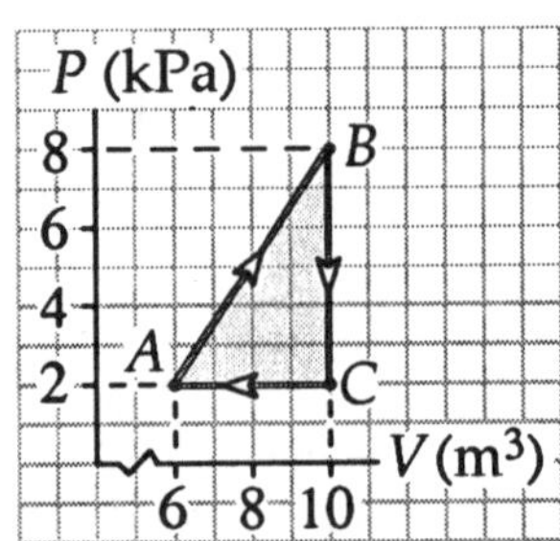

Figure P12.15

Solution

For a cyclic process, the net change in the internal energy of the system is zero. This is because the system returns to its initial thermodynamic state.

The first law of thermodynamics, then gives $\Delta U = Q - W = 0$ or $Q = W$ (i.e., the net heat transferred to the system during one cycle equals the net work done by the system during the cycle).

(a) When the system follows the path *ABCA* as indicated in Figure P12.15, work is done during processes *AB* and *CA* of the cycle. The work done during the constant-volume process *BC* is zero. Notice that during *AB*, the system is expanding and hence doing **positive** work on its surroundings. During process *CA*, the system is being compressed so the surroundings are doing work **on** the system (i.e., there is a **negative** quantity of work done **by** the system). The net work done by the system during the cycle *ABCA* is thus

$$W_{ABCA} = +|W_{AB}| - |W_{CA}| = +(\text{area under path } AB) - (\text{area under path } CA)$$

or $$W_{ABCA} = +(\text{area enclosed by the cyclic path})$$

The area enclosed by this cyclic path is triangular with a height $\Delta P = (8.0 - 2.0)\text{ kPa} = 6.0 \times 10^3\text{ Pa}$ and base $\Delta V = (10 - 6.0)\text{ m}^3 = 4.0\text{ m}^3$. The enclosed area is then

$$|W_{\text{net}}| = \tfrac{1}{2}(4.0\text{ m}^3)(6.0 \times 10^3\text{ Pa}) = 12 \times 10^3\text{ J} = 12\text{ kJ}$$

Thus, the net heat transferred to the system is $Q_{ABCA} = W_{ABCA} = +12\text{ kJ}$ ◊

(b) If the cycle is reversed (i.e., the system follows the path *ACBA*), the system is expanding and does **positive** work on the surroundings during the process *AC*, but is being compressed and doing **negative** work during the process *BA*. No work is done during the constant-volume process *CB*. Thus, the net work for this cycle is

$$W_{ACBA} = +|W_{AC}| - |W_{BA}| = (\text{area under path } AC) - (\text{area under path } BA)$$

or $$W_{ACBA} = -(\text{area enclosed by the cyclic path}) = -W_{ABCA} = -12\text{ kJ}$$

The first law of thermodynamics then gives $Q_{ACBA} = W_{ACBA} = -12\text{ kJ}$ ◊

19. One gram of water changes from liquid to solid at a constant pressure of one atmosphere and a constant temperature of 0 °C. In the process, the volume changes from 1.00 cm^3 to 1.09 cm^3. (a) Find the work done by the water, and (b) the change in the internal energy of the water.

Solution

(a) In a constant pressure process, the work done on its surroundings by a thermodynamic system is $W = P(\Delta V)$. In the process described, the pressure is $P = 1.00 \text{ atm} = 1.013 \times 10^5 \text{ Pa}$, and

$$\Delta V = V_f - V_i = \left(0.090 \text{ cm}^3\right)\left(\frac{1.00 \text{ m}^3}{1.00 \times 10^6 \text{ cm}^3}\right) = 9.00 \times 10^{-8} \text{ m}^3$$

The work done by the water is then

$$W = \left(1.013 \times 10^5 \text{ Pa}\right)\left(9.00 \times 10^{-8} \text{ m}^3\right) = 9.12 \times 10^{-3} \text{ J} \qquad \Diamond$$

(b) From the first law of thermodynamics, the change in internal energy of the system is $\Delta U = Q - W$, where Q is the heat energy **absorbed** by the system during the process. In this case, the heat of fusion must be **removed** (i.e., $Q < 0$) from the system to convert liquid water at 0 °C into a solid at 0 °C.

Thus, $$Q = -mL_f = -\left(1.00 \times 10^{-3} \text{ kg}\right)\left(3.33 \times 10^5 \text{ J/kg}\right) = -333 \text{ J}$$

and $$\Delta U = -333 \text{ J} - 9.12 \times 10^{-3} \text{ J} = -333 \text{ J} \qquad \Diamond$$

---

23. The efficiency of a Carnot engine is 30%. The engine absorbs 800 J of heat per cycle from a hot reservoir at 500 K. Determine (a) the heat expelled per cycle and (b) the temperature of the cold reservoir.

Solution

(a) The efficiency of a heat engine is defined as $e = W_{net}/Q_{input}$, where W_{net} is the net work done by the engine during its cycle and Q_{input} is the energy the engine takes in each cycle from the hot reservoir. Thus, for the engine described,

$$W_{net} = e \cdot Q_{input} = 0.30(800 \text{ J}) \quad \text{or} \quad W_{net} = 2.4 \times 10^2 \text{ J}$$

In the cyclic process of a heat engine, the change in internal energy per cycle is zero ($\Delta U_{cycle} = 0$). The first law of thermodynamics then gives

$$\Delta U_{cycle} = Q_{net} - W_{net} = 0 \quad \text{or} \quad Q_{net} = W_{net}$$

Thus, $W_{net} = Q_{input} - Q_{output}$ and the heat expelled each cycle is

$$Q_{output} = Q_{input} - W_{net} = 800 \text{ J} - 2.4 \times 10^2 \text{ J} = 5.6 \times 10^2 \text{ J} \qquad \Diamond$$

(b) The thermal efficiency of a heat engine operating on the Carnot cycle may be expressed in terms of the absolute temperatures of the hot and cold reservoirs used. This efficiency is:

$$e_c = (T_h - T_c)/T_h = 1 - T_c/T_h$$

Thus, if a Carnot engine uses a hot reservoir with a temperature of $T_h = 500$ K and has an efficiency of 30%, the temperature of the cold reservoir used must be

$$T_c = (1 - e_c)T_h = (1 - 0.30)(500 \text{ K}) = 350 \text{ K} \qquad \Diamond$$

31. A nuclear power plant has a power output of 1000 MW and operates with an efficiency of 33%. If excess heat is carried away from the plant by a river with a flow rate of 1.0×10^6 kg/s, what is the rise in temperature of the flowing water?

Solution

The net work output from this plant each second is

$$W_{net} = P(\Delta t) = \left(1000 \times 10^6 \text{ J/s}\right)(1.0 \text{ s}) = 1.0 \times 10^9 \text{ J}$$

and the efficiency of the plant is $e = W_{net}/Q_{input} = 0.33$. Thus, the heat input to the plant each second is

$$Q_{input} = \frac{W_{net}}{0.33} = \frac{1.0 \times 10^9 \text{ J}}{0.33} = 3.0 \times 10^9 \text{ J}$$

The excess heat expelled from the plant each second is therefore

$$Q_{output} = Q_{input} - W_{net} = 3.0 \times 10^9 \text{ J} - 1.0 \times 10^9 \text{ J} = 2.0 \times 10^9 \text{ J}$$

If the heat expelled each second is absorbed by 1.0×10^6 kg of water with a specific heat of 4186 J/kg·°C, the rise in the temperature of that water will be

$$\Delta T = \frac{Q_{output}}{m_{water}\, c_{water}} = \frac{2.0 \times 10^9 \text{ J}}{\left(1.0 \times 10^6 \text{ kg}\right)\left(4186 \text{ J/kg}\cdot°\text{C}\right)}$$

The water will therefore rise in temperature by $\Delta T = 0.48$ °C ◊

35. Two 2000-kg cars, both traveling at 20 m/s, undergo a head-on collision and stick together. Find the entropy change of the Universe during the collision if the temperature is 23°C.

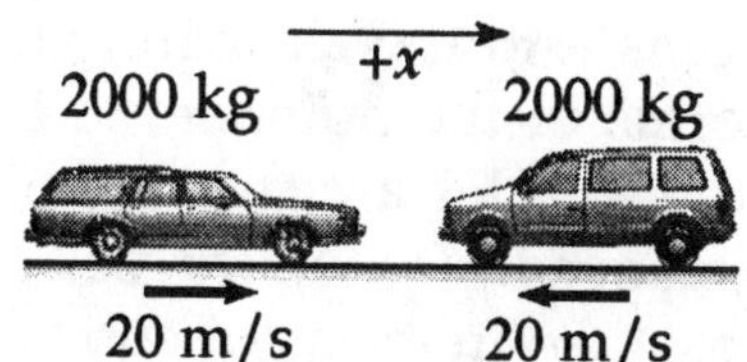

Solution Consider a system consisting of the two cars. The total mechanical energy of this system before impact is

$$E_i = (KE_{\text{total}})_i + (PE_{\text{total}})_i = (KE_{\text{total}})_i + 0 = (KE_{\text{total}})_i$$

where the zero potential energy level has been chosen at the roadway. Since the two cars have the same mass and initial speed, this energy is

$$E_i = 2\left(\tfrac{1}{2}mv_i^2\right) = (2000 \text{ kg})(20 \text{ m/s})^2 = 8.0\times10^5 \text{ J}$$

The collision is head-on and the two cars stick together and hence move with the same velocity, V, after impact. Conservation of momentum $(\mathbf{P}_{\text{after}} = \mathbf{P}_{\text{before}})$ then gives

$$(m_1 + m_2)V = m_1 v_{1i} + m_2 v_{2i} = (2000 \text{ kg})(+20 \text{ m/s} - 20 \text{ m/s}) = 0$$

or the velocity of the wreckage immediately after impact is $V = 0$. The mechanical energy of the system immediately after impact is then

$$E_f = KE_f + PE_f = \tfrac{1}{2}(m_1 + m_2)V^2 + (m_1 + m_2)g(0) = 0 + 0 = 0$$

The transformation of mechanical energy to heat energy resulting from the impact is $Q = E_i - E_f = +8.0\times10^5$ J. This heat energy is transferred to the surroundings which are at an **absolute** temperature of $T = 23 + 273 = 296$ K. The change in entropy generated by this process is

$$\Delta S = \frac{Q}{T} = \frac{+8.0\times10^5 \text{ J}}{296 \text{ K}} = +2.7\times10^3 \text{ J/K} \qquad \Diamond$$

37. Prepare a table like Table 12.1 for the following occurrence. You toss four coins into the air simultaneously. Record all the possible results of the toss in terms of the numbers of heads and tails that can result. (For example, HHTH and HTHH are two possible ways in which three heads and one tail can be achieved.) (a) On the basis of your table, what is the most probable result of a toss? (b) In terms of entropy, what is the most ordered state and (c) what is the most disordered?

Solution The desired table is shown below:

End Result	Possible Combinations	Total with Same Results
All Heads	HHHH	1
3 Heads, 1 Tail	THHH, HTHH, HHTH, HHHT	4
2 Heads, 2 Tails	TTHH, THTH, THHT, HTHT, HHTT, HTTH	6
1 Head, 3 Tails	HTTT, THTT, TTHT, TTTH	4
All Tails	TTTT	1

(a) The most probable result is seen to be 2 heads and 2 tails. Six combinations out of a total of 16 possible combinations will yield this result. The probability of obtaining 2 heads and 2 tails is thus 6/16. ◊

(b) The results with the most order are those in which all coins are oriented the same way (either all heads or all tails). These two states are equally probable, each having a probability of 1/16. ◊

(c) The result which allows the highest degree of disorder in the coin orientations is also the most probable state (2 heads and 2 tails). ◊

---

45. One object is at a temperature of T_h and another is at a lower temperature, T_c. Use the second law of thermodynamics to show that heat transfer can only occur from the hotter to the colder object. Assume a constant-temperature process.

Solution The second law of thermodynamics states that the change in the total entropy of an isolated system can never be negative. In any thermodynamic process, this total entropy must either remain constant or increase.

Assume that, in an isolated system, heat energy of magnitude Q flows from an object at absolute temperature T_c to a second object at absolute temperature $T_h > T_c$. Then, the heat transfers for each object would be $Q_h = +Q$ and $Q_c = -Q$, If the thermal capacity of each object is very large in comparison to Q, the temperatures of the objects remain essentially constant during this process. The entropy change associated with each object is then,

$$\Delta S_h = \frac{Q_h}{T_h} = \frac{+Q}{T_h} \quad \text{and} \quad \Delta S_c = \frac{Q_c}{T_c} = \frac{-Q}{T_c}$$

The total change in entropy for this isolated system would be

$$\Delta S_{\text{total}} = \Delta S_h + \Delta S_c \quad \text{or} \quad \Delta S_{\text{total}} = \left(\frac{+Q}{T_h}\right) + \left(\frac{-Q}{T_c}\right) = Q\left(\frac{T_c - T_h}{T_h T_c}\right)$$

Since absolute temperatures are never negative, the product $T_h T_c$ is positive. Since $T_h > T_c$, the factor $(T_c - T_h)/T_h T_c$ is negative. Therefore, the only way this process can be in agreement with the second law $(\Delta S_{\text{total}} \geq 0)$ is for the factor Q to also be negative. However, if $Q < 0$, then $Q = -|Q|$, and the heat transfers for each object are

$$Q_h = +Q = -|Q| < 0 \quad \text{and} \quad Q_c = -Q = -(-|Q|) = +|Q| > 0.$$

That is, the second law of thermodynamics requires that the actual flow of heat energy be **from** the hotter body **to** the cooler body . ◊

---

53. The surface of the Sun is approximately 5700 K, and the temperature of the Earth's surface is approximately 290 K. What entropy change occurs when 1000 J of thermal energy is transferred from the Sun to the Earth?

Solution Because of their very large thermal capacities, the temperatures of the Sun and Earth change by negligible amounts when 1000 J of energy is transferred from one body to the other. In this transfer, the Sun **loses** 1000 J of energy while the Earth **gains** an equal amount. Thus, $Q_{\text{Sun}} = -1000\text{ J}$ and $Q_{\text{Earth}} = +1000\text{ J}$. The change in entropy for the Sun and Earth are

$$\Delta S_{\text{Sun}} = \frac{Q_{\text{Sun}}}{T_{\text{Sun}}} = \frac{-1000\text{ J}}{5700\text{ K}} = -0.175\text{ J/K} \qquad \Delta S_{\text{Earth}} = \frac{Q_{\text{Earth}}}{T_{\text{Earth}}} = \frac{+1000\text{ J}}{290\text{ K}} = +3.45\text{ J/K}$$

The net change in entropy of the universe during this process is then

$\Delta S_U = \Delta S_{\text{Sun}} + \Delta S_{\text{Earth}}$ or $\Delta S_U = -0.175\text{ J/K} + 3.45\text{ J/K} = +3.27\text{ J/K}$ ◊

Note that $\Delta S_U \geq 0$, in agreement with the Clausius statement of the second law of thermodynamics.

---

57. One mole of neon gas is heated from 300 K to 420 K at constant pressure. Calculate (a) the heat energy transferred to the gas, (b) the change in the internal energy of the gas, and (c) the work done by the gas. Note that neon has a specific heat of $c = 20.79$ J/mol·K.

Solution (a) The heat energy transferred to this gas is $Q = mc(\Delta T)$

$$Q = (1.00\text{ mol})(20.79\text{ J/mol}\cdot\text{K})(420\text{ K} - 300\text{ K}) = 2.49 \times 10^3\text{ J} \quad ◊$$

(b) Neon is a noble gas, so it is monatomic, has negligible intermolecular forces, and closely approximates an ideal gas. The internal energy of the gas is thus $U = \frac{3}{2}PV = \frac{3}{2}nRT$ (see Problem 12.1). The gas' internal energy changes by:

$$\Delta U = U_f - U_i = \tfrac{3}{2}nRT_f - \tfrac{3}{2}nRT_i = \tfrac{3}{2}nR(\Delta T)$$

or $$\Delta U = \tfrac{3}{2}(1.00\text{ mol})(8.31\text{ J/mol}\cdot\text{K})(420\text{ K} - 300\text{ K}) = 1.50 \times 10^3\text{ J} \quad ◊$$

(c) The first law of thermodynamics, $\Delta U = Q - W$, then gives the work done by the gas as $W = Q - \Delta U = 2.49 \times 10^3\text{ J} - 1.50 \times 10^3\text{ J} = 990\text{ J}$ ◊

CHAPTER SELF-QUIZ

1. An ideal gas undergoes an adiabatic process in doing 25 J of work on its environment. What is its change in internal energy?
   a. 50 J
   b. 25 J
   c. Zero
   d. −25 J

2. A 3-mol ideal gas system is maintained at a constant volume of 4 liters; if 20 J of heat are added, what is the work done by the system?
   a. Zero
   b. 5.0 J
   c. 6.7 J
   d. 20 J

3. An ideal gas is maintained at a constant pressure of 200 N/m^2 during an isobaric process while its volume decreases by 0.2 m^3. What work is done by the system on its environment?
   a. 40 J
   b. 1000 J
   c. −40 J
   d. −1000 J

4. What is the work done by the gas as it expands from pressure P_1 and volume V_1 to pressure P_2 and volume V_2 along the indicated straight line?

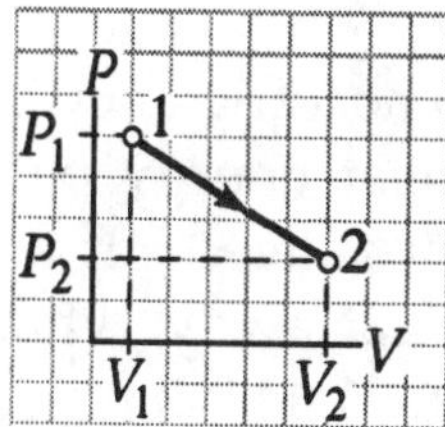

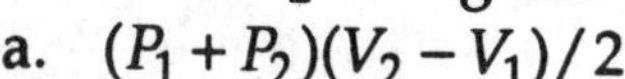

   a. $(P_1 + P_2)(V_2 - V_1)/2$
   b. $(P_1 + P_2)(V_2 - V_1)$
   c. $(P_1 - P_2)(V_2 - V_1)/2$
   d. $(P_1 - P_2)(V_2 + V_1)$

5. A 3-mol ideal gas system is maintained at a constant volume of 4 liters; if 20 J of heat are added, what is the change in internal energy of the system?
   a. Zero
   b. 5.0 J
   c. 6.7 J
   d. 20 J

6. A boulder of mass 600 kg tumbles down a mountainside and stops 160 m below. If the temperature of the boulder, mountain, and surrounding air are all at 300 K, what is the change in entropy of the Universe?
   a. 320 J/K
   b. 3100 J/K
   c. 1200 J/K
   d. 5200 J/K

7. An ideal gas at point O with pressure, volume, and temperature, P_0, V_0, and T_0, respectively, is heated to point A, allowed to expand to point B at constant temperature $2T_0$, and then returned to the original conditions at point O. The internal energy decreases by $3P_0V_0/2$ going from point B to Point O. How much heat left the gas from point B to point O?
   a. 0
   b. $5P_0V_0/2$
   c. $3P_0V_0/2$
   d. $P_0V_0/2$

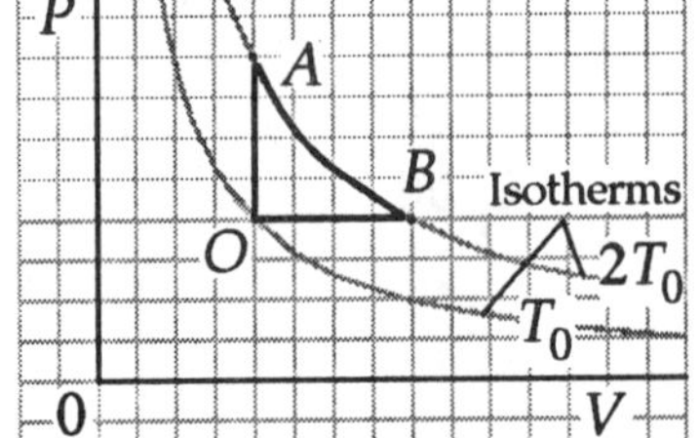

8. The efficiency of a 1000-MW nuclear power plant is 33 %: 2000 MW of heat energy per second is rejected for every 1000 MW of electrical power produced. If a river of flow rate 10^6 kg/s is used to transport the excess heat away, what is the average temperature increase of the river?
   a. 1.44 °C
   b. 0.96 °C
   c. 0.48 °C
   d. 0.24 °C

9. A bottle containing an ideal gas has a volume of 2 m^3 and a pressure of 1×10^5 N/m^2 at a temperature of 300 K. The bottle is placed against a metal block that is maintained at 900 K and the gas expands as the pressure remains constant until the temperature of the gas reaches 900 K. How much work is done by the gas?
   a. 0
   b. 2×10^5 J
   c. 4×10^5 J
   d. 6×10^5 J

10. A gasoline engine absorbs 2500 J of heat energy and performs 500 J of mechanical work in each cycle. The efficiency of the engine is
   a. 80 %
   b. 40 %
   c. 60 %
   d. 20 %

11. During each cycle of operation a refrigerator absorbs 55 cal of heat energy from the freezer compartment and expels 85 cal to the room. How much work is required top operate the system?
   a. 356 J
   b. 125 J
   c. 146 J
   d. 565 J

12. Calculate the entropy change when 1 mol (18 grams) of water at 100 °C is converted into steam. (The heat of vaporization of water is 540 cal/gm).
   a. 26.1 cal/K
   b. 35.6 cal/K
   c. 97.2 cal/K
   d. 174.9 cal/K

Chapter 13

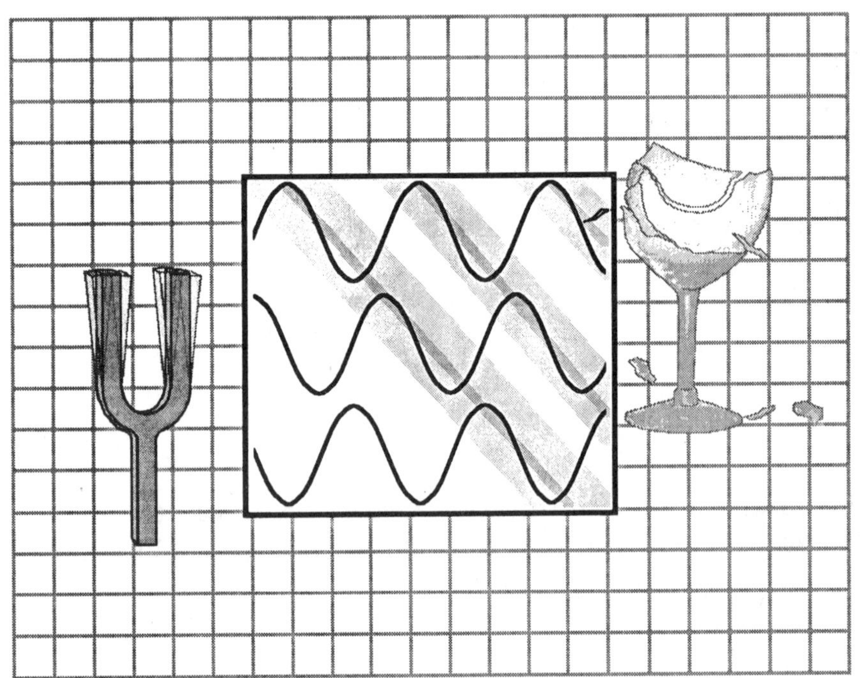

VIBRATIONS AND WAVES

Chapter 13

VIBRATIONS AND WAVES

During your study of this chapter, you will have an opportunity to use many of the concepts which were developed in the chapter on mechanics. You will examine various forms of **periodic motion,** concentrating especially on motion that occurs when the force on an object is proportional to the displacement of the object from its equilibrium position. When such a force acts only toward the equilibrium position, and has a magnitude which is proportional to the displacement from equilibrium, the result is a back-and-forth motion called simple harmonic motion--oscillation, or vibration, between two extreme positions for an indefinite period of time with no loss of energy. The terms **harmonic motion** and **periodic motion** are used interchangeably in this chapter. Both refer to back-and-forth motion.

Since vibrations can move through a medium, we also study wave motion in this chapter. Many kinds of waves occur in nature, including sound waves, seismic waves, and electromagnetic waves. We end this chapter with a brief discussion of some terms and concepts that are common to all types of waves, and in later chapters we shall focus our attention on specific categories of waves.

NOTES FROM SELECTED CHAPTER SECTIONS

13.1 Hooke's Law
13.2 Elastic Potential Energy

Simple harmonic motion occurs when the net force along the direction of motion is a Hooke's law type of force; that is, when the net force is proportional to the displacement and in the opposite direction. It is necessary to define a few terms relative to harmonic motion:

1. The amplitude, A, is the **maximum distance that an object moves away from its equilibrium position**. In the absence of friction, an object will continue in simple harmonic motion and reach a maximum displacement equal to the amplitude on each side of the equilibrium position during each cycle.

2. The period, T, is **the time it takes the object to execute one complete cycle of the motion.**

3. The frequency, f, **is the number of cycles or vibrations per unit of time.**

Oscillatory motions are exhibited by many physical systems such as a mass attached to a spring, a pendulum, atoms in a solid, stringed musical instruments, and electrical circuits driven by a source of alternating current. **Simple harmonic motion** of a mechanical system corresponds to the oscillation of an object between two points for an indefinite period of time, with no loss in mechanical energy.

An object exhibits simple harmonic motion if the net external force acting on it is a **linear restoring force.**

13.5 Position, Velocity, and Acceleration as a Function of Time

The position (x), velocity (v), and acceleration (a) of an object moving with simple harmonic motion are shown in the three graphs below. In this particular case, the object was released from rest when it was a maximum distance (amplitude) from the equilibrium position.

$$x = A\cos(2\pi f t)$$

Shown here, from top to bottom, are graphs of displacement, velocity, and acceleration versus time for an object moving with simple harmonic motion under the initial conditions that $x_0 = A$ and $v_0 = 0$ at $t = 0$.

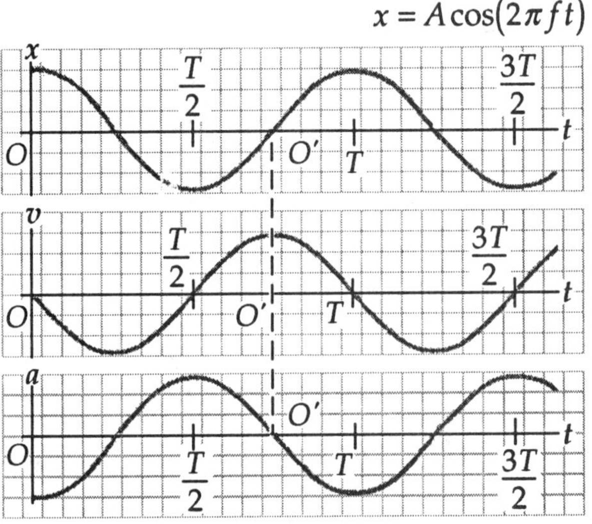

The most common system which undergoes simple harmonic motion is the mass-spring system shown in the figure at the right. The mass is assumed to move on a horizontal, **frictionless** surface. The point $x = 0$ is the equilibrium position of the mass; that is, the point where the mass would reside if left undisturbed. In this position, there is no horizontal force on the mass. When the mass is displaced a distance x from its equilibrium position, the spring produces a linear restoring force given by Hooke's law, $F = -kx$, where k is the force constant of the spring, and has SI units of N/m. The minus sign means that F is to the **left** when the displacement x is positive, whereas F is to the **right** when x is negative. In other words, the direction of the force F is **always** towards the equilibrium position.

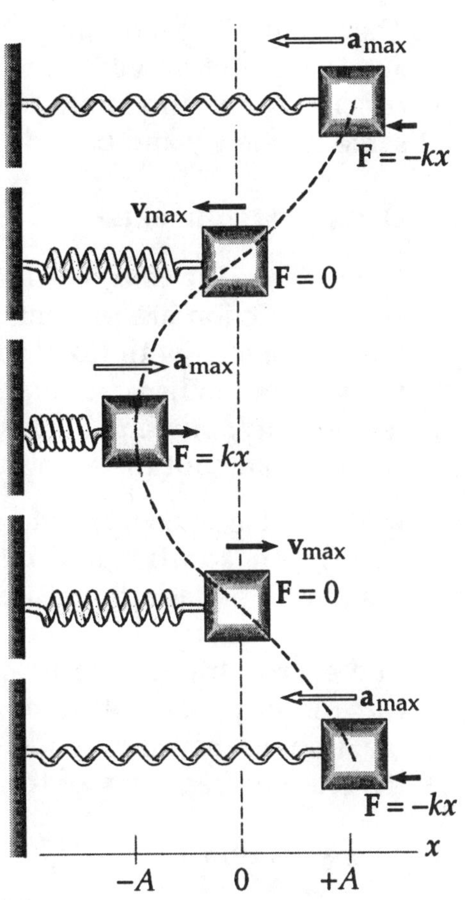

13.6 Motion of a Pendulum

A **simple pendulum consists** of a mass m attached to a light string of length L as shown in the figure. When the angular displacement θ is small during the entire motion (less than about 15°), the pendulum exhibits simple harmonic motion. In this case, the resultant force acting on the mass m equals the component of weight **tangent** to the circle, and has a magnitude $mg\sin\theta$. Since this force is always directed towards $\theta = 0$, it corresponds to a restoring force.

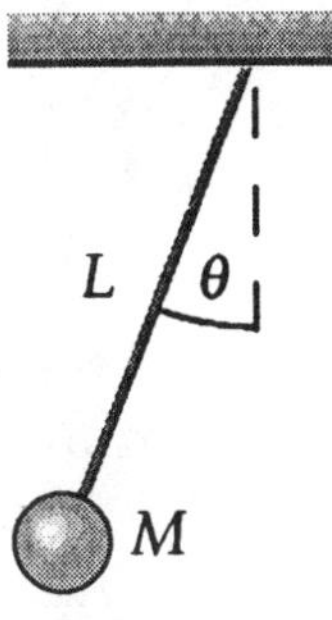

The period depends only on the length of the pendulum and the acceleration of gravity. The period **does not** depend on mass, so we conclude that **all** simple pendula of equal length oscillate with the same frequency and period.

13.7 Damped Oscillations

Damped oscillations occur in realistic systems in which retarding forces such as friction are present. These forces will reduce the amplitudes of the oscillations with time, since mechanical energy is continually lost by the system. When the retarding force is assumed to be proportional to the velocity, but small compared to the restoring force, the system will still oscillate, but the amplitude will decrease exponentially with time.

It is possible to compensate for the energy lost in a damped oscillator by adding an additional driving force that does positive work on the system. This additional energy supplied to the system must at least equal the energy lost due to friction to maintain constant amplitude. The energy transferred to the system is a maximum when the driving force is in phase with the velocity of the system. The amplitude is a maximum when the frequency of the driving force matches the natural (resonance) frequency of the system.

13.8 Wave Motion

The production of **mechanical waves** require: (1) an **elastic medium** which can be disturbed, (2) an **energy source** to provide a disturbance or deformation in the medium, and (3) a physical mechanism by way of which adjacent portions on the medium can **influence** each other. The three parameters important in characterizing waves are (1) wavelength, (2) frequency, and (3) wave velocity.

13.9 Types of Waves

Transverse waves are those in which particles of the disturbed medium move along a direction which is perpendicular to the direction of the wave velocity. For **longitudinal waves,** the particles of the medium undergo displacements which are parallel to the direction of wave motion.

13.10 Frequency, Amplitude, and Wavelength

Consider a wave traveling in the x direction on a very long string. Each particle along the string oscillates vertically in the y direction, in simple harmonic motion, with the same frequency as the source that vibrates the string. The maximum distance the string is raised above or below the equilibrium value is called the **amplitude,** A, of the wave. The distance between two successive points along the string that behave identically is called the **wavelength,** λ. The wave will advance along the string a distance of one wavelength in a time interval equal to one period of the vibration.

13.11 The Speed of Waves on Strings

For linear waves, the **velocity** of **mechanical waves** depends only on the physical properties of the medium through which the disturbance travels. In the case of waves on a **string,** the velocity depends on the tension in the string and the mass per unit length (linear mass density).

13.12 Superposition and Interference of Waves

If two or more waves are moving through a medium, the **resultant wave function** is the **algebraic sum** of the wave functions of the individual waves. Two traveling waves can pass through each other without being destroyed or altered.

13.13 Reflection of Waves

Whenever a traveling wave reaches a boundary, part or all of the wave is reflected. If the wave is traveling along a string and is reflected from a "fixed" end, the reflected pulse is inverted. By contrast, a pulse is reflected without inversion when it strikes a "free" end of a string.

EQUATIONS AND CONCEPTS

The force exerted by a spring on a mass attached to the spring and displaced a distance x from the unstretched position is given by Hooke's law. The force constant, k, is always positive and has a value which corresponds to the relative stiffness of the spring. The negative sign means that the force exerted on the mass is always directed opposite the displacement; the force is a restoring force, always directed toward the equilibrium position.

$$F_s = -kx \tag{13.1}$$

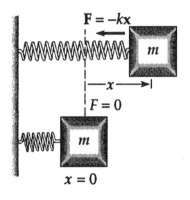

An object exhibits simple harmonic motion when the net force along the direction of motion is proportional to the displacement and oppositely directed.

Comment on simple harmonic motion.

This equation gives the acceleration of an object in simple harmonic motion as a function of position. Note that when the oscillating mass is at the equilibrium position ($x = 0$), the acceleration $a = 0$. The acceleration has its maximum magnitude when the displacement of the mass is maximum, $x = A$ (amplitude).

$$a = -\left(\frac{k}{m}\right)x \tag{13.2}$$

Work must be done by an external applied force in order to stretch or compress a spring. This work results in energy, called elastic potential energy, being stored in the spring.

$$PE_s \equiv \frac{1}{2}kx^2 \qquad (13.3)$$

The spring force is conservative; hence, in the absence of friction or other nonconservative forces, the total mechanical energy of the spring-mass system remains constant.

$$(KE + PE_g + PE_s)_i = (KE + PE_g + PE_s)_f \qquad (13.4)$$

The speed of an object in simple harmonic motion is a maximum at $x = 0$; the speed is zero when the mass is at the points of maximum displacement $(x = \pm A)$.

$$v = \pm\sqrt{\frac{k}{m}(A^2 - x^2)} \qquad (13.6)$$

The period of an object in simple harmonic motion is the time required to complete a full cycle of its motion.

$$T = 2\pi\sqrt{\frac{m}{k}} \qquad (13.8)$$

The frequency, the number of complete cycles per unit time, is the reciprocal of the period. The units of frequency are hertz (Hz).

$$f = \frac{1}{T} \qquad (13.9)$$

$$f = \frac{1}{2\pi}\sqrt{\frac{k}{m}} \qquad (13.10)$$

These equations represent the position of an object moving in simple harmonic motion as a function of time.

$$x = A\cos(\omega t) \qquad (13.11)$$

$$x = A\cos(2\pi f t) \qquad (13.13)$$

Notice that since $\cos(0) = 1$, when $t = 0$, $x = A$. Therefore, the particular form of the position equations shown here assumes that the vibrating object was at the point of maximum displacement when $t = 0$.

Comment on Equation 13.11 and Equation 13.13.

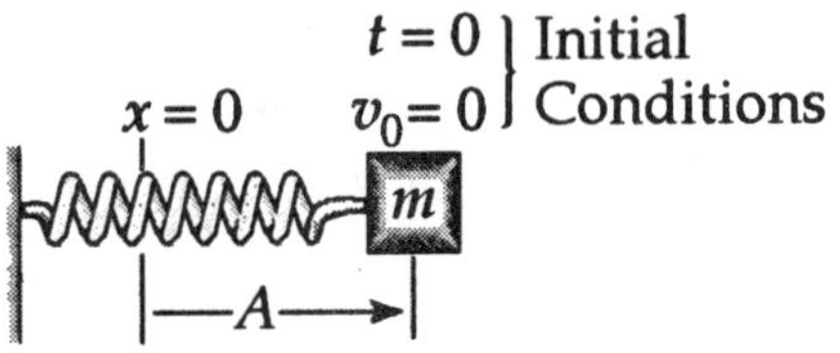

$x = 0$; $t = 0$, $v_0 = 0$ Initial Conditions; m; A

ω is the angular frequency (rad/s) of the object in simple harmonic motion and f is number of oscillations completed per unit time measured in hertz (Hz).

$$\omega = 2\pi f \qquad (13.13)$$

The period of oscillation of a simple pendulum depends only on its length, L, and the acceleration due to gravity, g. The period does not depend on the mass; and to a good approximation, the period does not depend on the amplitude, θ, within the range of small amplitudes.

$$T = 2\pi\sqrt{\frac{L}{g}} \qquad (13.15)$$

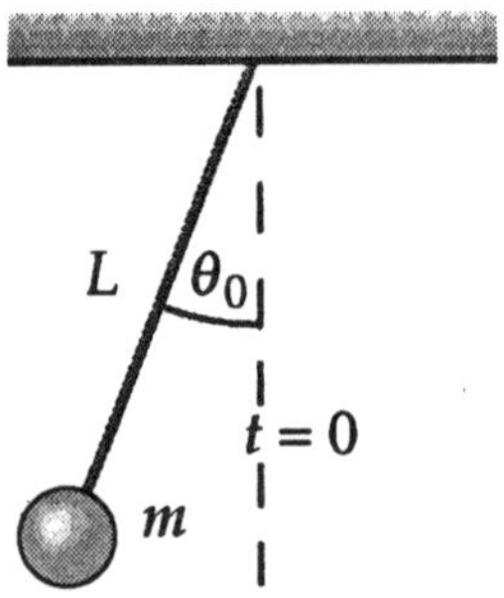

L; θ_0; $t = 0$; m

The following definitions of terms used to describe waves and wave motion will be important in discussing the remaining equations in this chapter:

Essential requirements of mechanical waves

(1) A source of disturbance.

(2) A material or medium which can undergo a disturbance.

(3) Some physical mechanism via which adjacent parts of the medium can influence each other. This allows the disturbance (pulse) to be propagated (travel along the medium).

A transverse wave is one in which the particles of the disturbed medium oscillate back and forth along a direction perpendicular to the direction of the wave velocity.

transverse wave

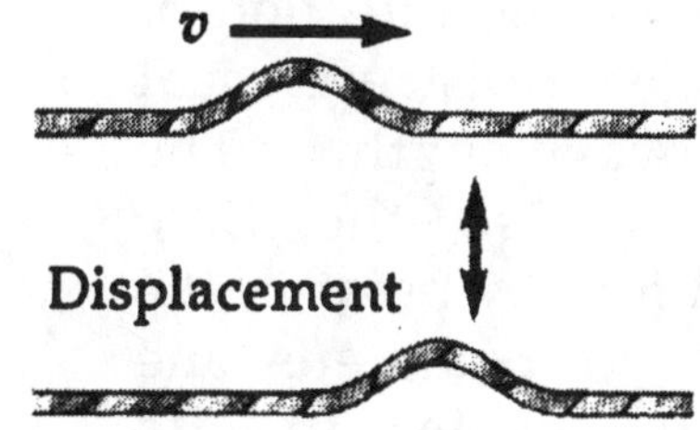

A longitudinal wave is one in which the particles of the medium undergo a displacement (oscillate back and forth) along a direction parallel to the direction of the wave velocity (direction along which the pulse travels).

longitudinal wave

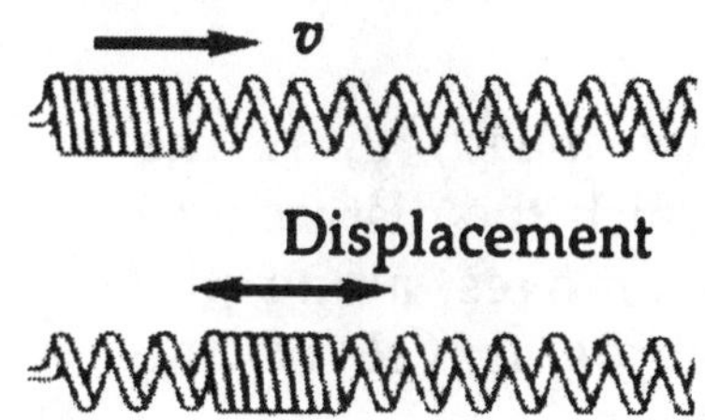

The wave amplitude, A, is the maximum possible value of the displacement of a particle of the medium away from its equilibrium position.

Amplitude, A

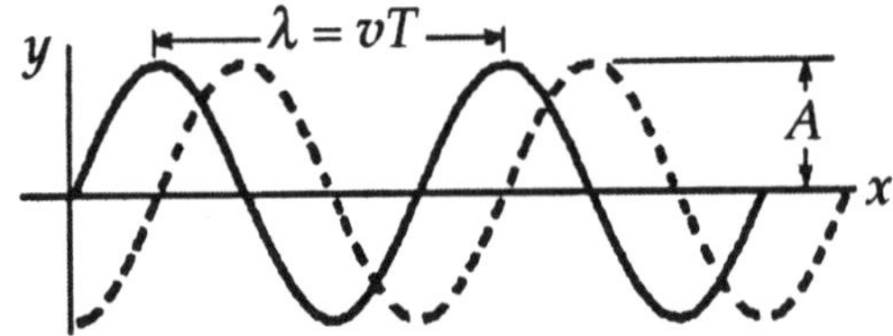

The wavelength, λ, is the minimum distance between two points which are the same distance from their equilibrium positions and are moving in the same direction (such a pair of points are said to be in phase).

wave length, λ

The period of the wave is the time required for the disturbance (or pulse) to travel along the direction of propagation a distance equal to the wavelength. The period is also the time required for any point in the medium to complete one complete cycle in its harmonic motion about its equilibrium point.

period, T

The wave speed, v, is the rate at which the disturbance or pulse moves along the direction of travel of the wave.

$$v = f\lambda \tag{13.16}$$

The wave speed in a stretched string depends on the tension in the string and the linear density (mass per unit length). For any mechanical wave, the speed depends only on the properties of the medium through which the wave travels.

$$v = \sqrt{\frac{F}{\mu}} \tag{13.17}$$

REVIEW CHECKLIST

▷ Describe the general characteristics of a system in simple harmonic motion; and define amplitude, period, frequency, and displacement.

▷ Define the following terms relating to wave motion: frequency, wavelength, velocity, and amplitude; and express a given harmonic wave function in several alternative forms involving different combinations of the wave parameters: wavelength, period, phase velocity, angular frequency, and harmonic frequency.

▷ Given a specific wave function for a harmonic wave, obtain values for the characteristic wave parameters: A, λ, k, ω, and f.

▷ Make calculations which involve the relationships between wave speed and the inertial and elastic characteristics of a string through which the disturbance is propagating.

▷ Define and describe the following wave associated phenomena: superposition, phase, interference, and reflection.

SOLUTIONS TO SELECTED END-OF-CHAPTER PROBLEMS

3. A ball dropped from a height of 4.00 m makes a perfectly elastic collision with the ground. Assuming no energy lost due to air resistance, (a) show that the motion is periodic and (b) determine the period of the motion. (c) Is the motion simple harmonic? Explain.

Solution (a) Since the collision is perfectly elastic $\left(KE_{\text{after}} = KE_{\text{before}}\right)$ and the recoil of the Earth may be neglected, the speed of the ball as it leaves the ground after a collision is the same as its speed immediately before the collision. The ball rebounds to the height from which it was initially dropped, comes to rest, and then falls again. The ball repeats this motion continuously, and thus, undergoes a periodic motion.

(b) The period (time for one complete cycle) of this motion is the total elapsed time from when the ball starts at rest 4.00 m above ground until it returns to rest at 4.00 m high following a bounce. For the downward motion,

$$v_0 = 0, \quad \Delta y = -4.00 \text{ m}, \quad a_y = -g = -9.80 \text{ m/s}^2$$

Thus, $\Delta y = v_0 t + \frac{1}{2} a_y t^2$ gives the time for the ball to reach the ground:

$$t_1 = \sqrt{\frac{2(4.00 \text{ m})}{9.80 \text{ m / s}^2}} = 0.904 \text{ s}$$

The speed of the ball just before impact is

$$v_1 = v_0 + a_y t_1 = 0 + \left(-9.80 \text{ m/s}^2\right)(0.904 \text{ s}) = -8.85 \text{ m/s}$$

On the upward flight, the initial velocity is $v_2 = -v_1 = +8.85 \text{ m/s}$, $v_{\text{final}} = 0$, and $\Delta y = +4.00 \text{ m}$. Therefore, the average velocity for this part of the motion is $\overline{v} = \left(v_2 + v_{\text{final}}\right)/2 = 4.43 \text{ m/s}$, and the time to return to the initial height is $t_2 = \Delta y / \overline{v} = (4.00 \text{ m})/(4.43 \text{ m/s}) = 0.904 \text{ s}$. The total time for this complete cycle of the motion is then

$$T = t_1 + t_2 = 1.81 \text{ s} \qquad \Diamond$$

(c) Simple harmonic motion is motion produced by a Hooke's law type force. Such a force is proportional to the displacement from an equilibrium position and is in the direction opposite to the displacement. This ball has a constant downward force, $w = mg$, acting on it while it is in the air. When the ball is at ground level, an upward impulsive force acts on it. These forces are not Hooke's law type forces. Thus, the motion **is not** simple harmonic. ◊

---

7. A child's toy consists of a piece of plastic attached to a spring (Fig. P13.7). The spring is compressed against the floor a distance of 2.00 cm, and the toy is released. If the toy has a mass of 100 g and rises to a maximum height of 60.0 cm, estimate the force constant of the spring.

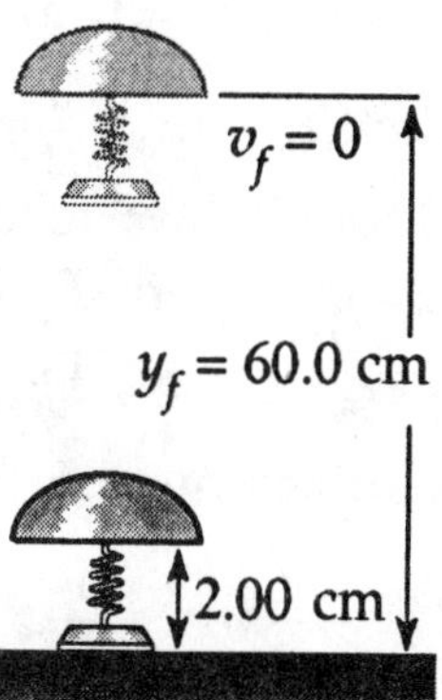

Solution When the toy is at rest against the floor with the spring compressed 2.00 cm, the total mechanical energy of the toy is

$$E_i = KE_i + PE_{g,i} + PE_{s,i} = \tfrac{1}{2}mv_i^2 + mgy_i + \tfrac{1}{2}kx_i^2 \quad \text{or} \quad E_i = 0 + 0 + \tfrac{1}{2}kx_i^2 = \tfrac{1}{2}kx_i^2$$

where the zero gravitational potential energy level has been chosen at the floor. When the toy comes to rest temporarily at its maximum height, the spring is no longer compressed and the total total mechanical energy is

$$E_f = KE_f + PE_{g,f} + PE_{s,f} = \tfrac{1}{2}mv_f^2 + mgy_f + \tfrac{1}{2}kx_f^2 = 0 + mgy_f + 0 = mgy_f$$

The forces acting on the toy during the intervening time interval are a gravitational force and a spring force, both conservative forces. Therefore, the mechanical energy is constant, or $E_i = E_f$. This gives $\tfrac{1}{2}kx_i^2 = mgy_f$, or the force constant is

$$k = \frac{2mgy_f}{x_i^2} = \frac{2(0.100\text{ kg})(9.80\text{ m/s}^2)(0.600\text{ m})}{(2.00\times10^{-2}\text{ m})^2} = 2.94\times10^3\text{ N/m} \quad ◊$$

---

13. A mass of 0.40 kg connected to a light spring with a spring constant of 19.6 N/m oscillates on a frictionless horizontal surface. If the spring is compressed 4.0 cm and released from rest, determine (a) the maximum speed of the mass, (b) the speed of the mass when the spring is compressed 1.5 cm, and (c) the speed of the mass when the spring is stretched 1.5 cm. (d) For what value of x does the speed equal one-half the maximum speed?

Solution Choosing the horizontal surface as the reference level $(y=0)$, $PE_g = mgy = 0$ at all times and the total energy at any point in the motion is $E = KE + PE_s = \frac{1}{2}mv^2 + \frac{1}{2}kx^2$. The only force doing work on the mass is a Hooke's law force (conservative). Therefore, the total energy is constant, or $E = E_0$ where E_0 is the energy when first released. Since the mass starts from rest, $E_0 = 0 + \frac{1}{2}kx_0{}^2 = \frac{1}{2}kx_0{}^2$ and the speed of the mass when it is at a displacement x from the equilibrium position is found from

$$\frac{1}{2}mv^2 + \frac{1}{2}kx^2 = \frac{1}{2}kx_0{}^2 \qquad \text{or} \qquad v = \sqrt{\left(x_0^2 - x^2\right)k/m} \qquad \textbf{[Equation 1]}$$

(a) Observing Equation 1, it is clear that the mass has maximum speed when $x = 0$ (i.e., at the equilibrium position). Thus,

$$v_{\max} = \sqrt{\frac{\left(x_0{}^2 - 0\right)k}{m}} = \sqrt{\frac{(0.040\text{ m})^2(19.6\text{ N/m})}{0.40\text{ kg}}} = 0.28\text{ m/s} = 28\text{ cm/s} \qquad \Diamond$$

(b) When the spring is compressed 1.5 cm, $x = -0.015$ m and Equation 1 gives the speed as

$$v = \sqrt{\left[(0.040\text{ m})^2 - (-0.015\text{ m})^2\right]\frac{19.6\text{ N/m}}{0.40\text{ kg}}} = 0.26\text{ m/s} = 26\text{ cm/s} \qquad \Diamond$$

(c) When the spring is stretched 1.5 cm, then $x = +0.015$ m and

$$v = \sqrt{\left[(0.040\text{ m})^2 - (+0.015\text{ m})^2\right]\frac{19.6\text{ N/m}}{0.40\text{ kg}}} = 0.26\text{ m/s} = 26\text{ cm/s} \qquad \Diamond$$

(d) When $v = \frac{1}{2}v_{\text{max}}$, Equation 1 gives $\sqrt{(x_o^2 - x^2)k/m} = \frac{1}{2}\sqrt{x_o^2 k/m}$

or

$$4(x_o^2 - x^2) = x_o^2 \quad \text{which reduces to} \quad x = \sqrt{\frac{3x_o^2}{4}} = \frac{\sqrt{3}(4.0\text{ cm})}{2} = 3.5\text{ cm} \quad \Diamond$$

---

19. A 200-g mass is attached to a spring and executes simple harmonic motion with a period of 0.250 s. If the total energy of the system is 2.00 J, find (a) the force constant of the spring and (b) the amplitude of the motion.

Solution

(a) When an object of mass m is attached to a spring of force constant k, it undergoes simple harmonic motion with a period given by $T = 2\pi/\omega$ where $\omega = \sqrt{k/m}$ is the angular frequency of the motion. In this case,

$$\omega = 2\pi/T = 2\pi/0.250\text{ s} = 8\pi\text{ rad/s}$$

Thus, the force constant of the spring is

$$k = m\omega^2 = (0.200\text{ kg})(8\pi\text{ rad/s})^2 = 126\text{ N/m} \quad \Diamond$$

(b) The energy of a mass undergoing simple harmonic motion on the end of a horizontal spring is $E = KE + PE_s = \frac{1}{2}mv^2 + \frac{1}{2}kx^2$. When the displacement from equilibrium, x, has it maximum value of $x = A =$ amplitude, the mass is in the process of reversing directions and has zero velocity. Thus, the relation between the energy and the amplitude is $E = 0 + \frac{1}{2}kA^2$, or $A = \sqrt{2E/k}$. The amplitude for this motion is then

$$A = \sqrt{\frac{2(2.00\text{ J})}{126\text{ N/m}}} = 0.178\text{ m} = 17.8\text{ cm} \quad \Diamond$$

---

25. A spring of negligible mass stretches 3.00 cm from its relaxed length when a force of 7.50 N is applied. A 0.500-kg particle rests on a frictionless horizontal surface and is attached to the free end of the spring. The particle is pulled horizontally so that it stretches the spring 5.00 cm and is then released from rest at $t = 0$. (a) What is the force constant of the spring? (b) What are the period, frequency, and angular frequency (ω) of the motion? (c) What is the total energy of the system? (d) What is the amplitude of the motion? (e) What are the maximum velocity and the maximum acceleration of the particle. (f) Determine the displacement, x, of the particle from the equilibrium position at $t = 0.500$ s.

Solution (a) According to Hooke's law, the force required to stretch a spring is directly proportional to the amount the spring is stretched, i.e., $F = kx$. Thus, if a force of 7.50 N stretches the spring by 3.00 cm, the force constant of the spring is

$$k = \frac{F}{x} = \frac{7.50 \text{ N}}{3.00 \times 10^{-2} \text{ m}} = 250 \text{ N/m} \qquad \Diamond$$

(b) The angular frequency, ω, of the motion is defined as $\omega = \sqrt{k/m}$ where k is the force constant of the spring and m is the mass attached to the spring. For the system described,

$$\omega = \sqrt{\frac{250 \text{ N/m}}{0.500 \text{ kg}}} = 22.4 \text{ rad/s} \qquad \Diamond$$

The period is then $$T = \frac{2\pi}{\omega} = \frac{2\pi \text{ rad}}{22.4 \text{ rad/s}} = 0.281 \text{ s} \qquad \Diamond$$

and the frequency is $$f = \frac{1}{T} = \frac{1}{0.281 \text{ s}} = 3.56 \text{ Hz} \qquad \Diamond$$

(c) The total energy of the system is constant. Therefore, the energy at any time is the same as the energy when it is released at $t = 0$,

$$E = E_0 = \tfrac{1}{2}mv_0^2 + \tfrac{1}{2}kx_0^2 = 0 + \tfrac{1}{2}(250 \text{ N/m})\left(5.00 \times 10^{-2} \text{ m}\right)^2 = 0.313 \text{ J} \qquad \Diamond$$

(d) When the object is at its maximum displacement from equilibrium $(x = A = \text{amplitude})$, it is momentarily at rest and all the energy of the system is stored as elastic potential energy, i.e., $E = 0 + \frac{1}{2}kA^2$. The amplitude is then

$$A = \sqrt{\frac{2E}{k}} = \sqrt{\frac{2(0.313\text{ J})}{250\text{ N/m}}} = 0.050\text{ m} = 5.00\text{ cm} \qquad \lozenge$$

(e) The maximum velocity will occur at the equilibrium position $(x = 0)$ where all the energy of the system is in the form of kinetic energy. Thus,

$$E = \frac{1}{2}mv_{\max}^2 + 0 \quad \text{or} \quad v_{\max} = \sqrt{\frac{2E}{m}} = \sqrt{\frac{2(0.313\text{ J})}{0.500\text{ kg}}} = 1.12\text{ m/s} \qquad \lozenge$$

From Newton's second law, the maximum acceleration of the particle occurs when the spring exerts maximum force on it. This occurs when the spring is stretched or compressed the maximum amount, at $|x| = A$. Thus,

$$a_{\max} = \frac{F_{\max}}{m} = \frac{kA}{m} = \frac{(250\text{ N/m})(0.050\text{ m})}{0.500\text{ kg}} = 25.0\text{ m/s}^2 \qquad \lozenge$$

(f) Since $\omega = \sqrt{k/m}$, and the object starts from rest at $x = A$ when $t = 0$, the displacement at any time $t > 0$ is

$$x = A\cos(\omega t) = A\cos\left(t\sqrt{k/m}\right)$$

At $t = 0.500$ s

$$x = (5.00\text{ cm})\cos\left((0.500\text{ s})\sqrt{\frac{250\text{ N/m}}{0.500\text{ kg}}}\right) = 0.919\text{ cm} \qquad \lozenge$$

Note: Your calculator must be set for radians, rather than degrees, to properly perform this calculation.

---

31. The acceleration due to gravity on Mars is 3.7 m/s². (a) What length pendulum has a period of one second on Earth? What length pendulum would have a one second period on Mars? (b) A mass is suspended from a spring with force constant 10 N/m. Find the mass suspended from this spring that would result in a one second period on Earth and on Mars.

Solution (a) The period of a simple pendulum is $T = 2\pi\sqrt{L/g}$ where L is the length of the pendulum and g is the acceleration due to gravity at the pendulum's location. Thus, if a pendulum has a period of $T = 1.0$ s on Earth where $g = 9.8\ \text{m/s}^2$, its length is

$$L_{\text{Earth}} = \frac{g_{\text{Earth}}\,T^2}{4\pi^2} = \frac{\left(9.8\ \text{m/s}^2\right)\left(1.0\ \text{s}\right)^2}{4\pi^2} = 0.25\ \text{m} = 25\ \text{cm} \qquad \Diamond$$

On Mars, where $g = 3.7\ \text{m/s}^2$, the length of a pendulum with a 1.0 s period is

$$L_{\text{Mars}} = \frac{g_{\text{Mars}}\,T^2}{4\pi^2} = \frac{\left(3.7\ \text{m/s}^2\right)\left(1.0\ \text{s}\right)^2}{4\pi^2} = 0.094\ \text{m} = 9.4\ \text{cm} \qquad \Diamond$$

(b) The period of vibration for an object suspended from a spring is $T = 2\pi\sqrt{m/k}$ where k is the force constant of the spring and m is the mass of the object. Thus, to have a period of $T = 1.0$ s when suspended from a spring with a force constant $k = 10$ N/m, the required mass is

$$m = \frac{kT^2}{4\pi^2} = \frac{\left(10\ \text{N/m}\right)\left(1.0\ \text{s}\right)^2}{4\pi^2} = 0.25\ \text{kg}$$

Both k and m are constants of the system and do not depend on the location of the system. Therefore, the same mass is needed on Earth and Mars, or

$$m_{\text{Earth}} = m_{\text{Mars}} = 0.25\ \text{kg} \qquad \Diamond$$

37. A harmonic wave is traveling along a rope. It is observed that the oscillator that generates the wave completes 40.0 vibrations in 30.0 s. Also, a given maximum travels 425 cm along the rope in 10.0 s. What is the wavelength?

Solution

The frequency of the oscillator, and the frequency that wave-crests (maxima) start down the rope, is

$$f = \frac{40.0 \text{ vibrations}}{30.0 \text{ s}} = \frac{4}{3} \text{ Hz}$$

The speed of the wave in the rope is

$$v = \frac{425 \text{ cm}}{10.0 \text{ s}} = 42.5 \text{ cm/s}$$

The wave's wavelength, frequency, and wave speed are related by $v = \lambda f$. Therefore, the wavelength is

$$\lambda = \frac{v}{f} = \frac{42.5 \text{ cm/s}}{4/3 \text{ s}^{-1}} = 31.9 \text{ cm} \ \Diamond$$

---

41. A string is 50.0 cm long and has a mass of 3.00 g. A wave travels at 5.00 m/s along this string. A second string has the same length but half the mass of the first. If the two strings are under the same tension, what is the speed of a wave along the second string?

Solution

The speed of transverse waves in a string is $v = \sqrt{F/\mu}$ where F is the tension in the string and μ is the mass per unit length of the string. The second string has the same tension and length as the first string, but only half the mass $\left(\text{i.e., } F_2 = F_1,\ L_2 = L_1, \text{and } m_2 = m_1/2\right)$. Therefore,

$$\mu_2 = \frac{m_2}{L_2} = \frac{m_1/2}{L_1} = \frac{1}{2}\left(\frac{m_1}{L_1}\right) = \frac{\mu_1}{2}$$

and the ratio of the wave speeds in the two strings is

$$\frac{v_2}{v_1} = \sqrt{\frac{F_2}{\mu_2}} \Big/ \sqrt{\frac{F_1}{\mu_1}} = \sqrt{\left(\frac{F_2}{F_1}\right)\left(\frac{\mu_1}{\mu_2}\right)} = \sqrt{(1)(2)} = \sqrt{2}$$

Thus, if $v_1 = 5.00$ m/s, the wave speed in the second string is

$$v_2 = v_1\sqrt{2} = (5.00 \text{ m/s})\sqrt{2} = 7.07 \text{ m/s} \qquad \Diamond$$

---

45. A wave of amplitude 0.30 m interferes with a second wave of amplitude 0.20 m traveling in the same direction. What are the (a) largest and (b) smallest resultant amplitudes that can occur, and under what conditions will these maxima and minima occur?

Solution When two waves meet at a point, the resultant displacement at that point for that instant in time is found by adding the displacements of the individual waves at that point and time.

(a) The largest magnitude displacement at the point will occur if the two waves each have maximum displacements (equal to the amplitudes of the waves) **in the same direction** at the instant they meet. That is, when the two waves meet in phase and interfere constructively. In this case the magnitude of the displacement at the point is

$$|\text{resultant displacment}| = A_1 + A_2 = 0.30 \text{ m} + 0.20 \text{ m} = 0.50 \text{ m} \qquad \Diamond$$

(b) The smallest magnitude displacement occurs at the point if the two waves each have maximum displacements **in opposite directions** at the instant they meet. That is, when the two waves meet out of phase and interfere destructively. In that case the magnitude of the displacement at the point is

$$|\text{resultant displacment}| = A_1 - A_2 = 0.30 \text{ m} - 0.20 \text{ m} = 0.10 \text{ m} \qquad \Diamond$$

---

49. A 3.00-kg mass is fastened to a light spring that passes over a pulley (Fig. P13.49). The pulley is frictionless, and its inertia may be neglected. The mass is released from rest when the spring is unstretched. If the mass drops 10.0 cm before stopping, find (a) the spring constant of the spring and (b) the speed of the mass when it is 5.00 cm below its starting point.

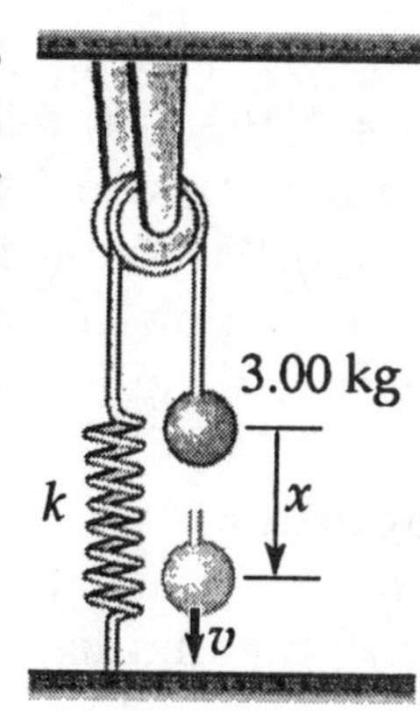

Solution Choose the zero gravitational potential energy level where the mass starts from rest and the spring is unstretched. Since the pulley is frictionless with negligible inertia, conservation of energy gives the energy after the spring is stretched a distance x as $E = E_0$ where E_0 is the energy at $x = 0$.

Thus, $KE + PE_g + PE_s = KE_0 + PE_{g,0} + PE_{s,0}$, or $\frac{1}{2}mv^2 - mgx + \frac{1}{2}kx^2 = 0+0+0$ and the speed of the mass after it has dropped a distance x is

$$v = \sqrt{x\left(2g - \frac{kx}{m}\right)} = \sqrt{x\left(19.6\ \text{m/s}^2 - \frac{kx}{m}\right)}$$

(a) Since the mass comes to rest when x = 10.0 cm = 0.100 m, the equation for the speed

$$0 = \sqrt{(0.100\ \text{m})\left[19.6\ \text{m/s}^2 - \frac{k(0.100\ \text{m})}{3.00\ \text{kg}}\right]}$$

gives the force constant of the spring: $k = \dfrac{(3.00\,\text{kg})(19.6\ \text{m/s}^2)}{0.100\ \text{m}} = 588\ \text{N/m}$ ◊

(b) When $x = 5.00\ \text{cm} = 5\times10^{-2}$ m, the speed equation gives

$$v = \sqrt{(5\times10^{-2}\ \text{m})\left[19.6\ \text{m/s}^2 - (588\ \text{N/m})(5\times10^{-2}\ \text{m})/(3.00\ \text{kg})\right]}$$

Thus, the speed of the mass at this point is $v = 0.700\ \text{m/s}$ ◊

57. A 2.00-kg block hangs without vibrating at the end of a spring (k = 500 N/m) that is attached to the ceiling of an elevator car. The car is rising with an upward acceleration of $g/3$ when the acceleration suddenly ceases (at t = 0). (a) What is the angular frequency of oscillation of the block after the acceleration ceases? (b) By what amount is the spring stretched during the time that the elevator car is accelerating?

Solution

(a) The angular frequency of oscillation of a mass attached to the end of a spring is given by $\omega = \sqrt{k/m}$ where k is the force constant of the spring and m is the mass. Thus, for this system, the angular frequency is

$$\omega = \sqrt{\frac{500 \text{ N/m}}{2.00 \text{ kg}}} = \sqrt{250 \text{ s}^{-2}} = 15.8 \text{ rad/s}$$ ◊

(b) The sketch at the right gives a free-body diagram of the block while the elevator (as well as its contents, including the block) is accelerating upward at $a_y = +g/3$. Newton's second law gives

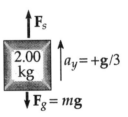

$$\sum F_y = F_s - mg = m(+g/3)$$

so the tension in the spring must be

$$F_s = \frac{4}{3}mg = \frac{4}{3}(2.00 \text{ kg})(9.80 \text{ m/s}^2) = 26.1 \text{ N}$$

From Hooke's law, the amount the spring is stretched is then

$$x = \frac{F_s}{k} = \frac{26.1 \text{ N}}{500 \text{ N/m}} = 5.23 \times 10^{-2} \text{ m} = 5.23 \text{ cm}$$ ◊

61. A light string of mass 10.0 g and length L = 3.00 m has its ends tied to two walls that are separated by the distance D = 2.00 m. Two masses, each of mass m = 2.00 kg, are suspended from the string as in Figure P13.61. If a wave pulse is sent from point A, how long does it take to travel to point B?

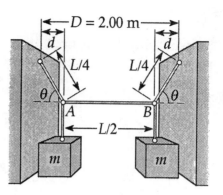

Figure P13.61

Solution

Observe from the geometry of this situation that $2d + L/2 = D = 2.00\text{ m}$. Since $L = 3.00\text{ m}$, this gives the distance each mass is from one of the walls:

$$d = \frac{2.00\text{ m} - 1.50\text{ m}}{2} = 0.250\text{ m}$$

The angle the inclined portions of the rope make with the horizontal is found from

$$\cos\theta = \frac{d}{L/4} = \frac{0.250\text{ m}}{0.750\text{ m}} = 0.333$$

so $\theta = 70.5°$

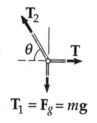

Now, consider the free-body diagram of the knot in the ropes at point A. This knot is in equilibrium and the first condition of equilibrium gives

$\mathbf{T}_1 = \mathbf{F}_g = m\mathbf{g}$

$$\sum F_x = T - T_2\cos\theta = 0 \quad \text{or} \quad T = T_2\cos 70.5° \qquad \textbf{[Equation 1]}$$

and

$$\sum F_y = T_2\sin\theta - mg = 0 \quad \text{or} \quad T_2 = \frac{mg}{\sin 70.5°} \qquad \textbf{[Equation 2]}$$

Substituting Equation 2 into Equation 1 gives the tension in the horizontal rope connecting points A and B as

$$T = \frac{mg}{\tan 70.5°}$$

or
$$T = \frac{(2.00 \text{ kg})(9.8 \text{ m/s}^2)}{\tan 70.5°} = 6.93 \text{ N}$$

The speed of a transverse wave in a rope is $v = \sqrt{F/\mu}$ where F is the tension in the rope and μ is the mass per unit length of the rope. The mass per unit length of these ropes is

$$\mu = \frac{10 \text{ g}}{3.00 \text{ m}} = \frac{0.0100 \text{ kg}}{3.00 \text{ m}}$$

or
$$\mu = 3.33 \times 10^{-3} \text{ kg/m}$$

The speed of waves in the section of rope between points A and B is then

$$v = \sqrt{T/\mu} = \sqrt{\frac{6.93 \text{ N}}{3.33 \times 10^{-3} \text{ kg/m}}} = 45.6 \text{ m/s}$$

and the time required for a pulse to travel from A to B is

$$t = \frac{L/2}{v} = \frac{1.5 \text{ m}}{45.6 \text{ m/s}} = 32.9 \times 10^{-3} \text{ s} = 32.9 \text{ ms}$$ ◊

CHAPTER SELF-QUIZ

1. A 0.2-kg object, suspended from a spring with a spring constant of k = 10 N/m, is moving in simple harmonic motion and has an amplitude of 0.08 m. What is its kinetic energy at the instant when its displacement is 0.04 m?
   a. 0.8×10^{-2} J
   b. zero
   c. 2.4×10^{-2} J
   d. 31.25 J

2. The tension in a guitar string is increased by a factor of four. By what factor does the wave velocity change?
   a. 4.0
   b. 2.0
   c. 0.5
   d. 0.25

3. A 0.4-kg mass hangs from a spring (k = 80 N/m), and is set into an up-and-down simple harmonic motion. What is the potential energy stored **in the spring alone** when the mass is displaced 0.1 m?
   a. zero
   b. 0.2 J
   c. 0.4 J
   d. 0.8 J

4. A runaway railroad car, with mass 30 x 10^4 kg, coasts across a level track at 1.5 m/s when it collides with a spring-loaded bumper at the end of the track. If the spring constant of the bumper is 2 x 10^6 N/m, what is the maximum displacement of the spring after the collision? (Assume collision is elastic.)
   a. 0.58 m
   b. 0.34 m
   c. 3.8 m
   d. 1.7 m

5. A 0.20-kg block rests on a frictionless level surface and is attached to a horizontally aligned spring with a spring constant of 40 N/m. The block is initially displaced 4 cm from the equilibrium point and then released to set up a simple harmonic motion. What is the frequency?
   a. 10.0 Hz
   b. 2.3 Hz
   c. 88.0 Hz
   d. 0.9 Hz

6. A mass on a spring vibrates in simple harmonic motion at a frequency of 4.0 Hz and an amplitude of 8.0 cm. If a timer is started when its displacement is a maximum (hence x = 8 cm when t = 0), what is the magnitude of its displacement when t = 3 s?
   a. zero
   b. 3.0 cm
   c. 6.7 cm
   d. 8.0 cm

7. A radio wave has a speed of 3×10^8 m/s and a frequency of 101 MHz. What is its wavelength?
   a. 3.0 m
   b. 45 m
   c. 0.10 m
   d. 2.97 m

8. Suppose a mass m is on a spring that has been compressed a distance h. How much further must the spring be compressed to triple the elastic potential energy?
   a. $2mgh$
   b. $2h$
   c. $1.7h$
   d. $0.73h$

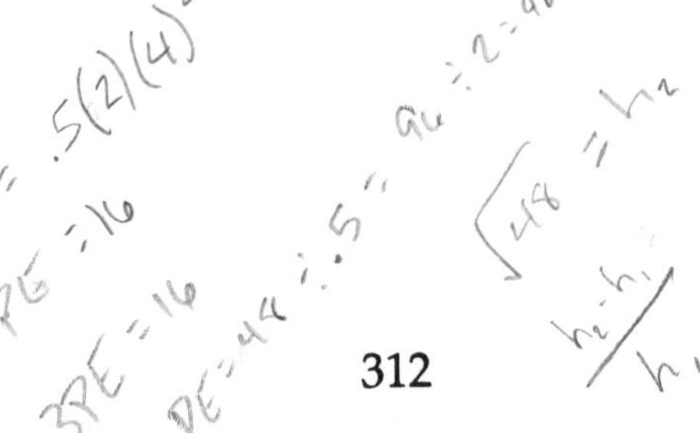

9. Ocean waves with a wavelength of 120 m are coming in at a rate of 8 per minute. What is their speed?
   a. 8 m/s
   b. 16 m/s
   c. 24 m/s
   d. 30 m/s

10. A piano string of mass 0.005 kg/m is under a tension of 1350 N. Find the velocity with which a wave travels on this string.
    a. 260 m/s
    b. 520 m/s
    c. 1040 m/s
    d. 2080 m/s

11. An earthquake emits both P-waves and S-waves which travel at different speeds through the Earth. A P-wave travels at 9000 m/s and an S-wave at 5000 m/s. If P-waves are received at a seismic station 1 minute before an S-wave arrives, how far is it to the earthquake center?
    a. 2420 km
    b. 1210 km
    c. 680 km
    d. 240 km

12. The speed of a sound wave in sea water is 1500 m/s. If this wave is transmitted at frequency 10 kHz, what is its wavelength?
    a. 5 cm
    b. 10 cm
    c. 15 cm
    d. 20 cm

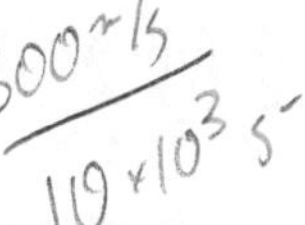

Chapter 14

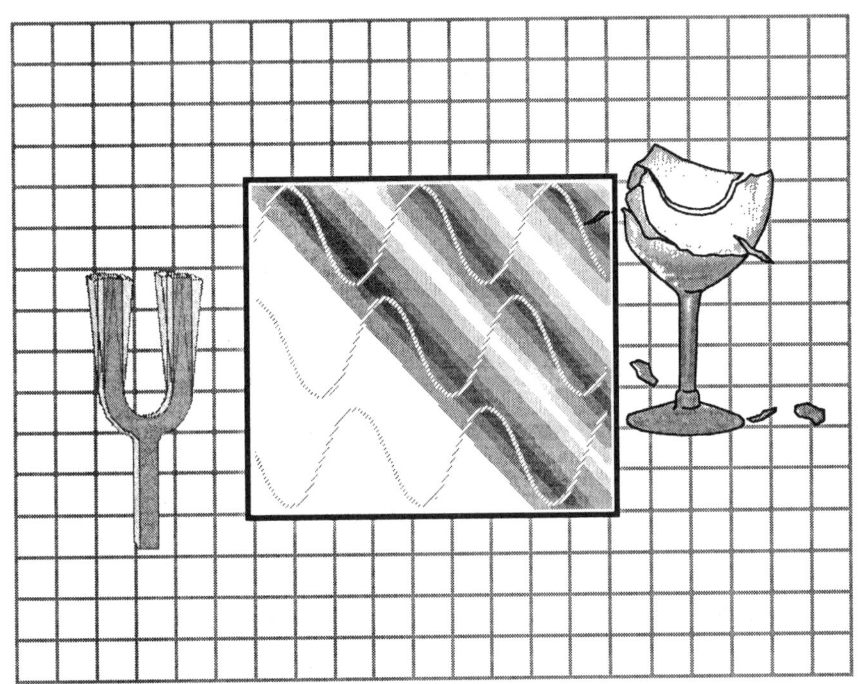

SOUND

Chapter 14

SOUND

Sound waves are the most important example of longitudinal waves. In this chapter we discuss the characteristics of sound waves—how they are produced, what they are, and how they travel through matter. We then investigate what happens when sound waves interfere with each other. The insights gained in this chapter will help you understand why we hear what we hear.

NOTES FROM SELECTED CHAPTER SECTIONS

14.1 Producing a SoundWave
14.2 Characteristics of Sound Waves

Sound waves, which have as their source vibrating objects, are longitudinal waves traveling through a medium such as air. The motion of the medium particles is **back and forth along the direction in which the wave travels.** This is in contrast to a transverse wave, in which the vibrations of the medium are **at right angles to the direction of travel of the wave.**

A sound wave traveling through air creates an alternating series of regions of molecular density and air pressure. Regions of high density and air pressure are called **compression** or **condensation,** while regions of lower-than-normal density are called **rarefaction**. A sinusoidal curve can be used to represent a sound wave. There are crests in the sinusoidal wave at the points where the sound wave has condensations, and troughs where the sound wave has rarefactions.

14.4 Energy and Intensity of Sound Waves

The **intensity** of a wave is the rate at which sound energy flows through a unit area perpendicular to the direction of travel of the wave. The faintest sounds the human ear can detect (about $1\times10^{-12}\ \mathrm{W/m^2}$) is called the **threshold of hearing**. The loudest sounds the ear can tolerate, at the **threshold of pain,** have an intensity of about $1\,W/m^2$. The sensation of loudness is approximately logarithmic in the human ear, and the relative intensity of a sound is called the **intensity level or decibel level.**

14.5 Spherical and Plane Waves

The intensity of a **spherical wave** produced by a point source is proportional to the average power emitted and inversely proportional to the square of the distance from the source.

14.6 The Doppler Effect

In general, a Doppler effect is experienced whenever there is relative motion between source and observer. When the source and observer are moving toward each other, the frequency heard by the observer is higher than the frequency of the source. When the source and observer are moving away from each other, the observer hears a frequency lower than the source frequency.

14.8 Standing Waves

Standing waves can be set up in a string by a continuous superposition of waves incident on and reflected from the ends of the string. The string has a number of natural patterns of vibration, called **normal modes**. Each normal mode has a **characteristic frequency**. The lowest of these frequencies is called the **fundamental frequency,** which together with the higher frequencies form a **harmonic series.**

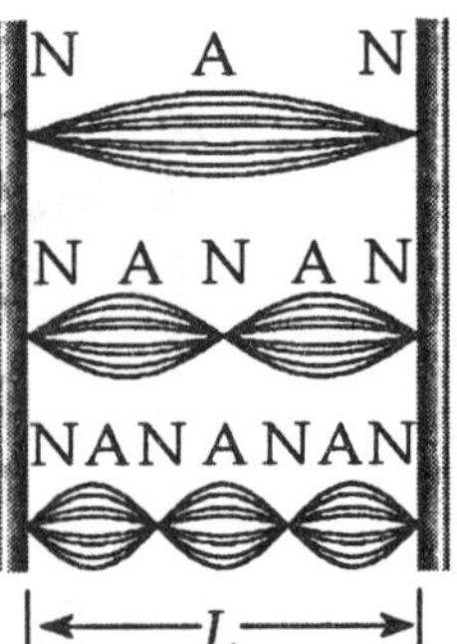

The figure on the right is a schematic representation of standing waves on a stretched string of length L, where the envelope represents many successive vibrations. The first three normal modes are shown. The points of zero displacement are called **nodes;** the points of maximum displacement are called **antinodes.**

14.9 Forced Vibrations and Resonance

Suppose a system has a natural frequency of vibration, f_0, and is driven or pushed back and forth with a periodic force whose frequency is f. This type of motion is referred to as a **forced vibration**. Its amplitude reaches a maximum when the frequency of the driving force equals the natural frequency of the system, f_0, called the **resonant frequency** of the system. Under this condition, the system is said to be in **resonance**.

14.10 Standing Waves in Air Columns

Standing waves are produced in strings by interfering **transverse** waves. Sound sources can be used to produce **longitudinal** standing waves in air columns. The phase relationship between incident and reflected waves depends on whether or not the reflecting end of the air column is open or closed. This gives rise to two sets of possible standing wave conditions:

> In a **pipe open at both ends,** the natural frequencies of vibration form a series in which all harmonics are present and are equal to integral multiples of the fundamental.
>
> In a **pipe closed at one end and open at the other,** only odd harmonics are present.

14.11 Beats

Consider a type of interference effect that results from the superposition of two waves with slightly different frequencies. In this situation, at some fixed point the waves are periodically in and out of phase, corresponding to an alternation in time between constructive and destructive interference. A listener hears an alternation in loudness, known as **beats**. The number of beats per second, or beat frequency, equals the difference in frequency between the two sources.

EQUATIONS AND CONCEPTS

A sound wave propagates through an elastic medium as a compressional wave. The speed of the sound wave depends on value of the bulk modulus, B (an elastic property), and the equilibrium density, ρ (an inertial property), of the material through which it is traveling.

$$v = \sqrt{\frac{B}{\rho}} \tag{14.1}$$

$$B = -\frac{\Delta P}{\Delta V / V} \tag{14.2}$$

The speed of sound (or any longitudinal wave) in a solid depends on the value of Young's modulus and the density of the material.

$$v = \sqrt{\frac{Y}{\rho}} \tag{14.3}$$

The velocity of sound depends on the temperature of the medium. Equation 14.4 shows the temperature dependence in air where T is the temperature in degrees Celsius and 331 m/s is the speed of sound in air at 0 °C.

$$v = (331\ \text{m/s})\sqrt{1 + \frac{T}{273\ ^\circ\text{C}}} \tag{14.4}$$

The intensity of a wave is the rate at which energy flows across a unit area, A, in a plane perpendicular to the direction of travel of the wave. The SI units of intensity, I, are watts per square meter, W/m^2.

$$I \equiv \frac{\text{power}}{\text{area}} = \frac{P}{A} \tag{14.6}$$

At a frequency of 1000 Hz, the faintest sound detectable by the human ear (threshold of hearing) has an intensity of 10^{-12} W/m^2. An intensity of 1 W/m^2 is the greatest intensity which the ear can tolerate (the threshold of pain).

Comment on sound intensity.

The decibel scale is a logarithmic intensity scale. On this scale, the unit of sound intensity is the decibel, dB. The constant I_0 is a reference intensity.

$$\beta \equiv 10 \log\left(\frac{I}{I_0}\right) \qquad (14.7)$$

$$I_0 = 1.00 \times 10^{-12} \text{ W/m}^2$$

In order to determine the decibel level corresponding to two different sources sounded simultaneously, first find the individual intensities I_1 and I_2 in W/m^2 and add these values to obtain the combined intensity $I = I_1 + I_2$. Finally, use Equation 14.7 to convert the intensity I to the decibel scale.

Comment on the decibel scale.

The intensity of a spherical wave is inversely proportional to the square of the distance from the source.

$$I = \frac{P_{av}}{4\pi r^2} \qquad (14.8)$$

The apparent change in frequency heard by an observer whenever there is relative motion between the source and the observer is called the Doppler effect. In Equation 14.15, the upper signs ($+v_0$ in numerator and $-v_s$ in the denominator) are used when there is motion of source or observer toward the other. The lower signs ($-v_0$ in numerator and $+v_s$ in denominator) are used when there is motion of source or observer away from the other. Also, it is important to remember that v_0 (velocity of the observer) and v_s (velocity of the source) are **each measured relative to the medium in which the sound travels.**

$$f' = f\left(\frac{v \pm v_o}{v \mp v_s}\right) \qquad (14.15)$$

(See Table, following page)

A series of standing wave patterns called normal modes can be excited in a stretched string. Each mode corresponds to a characteristic frequency and wavelength.

$$\lambda_n = \frac{2L}{n}$$

$$f_n = nf_1 = \frac{n}{2L}\sqrt{\frac{F}{\mu}} \qquad (4.20)$$

where $n = 1,2,3\ldots$

EXAMPLES OF DOPPLER EFFECT WITH OBSERVER/SOURCE IN MOTION

Observer	Source	Equation	Remark
O ⟶	S	$f' = f\left(\frac{v+v_o}{v}\right)$	Observer moving toward stationary source
⟵ O	S	$f' = f\left(\frac{v-v_o}{v}\right)$	Observer moving away from stationary source
O	⟵ S	$f' = f\left(\frac{v}{v-v_s}\right)$	Source moving toward stationary observer
O	S ⟶	$f' = f\left(\frac{v}{v+v_s}\right)$	Source moving away from stationary observer
O ⟶	S ⟶	$f' = f\left(\frac{v+v_o}{v+v_s}\right)$	Observer following moving source
⟵ O	⟵ S	$f' = f\left(\frac{v-v_o}{v-v_s}\right)$	Source following moving observer
⟵ O	S ⟶	$f' = f\left(\frac{v-v_o}{v+v_s}\right)$	Observer and source moving away from each other along opposite directions
O ⟶	⟵ S	$f' = f\left(\frac{v+v_o}{v-v_s}\right)$	Observer and source moving toward each other
O	S	$f' = f$	Observer and source both stationary

The frequency f_1 corresponding to $n = 1$, is the fundamental frequency and is the lowest frequency for which a standing wave is possible.

Comment on standing waves in strings.

The fundamental together with the other frequencies ($n = 1,2,3, \ldots$) form a harmonic series where f_2 is the second harmonic, f_3 is the third harmonic, and so on. The second harmonic is also called the first overtone; the third harmonic, the second overtone; and so on.

(See Figure below)

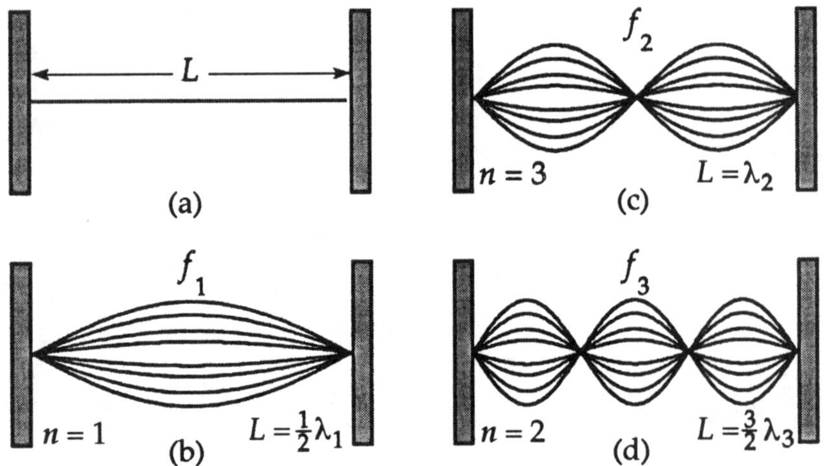

Standing waves in a stretched string of length L fixed at both ends. The normal frequencies of vibrations form a harmonic series: (b) the fundamental frequency, or first harmonic, (c) the second harmonic, and (d) the third harmonic.

Chapter 14

Standing waves are produced in strings by interfering transverse waves. Sound sources can also be used to produce longitudinal standing waves in air columns. The phase relationship between incident and reflected waves depends on whether the reflecting end of the air column is open or closed. This gives rise to two possible sets of standing wave conditions: those produced in an "open" pipe (open at both ends) and those produced in a "closed" pipe (open at only one end).

Comment on longitudinal standing waves in air columns.

In a pipe open at both ends, the natural frequencies of vibration form a series in which all harmonics (integer multiples of the fundamental) are present.

$$f_n = n\frac{v}{2L} \qquad (14.21)$$

$n = 1,2,3, \ldots$

In a pipe open at one end, only the odd harmonics (odd multiples of the fundamental) are possible.

$$f_n = n\frac{v}{4L} \qquad (14.22)$$

$n = 1,3,5, \ldots$

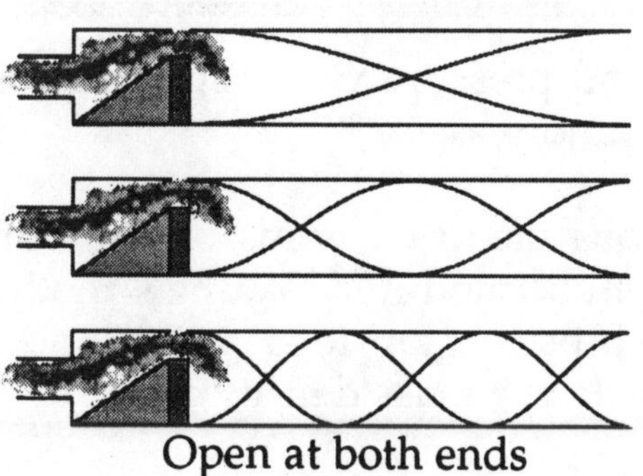

Open at both ends

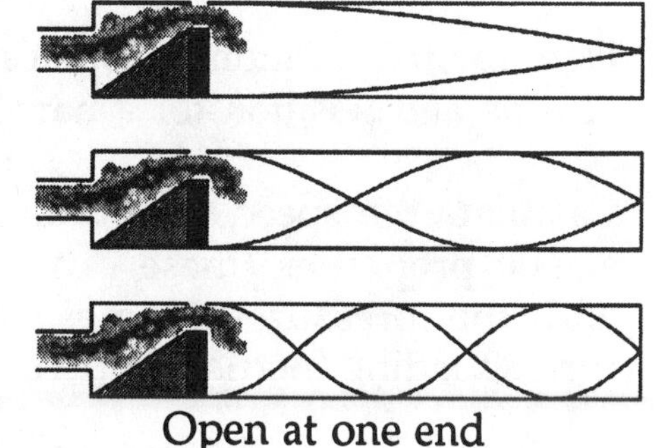

Open at one end

SUGGESTIONS, SKILLS, AND STRATEGIES

When making calculations using Equation 14.7 which defines the intensity of a sound wave on the decibel scale, the properties of logarithms must be kept clearly in mind.

In order to determine the decibel level corresponding to two sources sounded simultaneously, you must first find the intensity, I, of each source in W/m^2, add these values, and then convert the resulting intensity to the decibel scale. As an illustration of this technique, note that if two sounds of intensity 40 dB and 45 dB are sounded together, the intensity level of the combined sources **is** 46.2 dB (**not 85 dB**).

The most likely error in using Equation 14.15 to calculate the Doppler frequency shift due to relative motion between a sound source and an observer is due to using the incorrect algebraic sign for the velocity of either the observer or the source. These sign conventions are illustrated in the chart in the section on Equations and Concepts in which the directions of motion of the observer O and source S are indicated by arrows; and the correct choice of signs used in Equation 14.15,

$$f' = f\left(\frac{v \pm v_0}{v \mp v_s}\right)$$

where f is the true frequency of the source and f' is the apparent frequency as measured by the observer.

REVIEW CHECKLIST

▷ Describe the harmonic displacement and pressure variation as functions of time and position for a harmonic sound wave.

▷ Calculate the speed of sound in various media in terms of appropriate elastic properties (these can include bulk modulus, Young's modulus, and the pressure-volume relationships of an ideal gas) and the corresponding inertial properties (usually the mass density).

▷ Understand the basis of the logarithmic intensity scale (decibel scale) and convert intensity values (given in W/m[2]) to loudness levels on the dB scale. Determine the intensity ratio for two sound sources whose decibel levels are known. Calculate the intensity of a point source wave at a given distance from the source.

▷ Describe the various situations under which a Doppler shifted frequency is produced. Calculate the apparent frequency for a given actual frequency for each of the various possible relative motions between source and observer.

▷ Describe in both qualitative and quantitative terms the conditions which produce standing waves in a stretched string and in an open or closed air column pipe.

SOLUTIONS TO SELECTED END-OF-CHAPTER PROBLEMS

3. The range of human hearing extends from approximately 20 Hz to 20 000 Hz. Find the wavelengths of these extremes at a temperature of 27 °C.

Solution The speed of sound in air varies with temperature as

$$v = (331\text{ m/s})\sqrt{1 + T/273\ ^\circ\text{C}}$$

where T is the temperature in degrees **Celsius.** At $T = 27\ ^\circ\text{C}$, this gives

$$v = (331\text{ m/s})\sqrt{1 + 27\ ^\circ\text{C}/273\ ^\circ\text{C}} = 347\text{ m / s}$$

The wavelength λ, frequency f, and speed v of a wave are related by $v = \lambda f$. Thus, at 27 °C, the wavelengths corresponding to the extremes of the range of human hearing are:

$$\lambda_{\text{short}} = \frac{v}{f_{\text{high}}} = \frac{(347\text{ m/s})(100\text{ cm/m})}{2.0 \times 10^4\text{ Hz}} = 1.7\text{ cm} \qquad \lozenge$$

and $$\lambda_{\text{long}} = \frac{v}{f_{\text{low}}} = \frac{347\text{ m/s}}{20\text{ Hz}} = 17\text{ m} \qquad \lozenge$$

9. The intensity level of an orchestra is 85 dB. A single violin achieves a level of 70 dB. How does the intensity of the sound of the full orchestra compare with that of the violin's sound?

Solution

If a sound has intensity *I*, its decibel level is defined as

$$\beta \equiv 10\log\left(\frac{I}{I_0}\right)$$

where $I_0 = 1.0\times10^{-12}\ \mathrm{W/m^2}$ is the intensity of sound at the threshold of hearing. The difference in the decibel levels for the full orchestra and for the single violin is $\beta_{\text{orchestra}} - \beta_{\text{violin}} = 85\ \text{db} - 70\ \text{db}$, or

$$10\log\left(\frac{I_{\text{orchestra}}}{I_0}\right) - 10\log\left(\frac{I_{\text{violin}}}{I_0}\right) = 15\ \text{db}$$

Since $\log(A) - \log(B) = \log(A/B)$, this becomes

$$10\log\left(\frac{I_{\text{orchestra}}}{I_0}\bigg/\frac{I_{\text{violin}}}{I_0}\right) = 10\log\left(\frac{I_{\text{orchestra}}}{I_{\text{violin}}}\right) = 15\ \text{db}$$

Thus, $\log\left(\frac{I_{\text{orchestra}}}{I_{\text{violin}}}\right) = 1.5$, and taking the anti-logarithm gives

$$\frac{I_{\text{orchestra}}}{I_{\text{violin}}} = 10^{1.5} \text{ or } I_{\text{orchestra}} = \left(10^{1.5}\right)I_{\text{violin}} = 32I_{\text{violin}}$$

The intensity of sound from the full orchestra is 32 times the intensity of the sound from a single violin. ◊

13. A train sounds its horn as it approaches an intersection. The horn can just be heard at a level of 50 dB by an observer 10 km away. (a) What is the average power generated by the horn? (b) What intensity level of the horn's sound is observed by someone waiting at an intersection 50 m from the train? Treat the horn as a point source and neglect any absorption of sound by the air. Suppose the ground absorbs all sound that is radiated downward.

Solution (a) The intensity of a 50 db sound can be found from the defining equation for the decibel level, $\beta \equiv 10\log(I/I_0)$ where $I_0 = 1.0\times10^{-12}\text{ W/m}^2$. If $\beta = 50$ db, then $\log(I/I_0) = 5$ and $I = 10^5 I_0$, or

$$I = 10^5\left(1.0\times10^{-12}\text{ W/m}^2\right) = 1.0\times10^{-7}\text{ W/m}^2$$

If the train horn acts as a point source, the wavefronts are spherical with a surface area $A = 4\pi r^2$ at distance r from the horn. The average power emitted by a source can be expressed as $P = IA$ where I is the intensity of the wave and A is the surface area of the wavefront, both at the same distance r from the source. At $r = 10\text{ km} = 1.0\times10^4\text{ m}$ from the horn, $I = 1.0\times10^{-7}\text{ W/m}^2$ so the emitted power is

$$P = 4\pi\left(1.0\times10^4\text{ m}\right)^2\left(1.0\times10^{-7}\text{ W/m}^2\right) = 1.3\times10^2\text{ W} \qquad \Diamond$$

(b) At a distance of $r = 50$ m, the intensity of the sound is

$$I = \frac{P}{A} = \frac{P}{4\pi r^2} = \frac{1.3\times10^2\text{ W}}{4\pi(50\text{ m})^2} = 4.1\times10^{-3}\text{ W/m}^2$$

and the decibel level is

$$\beta = 10\log\left(\frac{I}{I_0}\right) = 10\log\left(\frac{4.1\times10^{-3}\text{ W/m}^2}{1.0\times10^{-12}\text{ W/m}^2}\right) = 10\log\left(4.1\times10^9\right) = 96\text{ db} \qquad \Diamond$$

21. An alert physics student stands beside the tracks as a train rolls slowly past. He notes that the frequency of the train whistle is 442 Hz when the train is approaching him and 441 Hz when the train is receding from him. From this he can find the speed of the train. What value does he find?

Solution According to the Doppler effect, a sound with a source frequency f will have an observed frequency of f':

$$f' = f\left(\frac{v \pm v_0}{v \mp v_s}\right)$$

Here, v is the speed of the sound wave, while v_0 and v_S are the speeds of the observer and the source respectively. All speeds are relative to the medium through which the wave is traveling. The upper set of signs is used when the source and observer are approaching (getting closer together) and the lower set is used when the source and observer are separating (getting farther apart).

Since the student is at rest beside the tracks, $v_0 = 0$. Without contrary information, $v = 345\text{ m/s}$ will be used for the speed of sound. Then, the Doppler effect equation gives

$$442\text{ Hz} = f\left(\frac{345\text{ m/s}}{345\text{ m/s} - v_s}\right) \qquad \textbf{[Equation 1]}$$

when the train is approaching the student. When the train is receding, the relation becomes

$$441\text{ Hz} = f\left(\frac{345\text{ m/s}}{345\text{ m/s} + v_s}\right) \qquad \textbf{[Equation 2]}$$

Dividing Equation 1 by Equation 2 gives

$$\frac{442}{441} = \frac{345\text{ m/s} + v_s}{345\text{ m/s} - v_s} \qquad \text{or} \qquad 442(345\text{ m/s}) - 442v_s = 441(345\text{ m/s}) + 441v_s$$

Solving for the train speed,

$$883v_s = 345\text{ m/s} \qquad \text{and} \qquad v_s = 0.391\text{ m/s}$$ ◊

27. A pair of speakers separated by 0.700 m are driven by the same oscillator at a frequency of 690 Hz. An observer, originally positioned at one of the speakers, begins to walk along a line perpendicular to the line joining the two speakers. (a) How far must the observer walk before reaching a relative maximum in intensity? (b) How far will the observer be from the speaker when the first relative minimum is detected in the intensity?

Solution The wavelength of the sound emitted by the speakers is

$$\lambda = \frac{v}{f} = \frac{345 \text{ m/s}}{690 \text{ Hz}} = 0.500 \text{ m}$$

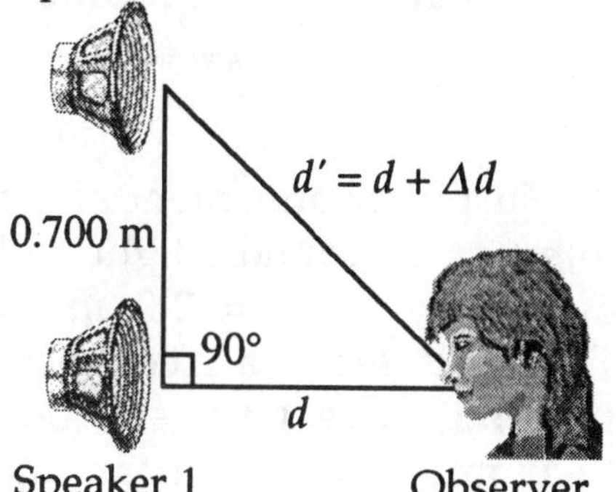

When the observer is distance d from the first speaker, he is distance $d' = d + \Delta d$ from the second as shown in the sketch. If a relative maximum is to occur, the difference in distances the two sound waves have traveled from the sources to the observer, $\Delta d = d' - d$, must be a whole number of wavelengths.

For a relative minimum, this path difference must be an odd number of half wavelengths. Note that when the observer is at Speaker 1 (i.e., when $d = 0$), the path difference is $\Delta d = 0.700$ m. As the observer moves away from this speaker, Δd decreases and approaches zero as d approaches infinity. Along this route, Δd ranges from 1.4λ to 0, so the observer will experience only one relative maximum (at $\Delta d = \lambda$) and one relative minimum (at $\Delta d = \lambda/2$).

(a) When the relative maximum is reached, $\Delta d = \lambda = 0.500\text{m}$. Applying the Pythagorean theorem to the right triangle shown in the sketch, $(d + 0.500 \text{ m})^2 = (0.700 \text{ m})^2 + d^2$. Expanding and simplifying this gives

$$(1.00 \text{ m})d + 0.250 \text{ m}^2 = 0.490 \text{ m}^2 \quad \text{or} \quad d = \frac{0.490 \text{ m}^2 - 0.250 \text{ m}^2}{1.00 \text{ m}}$$

Thus, the relative maximum is located at $d = 0.240 \text{ m} = 24.0 \text{ cm}$ in front of Speaker 1. ◊

(b) At the relative minimum, $\Delta = \lambda/2 = 0.250$ m and the Pythagorean theorem gives $(d + 0.250\text{ m})^2 = (0.700\text{ m})^2 + d^2$. Thus,

$$d = \frac{0.490\text{ m}^2 - 0.0625\text{ m}^2}{0.500\text{ m}} = 0.855\text{ m} = 85.5\text{ cm}$$

from Speaker 1 is the location of the relative minimum. ◊

---

33. In the arrangement shown in Figure P14.33, a mass, m = 5.0 kg, hangs from a cord around a light pulley. The length of the cord between point P and the pulley is L = 2.0 m. (a) When the vibrator is set to a frequency of 150 Hz, a standing wave with six loops is formed. What must be the linear mass density of the cord? (b) How many loops (if any) will result if m is changed to 45 kg? (c) How many loops (if any) will result if m is changed to 10 kg?

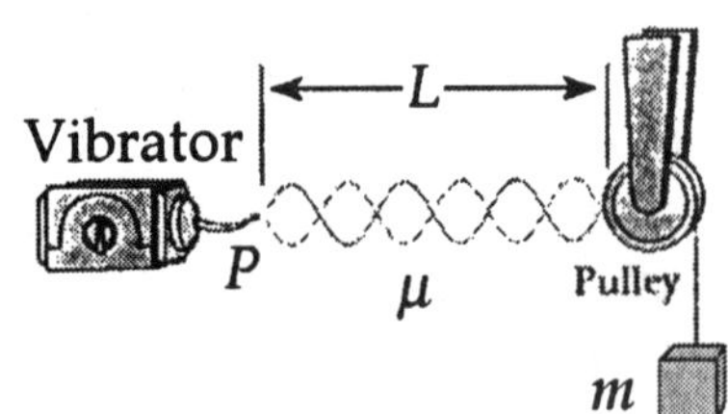

Figure P14.33

Solution (a) Each loop in the cord is one-half wavelength long. Thus, when a standing wave pattern with six loops forms in the cord, $L = 6(\lambda / 2)$ and the wavelength is $\lambda = L / 3 = (2.0\text{ m}) / 3 = 0.67\text{ m}$. The speed of the waves in the cord is

$$v = \lambda f = (0.67\text{ m})(150\text{ Hz}) = 100\text{ m/s}$$

This speed is also given by $v = \sqrt{F / \mu}$ where F is the tension in the cord and μ is the linear mass density. Since the 5.0-kg mass is in equilibrium, the tension in the cord must equal the weight of this mass, or $F = mg = 49\text{ N}$. Therefore, the mass per unit length of the cord is

$$\mu = \frac{F}{v^2} = \frac{49\text{ N}}{(100\text{ m/s})^2} = 4.9 \times 10^{-3}\text{ kg/m}$$ ◊

(b) If the mass is changed to the $m = 45$ kg, the tension in the cord will be

$$F = mg = (45\ \text{kg})(9.8\ \text{m/s}^2) = 4.4\times10^2\ \text{N}$$

and the speed of the waves is

$$v = \sqrt{\frac{F}{\mu}} = \sqrt{\frac{4.4\times10^2\ \text{N}}{4.9\times10^{-3}\ \text{kg/m}}} = 300\ \text{m/s}$$

The wavelength is then $\lambda = v/f = (300\ \text{m/s})/(150\ \text{Hz}) = 2.0\ \text{m}$. The number of loops that now fit in the length of the cord is

$$n = \frac{L}{\lambda/2} = \frac{2.0\ \text{m}}{1.0\ \text{m}} = 2$$

This is a whole number, so the cord forms a standing wave of 2 loops. ◊

(c) If $m = 10$ kg, then $F = mg = 98$ N and the speed is

$$v = \sqrt{\frac{98\ \text{N}}{4.9\times10^{-3}\ \text{kg/m}}} = 1.4\times10^2\ \text{m/s}$$

Therefore, the wavelength is $\lambda = v/f = (1.4\times10^2\ \text{m/s})/(150\ \text{Hz}) = 0.94\ \text{m}$. The number of half-wavelengths in the length of the cord is

$$n = \frac{L}{\lambda/2} = \frac{2.0\ \text{m}}{(0.94\ \text{m})/2} = 4.2$$

Since this is **not an integer,** resonance does not occur and no standing wave pattern is produced. ◊

35. A 5.0-kg mass connected to a spring is found to resonate when it is pushed at the frequencies 2.4 Hz, 1.2 Hz, 0.80 Hz, 0.60 Hz, ... Determine the spring constant for the spring.

Solution

To produce resonance, the pushes must be timed to always compliment what the spring is doing. Every time the spring has maximum compression, it is pushing the mass toward the equilibrium position. Resonance could be produced if an additional push directed toward the equilibrium position is applied to the mass every time it comes to this location (i.e., pushing at a frequency equal to the natural frequency, f_0, of the system).

However, resonance could also be produced by pushing on the mass every other time it comes to this position (i.e., pushing at a frequency $f = f_0/2$), or by pushing every third time it is at this position (i.e., at a frequency $f = f_0/3$), etc. In general, resonance occurs if the frequency of the pushes is given by $f = f_0/n$ where n is any positive integer.

Comparing the set of frequencies found to produce resonance in this system (2.4 Hz, 1.2 Hz, 0.80 Hz, 0.60 Hz, . . .) to f_0, $f_0/2$, $f_0/3$, $f_0/4$, . . ., it is clear that the natural frequency of vibration for this system is $f_0 = 2.4 \text{ Hz}$.

The natural frequency of vibration for a mass m attached to a spring of force constant k is

$$f_0 = \frac{1}{2\pi}\sqrt{\frac{k}{m}}$$

Thus, the force constant for this spring is

$$k = 4\pi^2 m f_0^{\,2} = 4\pi^2(5.0 \text{ kg})(2.4 \text{ Hz})^2 = 1.1\times 10^3 \text{ N/m} \qquad \Diamond$$

---

41. A pipe open at both ends has a fundamental frequency of 300 Hz when the temperature is 0 °C. (a) What is the length of the pipe? (b) What is the fundamental frequency at a temperature of 30 °C?

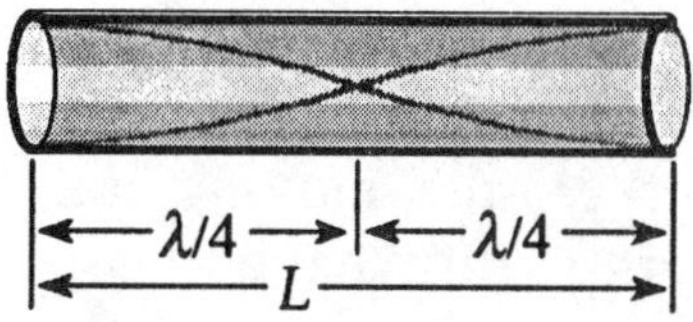

Solution A standing wave pattern in a pipe open at both ends must have an anti-node at each end. In the fundamental mode of vibration (i.e., for the lowest frequency and hence longest wavelength capable of producing resonance), the standing wave pattern is as shown in the sketch. Therefore, the length of the pipe is $L = \lambda/4 + \lambda/4 = \lambda/2$.

(a) The speed of sound in air when the temperature is T (Celsius) is $v = (331 \text{ m/s})\sqrt{1 + T/273 \text{ °C}}$. Thus, at 0 °C, $v = 331 \text{ m/s}$. If the fundamental frequency of the pipe is 300 Hz, then

$$\lambda = \frac{v}{f} = \frac{331 \text{ m/s}}{300 \text{ Hz}} = 1.10 \text{ m}$$

and the length of the pipe is $L = \dfrac{\lambda}{2} = 0.550 \text{ m}$ ◊

(b) At $T = 30 \text{ °C}$, the speed of sound is

$$v = (331 \text{ m/s})\sqrt{1 + 30/273} = 349 \text{ m/s}$$

The change in the length of the pipe as the temperature rises from 0 °C to 30 °C is negligible in comparison to the total length. Thus, the length of the pipe is still $L = 0.550 \text{ m}$ and, in the fundamental resonance mode, $\lambda = 2L = 1.10 \text{ m}$. The fundamental frequency at this temperature is then

$$f = \frac{v}{\lambda} = \frac{349 \text{ m/s}}{1.10 \text{ m}} = 317 \text{ Hz}$$ ◊

45. Two train whistles have identical frequencies of 180 Hz. When one train is at rest in the station, sounding its whistle, a beat frequency of 2 Hz is heard from a moving train. What two possible speeds and directions can the moving train have?

Solution

The beat frequency is equal to the difference in the detected frequencies of the two sounds. Thus, the frequency the observer detects for the sound from the whistle on the moving train must be either

$$f' = 180\text{ Hz} - 2\text{ Hz} = 178\text{ Hz} \quad \text{or} \quad f' = 180\text{ Hz} + 2\text{ Hz} = 182\text{ Hz}$$

The Doppler effect is responsible for the difference in the detected frequencies. The frequency a stationary observer detects from a moving sound source is

$$f' = f\left(\frac{v}{v \mp v_s}\right)$$

where v is the speed of sound, v_s is the speed of the source, and f is the frequency heard when the source is stationary relative to the observer. The upper sign is used when the source is approaching the observer and the lower sign when the source is moving away from the observer. Note that if the source is approaching the observer, $f' > f$, and $f' < f$ if the source is receding. In this case, $f = 180$ Hz and the speed of sound is assumed to be $v = 345$ m/s. The two possible situations are as follows:

(1) If the moving train is approaching the station, the Doppler effect equation becomes

$$182\text{ Hz} = (180\text{ Hz})\left(\frac{345\text{ m/s}}{345\text{ m/s} - v_s}\right) \quad \text{or} \quad 345\text{ m/s} - v_s = \left(\frac{180}{182}\right)(345\text{ m/s})$$

Thus, $$v_s = \left(1 - \frac{180}{182}\right)(345\text{ m/s}) = 3.8\text{ m/s} \quad \text{toward the station.}$$ ◊

(2) If the train is moving away from the station, the equation yields

$$178\text{ Hz} = (180\text{ Hz})\left(\frac{345\text{ m/s}}{345\text{ m/s} + v_s}\right) \quad \text{or} \quad 345\text{ m/s} + v_s = \left(\frac{180}{178}\right)(345\text{ m/s})$$

In this case, $v_s = \left(\frac{180}{178} - 1\right)(345\text{ m/s}) = 3.9\text{ m/s}$ away from the station. ◊

---

47. If a human ear canal can be thought of as resembling an organ pipe, closed at one end, that resonates at a fundamental frequency of 3000 Hz, what is the length of the canal? Use normal body temperature for your determination of the speed of sound in the canal.

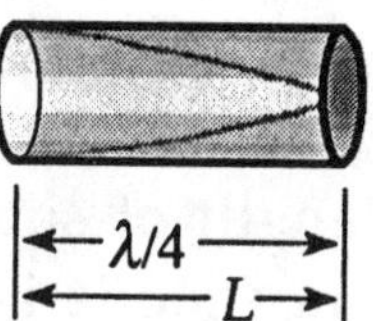

Solution A standing wave pattern in a pipe that is open at one end but closed at the other, must have an anti-node at the open end and a node at the closed end of the pipe. In the fundamental resonance mode (longest wavelength and lowest frequency that can produce a standing wave) for such a pipe, the wavelength of the sound is $\lambda = 4L$ as illustrated in the sketch. Normal body temperature is $T = 37.0\text{ °C}$ and the speed of sound in air at this temperature is

$$v = (331\text{ m/s})\sqrt{1 + \frac{37.0}{273}} = 353\text{ m/s}$$

The wavelength of a 3000 Hz sound wave at this temperature is

$$\lambda = \frac{v}{f} = \frac{353\text{ m/s}}{3000\text{ hz}} = 0.118\text{ m} = 11.8\text{ cm}$$

The length of the human ear canal that resonates at a fundamental frequency of 3000 Hz is therefore $L = \frac{\lambda}{4} = \frac{11.8\text{ cm}}{4} = 2.94\text{ cm}$ ◊

---

53. A flute is designed so that it plays a frequency of 261.6 Hz, middle C, when all the holes are covered and the temperature is 20.0 °C. (a) Consider the flute to be a pipe open at both ends, and find its length, assuming that the middle-C frequency is the fundamental. (b) A second player, nearby in a colder room, also attempts to play middle C on an identical flute. A beat frequency of 3.00 beats/s is heard. What is the temperature of the room?

Solution

Considering the flute to be a pipe open at both ends (as shown in the sketch), the wavelength of sound in the fundamental resonance mode is $\lambda = 2L$, where L is the length of the flute. Thus, the fundamental resonance frequency of the flute is $f = v/\lambda = v/2L$.

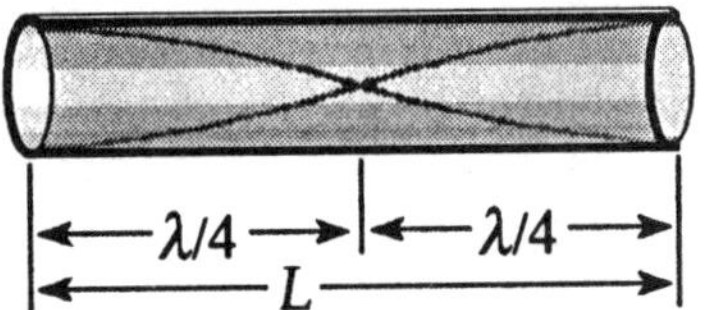

(a) The speed of sound at 20.0 °C is

$$v = (331 \text{ m/s})\sqrt{1 + \frac{20.0 \text{ °C}}{273 \text{ °C}}} = 343 \text{ m/s}$$

Thus, the length of a flute with a fundamental resonance frequency of 261.6 Hz is:

$$L = \frac{v}{2f} = \frac{343 \text{ m/s}}{2(261.6 \text{ Hz})} = 0.655 \text{ m} \qquad \Diamond$$

(b) Assuming a moderate difference in temperature in the two rooms, any difference in the lengths of the two flutes is negligible. Then, the ratio of the fundamental frequency of the flute in the cooler room to the fundamental frequency in the 20 °C room is

$$\frac{f_c}{f_{20}} = \frac{v_c/2L}{v_{20}/2L} = \frac{v_c}{v_{20}}$$

where v_c is the speed of sound in the cooler room and $v_{20} = 343\text{ m/s}$. Since the speed of sound decreases as temperature decreases, $v_c < v_{20}$ and hence $f_c < f_{20}$. The observed beat frequency is the difference in the fundamental frequencies in the two rooms. Therefore,

$$f_c = f_{20} - 3.00\text{ Hz} = 261.6\text{ Hz} - 3.00\text{ Hz} = 258.6\text{ Hz}$$

Hence, the speed of sound in the cooler room must be

$$v_c = \left(\frac{f_c}{f_{20}}\right)v_{20} = \left(\frac{258.6\text{ Hz}}{261.6\text{ Hz}}\right)(343\text{ m/s}) = 339\text{ m/s}$$

and the temperature in the cooler room may now be found from

$$339\text{ m/s} = (331\text{ m/s})\sqrt{1 + \frac{T}{273\text{ °C}}}$$

or

$$T = (273\text{ °C})\left[\left(\frac{339\text{ m/s}}{331\text{ m/s}}\right)^2 - 1\right] = 13.4\text{ °C}$$

◊

63. By proper excitation, it is possible to produce both longitudinal and transverse waves in a long metal rod. In a particular case, the rod is 150 cm long and 0.200 cm in radius and has a mass of 50.9 g. Young's modulus for the material is 6.80×10^{10} Pa. Determine the required tension in the rod so that the ratio of the speed of longitudinal waves to the speed of transverse waves is 8.

Solution

The speed of longitudinal waves in a solid is given by $v_L = \sqrt{Y/\rho}$ where Y is the Young's modulus and ρ is the density (mass per unit volume) of the solid. Transverse waves travel along a long, thin medium such as a rod with a speed $v_t = \sqrt{F/\mu}$ where F is the tension in this medium and μ is the mass per unit length. If $v_L = 8v_t$, then $\sqrt{Y/\rho} = 8\sqrt{F/\mu}$, or the required tension in the rod is $F = (Y/64)(\mu/\rho)$. The volume of a rod is $V = AL$ where A is the cross-sectional area of the rod and L is its length. Therefore, for a uniform cylindrical rod, the ratio of densities that appears in the equation for the tension is

$$\frac{\mu}{\rho} = \frac{(m/L)}{(m/V)} = \frac{V}{L} = \frac{AL}{L} = A = \pi r^2$$

Thus, the needed tension is $F = \pi r^2 Y/64$. This rod has a radius of $r = 2.00 \times 10^{-1}\text{ cm} = 2.00 \times 10^{-3}\text{ m}$ and is made from a material with a Young's modulus of $Y = 6.80 \times 10^{10}\text{ Pa} = 6.80 \times 10^{10}\text{ N/m}^2$.

Therefore, the required tension in the rod is

$$F = \frac{\pi\left(2.00 \times 10^{-3}\text{ m}\right)^2\left(6.80 \times 10^{10}\text{ N/m}^2\right)}{64} = 1.34 \times 10^4\text{ N} \qquad \Diamond$$

CHAPTER SELF-QUIZ

1. A series of ocean waves, each 6.0 m from crest to crest, moving past the observer at a rate of 2 waves per second, have what velocity?
   a. 1/3 m/s
   b. 3.0 m/s
   c. 8.0 m/s
   d. 12.0 m/s

2. A standing wave is set up in a 2.0-m length string fixed at both ends. The string vibrates in 5 distinct segments when driven by a 120 Hz source. What is the wave velocity in this string?
   a. 96 m/s
   b. 48 m/s
   c. 24 m/s
   d. 12 m/s

3. If a sound source with a 1000 Hz frequency is at rest, and a listener moves at a speed of 30.0 m/s away from the source, what is the apparent frequency heard by the listener? (The velocity of sound = 340 m/s.)
   a. 919 Hz
   b. 912 Hz
   c. 1090 Hz
   d. 1097 Hz

4. A low C (f = 65 Hz) is sounded on a piano. If the length of the piano wire is 2.0 m and its mass density is 5.0 g/m, what is the tension of the wire?
   a. 84 N
   b. 168 N
   c. 338 N
   d. 677 N

5. When I stand half way between two speakers, with one on my left and one on my right, a musical note from the speaker gives me constructive interference. How far to my left should I move to obtain destructive interference?
   a. one-fourth of a wavelength
   b. half a wavelength
   c. one wavelength
   d. one and a half wavelengths

6. For a standing wave on a string, the wavelength must equal
   a. the distance between adjacent nodes
   b. the distance between adjacent antinodes
   c. twice the distance between adjacent antinodes
   d. the distance between supports

7. An organ pipe, closed at one end, and a guitar string have the same fundamental frequency The frequency of the second overtone of the pipe corresponds to which overtone of the guitar string?
   a. first
   b. second
   c. third
   d. fourth

8. The sound level 5.0 m from a point source is 95 dB. At what distance will it be 75 dB?
   a. 50 m
   b. 75 m
   c. 225 m
   d. 500 m

9. A fireworks rocket explodes at a height of 100 m above the ground. An observer on the ground directly under the explosion experiences an average sound intensity of 7×10^{-2} W/m^2. What is the sound level in dB heard by the observer? ($I_0 = 10^{-12}$ W/m^2)
   a. 94.4 dB
   b. 100.0 dB
   c. 108.4 dB
   d. 119.4 dB

10. A clarinet behaves like a tube closed at one end. If its length is 80 cm, and the velocity of sound is 340 m/s, what is its fundamental frequency in Hz?
    a. 106 Hz
    b. 159 Hz
    c. 212 Hz
    d. 265 Hz

11. A bat, flying at 5.0 m/s, emits a chirp at 40 kHz. If this sound pulse is reflected by a wall, what is the frequency of the echo received by the bat?
    a. 41.2 kHz
    b. 40.9 kHz
    c. 40.6 kHz
    d. 40.3 kHz

12. Shortening a guitar string to one-third its initial length will change its natural frequency by what factor?
    a. 0.58
    b. 1.0
    c. 1.7
    d. 3.0

Chapter 15

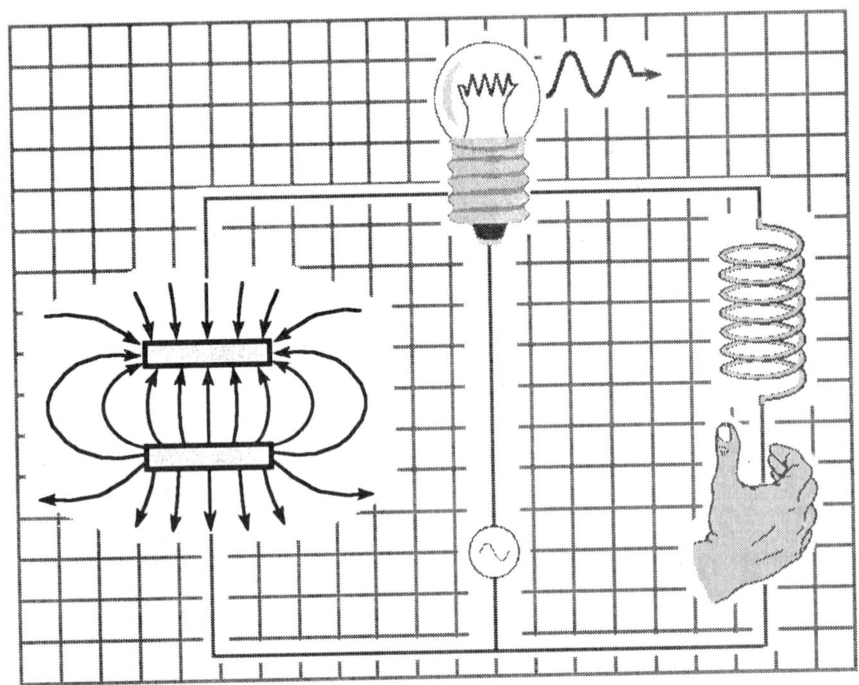

ELECTRIC FORCES AND ELECTRIC FIELDS

Chapter 15

ELECTRIC FORCES AND ELECTRIC FIELDS

In this chapter we use the effect, charging by friction, to begin an investigation of electric forces. We then discuss Coulomb's law, which is the fundamental law of force between any two charged particles. The concept of an electric field associated with charges is then introduced, and its effects on other charged particles described. We end with brief discussions of the Van de Graaff generator and the oscilloscope.

NOTES FROM SELECTED CHAPTER SECTIONS

15.1 Properties of Electric Charges

Electric charge has the following important properties:

1. There are two kinds of charges in nature, with the property that unlike charges attract one another and like charges repel one another.

2. The force between charges varies as the inverse square of their separation.

3. Charge is conserved.

4. Charge is quantized.

15.2 Insulators and Conductors

Conductors are materials in which electric charges move freely under the influence of an electric field; **insulators** are materials that do not readily transport charge.

15.3 Coulomb's Law

Experiments show that an **electric force** has the following properties:

1. It is inversely proportional to the square of the separation, r, between the two particles and is along the line joining them.
2. It is proportional to the product of the magnitudes of the charges, $|q_1|$ and $|q_2|$, on the two particles.
3. It is attractive if the charges are of opposite sign and repulsive if the charges have the same sign.

15.4 The Electric Field

The electric field vector **E** at some point in space is defined as the electric force **F** acting on a positive test charge placed at that point divided by the magnitude of the test charge q_0.

An electric field exists at some point if a test charge at rest placed at that point experiences an electrical force.

The total electric field at a point, due to a group of charges, equals the **vector sum** of the electric fields at that point due to each of the charges.

15.5 Electric Field Lines

A convenient aid for visualizing electric field patterns is to draw lines pointing in the same direction as the electric field vector at any point. These lines, called electric field lines, are related to the electric field in any region of space in the following manner:

1. The electric field vector **E** is **tangent** to the electric field line at each point.
2. The number of lines per unit area through a surface perpendicular to the lines is proportional to the strength of the electric field in that region. Thus, **E** is large when the field lines are close together and small when they are far apart.

The rules for drawing electric field lines for any charge distribution are as follows:

1. The lines must begin on positive charges and terminate on negative charges, or at infinity in the case of an excess of charge.
2. The number of lines drawn leaving a positive charge or approaching a negative charge is proportional to the magnitude of the charge.
3. No two field lines can cross.

15.6 Conductors in Electrostatic Equilibrium

A conductor in **electrostatic equilibrium** has the following properties:

1. The electric field is zero everywhere inside the conductor.
2. Any excess charge on an isolated conductor resides entirely on its surface.
3. The electric field just outside a charged conductor is perpendicular to the conductor's surface.
4. On an irregularly shaped conductor, the charge per unit area is greatest at locations where the curvature of the surface is greatest, that is, at sharp points.

15.10 Electric Flux and Gauss's Law

There is an important general relation between the net electric flux through a closed surface (often called a **gaussian surface**) and the charge **enclosed** by the surface. This is known as **Gauss's law,** and states that **the net electric flux through any closed gaussian surface is equal to the net charge inside the surface divided by ϵ_o.** The technique is useful to calculate the value of the electric field only in a limited number of situations where there is a high degree of symmetry in the charge distribution.

EQUATIONS AND CONCEPTS

The magnitude of the electrostatic force between two stationary point charges, q_1 and q_2, separated by a distance, r, is given by Coulomb's law. In calculations the approximate value of the Coulomb constant, k, may be used.

$$F = k_e \frac{|q_1||q_2|}{r^2} \quad (15.1)$$

$$k_e \approx 8.99 \times 10^9 \ \text{N} \cdot \text{m}^2 / \text{C}^2 \quad (15.3)$$

The smallest unit of electric charge known in nature is the charge on the electron or on the proton, represented by the symbol, e.

$$e = 1.6 \times 10^{-19} \ \text{C}$$

The direction of the electrostatic force on each charge is determined from the experimental observation that like sign charges experience forces of mutual repulsion and unlike sign charges attract each other. By virtue of Newton's third law, the magnitude of the force on each of the two charges is the same regardless of the relative magnitude of the values of q_1 and q_2.

Comment on Coulomb's law.

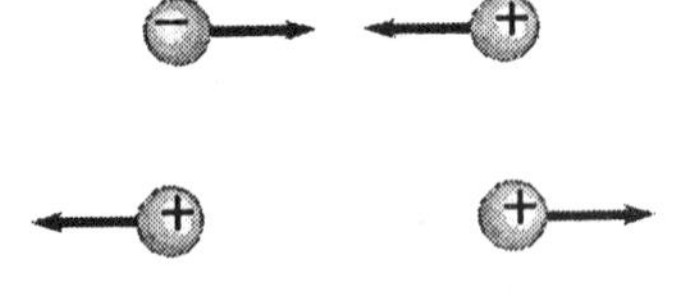

In cases where there are more than two charges present, the resultant force on any one charge is the vector sum of the forces exerted on that charge by the remaining individual charges present.

Comment on the principle of superposition.

The electric field, **E**, at any point in space is defined as the ratio of electric force per unit charge exerted on a small positive test charge, q_0, placed at the point where the field is to be determined.

$$E \equiv \frac{|\mathbf{F}|}{|q_0|} \qquad (15.4)$$

The direction of **E** at any point is defined to be the direction of the electric force that would be exerted on small positive charge if placed at the point in question. The SI units of the electric field are newtons per coulomb (N/C).

Comment on direction and units of the electric field.

The definition of the electric field combined with Coulomb's law leads to an expression for calculating the electric field a distance, r, from a point charge, q. The direction of the electric field is radially outward from a positive point charge and radially inward toward a negative point charge. The superposition principle holds when the electric field at a point is due to a number of point charges.

$$E = k_e \frac{|q|}{r^2} \qquad (15.5)$$

Comment on electric field lines.

Electric field lines are a convenient graphical representation of electric field patterns. These lines are related to the electric field in the following manner:

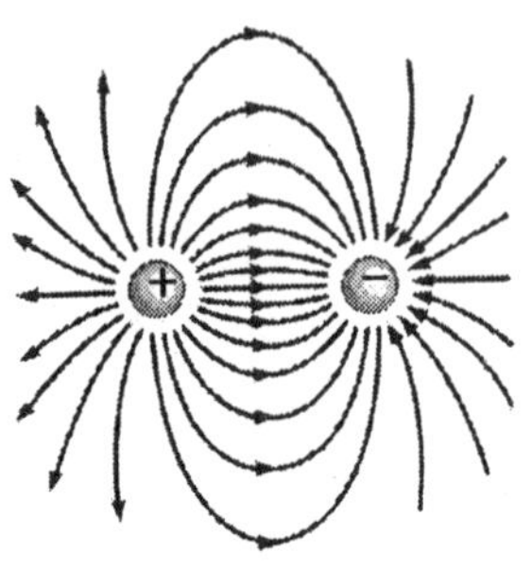

1. The electric field vector **E** is tangent to the electric field lines at each point.

2. The number of lines pre unit area through a surface perpendicular to the lines is proportional to the strength of the electric field in a given region. Thus, **E** is large when the field lines are lose together, and small when they are far apart.

Rules for drawing electric field lines

In drawing electric field lines,

1. The lines must begin on positive charges (or at infinity) and must terminate on negative charges or, in the case of an excess of charge, at infinity.

2. The number of lines drawn leaving a positive charge or approaching a negative charge is proportional to the magnitude of the charge.

3. No two field lines can cross each other.

A conductor in electrostatic equilibrium has the following properties:

Comment on conductors in electrostatic equilibrium.

(1) The electric field is zero everywhere inside the conductor;

(2) Any excess charge on an isolated conductor resides entirely on its surface;

(3) The electric field just outside a charged conductor is perpendicular to the surface of the conductor;

(4) On an irregularly shaped conductor, the charge tends to accumulate at locations where the radius of curvature of the surface is smallest—that is, at sharp points.

In the case of an electric field that is uniform in both magnitude and direction, the electric flux, which has SI units of $N \cdot m^2/C$, is proportional to the magnitude of the electric field, and depends on the angle between the normal to the surface and the direction of the electric field.

$$\Phi = EA \cos\theta$$

SUGGESTIONS, SKILLS, AND STRATEGIES

ELECTRIC FORCES AND FIELDS

1. **Units**: When performing calculations that involve the use of the Coulomb constant k that appears in Coulomb's law, charges must be in coulombs and distances in meters. If they are given in other units, you must convert them to SI.
2. **Applying Coulomb's law to point charges**: Use the superposition principle properly when dealing with a collection of interacting point charges. When several charges are present, the resultant force on any one of them is found by finding the individual force that every other charge exerts on it and then finding the vector sum of all these forces. The magnitude of the force that any charged object exerts on another is given by Coulomb's law. The direction of the force is found by noting that the forces are repulsive between like charges and attractive between unlike charges.
3. **Calculating the electric field of point charges**: The superposition principle can also be applied to electric fields, which are also vector quantities. To find the total electric field at a given point, first calculate the electric field at the point due to each individual charge. The resultant field at the point is the vector sum of the fields due to the individual charges.

REVIEW CHECKLIST

▷ Use Coulomb's law to determine the net electrostatic force on a point electric charge due to a known distribution of a finite number of point charges.

▷ Calculate the electric field **E** (magnitude and direction) at a specified location in the vicinity of a group of point charges.

▷ Describe the configuration of electric field lines as they are associated with various patterns of charge distribution such as (i) point charges, (ii) dipole, (iii) charged metallic sphere, (iv) etc.

▷ State and justify the conditions for charge distribution on conductors in electrostatic equilibrium.

SOLUTIONS TO SELECTED END-OF-CHAPTER PROBLEMS

3. Two identical conducting spheres are placed with their centers 0.30 m apart. One is given a charge of 12×10^{-9} C and the other a charge of -18×10^{-9} C. (a) Find the electrostatic force exerted on one sphere by the other. (b) The spheres are connected by a conducting wire. After equilibrium has occurred, find the electrostatic force between the two.

Solution

(a) The two spheres have unlike charges and attract one another. The magnitude of the force is given by Coulomb's law as $|F| = k|Q_1||Q_2|/r^2$, where $k = 8.99\times10^9\ \text{N}\cdot\text{m}^2/\text{C}^2$ and r is the distance between the centers of the two spheres. Then,

$$|F| = \frac{\left(8.99\times10^9\ \text{N}\cdot\text{m}^2/\text{C}^2\right)\left(12\times10^{-9}\ \text{C}\right)\left(18\times10^{-9}\ \text{C}\right)}{(0.30\ \text{m})^2} = 2.2\times10^{-5}\ \text{N} \qquad \Diamond$$

(b) Since the two spheres are identical, they share the total charge equally when connected by a conducting wire. Thus, the equilibrium charge on each sphere is

$$Q = \frac{Q_1 + Q_2}{2} = \frac{\left(12\times10^{-9}\ \text{C}\right) + \left(-18\times10^{-9}\ \text{C}\right)}{2} \quad \text{or} \quad Q = -3.0\times10^{-9}\ \text{C}$$

Because the spheres now have the same type charge (both negative) they repel one another with a force whose magnitude is

$$|F| = \frac{k|Q||Q|}{r^2} = \frac{\left(8.99\times10^9\ \text{N}\cdot\text{m}^2/\text{C}^2\right)\left(-3.0\times10^{-9}\ \text{C}\right)^2}{(0.30\ \text{m})^2} = 9.0\times10^{-7}\ \text{N} \qquad \Diamond$$

11. Three charges are arranged as shown in Figure P15.11. Find the magnitude and direction of the electrostatic force on the charge at the origin.

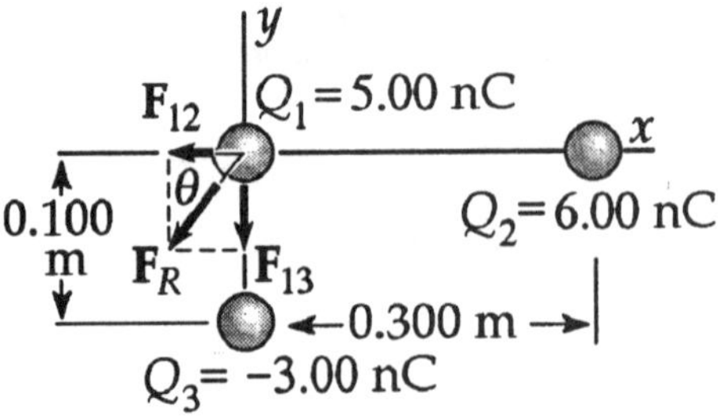

Figure P15.11

Solution The strategy to use when calculating the force exerted by a set of several point charges is to first find the force exerted by each of the point charges individually. The force exerted by the entire set of charges is the resultant of the forces exerted by the individual charges.

The charge Q_2 exerts a repulsive force, F_{12} on Q_1 as shown in the sketch. The distance separating these charges is $r_{12} = 0.300$ m so $F_{12} = k|Q_1||Q_2|/r_{12}^2$ becomes

$$F_{12} = \frac{\left(8.99\times10^{9}\ \text{N}\cdot\text{m}^2/\text{C}^2\right)\left(5.00\times10^{-9}\ \text{C}\right)\left(6.00\times10^{-9}\ \text{C}\right)}{(0.300\ \text{m})^2} = 3.00\times10^{-6}\ \text{N}$$

The negative charge Q_3 exerts an attractive force, F_{13} on Q_1. The distance between Q_1 and Q_3 is $r_{13} = 0.100$ m, so $F_{13} = k|Q_1||Q_3|/r_{13}^2$ becomes

$$F_{13} = \frac{\left(8.99\times10^{9}\ \text{N}\cdot\text{m}^2/\text{C}^2\right)\left(5.00\times10^{-9}\ \text{C}\right)\left(3.00\times10^{-9}\ \text{C}\right)}{(0.100\ \text{m})^2} = 1.35\times10^{-5}\ \text{N}$$

F_{12} and F_{13} can be combined with the Pythagorean theorem as

$$F_R = \sqrt{(F_{12})^2 + (F_{13})^2} = \sqrt{\left(3.00\times10^{-6}\ \text{N}\right)^2 + \left(1.35\times10^{-5}\ \text{N}\right)^2} = 1.38\times10^{-5}\ \text{N} \quad \lozenge$$

The angle that F_R makes with the horizontal is

$$\theta = \arctan\left(\frac{F_{13}}{F_{12}}\right) = \arctan\left(\frac{1.35\times10^{-5}\ \text{N}}{3.00\times10^{-6}\ \text{N}}\right) = 77.5^\circ \text{ below the } -x \text{ axis} \quad \lozenge$$

19. A piece of aluminum foil of mass 5.00 × 10^{-2} kg is suspended by a string in an electric field directed vertically upward. If the charge on the foil is 3.00 μC, find the strength of the field that will reduce the tension in the string to zero.

Solution Note that the foil has a positive charge. Thus, the force exerted on it by the field is in the direction of the field (i.e., vertically upward). The free-body diagram for the aluminum foil is shown in the diagram. Three forces act on the foil: the tension in the string, T; the force exerted by the electric field, F_e; and the weight of the foil, F_g. Since the foil is in equilibrium,

$$\sum F_y = +T + F_e - F_g = 0$$

and the tension in the cord is

$$T = F_g - F_e = mg - qE$$

Thus, if the tension in the cord is to be zero, it is necessary that $qE = mg$, and the required field strength is

$$E = \frac{mg}{q} = \frac{\left(5.00\times 10^{-2}\text{ kg}\right)\left(9.80\text{ m/s}^2\right)}{3.00\times 10^{-6}\text{ C}} = 1.63\times 10^5\text{ N/C}$$ ◊

23. Positive charges are situated at three corners of a rectangle, as shown in Figure P15.23. Find the electric field at the fourth corner.

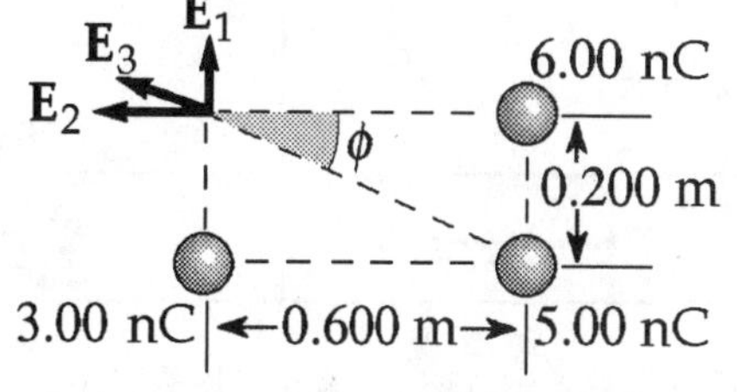

Figure P15.23

Solution The electric field at the fourth corner is the superposition of three contributions, one each from the charges on the other three corners. The contribution E_1, from the 3.00-nC charge, is directed upward as shown and has a magnitude

$$E_1 = \frac{kQ_1}{r_1^2} = \frac{\left(8.99\times 10^9\text{ N}\cdot\text{m}^2/\text{C}^2\right)\left(3.00\times 10^{-9}\text{ C}\right)}{(0.200\text{ m})^2} = 674\text{ N/C}$$

The 6.00-nC charge makes a contribution, E_2, which is directed to the left with a magnitude of

$$E_2 = \frac{kQ_2}{r_2^{\,2}} = \frac{\left(8.99\times10^9\ \text{N}\cdot\text{m}^2/\text{C}^2\right)\left(6.00\times10^{-9}\ \text{C}\right)}{(0.600\ \text{m})^2} = 150\ \text{N}/\text{C}$$

The contribution from the 5.00-nC charge is directed parallel to the diagonal of the rectangle as shown. The distance from the corner to this charge is the diagonal of the rectangle, so $r_3^{\,2} = (0.200\ \text{m})^2 + (0.600\ \text{m})^2 = 0.400\ \text{m}^2$

Thus, $$E_3 = \frac{kQ_3}{r_3^{\,2}} = \frac{\left(8.99\times10^9\ \text{N}\cdot\text{m}^2/\text{C}^2\right)\left(5.00\times10^{-9}\ \text{C}\right)}{0.400\ \text{m}^2} = 113\ \text{N}/\text{C}$$

The angle this contribution makes with the horizontal is:

$$\phi = \arctan\left(\frac{0.200\ \text{m}}{0.600\ \text{m}}\right) = \arctan(0.333) = 18.4^\circ$$

The horizontal and vertical components of the individual contributions and of the resultant field at the corner are as follows:

Contribution	Horizontal Component	Vertical Component
E_1	0	+674 N/C
E_2	−150 N/C	0
E_3	$-E_3\cos\phi = -107$ N/C	$+E_3\sin\phi = +35.7$ N/C
Resultant, E	$\sum E_x = -257$ N/C	$\sum E_y = +710$ N/C

The magnitude and direction of the resultant electric field at this corner are:

$$E = \sqrt{\left(\sum E_x\right)^2 + \left(\sum E_y\right)^2} = 755\ \text{N}/\text{C} \quad \text{and} \quad \tan\theta = \frac{\sum E_y}{\left|\sum E_x\right|} = 2.77 \quad \text{with} \quad \theta = 70.1^\circ$$

Thus, the electric field at the vacant corner of the rectangle is $E = 755$ N/C oriented at 70.1° clockwise from $-x$ axis. ◊

27. In Figure P15.27, determine the point (other than infinity) at which the total electric field is zero.

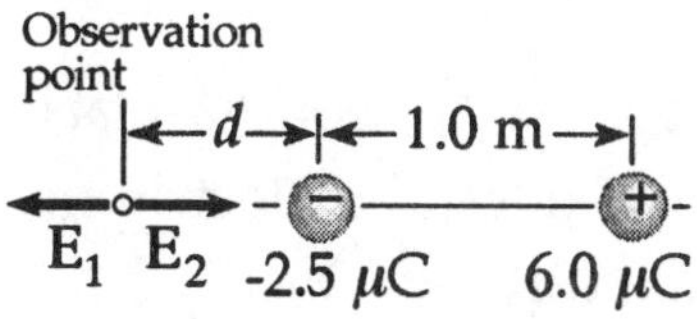

Observation point

$|\leftarrow d \rightarrow|\leftarrow 1.0\text{ m} \rightarrow|$

E_1 E_2 -2.5 μC 6.0 μC

Figure P15.27

Solution The electric field at any point will be the resultant of two contributions, one from each of the charges shown in the figure. The sum of two vectors can be zero only if the two vectors have the **same magnitude** and **opposite directions.**

The electric field due to a positive charge points away from the charge, while the field of a negative charge points toward the charge. If, at the observation point, the contributions E_1 and E_2 are to have opposite directions, it is necessary that the observation point lie somewhere on the line connecting the two charges. Also, the observation point cannot be between the charges since the two contributions have the same directions there. Since the two contributions must also have the same magnitude, the distance from the observation point to the smaller charge must be less than the distance to the larger charge. Therefore, the observation point must be located some distance d to the left of the $-2.5\ \mu$C charge as shown.

To solve for d, require that $E_2 = E_1$, or $\dfrac{k|-2.5\ \mu\text{C}|}{d^2} = \dfrac{k|6.0\ \mu\text{C}|}{(d+1.0\text{ m})^2}$

This reduces to $2.5(d+1.0\text{ m})^2 = 6.0d^2$ or $d+1.0\text{ m} = \pm\sqrt{\dfrac{6.0}{2.5}}d = \pm 1.55d$

Solving, $d = \dfrac{1.0\text{ m}}{+1.55-1} = 1.8\text{ m}$, and $d = \dfrac{1.0\text{ m}}{-1.55-1} = -0.39\text{ m}$

The negative solution is unacceptable because it refers to a point located between the two charges where the two contributions have equal magnitudes but are in the same direction. Thus, the only point (other than at a infinite distance from both charges) where the resultant field is zero is located on the line connecting the two charges and 1.8 m to the left of the -2.5 μC charge. ◊

29. (a) Sketch the electric field lines around an isolated point charge, $q > 0$. (b) Sketch the electric field pattern around an isolated negative point charge of magnitude $-2q$.

Solution

There are several points to keep in mind when sketching the electric field patterns for the given charge distributions. These are:

(1) The field lines for a point charge exhibit radial symmetry. That is, they are either radially outward away from the charge or are radially inward toward the charge.

(2) The lines originate on positive charges and terminate on negative charges. Thus, the pattern of the lines is radially outward for a positive point charge and radially inward for a negative point charge.

(3) The number of lines drawn leaving a positive charge or approaching a negative charge is proportional to the magnitude of the charge. Therefore, the number of lines drawn approaching the negative charge of magnitude $2q$ in part (b) must be twice the number drawn leaving the positive charge of magnitude q in part (a).

The patterns with all of these properties are shown above for each case. ◊

35. If the electric field strength in air exceeds 3.0×10^6 N/C, the air becomes a conductor. Using this fact, determine the maximum amount of charge that can be carried by a metal sphere 2.0 m in radius. (See the hint in Problem 34.)

Solution

In electrostatic equilibrium, the excess charge on a conducting sphere is uniformly distributed over the outer surface of that sphere. Also, at all points **outside** a spherically symmetric charge distribution, the electric field is identical to that which would exist if the total charge was concentrated as a point charge located at the center of the sphere. Thus, the electric field strength at any point outside the charged metal sphere is given by $E = kQ/r^2$, where Q is the net charge on the sphere and r is the distance to this point from the center of the sphere. Notice that the field has the greatest strength where r has its minimum allowed value. Since the observation point must be **outside** the spherical charge distribution, the minimum allowed value is $r = R$ where R is the radius of the sphere. Therefore, the field is strongest just outside the surface of the sphere and, as the charge on the sphere is increased, electrical breakdown (i.e., the air will become conducting) will begin to occur when the electric field at the surface of the sphere reaches 3.0×10^6 N/m. The maximum charge that the sphere may have is found from

$$E\Big|_{r=R} = \frac{kQ_{max}}{R^2} = 3.0 \times 10^6 \text{ N/m}, \qquad \text{or} \qquad Q_{max} = \frac{\left(3.0 \times 10^6 \text{ N/m}\right)R^2}{k}$$

If $R = 2.0$ m,

$$Q_{max} = \frac{\left(3.0 \times 10^6 \text{ N/m}\right)\left(2.0 \text{ m}\right)^2}{8.99 \times 10^9 \text{ N} \cdot \text{m}^2/\text{C}^2} = 1.3 \times 10^{-3} \text{ C} \qquad \Diamond$$

39. A 40-cm diameter loop is rotated in a uniform electric field until the position of maximum electric flux is found. The flux in this position is measured to be 5.2 x 10^5 N·m^2/C. Calculate the electric field strength in this region.

Solution When an plane surface of area A is located in a uniform electric field E, the electric flux through the surface (i.e., the number of electric field lines that penetrate the surface) is given by $\Phi = EA\cos\theta$. Here, θ is the angle between the direction of the electric field and the line perpendicular to the plane surface. For constant values of E and A, the flux through the surface is a maximum when $\cos\theta = 1$. This occurs when $\theta = 0°$, and when the line perpendicular to the surface is parallel to the field. Hence, if the maximum flux that passes through an area as it is rotated in a uniform electric field is Φ_{max}, the strength of the field is given by $E = \Phi_{max}/(A\cos 0°) = \Phi_{max}/A$. The area enclosed by a 40-cm diameter circular loop is

$$A = \frac{\pi d^2}{4} = \frac{\pi(0.40\text{ m})^2}{4} = (0.040\pi)\text{ m}^2$$

Thus, if the maximum flux that passes through this area is $\Phi_{max} = 5.2\times10^5\text{ N}\cdot\text{m}^2/\text{C}$, the field strength is

$$E = \frac{\Phi_{max}}{A} = \frac{5.2\times10^5\text{ N}\cdot\text{m}^2/\text{C}}{(0.040\pi)\text{ m}^2} = 4.1\times10^6\text{ N/C}$$ ◊

43. An infinite plane conductor has charge spread out on its surface as shown in Figure P15.43. Use Gauss' law to show that the electric field at any point outside the conductor is given by $E = \sigma/\epsilon_0$, where σ is the charge per unit area on the conductor. (**Hint:** Choose a gaussian surface in the shape of a cylinder with one end inside the conductor and one end outside the conductor.)

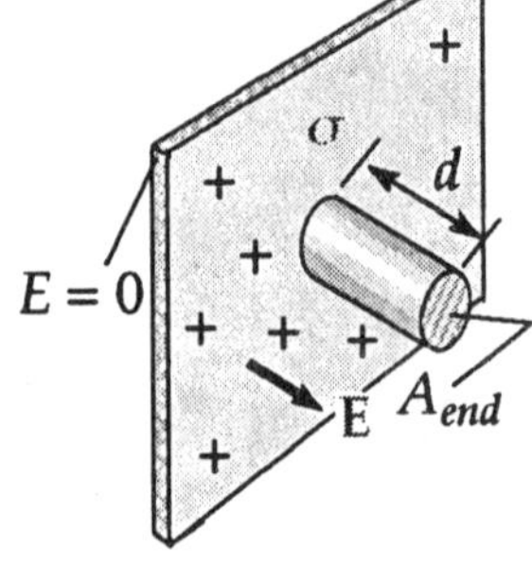

Figure P15.43

Solution At equilibrium, all excess charge on a conductor resides on the surface. For an infinite plane conductor (shown in an edge-on view in the figure), a uniform charge per unit area σ exists on each surface of the conductor. At points just outside a conducting surface, the electric field is perpendicular to the surface, and the field is zero at all points inside a conductor.

Gauss' law states: "The net electric flux through any closed gaussian surface is equal to the net charge enclosed by that surface divided by ϵ_o ". This may be used to determine the electric field strength just outside the surface of the plane conductor. Consider the cylindrical gaussian surface shown. The cylinder is oriented with its axis perpendicular to the plane conductor. One end of the cylinder is inside the conducting material and the other end is a distance d outside the conductor. The flux passing through any part of a gaussian surface is given by $\Phi = EA\cos\theta$ where A is the area of that part of the surface, E is the field strength at that point on the surface, and θ is the angle between the direction of the field and the line perpendicular to the surface at that point.

The net flux through the entire gaussian surface may be written as $\Phi_{\text{net}} = \Phi_{\text{left end}} + \Phi_{\text{cylindrical side}} + \Phi_{\text{right end}}$. Since the electric field is zero everywhere inside the conductor, $\Phi_{\text{left end}} = 0$. Outside the conductor, the field is parallel to the side of the cylinder, or the angle between the field and a line perpendicular to the cylindrical side is 90°. Hence, $\Phi_{\text{cylindrical side}} = EA_{\text{side}}\cos 90° = 0$. The field is perpendicular to the right end. Thus, $\theta = 0°$ here, and $\Phi_{\text{right end}} = EA_{\text{end}}\cos 0° = EA_{\text{end}}$. The charge enclosed by the gaussian surface is the charge on the circular area of the plane that lies inside the cylinder. This area equals the cross-sectional area of the cylinder and is the same as the area of the end, A_{end}. Therefore, $Q_{\text{enclosed}} = \sigma A_{\text{end}}$, and Gauss' law $(\Phi_{\text{net}} = Q_{\text{enclosed}}/\epsilon_o)$ gives

$$0 + 0 + EA_{\text{end}} = \frac{\sigma A_{\text{end}}}{\epsilon_o}$$

Thus, outside the infinite plane, $E = \sigma/\epsilon_o$ and is perpendicular to the plane. ◊

47. A small 2.00-g plastic ball is suspended by a 20.0-cm-long string in a uniform electric field, as shown in Figure P15.47. If the ball is in equilibrium when the string makes a 15.0° angle with the vertical as indicated, what is the net charge on the ball?

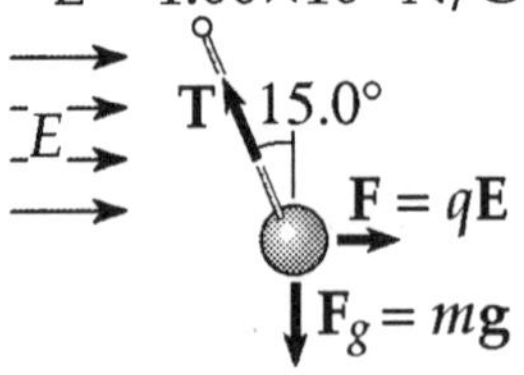

Figure P15.47

Solution Three forces act on the ball as shown in the free-body diagram. These forces are the tension in the string, T, the weight of the ball, w, and a force F exerted on the ball by the electric field. Note that for the string to be tilted as shown, the force F must be in the direction of the field. Thus, the ball has a net positive charge. Since the ball is in equilibrium, $\sum F_x = 0$ and $\sum F_y = 0$.

$\sum F_x = 0$: $\sum F_x = -T\sin 15.0° + F = 0$, or $F = T\sin 15.0°$ **[Equation 1]**

$\sum F_y = 0$: $T\cos 15.0° - F_g = 0$, or $mg = T\cos 15.0°$ **[Equation 2]**

Dividing Equation 1 by Equation 2: $\dfrac{F}{mg} = \dfrac{\sin 15.0°}{\cos 15.0°}$, or $F = mg\tan 15.0°$. The electrical force acting on the ball is the product of its charge and the electric field strength. Thus, the charge on the ball is $q = \dfrac{F}{E} = \dfrac{mg\tan 15.0°}{E}$, or

$$q = \frac{\left(2.00\times10^{-3}\text{ kg}\right)\left(9.80\text{ m/s}^2\right)\tan 15.0°}{1.00\times10^{3}\text{ N/C}} = 5.25\times10^{-6}\text{ C} = 5.25\ \mu\text{C} \qquad \Diamond$$

---

52. Two small silver spheres, each with a mass of 100 g, are separated by 1.00 m. Calculate the fraction of the electrons in one sphere that must be transferred to the other in order to produce an attractive force of 1.00×10^4 N (about a ton) between the spheres. (The number of electrons per atom of silver is 47, and the number of atoms per gram is Avogadro's number divided by the molar mass of silver, 107.87.)

Solution The net charge that must exist on each sphere if they are to attract each other with the specified force may be determined from Coulomb's law:

$$F = \frac{k|Q_1||Q_2|}{r^2}$$

Since the spheres were charged by transferring negative charge from one to the other, the magnitude of the positive charge on one is the same as the magnitude of the negative charge on the other (i.e., $|Q_1| = |Q_2| = |Q|$). Thus, $F = k|Q|^2/r^2$ and the magnitude of charge on either sphere is

$$|Q| = \sqrt{\frac{Fr^2}{k}} = \sqrt{\frac{(1.00 \times 10^4 \text{ N})(1.00 \text{ m})^2}{8.99 \times 10^9 \text{ N} \cdot \text{m}^2/\text{C}^2}} = 1.05 \times 10^{-3} \text{ C}$$

The number of electrons that must be transferred from a neutral object to give it a net positive charge of $Q = 1.05 \times 10^{-3}$ C is

$$n = \frac{Q}{e} = \frac{1.05 \times 10^{-3} \text{ C}}{1.60 \times 10^{-19} \text{ C}} = 6.59 \times 10^{15} \text{ electrons}$$

The total number of electrons contained in one of the spheres is given by

$$N = \left(\begin{matrix}\text{47 electrons} \\ \text{per silver atom}\end{matrix}\right)\left(\frac{6.02 \times 10^{23} \text{ atoms}}{\text{1 mole}}\right)\left(\frac{\text{1 mole of silver}}{\text{107.87 grams}}\right)\left(\begin{matrix}\text{100 grams} \\ \text{per sphere}\end{matrix}\right)$$

or $N = 2.62 \times 10^{25}$ electrons per sphere. The fraction of total electrons in one sphere that must be transferred to the other to achieve the desired attractive force is

$$\frac{n}{N} = \frac{6.59 \times 10^{15}}{2.62 \times 10^{25}} = 2.51 \times 10^{-10}$$

or approximately 2.5 out of every 10 billion electrons in the sphere. ◊

59. Two equal positive charges, q, are on the x axis at $x = a$ and $x = -a$. Show that the field along the positive y axis is in the y direction and is given by the relation $E_y = 2kqy(y^2 + a^2)^{-3/2}$

Solution As seen in the diagram, any point on the positive y axis is equidistant from the two charges. The Pythagorean theorem gives this distance to either charge as

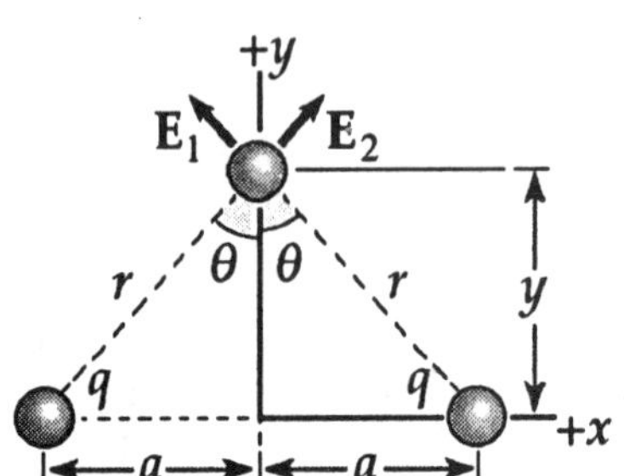

$$r = \sqrt{a^2 + y^2}$$

Since the charges are also equal, they make equal contributions, $E_1 = E_2 = kq/r^2$, to the total electric field at the observation point.

These two contributions must be added as vectors to find the resultant field. The x-component of the resultant field is $E_x = E_{1x} + E_{2x} = -E_1 \sin\theta + E_2 \sin\theta$. Since $E_1 = E_2$, the x-component of the resultant field is zero, and the resultant field is parallel to the y axis. The y-component of the resultant field is

$$E_y = E_{1y} + E_{2y} = +E_1 \cos\theta + E_2 \cos\theta = 2\left(\frac{kq}{r^2}\right)\cos\theta$$

From the diagram, observe that $\cos\theta = y/r$. Thus,

$$E_y = 2\left(kq/r^2\right)\left(y/r\right) = 2kqy/r^3$$

Since $r = \sqrt{a^2 + y^2}$, the resultant electric field at any point on the positive y axis is in the y direction and has a magnitude

$$E = E_y = \frac{2kqy}{\left(a^2 + y^2\right)^{3/2}} = 2kqy\left(a^2 + y^2\right)^{-3/2} \qquad \Diamond$$

CHAPTER SELF-QUIZ

1. A charge of +3 C is at the origin. When charge Q is placed at 2.00 m along the positive x axis, the electric field at 2.00 m along the negative x axis becomes zero. What is the value of Q?

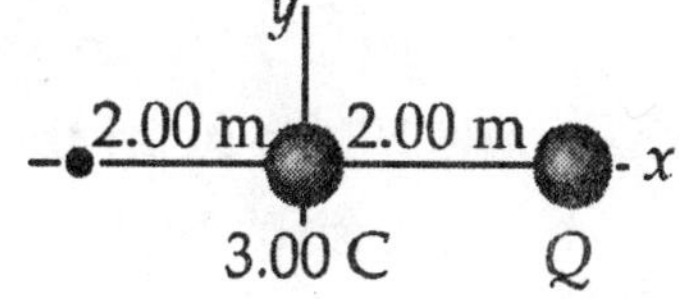

   a. −3 C
   b. −6 C
   c. −9 C
   d. −12 C

2. A charge, $+Q$, is placed inside a balloon and the balloon is blown up. As the radius, r, of the balloon increases, the number of field lines going through the surface of the balloon
   a. increases proportional to r^2
   b. increases proportional to r
   c. stays the same
   d. decreases as $1/r$

3. At what point is the electrical field associated with a uniformly-charged, hollow, metallic sphere greatest?
   a. center of the sphere
   b. at the sphere's inner surface
   c. at infinity
   d. at the sphere's outer surface

4. Two point charges are placed along a horizontal axis with the following values and positions: +3 μC at $x = 0$ cm and −7 μC at $x = 20$ cm. What is the magnitude of the electric field at the point midway between the two charges (at $x = 10$ cm)?
   a. 9.00×10^6 N/C
   b. 3.60×10^6 N/C
   c. 4.50×10^6 N/C
   d. 1.80×10^6 N/C

5. Initially, a net charge of –5.0 μC is placed on a 5.0-cm radius metallic hollow sphere. Next, a +10 μC charge is carefully inserted at the center through a hole in the latter's surface. What electric field is present at a point 10 cm from the center of the sphere? $(k = 8.99 \times 10^9 \text{ N} \cdot \text{m}^2 / \text{C}^2)$
   a. 2.30×10^6 N/C
   b. 4.50×10^6 N/C
   c. 9.00×10^6 N/C
   d. 18.0×10^6 N/C

6. Two point charges are separated by 4 cm and have charges of +2.0 μC and –2.0 μC, respectively. What is the electric field at a point midway between the two charges?
   a. 18.0×10^7 N/C
   b. 9.00×10^7 N/C
   c. 4.50×10^7 N/C
   d. Zero

7. The number of electric field lines that is incident normally on a unit cross-sectional area is indicative of which of the following?
   a. field direction
   b. charge density
   c. field strength
   d. charge motion

8. The electric field in a cathode ray tube is supposed to accelerate electrons from 0 to 1.60×10^7 m/s in a distance of 2 cm. What electric field is required? $(m_e = 9.10 \times 10^{-31}$ kg and $e = 1.60 \times 10^{-19}$ C$)$
   a. 9,000 N/m
   b. 18,200 N/m
   c. 36,400 N/m
   d. 72,800 N/m

9. A Van De Graaff generator has a spherical dome of radius 30 cm. Operating in dry air, where "atmospheric breakdown" is $E_{max} = 3.00 \times 10^6$ N/C what is the maximum charge that can be held on the dome? $(k = 8.99 \times 10^9 \text{ N} \cdot \text{m}^2/\text{C}^2)$
   a. 1.50×10^{-5} C
   b. 3.00×10^{-5} C
   c. 1.50×10^{-6} C
   d. 7.50×10^{-6} C

10. Two charges, $+Q$ and $-Q$, are located two meters apart and there is a point along the line that is equidistant from the two charges as indicated. Which vector best represents the direction of the electric field at that point?

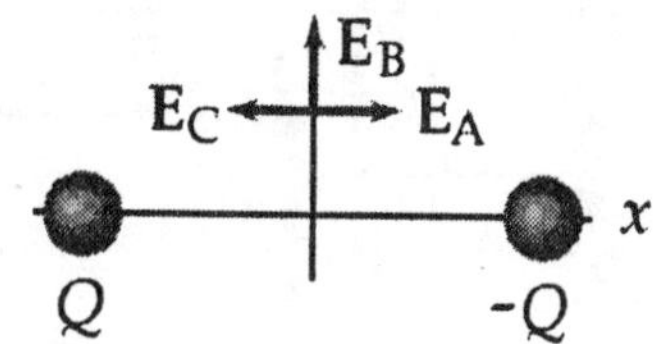

   a. Vector $\mathbf{E_A}$
   b. Vector $\mathbf{E_B}$
   c. Vector $\mathbf{E_C}$
   d. The electric field at that point is zero

11. A proton (mass = 1.67×10^{-27} kg, charge = $+1.60 \times 10^{-19}$ C) is accelerated from rest by an electric field of 400 N/C. Find its velocity after 10^{-6} s.
   a. 3.83×10^4 m/s
   b. 7.66×10^4 m/s
   c. 15.32×10^4 m/s
   d. 30.6×10^4 m/s

12. A glass rod rubbed with a silk cloth acquires a charge of $+7.40 \times 10^{-9}$ C. How many electrons were transferred from the glass to the silk?
   a. 4.62×10^{10}
   b. 2.10×10^{10}
   c. 0.90×10^9
   d. 1.30×10^9

Chapter 16

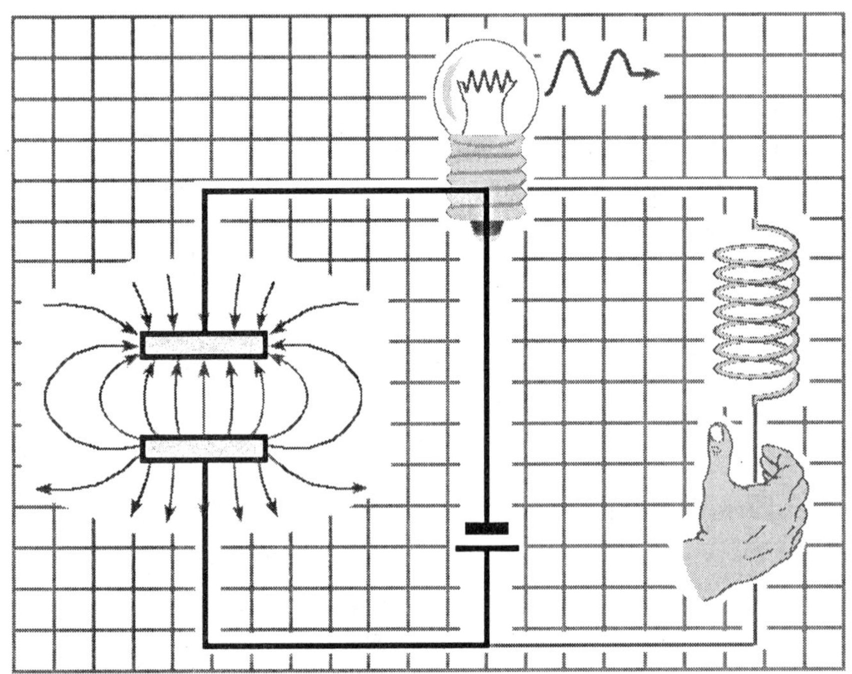

ELECTRICAL ENERGY AND CAPACITANCE

Chapter 16

ELECTRICAL ENERGY AND CAPACITANCE

The concept of potential energy was first introduced in Chapter 5. A potential energy function can be defined for any conservative force, such as the force of gravity. By using the principle of conservation of energy, we were often able to avoid working directly with forces when solving problems. In this chapter we discover that the energy concept is also useful in the study of electricity. Because the Coulomb force is conservative, we can define an electrical potential energy corresponding to the Coulomb force. This concept of potential energy is of value, but perhaps even more valuable is a quantity called electric potential, defined as potential energy per unit charge.

We take our first steps toward circuits with a discussion of electric potential, carried forward by an investigation of a common circuit element called a capacitor.

NOTES FROM SELECTED CHAPTER SECTIONS

16.1 Potential Difference and Electric Potential

The **electrostatic force is conservative**; therefore, it is possible to define an electric potential energy function associated with this force. The change in potential between two points is proportional to the change in potential energy of a charge as it moves between the two points.

Electrical potential difference (a scalar quantity) is the work done to move a charge from point A to point B divided by the magnitude of the charge. Thus, the SI units of potential are joules per coulomb, and are called volts (V).

A positive charge gains electrical potential energy when it is moved in a direction opposite the electric field. When a positive charge is placed in an electric field, it moves in the direction of the field, from a point of high potential to a point of lower potential.

16.2 Electric Potential and Potential Energy Due to Point Charges

In electric circuits, a point of zero potential is often defined by grounding (connecting to Earth) some point in the circuit. In the case of a point charge, the point of zero potential is taken to be at an infinite distance from the charge. The potential at a given point in space depends only on the quantity of charge on the object setting up the potential, and the distance r from the object to the specific point. A potential can exist at a point in space whether or not there is a charge at that point.

16.3 Potentials and Charged Conductors
16.4 Equipotential Surfaces

No work is required to move a charge between two points that are at the same potential. That is, $W = 0$ when $V_B = V_A$.

The electric potential is a constant everywhere on the surface of a charged conductor in equilibrium.

The electric potential is constant everywhere inside a conductor and equal to its value at the surface.

The electron volt is defined as the energy that an electron (or proton) gains when accelerated through a potential difference of 1 V.

16.6 The Definition of Capacitance

A capacitor is a device consisting of a pair of conductors separated by insulating material. A charged capacitor acts as a storehouse of charge and energy that can be reclaimed when needed for a specific application.

The capacitance, C, of a capacitor is defined as the ratio of the magnitude of the charge on either conductor to the magnitude of the potential difference between the conductors. Capacitance has SI units of coulombs per volt, called **farads** (F). The farad is a very large unit of capacitance. In practice, most capacitors have capacitances ranging from microfarads to picofarads.

16.7 Combinations of Capacitors

The potential difference across each capacitor in a parallel combination of capacitors is the same for each capacitor.

The equivalent capacitance of a parallel combination of capacitors is larger than any of the individual capacitances.

For a series combination of capacitors, the magnitude of the charge must be the same on all the plates.

In general, the potential difference across any number of capacitors (or other circuit elements) in series is equal to the sum of the potential differences across the individual capacitors.

16.10 Capacitors With Dielectrics

A dielectric is an insulating material, such as rubber, glass, or waxed paper. When a dielectric is inserted between the plates of a capacitor, the capacitance increases. If the dielectric completely fills the space between the plates, the capacitance is multiplied by the factor κ, called the **dielectric constant.**

The smallest plate separation for a capacitor is limited by the electric discharge that can occur through the dielectric material separating the plates. For any given plate separation, there is a maximum electric field that can be produced in the dielectric before it breaks down and begins to conduct. This maximum electric field is called the **dielectric strength**.

EQUATIONS AND CONCEPTS

The change in electric potential energy of an electric charge in moving between two points in an electric field is equal to the negative of the work done by the electric force. The change in electric potential energy can be expressed in terms of the charge, the magnitude of the field, and the distance moved parallel to the direction of the field.

$$\Delta PE = -W = -qEd \tag{16.1}$$

The electric potential difference between points A and B is defined as the change in electric potential energy per unit charge as a positive charge is moved from point A to point B.

$$\Delta V \equiv V_B - V_A = \frac{\Delta PE}{q} \tag{16.2}$$

The SI unit of electric potential is the volt, V. One joule of work must be done by an external force in order to move 1 coulomb of charge from point A to a second point B where the electric potential is 1 volt greater. Since the electric force is a conservative force, the work done is independent of the path taken from A to B.

$$1\ \text{V} \equiv 1\ \text{J/C} \tag{16.3}$$

$$1\ \text{N/C} = 1\ \text{V/m}$$

The electric potential in the vicinity of a point charge, q, is inversely proportional to the distance from the charge. This equation assumes that the potential at infinity is zero. Note that the sign of the potential depends on the sign of the charge q.

$$V = k\frac{q}{r} \tag{16.5}$$

The total electric potential at some point P, due to several point charges, is the algebraic sum of the potentials due to the individual charges.

Comment on a potential due to several charges.

The potential energy of a pair of charges separated by a distance, r, represents the minimum work required to assemble the charges from an infinite separation. The potential energy of the two charges is positive if the two charges have the same sign; and it is negative if the two charges are of opposite sign.

$$PE = k\frac{q_1 q_2}{r} \tag{16.6}$$

No work is required to move a charge between two points that are at the same potential.

$$W = -q(V_B - V_A) \tag{16.7}$$

The net charge on a conductor in electrostatic equilibrium resides entirely on the surface. The electric potential is constant everywhere on the surface. The electric potential is constant everywhere inside and equal to its value at the surface. Also, recall that the electric field inside a charged conductor is zero.

Comment on charged conductors in electrostatic equilibrium.

The electron volt is defined as that quantity of energy which an electron or proton gains when accelerated through a potential difference of 1 volt.

$$1\text{ eV} = 1.60 \times 10^{-19}\text{ J} \qquad (16.8)$$

The capacitance, C, of a capacitor is defined as the ratio of the charge on either plate (conductor) to the potential difference between the plates. (C is always positive.)

$$C \equiv \frac{Q}{\Delta V} \qquad (16.9)$$

The SI unit of capacitance is the farad (F). The farad is a very large unit of capacitance and in practice typical devices have capacitances ranging from picofarads (10^{-12} F) to microfarads (10^{-6} F).

$$1\text{ F} \equiv 1\text{ C/V}$$

The capacitance of a capacitor depends on the physical characteristics of the device (size, shape, plate separation, and the nature of the dielectric medium filling the region between the plates).

Comment on capacitors.

The capacitance of an air-filled parallel plate capacitor is proportional to the area of the plates and inversely proportional to the separation of the plates.

$$C = \epsilon_0 \frac{A}{d} \qquad (16.10)$$

The value for the permittivity constant for free space.

$$\epsilon_o = \frac{1}{4\pi k_e} = 8.85 \times 10^{-12}\ C^2 / N \cdot m^2$$

The equivalent capacitance of a parallel combination of capacitors is the sum of the values of the individual capacitors in the parallel group. The total charge stored by a group of capacitors connected in parallel is the sum of the charges stored on the individual capacitors.

$$C_{eq} = C_1 + C_2 + C_3 + \ldots \qquad (16.13)$$

$$Q_{total} = Q_1 + Q_2 + Q_3 + \ldots$$

(parallel combination)

The equivalent capacitance of a series combination of capacitors is smaller than the smallest individual value of capacitance in the group. The potential difference across the series group equals the sum of the values of potential difference across the individual capacitors.

$$\frac{1}{C_{eq}} = \frac{1}{C_1} + \frac{1}{C_2} + \frac{1}{C_3} + \ldots \qquad (16.16)$$

$$\Delta V = \Delta V_1 + \Delta V_2 + \Delta V_3 + \ldots$$

(series combination)

The electrostatic energy stored in the electric field of a charged capacitor is equal to the work done by a battery (or other source of emf) in charging the capacitor from $q = 0$ to $q = Q$.

$$W = \tfrac{1}{2} Q \Delta V \qquad (16.17)$$

$$\text{Energy stored} = \tfrac{1}{2} C \Delta V^2 = \frac{Q^2}{2C}$$

When a dielectric (insulating material) is inserted between the plates of a capacitor, the capacitance of the device increases.

$$C = \kappa \epsilon_o \left(\frac{A}{d} \right) \qquad (16.20)$$

SUGGESTIONS, SKILLS, AND STRATEGIES

A STRATEGY FOR PROBLEMS INVOLVING ELECTRIC POTENTIAL

1. When working problems involving electric potential, remember that electric potential is a **scalar quantity** (rather than a vector quantity like the electric field), so there are no components to worry about. Therefore, when using the superposition principle to evaluate the electric potential at a point due to a system of point charges, you simply take the algebraic sum of the electric potentials due to each charge. However, you must keep track of signs. The electric potential due to each positive charge is positive; the electric potential due to each negative charge is negative. The basic equation to use is $V = kq / r$.

2. As in mechanics, only changes in potential energy are significant; hence, the point where you choose the potential energy to be zero is arbitrary.

A PROBLEM–SOLVING STRATEGY FOR CAPACITORS

1. Be careful with your choice of units. To calculate the capacitance of a device in farads, make sure that distances are in meters and use the SI value of ϵ_0.

2. When two or more unequal capacitors are connected in **series,** they carry the same charge, but the potential differences across them are not the same. Their capacitances add as reciprocals, and the equivalent capacitance of the combination is always **less** than the smallest individual capacitor.

3. When two or more capacitors are connected in **parallel,** the potential difference across each is the same. The charge on each capacitor is proportional to its capacitance; hence, the capacitances add directly to give the equivalent capacitance of the parallel combination.

4. A complicated circuit consisting of capacitors can often be reduced to a simple circuit containing only one capacitor. To do so, examine your initial circuit and replace any capacitors in series or any in parallel using the rules of Steps 2 and 3 above. Draw a sketch of your new circuit after

these changes have been made. Examine this new circuit and replace any series or parallel combinations. Continue this process until a single, equivalent capacitor is found.

5. If the charge on, or the electric potential difference across, one of the capacitors in the complicated circuit is to be found, start with the final circuit found in Step 4 and gradually work your way back through the circuits using $C = Q / V$ and the rules given in Steps 2 and 3 above.

REVIEW CHECKLIST

▷ Understand that each point in the vicinity of a charge distribution can be characterized by a scalar quantity called the electric potential, V; and define the quantity, **electrical potential difference.**

▷ Calculate the electric potential difference between any two points in a uniform **electric field** and calculate the electric potential difference between any two points in the vicinity of a **group of point charges.**

▷ Calculate the electric **potential energy** associated with a group of point charges and define the unit of energy, **electron volt.**

▷ Justify the claims that (i) all points on the surface and within a charged conductor are at the same potential and (ii) the electric field within a charged conductor is zero.

▷ Define the quantity, **capacitance**; and evaluate the capacitance of a parallel plate capacitor of given area and plate separation.

▷ Determine the equivalent capacitance of a network of capacitors in series-parallel combination and calculate the final charge on each capacitor and the potential difference across each when a known potential is applied across the combination.

SOLUTIONS TO SELECTED END-OF-CHAPTER PROBLEMS

1. A proton moves 2.0 cm parallel to a uniform electric field of $E = 200$ N/C. (a) How much work is done on the proton by the field? (b) What change occurs in the potential energy of the proton? (c) What potential difference did the proton move through?

Solution (a) The uniform electric field exerts a constant force of magnitude $F = qE$ directed parallel to the field on the positively charged proton. The work done is $W = FS\cos\theta$ where θ is the angle between the force and displacement S. Since the proton moves parallel to the field, and hence parallel to the force, $\theta = 0°$. Thus, the work done by the field is $W = (qE)S\cos 0°$, or

$$W = \left(1.60\times10^{-19}\ \text{C}\right)\left(200\ \text{N/C}\right)\left(2.0\times10^{-2}\ \text{m}\right)(1.0) = 6.4\times10^{-19}\ \text{J} \qquad \Diamond$$

(b) The work done on the proton by the field equals the negative of the change in the proton's electrical potential energy. Therefore,

$$\Delta PE = -W = -6.4\times10^{-19}\ \text{J} \qquad \Diamond$$

(c) The potential difference the proton moved through is equal to the change in the electrical potential energy **per unit charge.** Hence,

$$\Delta V = V_{\text{final}} - V_{\text{initial}} = \frac{\Delta PE}{q} = \frac{-6.4\times10^{-19}\ \text{J}}{1.60\times10^{-19}\ \text{C}} = -4.0\ \text{V} \qquad \Diamond$$

---

7. A pair of oppositely charged, parallel plates are separated by 5.33 mm. A potential difference of 600 V exists between the plates. (a) What is the magnitude of the electric field strength between the plates? (b) What is the magnitude of the force on an electron between the plates? (c) How much work must be done on the electron to move it to the negative plate if it is initially positioned 2.90 mm from the positive plate?

Solution (a) A uniform electric field, directed perpendicular to the plates, exists in the region between the plates and exerts a constant force $F = qE$ on a charged particle in this region. When the particle moves a distance d parallel to the field, going from one plate to the other, the work done by the field is $W = Fd\cos 0° = qEd$. Thus, the magnitude of the change in the potential energy of the particle is $|\Delta PE| = W = qEd$ and the magnitude of the potential difference between the plates is

$$|\Delta V| = \frac{|\Delta PE|}{q} = \frac{qEd}{q} = Ed$$

For the given set of plates, $|\Delta V| = 600\text{ V}$ and $d = 5.33\times10^{-3}\text{ m}$. Therefore, the magnitude of the electric field between the plates is

$$E = \frac{|\Delta V|}{d} = \frac{600\text{ V}}{5.33\times10^{-3}\text{ m}} = 1.13\times10^{5}\text{ V/m} \quad \Diamond$$

(b) The magnitude of the force on an electron located between the plates is

$$F = |q|E = eE = \left(1.60\times10^{-19}\text{ C}\right)\left(1.13\times10^{5}\text{ V/m}\right) = 1.80\times10^{-14}\text{ N} \quad \Diamond$$

(c) The displacement of the electron is

$$S = 5.33\text{ mm} - 2.90\text{ mm} = 2.43\text{ mm}$$

toward the negative plate. An applied force with magnitude $F = eE = 1.80\times10^{-14}\text{ N}$ and directed toward the negative plate must be used to offset the influence of the field and move the electron without acceleration. The work done by the applied force as the electron moves to the negative plate is

$$W = FS\cos 0° = \left(1.80\times10^{-14}\text{ N}\right)\left(2.43\times10^{-3}\text{ m}\right)(1.00) = 4.37\times10^{-17}\text{ J} \quad \Diamond$$

15. Two point charges, $Q_1 = +5.00$ nC and $Q_2 = -3.00$ nC, are separated by 35.0 cm. (a) What is the electric potential at a point midway between the charges? (b) What is the potential energy of the pair of charges? What is the significance of the algebraic sign of your answer?

Solution

(a) The point midway between the charges lies a distance from each charge of $r_1 = r_2 = 17.5 \text{ cm} = 0.175 \text{ m}$. The electric potential at this point is the algebraic sum of two contributions, one due to each charge,

$$V = V_1 + V_2 = \frac{kQ_1}{r_1} + \frac{kQ_2}{r_2} = k\left(\frac{Q_1}{r_1} + \frac{Q_2}{r_2}\right)$$

or $$V = \left(8.99 \times 10^9 \text{ N} \cdot \text{m}^2/\text{C}^2\right)\left(\frac{5.00 \times 10^{-9} \text{ C}}{0.175 \text{ m}} - \frac{3.00 \times 10^{-9} \text{ C}}{0.175 \text{ m}}\right) = 103 \text{ V}$$ ◊

(b) Consider the charge Q_1 to be isolated in space. The electrical potential at a distance of $r_{12} = 35.0$ cm from Q_1 is then $V_1 = kQ_1/r_{12}$. By definition, this is the potential energy per unit charge a charged particle will possess when located at distance r_{12} from Q_1. Thus, when the charge Q_2 is placed in this position to complete construction of the specified charge distribution, the potential energy associated with the pair of charges is $PE = Q_2V_1 = kQ_1Q_2/r_{12}$:

$$PE = \frac{\left(8.99 \times 10^9 \text{ N} \cdot \text{m}^2/\text{C}^2\right)\left(5.00 \times 10^{-9} \text{ C}\right)\left(-3.00 \times 10^{-9} \text{ C}\right)}{0.350 \text{ m}} = -3.85 \times 10^{-7} \text{ J}$$ ◊

Note that the potential energy of the pair of charges is zero when r_{12} is very large (i.e., $PE \to 0$ as $r_{12} \to \infty$). The fact that the current potential energy of the pair is negative means that positive work must be done on this system to reach the zero potential energy level (to completely separate the charges). ◊

19. In Rutherford's famous scattering experiments that led to the "planetary model" of the atom, alpha particles (having charges of +2*e* and masses of 6.6×10^{-27} kg) were fired toward a "fixed" gold nucleus with charge +79*e*. An alpha particle, initially very far from the gold nucleus, is fired at 2.0×10^7 m/s directly toward the gold nucleus as in Figure P16.19. How close does the alpha particle get to the gold nucleus before turning around?

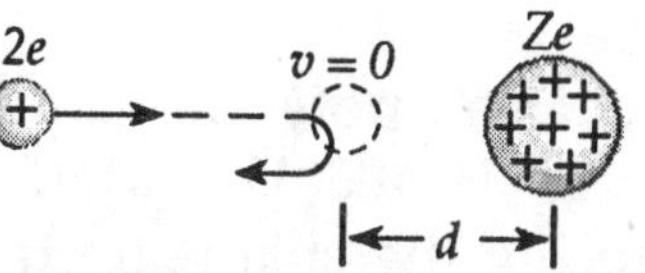

Figure P16.19

Solution As the alpha particle approaches the gold nucleus, the only force acting on it is the electrostatic force exerted by the nucleus. This is a conservative force. Hence, the total mechanical energy of the system remains constant $\left(KE_f + PE_f = KE_i + PE_i\right)$. Since the gold nucleus is "fixed" (i.e., is so massive in comparison to the alpha particle that its recoil may be ignored), the total kinetic energy is that of the alpha particle, $KE = \frac{1}{2}m_\alpha v_\alpha^2$. The potential energy of this pair of charges is $PE = kQ_1Q_2/r$, and the conservation of energy equation becomes

$$\frac{1}{2}m_\alpha v_f^2 + \frac{kQ_1Q_2}{r_f} = \frac{1}{2}m_\alpha v_i^2 + \frac{kQ_1Q_2}{r_i}$$

When the alpha particle is the minimum distance from the gold nucleus, $r_f = d$ and $v_f = 0$. Also, the alpha particle is initially "very far" from the nucleus, so $r_i \approx \infty$ and $v_i = 2.0 \times 10^7$ m/s. The energy equation is then

$$0 + \frac{kQ_1Q_2}{d} = \frac{1}{2}m_\alpha v_i^2 + 0$$

and the closest the alpha particle gets to the nucleus is $d = 2kQ_1Q_2/m_\alpha v_i^2$:

$$d = \frac{2k(2e)(79e)}{m_\alpha v_i^2} = \frac{316\left(8.99 \times 10^9 \text{ N}\cdot\text{m}^2/\text{C}^2\right)\left(1.60 \times 10^{-19} \text{ C}\right)^2}{\left(6.6 \times 10^{-27} \text{ kg}\right)\left(2.0 \times 10^7 \text{ m/s}\right)^2} = 2.8 \times 10^{-14} \text{ m} \ \Diamond$$

23. The potential difference between a pair of oppositely charged parallel plates is 400 V. (a) If the spacing between the plates is doubled without altering the charge on the plates, what is the new potential difference between the plates? (b) If the plate spacing is doubled and the potential difference between the plates is kept constant, what is the ratio of the final charge on one of the plates to the original charge?

Solution

By definition, the capacitance of a capacitor is $C = Q/\Delta V$. Here, Q is the charge stored when a potential difference ΔV is maintained between the plates. A pair of parallel plates, with a vacuum between them, form a capacitor with capacitance $C = \epsilon_o A/d$ where ϵ_o is a constant, A is the surface area of one of the plates, and d is the distance separating the plates.

(a) If the spacing between the plates is doubled, the ratio of the final capacitance to the initial is

$$\frac{C_f}{C_i} = \frac{\epsilon_o A/d_f}{\epsilon_o A/d_i} = \frac{d_i}{d_f} = \frac{1}{2}, \quad \text{or} \quad 2C_f = C_i$$

Thus, $2\left(Q_f/\Delta V_f\right) = Q_i/\Delta V_i$ and since the charge on the plates is unaltered $\left(Q_f = Q_i\right)$, the new potential difference is $\Delta V_f = 2\Delta V_i = 2(400 \text{ V}) = 800 \text{ V}$ ◊

(b) If the plate spacing is doubled, $2C_f = C_i$ and $2\left(Q_f/\Delta V_f\right) = Q_i/\Delta V_i$ as before. However, the potential difference between the plates is kept constant in this case $\left(\Delta V_f = \Delta V_i\right)$ and the ratio of the final charge on one of the plates to the original charge is

$$Q_f/Q_i = 1/2$$ ◊

29. (a) Find the equivalent capacitance of the group of capacitors in Figure P16.29. (b) Find the charge on and the potential difference across each.

Solution In the analysis of a capacitor network such as this, the rules for combining capacitors in series and parallel should be used to reduce the circuit to its simplest equivalent form as outlined in the following steps. First, consider the original circuit as shown in Figure 1 below.

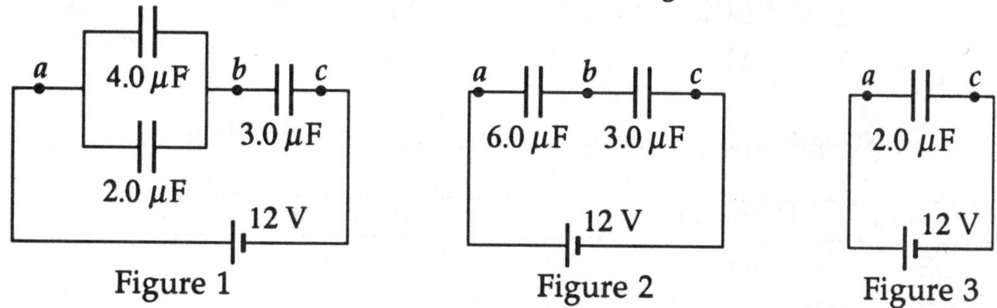

Figure 1 Figure 2 Figure 3

Note that the two capacitors between points a and b are in parallel and their equivalent capacitance is $C_{ab} = 4.0\ \mu\text{F} + 2.0\ \mu\text{F} = 6.0\ \mu\text{F}$ as shown in Figure 2. Now observe that the two capacitors between points a and c in Figure 2 are in series, so the equivalent capacitance is

$$\frac{1}{C_{ac}} = \frac{1}{6.0\ \mu\text{F}} + \frac{1}{3.0\ \mu\text{F}} = \frac{1+2}{6.0\ \mu\text{F}} \quad \text{or} \quad C_{ac} = 2.0\ \mu\text{F} \quad \text{as shown in Figure 3.}$$

(a) The equivalent capacitance of the group of capacitors in the original circuit is 2.0 μF as discussed above. ◊

(b) Now, the procedure is to start with the most simplified circuit and work back to the original circuit. The total charge stored between points a and c in Figure 3 is $Q_{ac} = C_{ac}\Delta V_{ac} = (2.0\ \mu\text{F})(12\ \text{V}) = 24\ \mu\text{C}$. This states that 24 μC of positive charge is stored on the plate next to point a and 24 μC of negative charge on the plate next to c. Now, return to Figure 2 and observe that this means $Q_{ab} = Q_{bc} = 24\ \mu\text{C}$. Therefore, the potential differences between these pairs of points are

$$\Delta V_{ab} = \frac{Q_{ab}}{C_{ab}} = \frac{24\ \mu\text{C}}{6.0\ \mu\text{F}} = 4.0\ \text{V} \quad \text{and} \quad \Delta V_{bc} = \frac{Q_{bc}}{C_{bc}} = \frac{24\ \mu\text{C}}{3.0\ \mu\text{F}} = 8.0\ \text{V}$$

Finally, in Figure 1, observe that the 24 μC stored between points a and b is divided between two capacitors which have the same potential difference, $\Delta V_{ab} = 4.0$ V, maintained across them. Thus, the charge stored on each capacitor is

$$Q_4 = C_4 \Delta V_{ab} = (4.0\ \mu\text{F})(4.0\ \text{V}) = 16\ \mu\text{C}$$

and
$$Q_2 = C_2 \Delta V_{ab} = (2.0\ \mu\text{F})(4.0\ \text{V}) = 8.0\ \mu\text{C}$$

The requested charges and potential differences are summarized as:

Capacitor	Stored Charge	Potential Difference
2.0 μF	8.0 μC	4.0 V
3.0 μF	24 μC	8.0 V
4.0 μF	16 μC	4 .0 V

◊ ◊ ◊

---

35. A 25.0-μF capacitor and a 40.0-μF capacitor are charged by being connected across separate 50.0-V batteries. (a) Determine the resulting charge on each capacitor. (b) The capacitors are then disconnected from their batteries and connected to each other, with each negative plate connected to the other positive plate. What is the final charge of each capacitor, and what is the final potential difference across the 40.0-μF capacitor?

Solution

(a) The charge stored in a capacitor of capacitance C when a potential difference ΔV is maintained between its plates is $Q = C\Delta V$. Thus, when the given capacitors are each connected across a 50.0-V battery, the charges stored are

$$Q_{25} = (25.0\ \mu\text{F})(50.0\ \text{V}) = 1.25 \times 10^3\ \mu\text{C}$$

and
$$Q_{40} = (40.0\ \mu\text{F})(50.0\ \text{V}) = 2.00 \times 10^3\ \mu\text{C}$$ ◊

(b) When the capacitors are carefully disconnected from the batteries, the capacitors are charged as shown in the "Before" figure below.

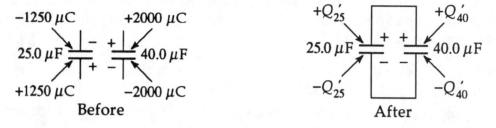

−1250 μC +2000 μC
25.0 μF 40.0 μF
+1250 μC −2000 μC
Before

$+Q'_{25}$ $+Q'_{40}$
25.0 μF 40.0 μF
$-Q'_{25}$ $-Q'_{40}$
After

When the capacitors are reconnected, with the positive plate of one connected to the negative plate of the other, total charge is conserved but redistributed between the two capacitors. The net positive charge that existed on the two upper plates ($+2000\ \mu\text{C} - 1250\ \mu\text{C} = +750\ \mu\text{C}$) is shared by these two plates as shown in the "After" figure. Likewise, the net negative charge ($-750\ \mu\text{C}$) that originally existed on the two lower plates is shared by those plates. Thus, after the capacitors are reconnected,

$$Q'_{40} + Q'_{25} = 750\ \mu\text{C} \qquad \textbf{[Equation 1]}$$

Also, after the capacitors are reconnected, they are in parallel and have the same potential difference across them. That is $\Delta V'_{40} = \Delta V'_{25}$, or

$$\frac{Q'_{40}}{40.0\ \mu\text{F}} = \frac{Q'_{25}}{25.0\ \mu\text{F}} \quad \text{which becomes} \quad Q'_{40} = 1.60\,Q'_{25} \qquad \textbf{[Equation 2]}$$

Substituting Equation 2 into 1, $\quad 2.60\,Q'_{25} = 750\ \mu\text{C}$

Thus, the final charge on the 25.0-μF capacitor is

$$Q'_{25} = 288\ \mu\text{C} \qquad \Diamond$$

Then, Equation 1 gives: $\quad Q'_{40} = 750\ \mu\text{C} - 288\ \mu\text{C} = 462\ \mu\text{C} \qquad \Diamond$

The final potential difference across each capacitor is:

$$\Delta V'_{25} = \Delta V'_{40} = \frac{Q'_{40}}{40.0\ \mu\text{F}} = \frac{462\ \mu\text{C}}{40.0\ \mu\text{F}} = 11.6\ \text{V} \qquad \Diamond$$

39. Two capacitors, $C_1 = 25\ \mu C$ and $C_2 = 5.0\ \mu C$, are connected in parallel and charged with a 100-V power supply. (a) Calculate the total energy stored in the two capacitors. (b) What potential difference would be required across the same two capacitors connected in **series** in order that the combination store the same energy as in (a)?

Solution (a) When the capacitors are connected in parallel, the total capacitance is $C_{eq} = C_1 + C_2 = 25\ \mu F + 5.0\ \mu F = 30\ \mu F$. Thus, when a 100-V power supply is connected across this combination, the total stored energy is

$$W = \frac{1}{2}C_{eq}(\Delta V)^2 = \frac{1}{2}\left(3.0\times10^{-5}\ F\right)(100\ V)^2 = 0.15\ J \qquad \lozenge$$

(b) When the two capacitors are connected in series, the total capacitance of the series combination is

$$1/C_{eq} = 1/C_1 + 1/C_2 = \left(1/25\ \mu F\right) + \left(1/5.0\ \mu F\right) \qquad \text{or} \qquad C_{eq} = (25/6)\ \mu F$$

To store the same energy as in the parallel combination of part (a), the required potential difference is $(\Delta V)^2 = 2W/C_{eq}$,

or $$\Delta V = \sqrt{\frac{2(0.15\ J)}{(25/6)\times10^{-6}\ F}} = \sqrt{7.2\times10^4}\ V = 2.7\times10^2\ V \qquad \lozenge$$

---

43. Determine (a) the capacitance and (b) the maximum voltage that can be applied to a Teflon-filled parallel-plate capacitor having a plate area of 175 cm^2 and insulation thickness of 0.0400 mm.

Solution (a) A parallel plate capacitor with a dielectric material between the plates has a capacitance of $C = \kappa\epsilon_o A/d$ where κ is the dielectric constant of the material, ϵ_o is the permittivity of vacuum, A is the area of one of the plates, and d is the distance separating the plates. From Table 16.1 in the text, the dielectric constant of Teflon is $\kappa = 2.1$, so the capacitance of this parallel plate capacitor is

$$C = (2.1)\left(8.85\times10^{-12}\ F/m\right)\left(\frac{175\ cm^2(1.0\ m^2/10^4\ cm^2)}{0.0400\ \times10^{-3}\ m}\right) = 8.1\times10^{-9}\ F = 8.1\ nF \quad \lozenge$$

(b) The dielectric strength of a material is the maximum electric field that material can withstand before it breaks down and begins to conduct. From Table 16.1, the dielectric strength for Teflon is 60×10^6 V/m. The electric field strength between the plates of a parallel plate capacitor is $E = \Delta V/d$ where ΔV is the applied voltage and d is the distance between the plates. Thus, the maximum voltage that can be applied is

$$\Delta V_{max} = E_{max} d = \left(60 \times 10^6 \text{ V/m}\right)\left(0.0400 \times 10^{-3} \text{ m}\right)$$

or $$\Delta V_{max} = 2.4 \times 10^3 \text{ V} = 2.4 \text{ kV}$$ ◊

47. Three parallel-plate capacitors are constructed, each having the same plate area A, and with C_1 having plate spacing d_1, C_2 having d_2, and C_3 having d_3. Show that the total capacitance C of these three capacitors connected in series is the same as a capacitor of plate area A and with plate spacing $d = d_1 + d_2 + d_3$.

Solution The capacitance of a parallel-plate capacitor, with vacuum between the plates, is $C = \epsilon_o A/d$, where ϵ_o is a constant, A is the area of one of the plates, and d is the distance between the plates. When three capacitors (having capacitances of C_1, C_2, and C_3) are connected in series, the total capacitance of the combination is

$$\frac{1}{C_{eq}} = \frac{1}{C_1} + \frac{1}{C_2} + \frac{1}{C_3}$$

If the capacitors all have the same plate area A, and plate spacings of d_1, d_2, and d_3 respectively, this becomes

$$\frac{1}{C_{eq}} = \frac{d_1}{\epsilon_o A} + \frac{d_2}{\epsilon_o A} + \frac{d_3}{\epsilon_o A} = \frac{d_1 + d_2 + d_3}{\epsilon_o A} \quad \text{or} \quad C_{eq} = \epsilon_o \frac{A}{d_1 + d_2 + d_3}$$

Comparing this result to the general expression for the capacitance of a parallel-plate capacitor, $C = \epsilon_0 A/d$, it is observed that the total capacitance of the series combination is the same as that of a single capacitor of plate area A and plate spacing $d = d_1 + d_2 + d_3$. ◊

---

52. An isolated capacitor of unknown capacitance has been charged to a potential difference of 100 V. When the charged capacitor is disconnected from the battery and then connected in parallel to an uncharged 10.0-μF capacitor, the voltage across the combination is measured to be 30.0 V. Calculate the unknown capacitance.

Solution If the unknown capacitance is C_x, the charge stored when it is connected to a potential difference of ΔV_0 is $Q = C_x \Delta V_0$. When the charged capacitor is carefully disconnected from the battery and connected in parallel to an uncharged 10.0-μF capacitor, the charge Q is shared by the two capacitors with Q_x' remaining on C_x and some amount Q_{10} moving to the 10.0-μF capacitor. Since charge is conserved, the relation between these charges is:

$$Q_x' + Q_{10} = Q = C_x \Delta V_0 \qquad \textbf{[Equation 1]}$$

Capacitors connected in parallel have the same potential difference across them. If the final potential difference across the parallel combination is $\Delta V'$, the final charge on each is $Q_x' = C_x \Delta V'$ and $Q_{10} = (10.0\ \mu\text{F})\Delta V'$. Substituting these results into Equation 1 above gives $C_x \Delta V' + (10.0\ \mu\text{F})\Delta V' = C_x \Delta V_0$, or

$$C_x = \frac{(10.0\ \mu\text{F})\Delta V'}{\Delta V_0 - \Delta V'}$$

If $\Delta V_0 = 100$ V and $\Delta V'$ is measured to be 30.0 V, the unknown capacitance is found to be

$$C_x = \frac{(10.0\ \mu\text{F})(30.0\ \text{V})}{100\ \text{V} - 30.0\ \text{V}} = 4.29\ \mu\text{F}$$ ◊

---

57. Two capacitors when connected in parallel give an equivalent capacitance of 9.00 pF and an equivalent capacitance of 2.00 pF when connected in series. What is the capacitance of each capacitor?

Solution When the two capacitors are connected in series, the equivalent capacitance is

$$\frac{1}{C_{\text{series}}}=\frac{1}{C_1}+\frac{1}{C_2}=\frac{1}{2.00\text{ pF}} \qquad \textbf{[Equation 1]}$$

When they are connected in parallel, the equivalent capacitance is $C_{\text{parallel}}=C_1+C_2=9.00\text{ pF}$, which may be written as $C_2=9.00\text{ pF}-C_1$. Substituting this result into Equation 1 yields

$$\frac{1}{C_1}+\frac{1}{9.00\text{ pF}-C_1}=\frac{1}{2.00\text{ pF}}$$

Finding a common denominator gives

$$\frac{(9.00\text{ pF}-C_1)+C_1}{C_1(9.00\text{ pF}-C_1)}=\frac{1}{2.00\text{ pF}} \quad \text{or} \quad (9.00\text{ pF})(2.00\text{ pF})=C_1(9.00\text{ pF}-C_1)$$

Expanding and rearranging yields the equation $C_1^2-(9.00\text{ pF})C_1+18.0\text{ pF}^2=0$ which may be solved with the quadratic formula:

$$C_1=\frac{(9.00\text{ pF})\pm\sqrt{(9.00\text{ pF})^2-72.0\text{ pF}^2}}{2}=\frac{(9.00\text{ pF})\pm(3.00\text{ pF})}{2}$$

Thus, there are two possible values for C_1: $C_1=6.00\text{ pF}$ or $C_1=3.00\text{ pF}$. Then, using $C_2=9.00\text{ pF}-C_1$ from above, two possible values for C_2 are found: $C_2=3.00\text{ pF}$ or $C_2=6.00\text{ pF}$. The conclusion is that one of the unknown capacitances is 3.00 pF and the other is 6.00 pF. ◊

CHAPTER SELF-QUIZ

1. At which location will the magnitude of the electric field between the two parallel plates of a charged capacitor be the greatest?
   a. near the positive plate
   b. near the negative plate
   c. midway between the two plates
   d. electric field is constant throughout space between plates

2. When moving an electrical charge from one point to another in the presence of an electrical field, which quantity depends on the size of the charge that is moved?
   a. the electric field
   b. the work done
   c. the potential difference
   d. the distance moved

3. Two capacitors with capacities of 1.00 μF and 0.50 μF, respectively, are connected in parallel. The system is connected to a 100-V battery. What charge accumulates on the 1.00-μF capacitor?
   a. 150 μC
   b. 100 μC
   c. 50 μC
   d. 33 μC

4. A 200-V battery is connected to a 0.50 μF, parallel plate, air-filled capacitor. Now, the battery is disconnected, with care taken not to discharge the plates. Some Pyrex glass is next inserted between the two plates, completely filling up the space. What is the final potential difference between the plates? (For Pyrex, dielectric constant = 5.60.)
   a. 36 V
   b. 200 V
   c. 560 V
   d. 1120 V

5. A parallel-plate capacitor has dimensions 2.00 cm × 3.00 cm. The plates are separated by a 1.00 mm thickness of paper (dielectric constant $\kappa = 3.70$). What is the charge that can be stored on this capacitor, when connected to a 9.00-volt battery? $(\epsilon_0 = 8.85 \times 10^{-12}\ C^2/N \cdot m)$
   a. 19.6×10^{-12} C
   b. 4.75×10^{-12} C
   c. 4.75×10^{-11} C
   d. 1.76×10^{-10} C

6. If $C_1 = 15.0\ \mu F$, $C_2 = 10.0\ \mu F$, $C_3 = 20.0\ \mu F$, and $\Delta V_0 = 18.0$ V, determine the energy stored by C_2.
   a. 0.72 mJ
   b. 0.32 mJ
   c. 0.50 mJ
   d. 0.18 mJ

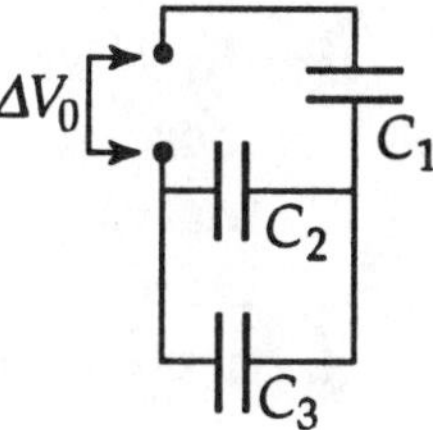

7. Very large capacitors have been considered as a means for storing electrical energy. If we constructed a very large parallel plate capacitor of plate area 1.00 m² using Pyrex ($k = 5.60$) of thickness 2.00 mm as a dielectric, how much electrical energy would it store at a plate voltage of 6000 V? $(\epsilon_0 = 8.85 \times 10^{-12}\ C^2/N \cdot m)$
   a. 0.45 J
   b. 90 J
   c. 9,000 J
   d. 45, 000 J

8. There is a hollow, conducting, uncharged sphere with a charge $+Q$ inside the sphere. Consider the potential and the electrical field at a point P_1 inside the metal of the sphere. At this point,
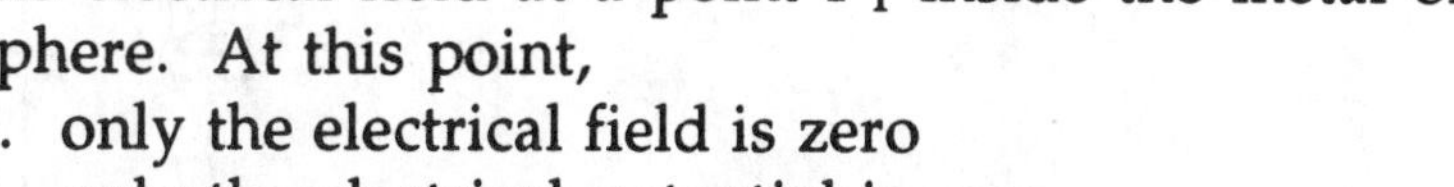
   a. only the electrical field is zero
   b. only the electrical potential is zero
   c. both the electrical field and electric potential are zero
   d. neither the electrical field nor the electrical potential are zero

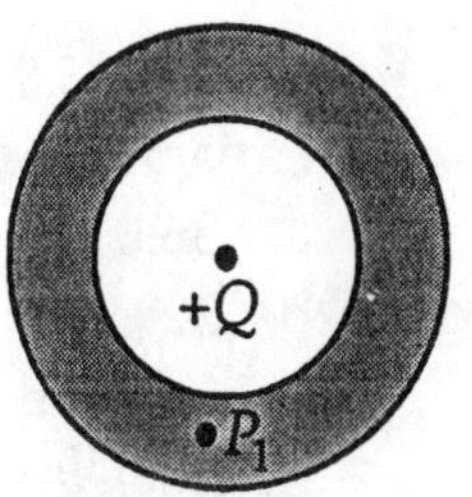

9. What is the equivalent capacitance of the combination shown?
    a. 20.0 μF
    b. 90.0 μF
    c. 22.0 μF
    d. 4.60 μF

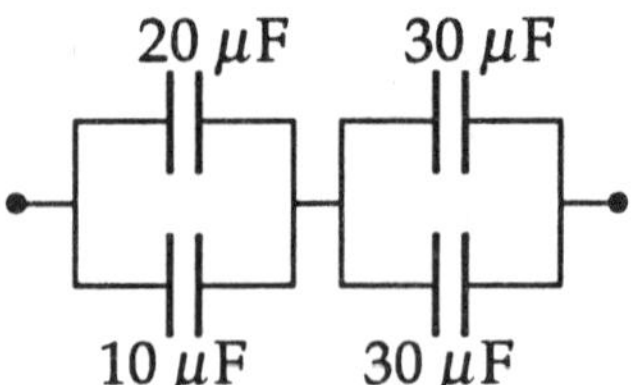

10. Inserting a dielectric material between two charged parallel conducting plates, originally separated by air and disconnected from a battery, will produce what effect on the capacitor?
    a. increase charge
    b. increase voltage
    c. increase capacitance
    d. decrease capacitance

11. A uniform electric field, with a magnitude of 5×10^2 N/C, is directed parallel to the positive x axis. If the potential at x = 5.00 m is 2500 V, what is the potential at x= 2.00 m?
    a. 1000 V
    b. 2000 V
    c. 4000 V
    d. 1500 V

12. If $C = 45.0\ \mu\text{F}$, determine the equivalent capacitance for the combination shown.
    a. 36.0 μF
    b. 32.0 μF
    c. 34.0 μF
    d. 30.0 μF

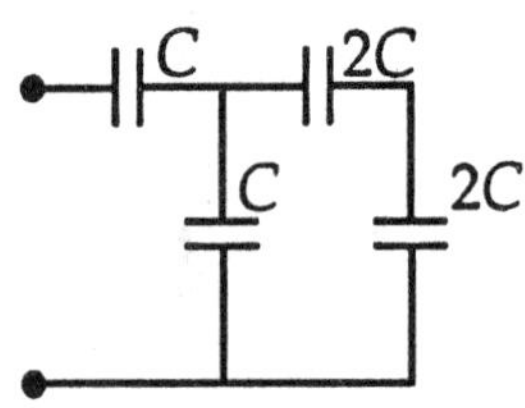

Chapter 17

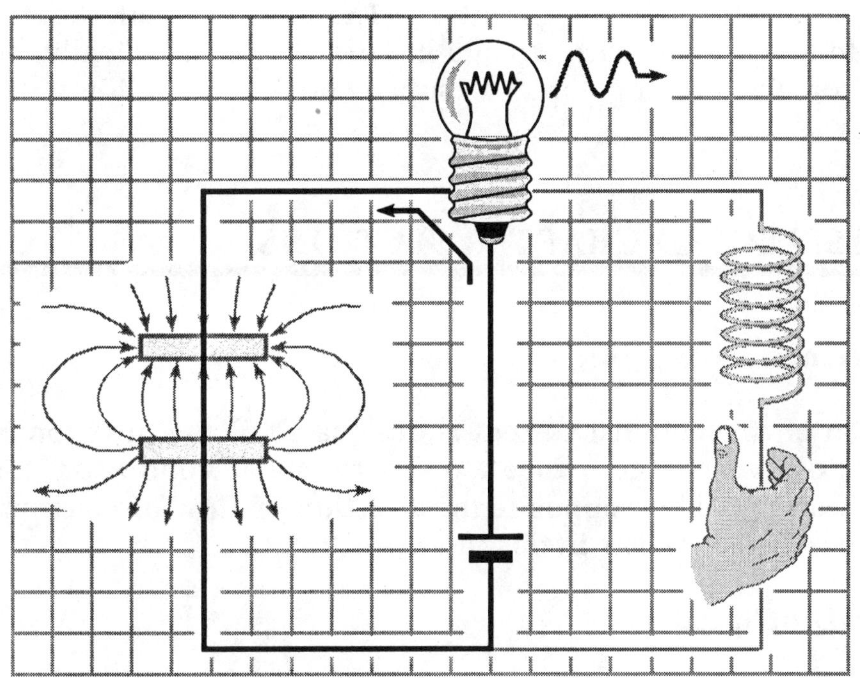

CURRENT AND RESISTANCE

Chapter 17

CURRENT AND RESISTANCE

Many practical applications and devices are based on the principles of static electricity, but electricity truly became an inseparable part of our daily lives when scientists learned how to control the **flow of electric charges**.

In this chapter we define current and discuss some of the factors that contribute to the resistance to flow of charge in conductors. We also discuss energy transformations in electric circuits. (These topics will be the foundation for additional work with circuits in later chapters.) Finally, an interesting essay on the exciting topic of superconductivity follows this chapter.

NOTES FROM SELECTED CHAPTER SECTIONS

17.1 Electric Current

The direction of conventional current is designated as the direction of motion of positive charge. In an ordinary metal conductor, the direction of current will be **opposite** the **direction of flow of electrons** (which are the charge carriers in this case).

17.2 Current and Drift Speed

In the classical model of electronic conduction in a metal, electrons are treated like molecules in a gas and, in the absence of an electric field, have a **zero average velocity.**

Under the influence of an electric field, the electrons move along a direction opposite the direction of the applied field with a **drift velocity** which is proportional to the average time between collisions with atoms of the metal and inversely proportional to the number of free electrons per unit volume.

17.3 Resistance and Ohm's Law

When a voltage (potential difference) ΔV is applied across the ends of a metallic conductor, the current in the conductor is found to be proportional to the applied voltage. If the proportionality is exact, we can write $\Delta V = IR$ where the proportionality constant R is called the resistance of the conductor. In fact, we define this resistance as the ratio of the voltage across the conductor to the current it carries. For many materials, including most metals, experiments show that **the resistance is constant over a wide range of applied voltages.** This statement is known as Ohm's law.

Ohm's law is not a fundamental law of nature, but an empirical relationship that is valid only for certain materials. Materials that obey Ohm's law, and hence have a constant resistance over a wide range of voltages, are said to be **ohmic**. Materials that do not obey Ohm's law **are nonohmic.**

Resistance has the SI units volts per ampere, called ohms (Ω). Thus, if a potential difference of 1 V across a conductor produces a current of 1 A, the resistance of the conductor is 1 Ω.

17.4 Resistivity

For ohmic materials, the ratio of the current density and the electric field which gives rise to the current is equal to a constant, σ, which is the conductivity of the material. The reciprocal of the conductivity is called the resistivity, ρ. Each ohmic material has a characteristic resistivity which depends only on the properties of the specific material and is a function of temperature.

17.5 Temperature Variation of Resistance

The resistivity, and hence the resistance, of a conductor depends on a number of factors. One of the most important is the temperature of the metal. For most metals, resistivity increases with increasing temperature. For most metals, resistivity increases approximately linearly with temperature over a limited temperature range.

EQUATIONS AND CONCEPTS

Under the influence of an electric field, electric charges will move through conducting gases, liquids, and solids. The electric current is defined as the rate at which charge flows through a cross section of a conductor.

$$I \equiv \frac{\Delta Q}{\Delta t} \tag{17.1}$$

The SI unit of current is the ampere (A).

$$1\text{ A} = 1\text{ C/s} \tag{17.2}$$

It is conventional to choose the direction of current to be in the direction of flow of positive charge. In a solid conductor, the current is due to the motion of negatively charged electrons. In such conductors, the direction of the current will be opposite the direction of flow of electrons.

Comment on direction of current.

The velocity of the charge carriers in a conductor is actually an average value of the individual velocities and is called the drift velocity, v_d. The current can be expressed in terms of the drift velocity and the number of mobile charge carriers per unit volume of conductor.

$$I = nqv_dA \tag{17.3}$$

For many practical applications, the resistance of a given conductor can be more conveniently stated as the ratio of the potential difference across a conductor to the value of the current in the conductor. This equation is usually referred to as Ohm's law.

$$\Delta V = IR \tag{17.5}$$

The resistance of a given conductor made of a homogeneous material of uniform cross section can be expressed in terms of the dimensions of the conductor and an intrinsic property of the material of which the conductor is made called its resistivity. The value of the resistivity of a given material depends on the electronic structure of the material and on the temperature.

$$R = \rho\frac{\ell}{A} \tag{17.6}$$

The symbol ρ used for resistivity should not be confused with the same symbol used earlier in the book for density. Very often, a single symbol is used to represent different quantities.

The SI unit of resistance is the ohm. If a potential difference of 1 V across a conductor produces a current of 1 A, the resistance of the conductor is 1 ohm.

$$1\,\Omega = 1\text{ V/A}$$

The resistivity, and therefore the resistance, of a conductor varies with temperature. Over a limited range of temperatures, this variation is approximately linear. In these equations, the parameter, α, is the temperature coefficient of resistivity, and T_0 is a reference temperature usually taken to be 20.0 °C. Some materials (for example, carbon) have a negative temperature of resistivity and in these cases the resistance decreases as the temperature increases.

$$\rho = \rho_0[1 + \alpha(T - T_0)] \tag{17.7}$$

$$R = R_0[1 + \alpha(T - T_0)] \tag{17.8}$$

Joule's law can be used to calculate the power delivered to a resistor or other device carrying a current, I, and having a potential difference, ΔV, between its terminals.

$$P = I\Delta V \tag{17.9}$$

When the device obeys Ohm's law (i.e. a resistor), the power can be expressed in either of two alternative forms.

$$P = I^2 R = \frac{(\Delta V)^2}{R} \tag{17.10}$$

The SI unit of power is the watt (W).

$$1\ \text{W} = 1\ \text{J/s} = 1\ \text{V}\cdot\text{A}$$

The kilowatt-hour is the quantity of energy consumed in one hour at a constant use rate (or power) of 1 kW.

$$1\ \text{kWh} = 3.60 \times 10^6\ \text{J} \tag{17.11}$$

REVIEW CHECKLIST

- ▷ Define the term, electric current, in terms of rate of charge flow, and its corresponding unit of measure, the ampere. Calculate electron drift velocity, and quantity of charge passing a point in a given time interval in a specified current-carrying conductor.
- ▷ Determine the resistance of a conductor using Ohm's law. Also, calculate the resistance based on the physical characteristics of a conductor. Distinguish between ohmic and nonohmic conductors.
- ▷ Make calculations of the variation of resistance with temperature, which involves the concept of the temperature coefficient of resistivity.
- ▷ Sketch a simple single-loop circuit to illustrate the use of basic circuit element symbols and direction of conventional current.
- ▷ Use Joule's law to calculate the power dissipated in a resistor.

SOLUTIONS TO SELECTED END-OF-CHAPTER PROBLEMS

5. In the Bohr model of the hydrogen atom, an electron in the lowest energy state moves at a speed of 2.19×10^6 m/s in a circular path having a radius of 5.29×10^{-11} m. What is the effective current associated with this orbiting electron?

Solution The effective current is the amount of charge passing a fixed point in the orbit each second. A charge of magnitude $e = 1.60 \times 10^{-19}$ C passes this point every time the electron completes a trip around the orbit. The electron must travel a distance $d = 2\pi r$ in a full trip around the orbit. The time required to complete this trip is $T = d/v = 2\pi r / v$, where v is the speed of the orbiting electron. Thus, the electron has an orbital frequency of $f = 1/T = v/2\pi r$, and a effective current of

$$I = ef = \frac{ev}{2\pi r} = \frac{(1.60 \times 10^{-19}\text{ C})(2.19 \times 10^{6}\text{ m/s})}{2\pi(5.29 \times 10^{-11}\text{ m})} = 1.05 \times 10^{-3}\text{ C/s} = 1.05\text{ mA} \quad \lozenge$$

9. An aluminum wire with a cross-sectional area of 4.0×10^{-6} m^2 carries a current of 5.0 A. Find the drift speed of the electrons in the wire. The density of aluminum is 2.7 g/cm^3. (Assume that one electron is supplied by each atom.)

Solution

In terms of the speed v_d of the drifting electrons, the current in a metallic conductor is $I = neAv_d$. Here, n is the number of free electrons per unit volume, and A is the cross-sectional area of the conductor. Since it is assumed that each atom supplies one free electron, n is the same as the number of atoms per unit volume. This may be found from

$$n = \frac{\text{mass per unit volume}}{\text{mass per atom}} = \frac{\text{density}}{\text{mass per atom}}$$

The mass of a single atom is $m_{\text{atom}} = \dfrac{\text{mass per mole}}{\text{atoms per mole}} = \dfrac{\text{molecular wt}}{\text{Avogadro's number}}$

For aluminum, this gives $\quad m_{\text{atom}} = \dfrac{27 \text{ g}}{6.02 \times 10^{23}} = 4.5 \times 10^{-23} \text{ g}$

The density of free electrons is

$$n = \frac{2.7 \text{ g/cm}^3}{4.5 \times 10^{-23} \text{ g}} = \left(\frac{6.0 \times 10^{22}}{\text{cm}^3}\right)\left(\frac{10^6 \text{cm}^3}{1 \text{ m}^3}\right) = 6.0 \times 10^{28} \text{ m}^{-3}$$

The drift speed of the electrons in this wire is then $v_d = I/neA$:

$$v_d = \frac{5.0 \text{ C/s}}{\left(6.0 \times 10^{28} \text{ m}^{-3}\right)\left(1.6 \times 10^{-19} \text{ C}\right)\left(4.0 \times 10^{-6} \text{ m}^2\right)} = 1.3 \times 10^{-4} \text{ m/s} \qquad \Diamond$$

14. A potential difference of 12 V is found to produce a current of 0.40 A in a 3.2-m length of wire with a uniform radius of 0.40 cm. What is (a) the resistance of the wire? (b) the resistivity of the wire?

Solution

(a) The resistance of a conductor is the ratio of the potential difference maintained across the conductor to the current this potential difference causes to flow through the conductor, $R = \Delta V/I$. In this wire, a potential difference of 12 V produces a current of 0.40 A. Thus, the resistance of the wire is

$$R = \frac{12 \text{ V}}{0.40 \text{ A}} = 30 \ \Omega \qquad \Diamond$$

(b) The resistance of an ohmic conductor may be expressed as $R = \rho\frac{\ell}{A}$ where ℓ is the length of the conductor, A is the cross-sectional area, and ρ is the resistivity of the material from which the conductor is made.

The resistivity of the material in this wire is

$$\rho = \frac{RA}{\ell} = \frac{R(\pi r^2)}{\ell} = \frac{(30 \ \Omega)\pi\left(0.40 \times 10^{-2} \text{ m}\right)^2}{3.2 \text{ m}}$$

Thus, the resistivity is

$$\rho = 4.7 \times 10^{-4} \ \Omega \cdot \text{m} \qquad \Diamond$$

19. An 1050-W toaster operates on a 120-V household circuit and a 4.00-m length of nichrome wire as its heating element. The operating temperature of this element is 320 °C. What is the cross-sectional area of the wire.

Solution The voltage across a conductor is the work per unit charge the electric field does on charges passing through the conductor. The current is the amount of charge passing through the conductor each second. Thus, the power dissipation (rate of doing work) in the conductor is

$$P = \begin{pmatrix}\text{work per} \\ \text{unit charge}\end{pmatrix}\begin{pmatrix}\text{charge per} \\ \text{second}\end{pmatrix} = (\Delta V)I$$

The current flowing in the heating element of this toaster must be

$$I = \frac{P}{\Delta V} = \frac{1050 \text{ W}}{120 \text{ V}} = 8.75 \text{ A}$$

The resistance of this heating element, at its operating temperature of 320 °C, is given by Ohm's law as

$$R = \frac{\Delta V}{I} = \frac{120 \text{ V}}{8.75 \text{ A}} = 13.7 \ \Omega$$

The resistivity of nichrome at this temperature may be found from $\rho = \rho_0\left[1 + \alpha(T - T_0)\right]$. From Table 17.1 in the text, $\rho_0 = 150 \times 10^{-8} \ \Omega \cdot \text{m}$ and $\alpha = 0.4 \times 10^{-3} \ (°\text{C})^{-1}$, all at a reference temperature of $T_0 = 20.0 \ °\text{C}$. Thus, at a temperature of $T = 320 \ °\text{C}$,

$$\rho = \left(150 \times 10^{-8} \ \Omega \cdot \text{m}\right)\left[1 + \left(0.4 \times 10^{-3} \ °\text{C}^{-1}\right)(320 \ °\text{C} - 20 \ °\text{C})\right] = 1.68 \times 10^{-6} \ \Omega \cdot \text{m}$$

and from $R = \rho \ell / A$ the cross-sectional area of the wire must be:

$$A = \rho \ell / R = \frac{\left(1.68 \times 10^{-6} \ \Omega \cdot \text{m}\right)(4.00 \text{ m})}{13.7 \ \Omega} = 4.91 \times 10^{-7} \text{ m}^2 \qquad \lozenge$$

27. A 100-cm-long copper wire 0.50 cm in radius has a potential difference across it sufficient to produce a current of 3.0 A at 20 °C. (a) What is the potential difference? (b) If the temperature of the wire is increased to 200 °C, what potential difference is now required to produce a current of 3.0 A?

Solution (a) The resistance of an ohmic conducting element of length ℓ and cross-sectional area A may be written as $R = \rho\ell/A$. where ρ is the resistivity of the material making up this element. From Table 17.1 in the textbook, the resistivity of copper at 20 °C is seen to be $1.7\times10^{-8}\ \Omega\cdot\text{m}$. Hence, the resistance of this copper wire at 20 °C is

$$R = \rho\ell/\pi r^2 = \frac{\left(1.7\times10^{-8}\ \Omega\cdot\text{m}\right)(1.00\ \text{m})}{\pi\left(5.0\times10^{-3}\ \text{m}\right)^2} = 2.2\times10^{-4}\ \Omega$$

The potential difference that must be maintained across this wire to produce a 3.0 A current is then

$$\Delta V = IR = (3.0\ \text{A})\left(2.2\times10^{-4}\ \Omega\right) = 6.6\times10^{-4}\ \text{V} = 0.66\ \text{mV} \qquad \lozenge$$

(b) The resistance of a resistor varies with temperature according to the relation $R = R_0\left[1+\alpha(T-T_0)\right]$. Here, R_0 is the resistance of the resistor at the reference temperature T_0, and α is the temperature coefficient of resistivity for the conducting material in the resistor. From part (a), the resistance of this wire at $T_0 = 20\ °\text{C}$ is $R_0 = 2.2\times10^{-4}\ \Omega$. From Table 17.1, $\alpha = 3.9\times10^{-3}\ (°\text{C})^{-1}$ for copper. The resistance of this wire at 200 °C is then

$$R = \left(2.2\times10^{-4}\ \Omega\right)\left[1+\left(3.9\times10^{-3}\ °\text{C}\right)(200\ °\text{C} - 20\ °\text{C})\right] \quad \text{or} \quad R = 3.7\times10^{-4}\ \Omega$$

The potential difference now required to produce a current of 3.0 A is

$$\Delta V = IR = (3.0\ \text{A})\left(3.7\times10^{-4}\ \Omega\right) = 1.1\times10^{-3}\ \text{V} = 1.1\ \text{mV} \qquad \lozenge$$

33. The power supplied to a typical black-and-white television set is 90 W when the set is connected to 120 V. (a) How much electric energy does this set consume in one hour? (b) A color television set draws about 2.5 A when connected to 120 V. How much time is required for it to consume the same energy as the black-and-white model consumes in one hour?

Solution (a) The black-and-white TV is using energy at the rate of $P = 90\text{ W} = 90\text{ J/s}$. The total energy it will consume in one hour is

$$E = Pt = (90\text{ J/s})(1.0\text{ h})(3600\text{ s}/1.0\text{ h}) = 3.2\times10^5\text{ J} \qquad \lozenge$$

(b) The rate at which the color TV uses energy is given by $P = (\Delta V)I$ or $P = (120\text{ V})(2.5\text{ A}) = 300\text{ W}$. The time required to consume $3.2\times10^5\text{ J}$ of energy is

$$t = \frac{E}{P} = \frac{3.2\times10^5\text{ J}}{300\text{ J/s}} = (1.1\times10^3\text{ s})\left(\frac{1.0\text{ min}}{60\text{ s}}\right) = 18\text{ min} \qquad \lozenge$$

---

37. A small motor draws a current of 1.75 A from a 120-V line. The output power of the motor is 0.20 hp. (a) At a rate of $0.060/kWh, what is the cost of operating the motor for 4.0 h? (b) What is the efficiency of the motor?

Solution (a) The power input to the motor is $P = (\Delta V)I = (120\text{ V})(1.75\text{ A})$, or $P = 210\text{ W} = 0.210\text{ kW}$. The kilowatt-hour (kWh) is the quantity of energy consumed in 1.0 hour when energy is used at a rate of 1.0 kW. The energy used by this motor in 4.0 h, expressed in kWh, is therefore $E = Pt = (0.210\text{ kW})(4.0\text{ h}) = 0.84\text{ kWh}$. The cost of operating this motor for 4.0 h is:

$$\text{cost} = (\text{energy used})(\text{rate}) = (0.84\text{ kWh})\left(\frac{\$0.060}{\text{kWh}}\right) = \$0.050 = 5.0¢ \qquad \lozenge$$

(b) The output power and the efficiency of the motor are

$$P_{output} = 0.20\text{ hp}\left(\frac{0.746\text{ kW}}{1.0\text{ hp}}\right) = 0.15\text{ kW} \qquad \text{eff} = \frac{P_{output}}{P_{input}} = \frac{0.15\text{ kW}}{0.210\text{ kW}} = 71\% \qquad \lozenge$$

---

43. The heating coil of a hot water heater has a resistance of 20 Ω and operates at 210 V. If electrical energy costs \$0.080/kWh, what does it cost to raise the 200 kg of water in the tank from 15.0 °C to 80.0 °C? (See Chapter 11.)

Solution

The kilowatt-hour is a measure of energy equal to

$$1\text{ kWh} = (1.00\text{ kW})(1.00\text{ h}) = (1000\text{ J/s})(3600\text{ s}) = 3.60\times10^{6}\text{ J}$$

The heat energy needed to raise the temperature of 200 kg of water from 15.0 °C to 80.0 °C is

$$Q = mc(\Delta T) = (200\text{ kg})(4186\text{ J/kg·°C})(65.0\text{°C}) = 5.44\times10^{7}\text{ J}$$

where the specific heat of water is from Table 11.1 in the textbook. In units of kWh,

$$Q = (5.44\times10^{7}\text{ J})\left(\frac{1.00\text{ kWh}}{3.60\times10^{6}\text{ J}}\right) = 15.1\text{ kWh}$$

and the cost of operating the heater to produce this quantity of heat energy is

$$\text{cost} = (\text{energy used})(\text{rate}) = (15.1\text{ kWh})(\$0.080/\text{kWh}) = \$1.21 \qquad \Diamond$$

---

47. Birds resting on high-voltage power lines are a common sight. The copper wire on which a bird stands is 2.2 cm in diameter and carries a current of 50 A. If the bird's feet are 4.0 cm apart, calculate the potential difference across its body.

Solution

The resistivity of copper is $\rho = 1.7 \times 10^{-8}\ \Omega \cdot \text{m}$ (Table 17.1 in the textbook).

Taking the diameter of the wire to be $d = 2.2\ \text{cm} = 2.2 \times 10^{-2}\ \text{m}$, the cross-sectional area of the copper wire is $A = \pi d^2/4$. Thus,

$$A = \frac{\pi\left(2.2 \times 10^{-2}\ \text{m}\right)^2}{4} = 3.8 \times 10^{-4}\ \text{m}^2$$

and the resistance of the length of wire between the bird's feet is

$$R = \rho \ell / A = \frac{\left(1.7 \times 10^{-8}\ \Omega \cdot \text{m}\right)\left(4.0 \times 10^{-2}\ \text{m}\right)}{3.8 \times 10^{-4}\ \text{m}^2}$$

or $\quad R = 1.8 \times 10^{-6}\ \Omega$

When the wire carries a current of 50 A, the potential difference between the two points where the bird's feet touch the wire is

$$\Delta V = IR = (50\ \text{A})\left(1.8 \times 10^{-6}\ \Omega\right) = 9.0 \times 10^{-5}\ \text{V} = 90 \times 10^{-6}\ \text{V} = 90\ \mu\text{V} \quad \Diamond$$

51. A length of metal wire has a radius of 5.00×10^{-3} m and a resistance of 0.100 Ω. When the potential difference across the wire is 15.0 V, the electron drift speed is found to be 3.17×10^{-4} m/s. Based on these data, calculate the density of free electrons in the wire.

Solution In terms of the drift speed of the electrons, the current flowing in a conductor is given by $I = nqAv_d$ where n is the density of free electrons in the conducting material, $q = e$ is the magnitude of the charge of the charge carriers (electrons), A is the cross-sectional area of the conductor, and v_d is the drift speed. Therefore, the density of free electrons may be expressed as $n = I/eAv_d$. The cross-sectional area of this wire is

$$A = \pi r^2 = \pi\left(5.00\times10^{-3}\ \text{m}\right)^2 = 7.85\times10^{-5}\ \text{m}^2$$

and the current that flows in the wire when a potential difference of 15.0 V is maintained across it is $I = \Delta V/R = 15.0\ \text{V}/0.100\ \Omega = 150\ \text{A}$. The density of free electrons in this wire is then

$$n = \frac{150\ \text{A}}{\left(1.60\times10^{-19}\ \text{C}\right)\left(7.85\times10^{-5}\ \text{m}^2\right)\left(3.17\times10^{-4}\ \text{m/s}\right)} = 3.77\times10^{28}\ \frac{\text{electrons}}{\text{m}^3} \quad \Diamond$$

53. A carbon wire and a nichrome wire are connected one after the other. If the combination has a total resistance of 10.0 kΩ at 20 °C, what is the resistance of each wire at 20 °C so that the resistance of the combination does not change with temperature?

Solution The total resistance of the two wires connected one after the other is $R = R_c + R_n$ where R_c is the resistance of the carbon wire and R_n is that of the nichrome wire. Note in Table 17.1 from the textbook that the temperature coefficient of resistivity is negative for carbon but positive for nichrome. Thus, as the temperature increases, the resistance of the carbon wire will decrease while the resistance of the nichrome wire will increase. It is desired to choose the values of R_c and R_n, at 20 °C, so the total resistance is constant at $R = 10.0\ \text{k}\Omega$ as temperature varies.

At temperature T, the resistance of each wire is given by $R_c = R_{0c}[1+\alpha_c(T-20\ °C)]$ and $R_n = R_{0n}[1+\alpha_n(T-20\ °C)]$. Here R_{0c} and R_{0n} are the resistances of the wires at 20 °C, while α_c and α_n are the temperature coefficients of resistivity for carbon and nichrome, respectively. The total resistance of the combination at temperature T is then

$$R = R_{0c}[1+\alpha_c(T-20\ °C)] + R_{0n}[1+\alpha_n(T-20\ °C)]$$

which may be written as

$$R = (R_{0c}+R_{0n}) - (R_{0c}\alpha_c + R_{0n}\alpha_n)(20\ °C) + (R_{0c}\alpha_c + R_{0n}\alpha_n)T$$

The only way this can remain constant as the temperature varies is for the multiplier of T in the last term be zero. That is, it is necessary that:

$$R_{0c}\alpha_c + R_{0n}\alpha_n = 0 \qquad \textbf{[Equation 1]}$$

A second equation relating the resistances at 20 °C is the requirement that $R = 10.0\ k\Omega$ at $T = 20\ °C$. This gives:

$$R_{0c} + R_{0n} = 10.0\ k\Omega \qquad \textbf{[Equation 2]}$$

From Equation 2, $R_{0n} = 10.0\ k\Omega - R_{0c}$. Substituting this into Equation 1 and simplifying gives $R_{0c} = \alpha_n(10.0\ k\Omega)/(\alpha_n - \alpha_c)$. From Table 17.1, $\alpha_n = 0.40\times10^{-3}\ (°C)^{-1}$ and $\alpha_c = -0.50\times10^{-3}\ (°C)^{-1}$. Therefore, the resistance of the carbon wire at 20 °C must be

$$R_{0c} = \frac{(0.40\times10^{-3}\ °C^{-1})(10.0\ k\Omega)}{(0.40\times10^{-3}\ °C^{-1}) - (-0.50\times10^{-3}\ °C^{-1})} = 4.4\ k\Omega \qquad \Diamond$$

Then, the required resistance of the nichrome wire at 20 °C is

$$R_{0n} = 10.0\ k\Omega - 4.4\ k\Omega = 5.6\ k\Omega \qquad \Diamond$$

60. (a) A sheet of copper ($\rho = 1.7 \times 10^{-8}$ Ω·m) is 2.0 mm thick and has surface dimensions of 8.0 cm × 24 cm. If the long edges are joined to form a tube 24 cm in length, what is the resistance between the ends? (b) What mass of copper is required to manufacture a 1500-m-long spool of copper cable with a total resistance of 4.5 Ω?

Solution (a) When the sheet of copper, as shown in part (A) of the figure, is joined along the 24-cm edges, it forms a hollow cylindrical shell as shown in (B). The shell has a circumference of 8.0 cm, length of 24 cm and a thickness of 2.0 mm. The cross-sectional area of the material in the shell is the same as the area of the upper end of the sheet. Thus,

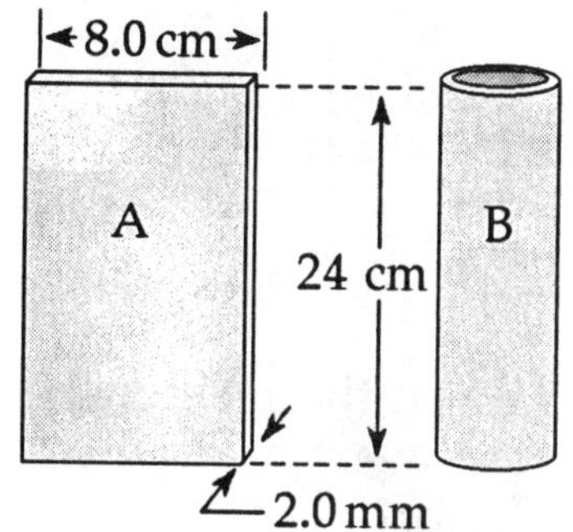

$$A = \left(2.0 \times 10^{-3} \text{ m}\right)\left(8.0 \times 10^{-2} \text{ m}\right) = 1.6 \times 10^{-4} \text{ m}$$

The resistivity of copper is $\rho = 1.7 \times 10^{-8}\ \Omega \cdot \text{m}$, so the resistance between the ends of the shell is

$$R = \rho \ell / A = \left(1.7 \times 10^{-8}\ \Omega \cdot \text{m}\right)\left(\frac{24 \times 10^{-2} \text{ m}}{1.6 \times 10^{-4} \text{ m}^2}\right) = 2.6 \times 10^{-5}\ \Omega \qquad \Diamond$$

(b) A solid cylindrical copper cable is to be 1500 m long and have a resistance of 4.5 Ω. The volume of copper needed will be $V = A\ell$ where A is the cross sectional area of the cable and $\ell = 1500$ m is the length. Thus, the area may be written as $A = V/\ell$, and the resistance $R = \rho\ell/A$ becomes $R = \rho\ell^2/V$. The mass of copper in the cable is $m = (\text{volume})(\text{density})$. Thus, the volume is $V = m/\text{density}$, and the resistance may be written as $R = \rho\ell^2(\text{density})/m$. The density of copper is $8.92 \times 10^3 \text{ kg/m}^3$ (see Table 9.2 in the textbook) and the mass of copper needed to make the specified cable is:

$$m = \rho\ell^2 \frac{(\text{density})}{R} = \left(1.7 \times 10^{-8}\ \Omega \cdot \text{m}\right)(1500 \text{ m})^2 \frac{\left(8.92 \times 10^3 \text{ kg/m}^3\right)}{(4.5\ \Omega)} = 76 \text{ kg} \qquad \Diamond$$

CHAPTER SELF-QUIZ

1. If a 6.00-V battery, with negligible internal resistance, and a 12.0-ohm resistor are connected in series, what is the amount of electrical energy transformed to heat per coul of charge that flows through the circuit?
   a. 0.50 J
   b. 3.00 J
   c. 6.00 J
   d. 72.0 J

2. When the potential difference between the ends of a conductor is tripled, the average drift velocity of mobile charge carriers is changed by what factor?
   a. 0.33
   b. 1.00
   c. 3.00
   d. 9.00

3. The heating coil of a hot water heater has a resistance of 20.0 ohm and operates at 210 V. What is its power rating?
   a. 8.80×10^5 W
   b. 2205 W
   c. 10.5 W
   d. 2940 W

4. If a certain resistor obeys Ohm's law, its resistance will change
   a. as the voltage across the resistor changes
   b. as the current through the resistor changes
   c. as the energy given off by the electrons in their collisions changes
   d. none of the above, since resistance is a constant for a given resistor

5. A certain material is in a room at 27.0 °C. If the absolute temperature (K) of the material is doubled, its resistance also doubles. (Water freezes at 273.0 K.) What is the value for α, the temperature coefficient of resistivity?
   a. 1 $°C^{-1}$
   b. 2 $°C^{-1}$
   c. 0.0034 $°C^{-1}$
   d. 0.038 $°C^{-1}$

6. A metal wire has a radius of 0.30 mm, is 1.00 m long, and has a resistance of 20.0 ohms. What must be the length of a second wire made of the same type of metal if its radius is 0.10 mm and also has a resistance of 20.0 ohms?
   a. 0.11 m
   b. 0.33 m
   c. 3.00 m
   d. 9.00 m

7. A platinum wire is utilized to determine the melting point of indium. The resistance of the platinum wire is 2.00 Ω at 20.0 °C and increases to 3.072 Ω as the indium starts to melt. $\alpha_{\text{platinum}} = 3.92 \times 10^{-3}\ °C^{-1}$. What is the melting temperature of indium?
   a. 137 °C
   b. 157 °C
   c. 351 °C
   d. 731 °C

8. Two wires made of the same metal and of the same length are connected across the same voltage. If one wire has three times the radius of the other, what is the ratio of currents, large wire to small wire?
   a. 1/9
   b. 1/3
   c. 3/1
   d. 9/1

9. A copper cable is to be designed to carry a current of 300 A with a power loss of only 2 watts per meter. What is the required radius of the copper cable? (The resistivity of copper is $1.70 \times 10^{-8}\ \Omega \cdot \text{m}$.)
   a. 0.80 cm
   b. 1.60 cm
   c. 3.20 cm
   d. 4.00 cm

10. A color television set draws about 2.50 A when connected to 120 V. What is the cost (with electrical energy at 6 cents/kWh) of running the color TV for 8 hours?
    a. 1.4 cents
    b. 3.0 cents
    c. 14.4 cents
    d. 30.0 cents

11. An electric car is designed to run off a bank of 12.0-V batteries with total energy storage of 2.00×10^{7} J. If the electric motor draws 8000 W in moving the car at a steady speed of 20 m/s, how far will the car go before it is "out of juice"?
    a. 25 km
    b. 50 km
    c. 100 km
    d. 150 km

12. Suppose you have a 10-gram copper cylinder. It is to be drawn down into a wire that will have a resistance of 1.00 Ω. What is the length of such a wire? (The resistivity of copper is $1.70 \times 10^{-8}\ \Omega \cdot \text{m}$.) (The density of copper is $8.90 \times 10^{3}\ \text{kg/m}^{3}$.)
    a. 2.00 m
    b. 4.00 m
    c. 8.00 m
    d. 16.0 m

Chapter 18

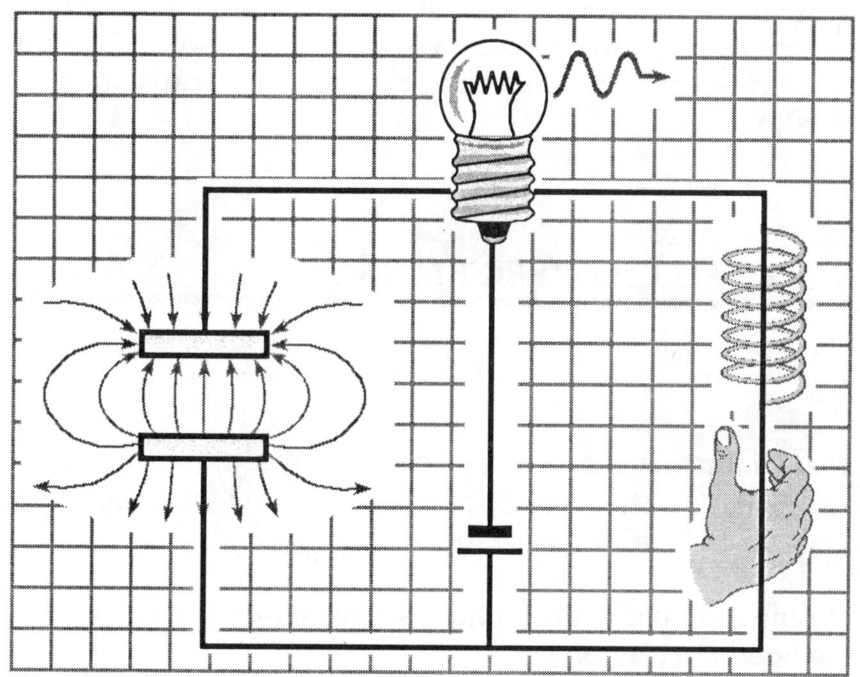

DIRECT CURRENT CIRCUITS

Chapter 18

DIRECT CURRENT CIRCUITS

This chapter analyzes some simple circuits whose elements include batteries, resistors, and capacitors in varied combinations. Such analysis is simplified by the use of two rules known as Kirchhoff's rules, which follow from the principles of conservation of energy and the law of conservation of charge. Most of the circuits are assumed to be in **steady state,** which means that the currents are constant in magnitude and direction. We close the chapter with a discussion of circuits containing resistors and capacitors, in which current varies with time.

NOTES FROM SELECTED CHAPTER SECTIONS

18.2 Sources of EMF

The source that maintains the constant current in a closed circuit is called a source of "emf." The emf, $\mathcal{E}$, of a source is the work done per unit charge, and the SI unit of emf is the volt.

Emf is equal to the terminal voltage of the source **when the current is zero,** called the **open-circuit voltage.**

18.2 Resistors in Series
18.3 Resistors in Parallel

The **current** must be the same for each of a group of resistors connected in **series.**

The **potential difference** must be the same across each of a group of resistors in **parallel.**

18.4 Kirchhoff's Rules and Complex DC Circuits

1. The sum of the currents entering any junction must equal the sum of the currents leaving that junction. (A junction is any point in the circuit where the current can split.)

2. The algebraic sum of the changes in potential around any closed circuit loop must be **zero.**

The first rule is a statement of **conservation of charge;** the second rule follows from the **conservation of energy.**

18.5 RC Circuits

Consider an uncharged capacitor in series with a resistor, a battery, and a switch. In the charging process, charges are transferred from one plate of the capacitor to the other moving along a path **through the resistor, battery, and switch.** The charges **do not move across the gap between the plates of the capacitor.**

The battery does work on the charges to increase their electrostatic potential energy as they move from one plate to the other.

EQUATIONS AND CONCEPTS

When a battery is providing a current to an external circuit, the **terminal voltage** of the battery will be less than the emf due to **internal resistance** of the battery.

$$\Delta V = \mathcal{E} - Ir \qquad (18.1)$$

The total or equivalent resistance of a series combination of resistors is equal to the sum of the resistances of the individual resistors.

$$R_{eq} = R_1 + R_2 + R_3 + \ldots \qquad (18.4)$$

(series combination)

A group of resistors connected in parallel has an equivalent resistance which is less than the smallest individual value of resistance in the group.

$$\frac{1}{R_{eq}} = \frac{1}{R_1} + \frac{1}{R_2} + \frac{1}{R_3} + \ldots \quad (18.6)$$

(parallel combination)

Resistors in series are connected so that they have only one common circuit point per pair; there is a common current through each resistor in the group.

Resistors in parallel are connected so that each resistor in the group has two circuit points in common with each of the other resistors; there is a common potential difference across each resistor in the group.

Comment on series and parallel connection of resistors.

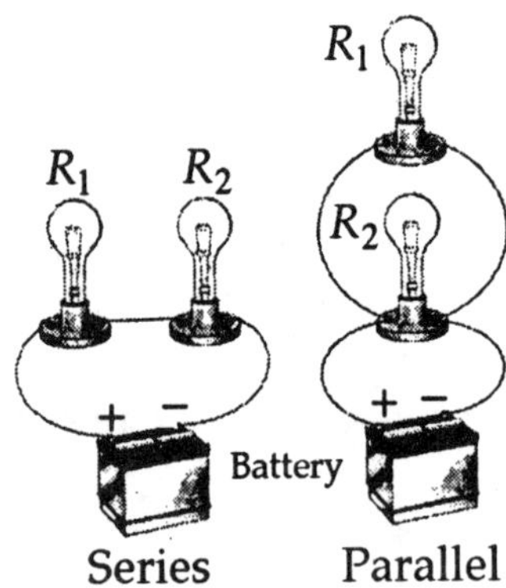

Many circuits which contain several resistors can be reduced to an equivalent single-loop circuit by successively combining groups of resistors in series and parallel. However, in the most general case, such a reduction is not possible and you must solve a true multiloop circuit by use of Kirchhoff's rules. Review the procedure suggested in the following section to apply Kirchhoff's rules.

When a battery is used to charge a capacitor in series with a resistor, a quantity τ, called the time constant of the circuit, is used to describe the manner in which the charge on the capacitor varies with time. The charge on the capacitor increases from zero to 63 percent of its maximum value in a time interval equal to one time constant. Also, during one time constant, the charging current decreases from its initial maximum value of $I_0 = \mathcal{E}/R$ to 37 percent of I_0.

$$q = Q(1 - e^{-t/RC}) \quad (18.7)$$

$$\tau = RC \quad (18.8)$$

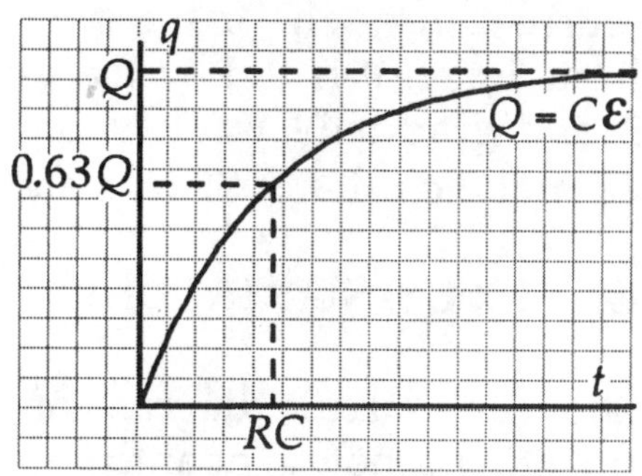

SUGGESTIONS, SKILLS, AND STRATEGIES

PROBLEM-SOLVING STRATEGY FOR RESISTORS

1. When two or more unequal resistors are connected in **series**, they carry the same current, but the potential differences across them are not the same. The resistors add directly to give the equivalent resistance of the series combination.

2. When two or more unequal resistors are connected in **parallel**, the potential differences across them are the same. Since the current is inversely proportional to the resistance, the currents through them are not the same. The equivalent resistance of a parallel combination of resistors is found through reciprocal addition, and the equivalent resistor is always **less** than the smallest individual resistor.

3. A complicated circuit consisting of resistors can often be reduced to a simple circuit containing only one resistor. To do so, examine the initial circuit and replace any resistors in series or any in parallel using the procedures outlined in Steps 1 and 2. Sketch the new circuit after these

changes have been made. Examine the new circuit and replace any series or parallel combinations. Continue this process until a single equivalent resistance is found.

4. If the current through, or the potential difference across, a resistor in the complicated circuit is to be identified, start with the final circuit found in Step 3 and gradually work your way back through the circuits, using $\Delta V = IR$ and the rules of Steps 1 and 2.

STRATEGY FOR USING KIRCHHOFF'S RULES

1. First, draw the circuit diagram and assign labels and symbols to all the known and unknown quantities. **You must assign directions to the currents** in each part of the circuit. Do not be alarmed if you guess the direction of a current incorrectly; the resulting value will be negative, but its magnitude will be correct. Although the assignment of current directions is arbitrary, you must stick with your guess throughout as you apply Kirchhoff's rules.

2. Apply the junction rule to any junction in the circuit. The junction rule may be applied as many times as a new current (one not used in a previous application) appears in the resulting equation. In general, the number of times the junction rule can be used is one fewer than the number of junction points in the circuit.

3. Now apply Kirchhoff's loop rule to as many loops in the circuit as are needed to solve for the unknowns. Remember you must have as many equations as there are unknowns (I's, R's, and $\mathcal{E}$'s). **For each loop, you must first choose a direction to sum the voltage changes, clockwise or counterclockwise. Then,** you must correctly identify the change in potential as you cross each element in traversing the closed loop. Watch out for signs!

4. Finally, you must solve the equations simultaneously for the unknown quantities. Be careful in your algebraic steps, and check your numerical answers for consistency.

Convenient "rules of thumb" which you may use to determine the increase or decrease in potential as you cross a resistor or seat of emf in traversing a circuit loop are illustrated in the following figure. Notice that the potential **decreases** (changes by $-IR$) when the resistor is traversed **in the direction of the current.** There is an **increase** in potential of $+IR$ if the direction of travel is **opposite** the direction of current. If a seat of emf is traversed **in** the direction of the emf (from – to + on the battery), the potential **increases** by $\mathcal{E}$. If the direction of travel is from + to –, the potential **decreases** by $\mathcal{E}$ (changes by $-\mathcal{E}$).

$$\Delta V = V_b - V_a = -IR$$

$$\Delta V = V_b - V_a = IR$$

$$\Delta V = V_b - V_a = \mathcal{E}$$

$$\Delta V = V_b - V_a = -\mathcal{E}$$

As an illustration of the use of Kirchhoff's rules, consider a three-loop circuit which has the **general form** shown in figure at the right. In this illustration, the actual circuit elements, R's and $\mathcal{E}$'s are not shown but assumed known. There are six possible different values of I in the circuit; therefore you will need six independent equations to solve for the six values of I. There are four junction points in the circuit (at points $a, d, f,$ and h). The first rule applied at **any three** of these points will yield three equations. The circuit can be thought of as a group of three "blocks" as shown in the following figure on the right. Kirchhoff's second law, when applied to each of these loops (*abcda*, *ahfga*, and *defhd*), will yield three additional equations. You can then solve the total of six equations simultaneously for the six values of I_1, I_2, I_3, I_4, I_5, and I_6. You can, of course, expect that the sum of the changes in potential difference around **any other closed loop** in the circuit will be zero (for example, *abcdefga* or *ahfedcba*); however the equations found by applying Kirchhoff's second rule to these additional loops **will not be independent** of the six equations found previously.

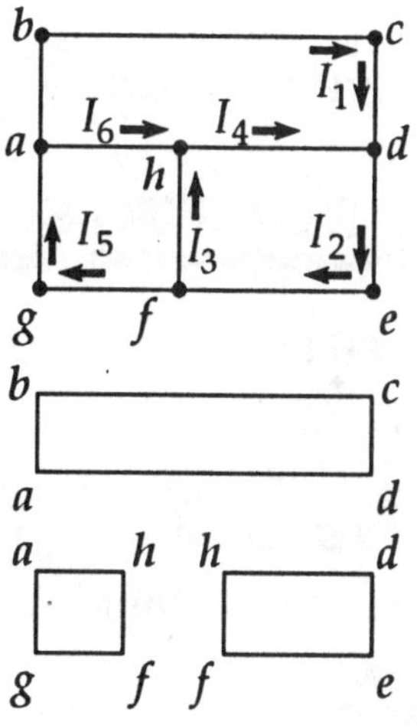

REVIEW CHECKLIST

▷ Calculate the current in a single-loop circuit and the potential difference between any two points in the circuit; and calculate the equivalent resistance of a group of resistors in parallel, series, or series-parallel combination.

▷ Use Ohm's law to calculate the current in a circuit and the potential difference between any two points in a circuit which can be reduced to an equivalent single-loop circuit.

▷ Use Joule's law to calculate the power dissipated by any resistor or group of resistors in a circuit.

▷ Apply Kirchhoff's rules to solve multiloop circuits; that is, find the currents at any point and the potential difference between any two points.

▷ Describe in qualitative terms the manner in which charge accumulates on a capacitor or current flow changes through a resistor with time in a series circuit with battery, capacitor, resistor, and switch.

SOLUTIONS TO SELECTED END-OF-CHAPTER PROBLEMS

5. A 9.0-Ω resistor and a 6.0-Ω resistor are connected in series with a power supply. (a) The voltage drop across the 6.0-Ω resistor is measured to be 12 V. Find the voltage output of the power supply. (b) The two resistors are connected in parallel across a power supply, and the current through the 9.0-Ω resistor is found to be 0.25 A. Find the power supply's voltage setting.

Solution (a) Since the voltage drop across the 6.0-Ω resistor is $\Delta V_{bc} = 12\text{ V}$, the current in this circuit is

$$I = \Delta V_{bc}/R_{bc} = 12\text{ V}/6.0\ \Omega = 2.0\text{ A}$$

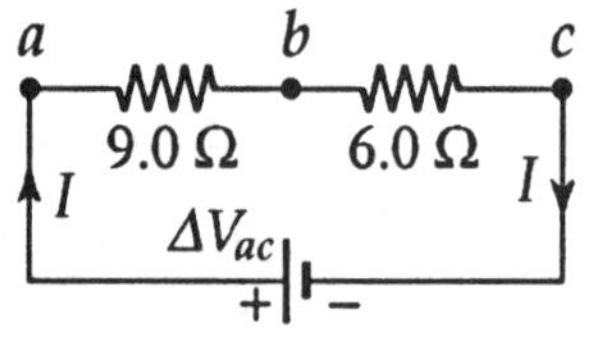

Figure 1

The series combination between points a and c in Figure 1 has a total resistance of $R_{ac} = 9.0\ \Omega + 6.0\ \Omega = 15\ \Omega$. Thus, the voltage output of the power supply is: $\Delta V_{ac} = IR_{ac} = (2.0\text{ A})(15\ \Omega) = 30\text{ V}$ ◊

(b) The two resistors are now connected in parallel with the power supply as shown in Figure 2. The power supply maintains the same potential difference, ΔV_{ab} across each of the resistors. Since the current through the 9.0 Ω resistor is known to be $I_9 = 0.25$ A, the potential difference across this resistor, and hence the voltage setting of the power supply, is

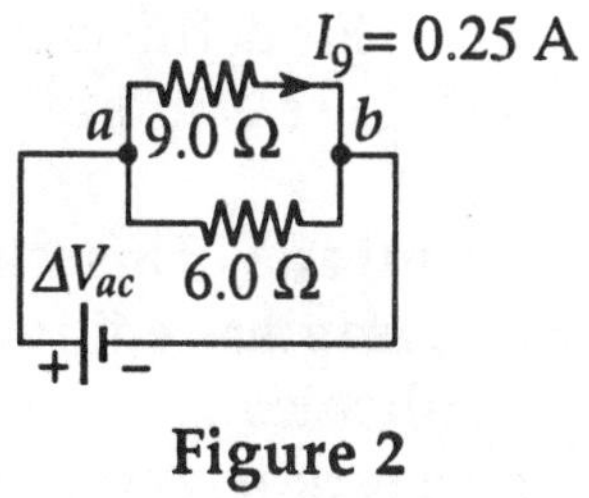

Figure 2

$$\Delta V_{ab} = I_9 R_9 = (0.25\ \text{V})(9.0\ \Omega) = 2.3\ \text{V}$$ ◊

---

11. Two resistors, A and B, are connected in parallel across a 6.0-V battery. The current through B is found to be 2.0 A. When the two resistors are connected in series to the 6.0-V battery, a voltmeter connected across resistor A measures a voltage of 4.0 V. Find the resistances of A and B.

Solution When the two resistors are connected in parallel across a 6.0-V battery, the battery maintains a potential difference of 6.0 V across each resistor. If this potential difference produces a 2.0-A current in B, the resistance of B is:

$$R_B = \frac{\Delta V_B}{I_B} = \frac{6.0\ \text{V}}{2.0\ \text{A}} = 3.0\ \Omega$$ ◊

When the resistors are connected in series to the 6.0-V battery, the voltage drop across the series combination is $\Delta V_A + \Delta V_B = 6.0\ \text{V}$. If $\Delta V_A = 4.0\ \text{V}$, then $\Delta V_B = 2.0\ \text{V}$, and the current through the series combination is

$$I = \frac{\Delta V_B}{R_B} = \frac{2.0\ \text{V}}{3.0\ \Omega} = 0.67\ A$$

Therefore, $$R_A = \frac{\Delta V_A}{I} = \frac{4.0\ \text{V}}{0.67\ A} = 6.0\ \Omega$$ ◊

---

14. Find the current in the 12-Ω resistor in Figure P18.14.

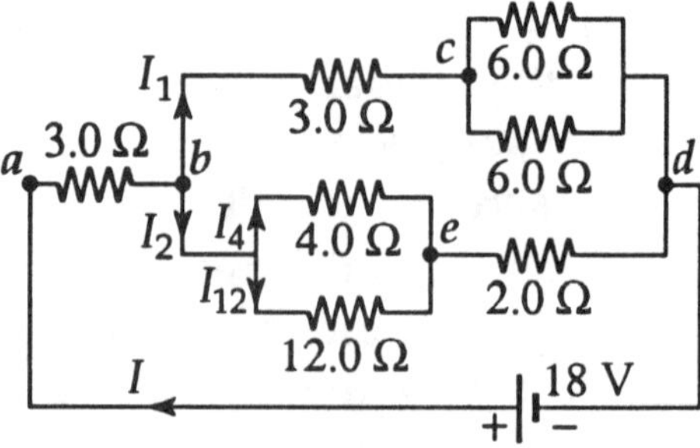

Figure P18.14

Solution This circuit can be reduced to its simplest equivalent form by successive applications of the rules for combining resistors in series and parallel as shown in the figures to the right. The strategy is to start with the simplest of these equivalent circuits and work back to the original circuit, gathering more information at each step on the way. From Figure 5, the total resistance of the circuit is seen to be $R_{ad} = 63/11\ \Omega$. Therefore, the total current supplied by the battery must be

$$I = \frac{\Delta V_{ad}}{R_{ad}} = \frac{18\text{ V}}{63/11\ \Omega} = 3.14\text{ A}$$

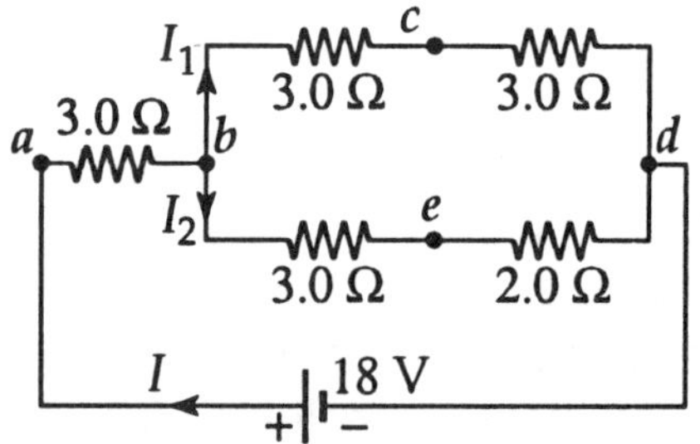

Figure 2

The potential difference between points b and d can now be found from Figure 4 as

$$\Delta V_{bd} = IR_{bd} = (3.14\text{ A})(30/11\ \Omega) = 8.57\text{ V}$$

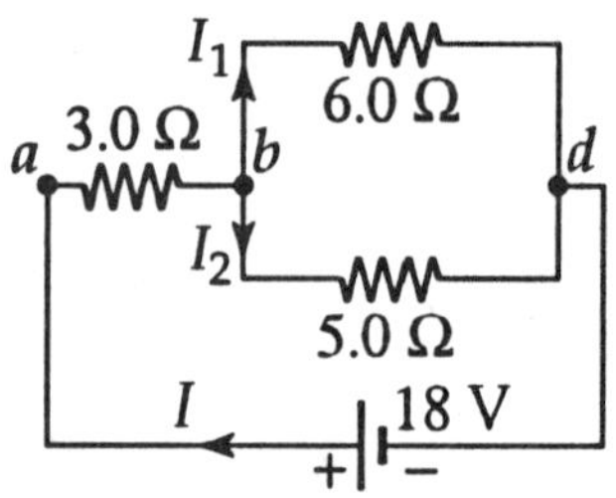

Figure 3

From Fig. 3, $\quad I_2 = \dfrac{\Delta V_{bd}}{5.0\ \Omega} = \dfrac{8.57\text{ V}}{5.0\ \Omega} = 1.71\text{ A}$

Going back to Figure 2, the potential difference between points b and e is seen to be $\Delta V_{be} = I_2 R_{be} = (1.71\text{ A})(3.0\ \Omega) = 5.13\text{ V}$. With this knowledge, the current in the 12-Ω resistor can be found from Figure 1 as:

$$I_{12} = \frac{\Delta V_{be}}{12\ \Omega} = \frac{5.13\text{ V}}{12\ \Omega} = 0.428\text{ A} \qquad \lozenge$$

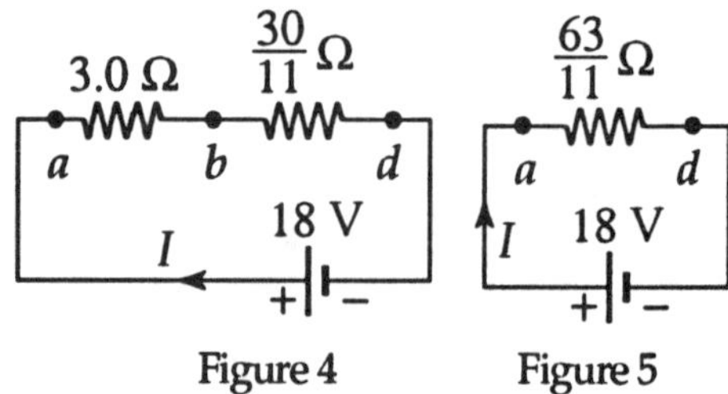

Figure 4 Figure 5

17. In the circuit of Figure P18.17, the current I_1 is 3.0 A and the value of $\mathcal{E}$ and R are unknown. What are the currents I_2 and I_3?

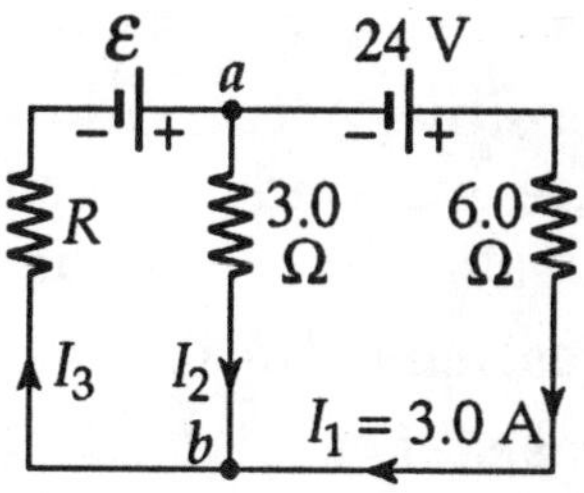

Figure P18.17

Solution

Follow the path taken by the current I_1 from point a to point b, keeping track of the changes in potential that occur. The change in potential along this path is found to be

$$\Delta V_{ab} = +24 \text{ V} - (3.0 \text{ A})(6.0 \ \Omega) = +24 \text{ V} - 18 \text{ V} = +6.0 \text{ V}$$

The same change in potential must occur as one goes from point a to point b along the path followed by the current I_2. Thus,

$$\Delta V_{ab} = -I_2(3.0 \ \Omega) = +6.0 \text{ V}, \quad \text{or} \quad I_2 = -2.0 \text{ A}$$

The negative sign tells us that I_2 actually flows from point b toward a rather than in the direction assumed by the arrow in the circuit diagram. Now that I_2 is known, application of Kirchoff's junction rule at point b gives

$$I_3 = I_1 + I_2 = (+3.0 \text{ A}) + (-2.0 \text{ A}) = +1.0 \text{ A}$$

The positive sign tells us that I_3 actually flows in the direction assumed in the circuit diagram. The results are:

I_2 = 2.0 A flowing from b toward a ◊

and I_3 = 1.0 A, also flowing from b toward a ◊

21. Two 1.50-V batteries—with their positive terminals in the same direction—are inserted in series into the barrel of a flashlight. One battery has an internal resistance of 0.255 Ω, the other an internal resistance of 0.153 Ω. When the switch is closed, a current of 0.600 A passes through the lamp. (a) What is the lamp's resistance? (b) What fraction of the power dissipated is dissipated in the batteries?

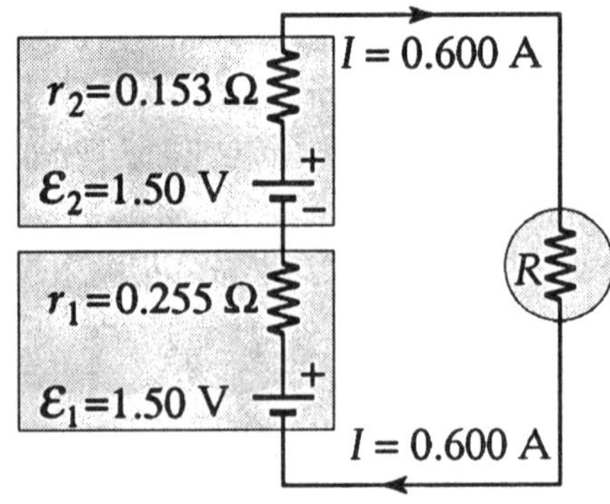

Solution (a) The circuit is shown in the diagram. Applying Kirchoff's loop rule to the only closed loop in this circuit, going around the circuit in the direction of the current, gives

$$+\mathcal{E}_1 - Ir_1 + \mathcal{E}_2 - Ir_2 - IR = 0$$

The resistance of the bulb is then $R = \dfrac{\mathcal{E}_1 + \mathcal{E}_2}{I} - r_1 - r_2$

or $$R = \frac{3.00 \text{ V}}{0.600 \text{ A}} - 0.153\ \Omega - 0.255\ \Omega = 4.59\ \Omega$$ ◊

(b) The total power input from the emfs of the batteries is:

$$P_{\text{in}} = (\mathcal{E}_1 + \mathcal{E}_2)I = (3.00 \text{ V})(0.600 \text{ A}) = 1.80 \text{ W}$$

The power dissipated within the batteries is:

$$P_{\text{loss}} = I^2(r_1 + r_2) = (0.600 \text{ A})^2(0.153\ \Omega + 0.255\ \Omega) = 0.147 \text{ W}$$

The fraction of the total power input represented by this internal loss is

$$\text{fractional loss} = \frac{P_{\text{loss}}}{P_{\text{in}}} = \frac{0.147 \text{ W}}{1.80 \text{ W}} = 0.0820 \quad \text{or} \quad 8.20\%$$ ◊

23. Calculate each of the unknown currents I_1, I_2, and I_3 for the circuit of Figure P18.23.

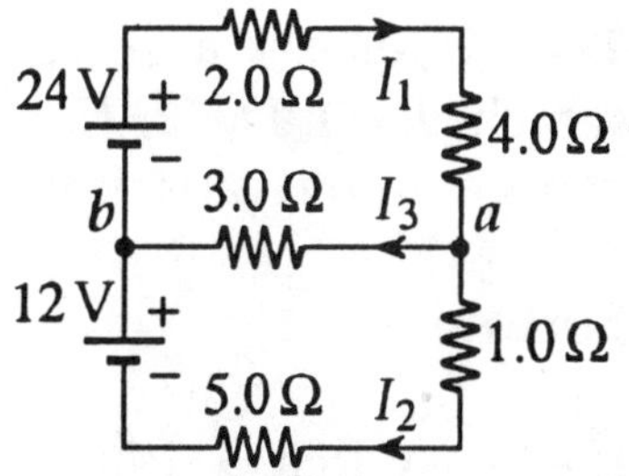

Figure P18.23

Solution Since there are three unknown currents to find, three equations are needed. One equation involving the unknown currents is obtained by applying Kirchoff's junction rule at point a. This gives:

$$I_1 = I_2 + I_3 \qquad \textbf{[Equation 1]}$$

Applying Kirchoff's loop rule to the upper loop in the circuit diagram, starting at the 3.0-Ω resistor and going clockwise, gives $-(3.0\ \Omega)I_3 + 24\ \text{V} - (2.0\ \Omega)I_1 - (4.0\ \Omega)I_1 = 0$, or reducing to the simplest form:

$$2I_1 + I_3 = 8.0\ \text{A} \qquad \textbf{[Equation 2]}$$

Then, applying Kirchoff's loop rule to the lower loop, starting at the 1.0-Ω resistor and going clockwise, gives $-(1.0\ \Omega)I_2 - (5.0\ \Omega)I_2 + 12\ \text{V} + (3.0\ \Omega)I_3 = 0$. The last term is positive because we were going against the flow of the current as we passed through the 3 Ω resistor. Since conventional current always flows from higher to lower potential, you encounter a drop in potential when going through a resistor in the direction of the current and a rise in potential when going through a resistor opposite to the direction of the current. This equation reduces to:

$$I_3 = 2I_2 - 4.0\ \text{A} \qquad \textbf{[Equation 3]}$$

Substituting Equation 1 into Equation 2 gives $2I_2 + 3I_3 = 8.0\ \text{A}$, and substituting Equation 3 into this result gives $2I_2 + 3(2I_2 - 4.0\ \text{A}) = 8.0\ \text{A}$, or $8I_2 = +20\ \text{A}$ and $I_2 = +2.5\ \text{A}$. Thus, I_2 flows in the direction assumed in the circuit diagram and has a magnitude of 2.5 A. Equation 3 then gives $I_3 = 2(+2.5\ \text{A}) - 4.0\ \text{A} = +1.0\ \text{A}$, and Equation 1 now yields $I_1 = (+2.5\ \text{A}) + (+1.0\ \text{A}) = +3.5\ \text{A}$. The currents in the circuit are: $I_1 = 3.5\ \text{A}$, $I_2 = 2.5\ \text{A}$, and $I_3 = 1.0\ \text{A}$, with the directions indicated by the arrows in the circuit diagram. ◊

31. Consider a series *RC* circuit (see Fig. 18.13) for which $R = 1.0\ \text{M}\Omega$, $C = 5.0\ \mu\text{F}$, and $\mathcal{E} = 30\ \text{V}$. Find the charge on the capacitor 10 s after the switch is closed.

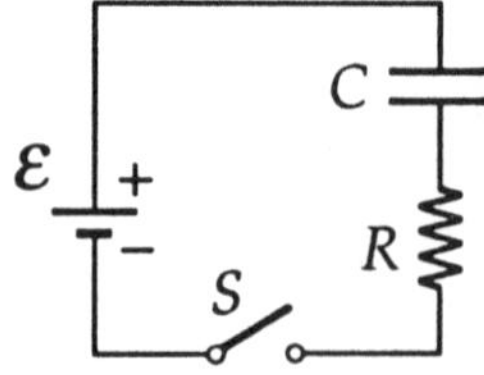

Figure 18.13

Solution

If the switch S is closed at $t = 0$, the charge on the capacitor at any time later is $q = Q_{max}\left(1 - e^{-t/\tau}\right)$. Here, $\tau = RC$ is the time constant of the circuit. For the given circuit,

$$\tau = (1\ \text{M}\Omega)(5.0\ \mu\text{F}) = \left(1.0\times10^{6}\ \Omega\right)\left(5.0\times10^{-6}\ \text{F}\right) = 5.0\ \text{s}$$

The charge stored in the capacitor when it is fully charged (i.e., when current has ceased to flow in the circuit) is Q_{max}. Note that if Kirchoff's loop rule is applied to this circuit, with $I = 0$, the result is $+\mathcal{E} - \Delta V_c - R(0) = 0$, or when the capacitor is fully charged, the voltage across the capacitor is $\Delta V_c = Q_{max}/C = \mathcal{E}$. Thus, $Q_{max} = C\mathcal{E} = (5.0\ \mu\text{F})(30\ \text{V}) = 150\ \mu\text{C}$. At $t = 10\ \text{s}$ after the switch is closed, the charge on the capacitor is:

$$q = (150\ \mu\text{C})\left[1 - e^{-(10\ \text{s})/(5.0\ \text{s})}\right] = (150\ \mu\text{C})\left[1 - e^{-2}\right] = 130\ \mu\text{C}$$ ◊

A **Voltswagon** towing a mobile **Ohm.**

35. A lamp (R = 150 Ω), an electric heater (R = 25 Ω), and a fan (R = 50 Ω) are connected in parallel across a 120-V line. (a) What total current is supplied to the circuit? (b) What is the voltage across the fan? (c) What is the current in the lamp? (d) What power is expended in the heater?

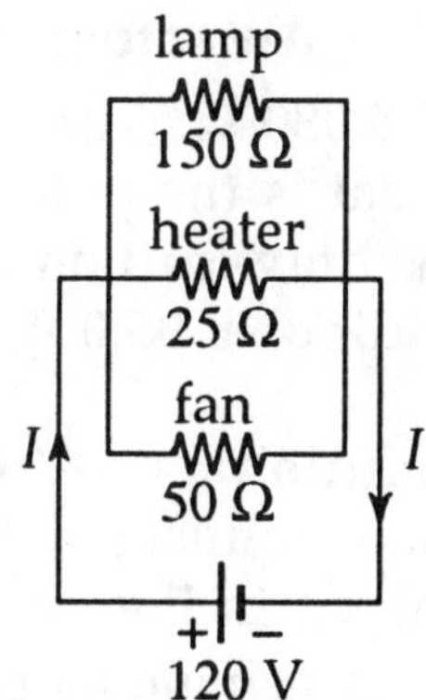

Solution

(a) The total resistance of this combination of three parallel resistances is

$$\frac{1}{R_{eq}} = \frac{1}{150\ \Omega} + \frac{1}{25\ \Omega} + \frac{1}{50\ \Omega} = \frac{10}{150\ \Omega} \quad \text{or} \quad R_{eq} = 15\ \Omega$$

Therefore, the total current supplied to the circuit by the power source is

$$I = \Delta V / R_{eq} = 120\ \text{V}/15\ \Omega = 8.0\ \text{A} \qquad \Diamond$$

(b) When a parallel combination of components is connected directly to a power source as in this circuit, the full voltage of the power source is maintained across each component. Thus,

$$\Delta V_{\text{fan}} = 120\ \text{V} \qquad \Diamond$$

(c) The current in the lamp is

$$I_{\text{lamp}} = \Delta V_{\text{lamp}} / R_{\text{lamp}} = 120\ \text{V}/150\ \Omega = 0.80\ \text{A} \qquad \Diamond$$

(d) The power expended in the heater is:

$$P_{\text{heater}} = \Delta V_{\text{heater}} I_{\text{heater}} = \frac{\Delta V_{\text{heater}}^2}{R_{\text{heater}}} = \frac{(120\ \text{V})^2}{25\ \Omega} = 576\ \text{W} \qquad \Diamond$$

43. An automobile battery has an emf of 12.60 V and an internal resistance of 0.080 Ω. The headlights have total resistance 5.00 Ω (assumed constant). What is the potential difference across the headlight bulbs (a) when they are the only load on the battery? (b) when the starter motor is operated, taking an additional 35.0 A from the battery?

Solution (a) When the headlights are the only load connected to the battery, the circuit is as shown in Figure 1. Applying Kirchoff's loop rule to this single loop circuit gives $+\mathcal{E} - Ir - IR_{\text{lights}} = 0$, or the current in the circuit is

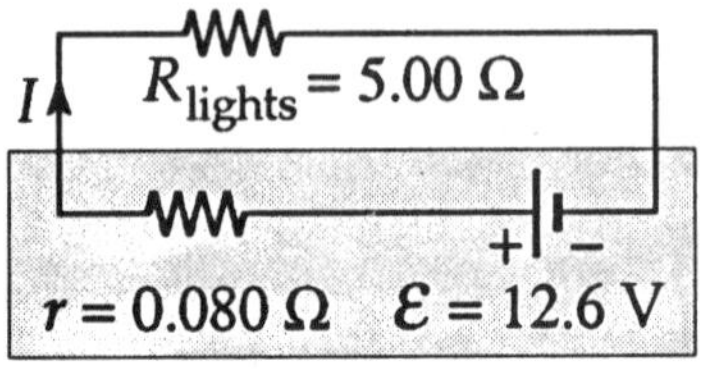

$$I = \frac{\mathcal{E}}{r + R_{\text{lights}}} = \frac{12.6 \text{ V}}{5.08 \text{ Ω}} = 2.48 \text{ A}$$

I $R_{\text{lights}} = 5.00 \text{ Ω}$ + − $r = 0.080 \text{ Ω}$ $\mathcal{E} = 12.6 \text{ V}$

Figure 1

The potential difference across the headlight bulbs is then

$$\Delta V_{\text{lights}} = IR_{\text{lights}} = (2.48 \text{ A})(5.00 \text{ Ω}) = 12.4 \text{ V} \quad \Diamond$$

(b) When the starter motor is turned on, the circuit is as shown in Figure 2. Using Kirchoff's junction rule at point a gives $I = I_{\text{lights}} + 35.0 \text{ A}$, and applying Kirchoff's loop rule to the lower loop of this circuit yields

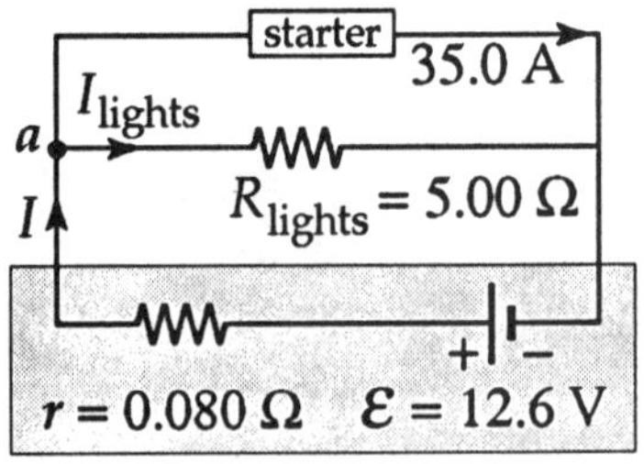

$$+\mathcal{E} - Ir - I_{\text{lights}} R_{\text{lights}} = 0$$

starter 35.0 A I_{lights} a I $R_{\text{lights}} = 5.00 \text{ Ω}$ + − $r = 0.080 \text{ Ω}$ $\mathcal{E} = 12.6 \text{ V}$

Combining these two equations, one obtains

$$+\mathcal{E} - \left(I_{\text{lights}} + 35.0 \text{ A}\right)r - I_{\text{lights}} R_{\text{lights}} = 0$$

or

$$I_{\text{lights}} = \frac{\mathcal{E} - (35.0 \text{ A})r}{r + R_{\text{lights}}} = \frac{12.6 \text{ V} - (35.0 \text{ A})(0.080 \text{ Ω})}{5.00 \text{ Ω} + 0.080 \text{ Ω}} = 1.93 \text{ A}$$

The potential difference that now exists across the headlight bulbs is

$$\Delta V_{\text{lights}} = I_{\text{lights}} R_{\text{lights}} = (1.93 \text{ A})(5.00 \text{ Ω}) = 9.65 \text{ V} \quad \Diamond$$

47. The resistance between points a and b in Figure P18.47 drops to one-half its original value when switch S is closed. Determine the value of R.

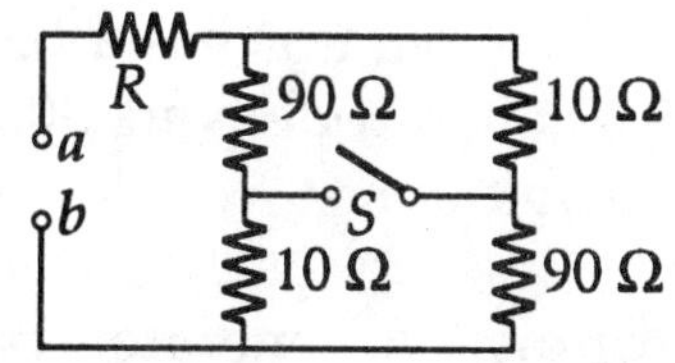

Figure P18.47

Solution With switch S open, the circuit is equivalent to that shown in Figure 1. Successive applications of the rules for combining resistors in series and parallel reduces this network to a single resistor as shown in Figures 2 through 4 below:

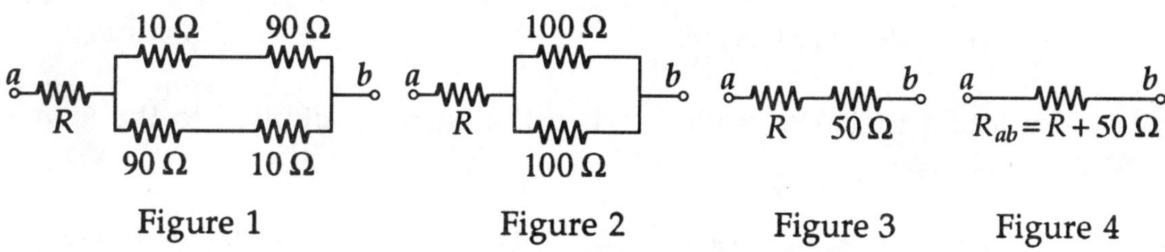

Figure 1 Figure 2 Figure 3 Figure 4

When the switch is closed, the resulting circuit is equivalent to that shown in Figure 5 below. Combining resistors in series and parallel then reduces it to a single resistor as shown in Figures 6 and 7.

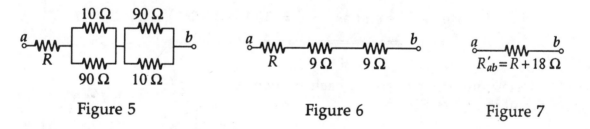

Figure 5 Figure 6 Figure 7

It is given that, when the switch is closed, the resistance between points a and b is one-half the value it had with the switch open. Therefore,

$$R'_{ab} = \tfrac{1}{2} R_{ab} \quad \text{or} \quad R + 18\ \Omega = \tfrac{1}{2}(R + 50\ \Omega), \text{ which yields } R = 14\ \Omega \qquad \Diamond$$

50. A generator has a terminal voltage of 110 V when it delivers 10.0 A, and 106 V when it delivers 30.0 A. Calculate the emf and the internal resistance of the generator.

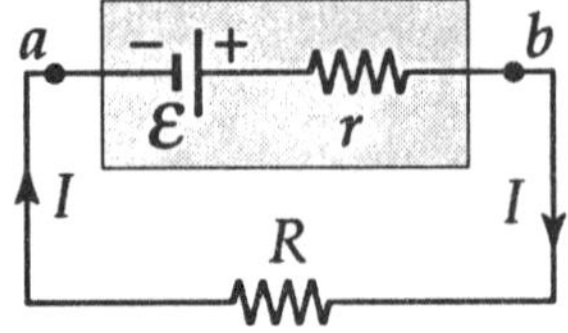

Solution A generator, battery, or other power source may be considered to consist of a seat of emf, $\mathcal{E}$, and internal resistance r as shown within the dotted outline. When this source supplies current I to a load, the voltage between the terminals of the source is $\Delta V_{ab} = \mathcal{E} - Ir$. If the terminal voltage is $\Delta V_{ab} = 110\text{ V}$ when the source delivers a current of $I = 10.0\text{ A}$, then

$$110\text{ V} = \mathcal{E} - (10.0\text{ A})r \qquad \textbf{[Equation 1]}$$

If it is also found that $\Delta V_{ab} = 106\text{ V}$ when the current supplied to the load is $I = 30.0\text{ A}$, one has:

$$106\text{ V} = \mathcal{E} - (30.0\text{ A})r \qquad \textbf{[Equation 2]}$$

Subtracting Equation 2 from Equation 1 gives $4.00\text{ V} = (20.0\text{ A})r$, so the internal resistance of the source is $r = (4.00\text{ V})/(20.0\text{ A}) = 0.200\ \Omega$ ◊

Then, Equation 1 gives the emf as $\mathcal{E} = 110\text{ V} + (10.0\text{ A})(0.200\ \Omega) = 112\text{ V}$ ◊

---

55. Determine the current in each branch of the circuit in Figure P18.55.

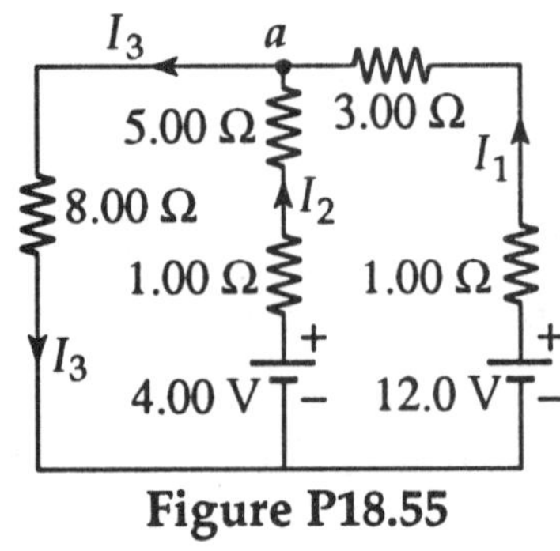

Figure P18.55

Solution Three distinct currents are present in this circuit. These are labeled I_1, I_2, and I_3 and will be assumed to flow in the directions indicated by the arrows in the circuit diagram. Application of Kirchoff's junction rule at point a gives:

$$I_3 = I_1 + I_2 \qquad \textbf{[Equation 1]}$$

Going counterclockwise around the leftmost loop, starting at point a, Kirchoff's loop rule gives $-(8.00\ \Omega)I_3 + 4.00\ \text{V} - (1.00\ \Omega)I_2 - (5.00\ \Omega)I_2 = 0$, which simplifies to:

$$(3.00)I_2 + (4.00)I_3 = 2.00\ \text{A} \qquad \textbf{[Equation 2]}$$

Starting at a and going clockwise around the rightmost loop, the loop rule yields

$$+(3.00\ \Omega)I_1 + (1.00\ \Omega)I_1 - 12.00\ \text{V} + 4.00\ \text{V} - (1.00\ \Omega)I_2 - (5.00\ \Omega)I_2 = 0$$

which, when solved for I_1, is: $\quad I_1 = 2.00\ \text{A} + (1.50)I_2 \qquad$ **[Equation 3]**

Upon substituting Equation 1 into Equation 2, one obtains $(7.00)I_2 + (4.00)I_1 = 2.00\ \text{A}$, and substitution of Equation 3 into this result gives $(13.0)I_2 + 8.00\ \text{A} = 2.00\ \text{A}$, which yields I_2 in the center branch:

$$I_2 = -\left(\frac{6.00}{13.0}\right)\text{A upward, or } I_2 = \frac{6.00}{13.0}\ \text{A downward in the center branch} \quad \Diamond$$

Thus, the current in the center branch of this circuit has a magnitude of $(6.00/13.0)$ A, but flows in the direction opposite to what was assumed. Now that we know the value of I_2, Equation 3 gives I_1, flowing upward in the rightmost branch:

$$I_1 = 2.00\ \text{A} + (1.50)\left(-\frac{6.00}{13.0}\ \text{A}\right) = +\frac{17.0}{13.0}\ \text{A} \qquad \Diamond$$

Then, substituting the results found for I_1 and I_2 into Equation 1, the remaining current is found to be flowing downward in the leftmost branch:

$$I_3 = \frac{17.0}{13.0}\ \text{A} - \frac{6.00}{13.0}\ \text{A} = +\frac{11.0}{13.0}\ \text{A} \qquad \Diamond$$

Since positive values are found for I_1 and I_3, these currents flow in the assumed directions indicated by arrows in the circuit diagram.

CHAPTER SELF-QUIZ

1. The following three appliances are connected to a 120-volt house circuit: (i) toaster, 1200 W; (ii) coffee pot, 750 W; and (iii) microwave, 600 W. If all were operated at the same time, what total current would they draw?
   a. 3.00 A
   b. 5.00 A
   c. 10.0 A
   d. 21.0 A

2. A circuit contains a 3.00-volt battery, a 2.00-ohm resistor, a 0.30-microfarad capacitor, an ammeter, and a switch, all in series. What will be the current reading 10 min after the switch is closed?
   a. zero
   b. 0.75 A
   c. 1.50 A
   d. 10^7 A

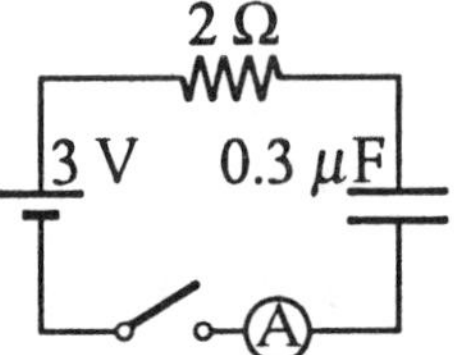

3. What is the current through the 4.00-ohm resistor?
   a. 1.00 A
   b. 0.50 A
   c. 1.50 A
   d. 2.00 A

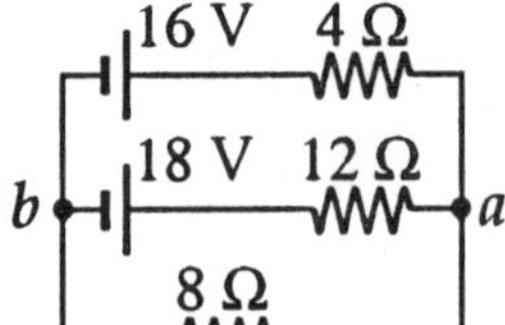

4. A circuit contains a 3.00-volt battery, a 2.00-ohm resistor, a 0.30-microfarad capacitor, an ammeter, and a switch in series. The sum of all the potential differences taken for all the components going completely around the loop is which of the following?
   a. zero
   b. 3.00 V
   c. -3.00 V
   d. 6.00 V

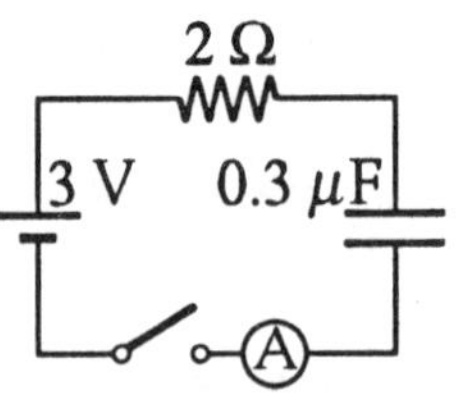

5. Three resistors, with values of 3.00, 6.00, and 12.0 ohms, respectively, are connected in parallel. What is the overall resistance of this combination?
   a. 0.58 ohms
   b. 1.71 ohms
   c. 7.00 ohms
   d. 21.0 ohms

6. Two resistors of values 6.00 and 12.00 ohms are connected in parallel. This combination in turn is hooked in series with a 4.00-ohm resistor and a 24.0-V battery. What is the current in the 6.00-ohm resistor?
   a. 2.00 A
   b. 3.00 A
   c. 6.00 A
   d. 12.0 A

7. Which two resistors are in parallel with each other?
   a. R and R_4
   b. R_2 and R_3
   c. R_2 and R_4
   d. R and R_1

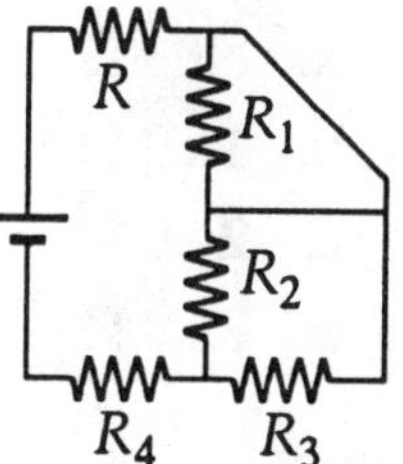

8. Which resistor in Question 7 above is in series with resistor R?
   a. R_1
   b. R_2
   c. R_3
   d. R_4

9. What is the current flowing through the 8.00-ohm resistor?

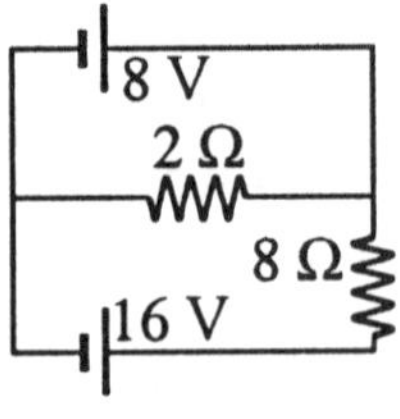

a. 1.00 A
b. 2.00 A
c. 3.00 A
d. 6.00 A

10. What is the equivalent resistance for these resistors?

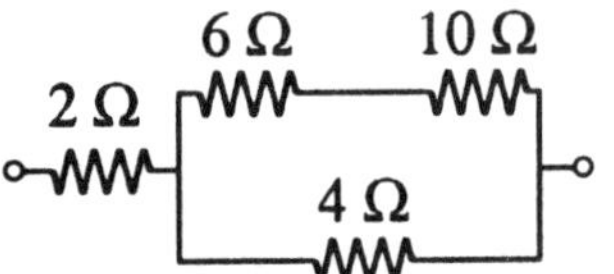

a. 2.30 ohms
b. 5.20 ohms
c. 13.0 ohms
d. 33.0 ohms

11. In a circuit, a current of 1.00 A is drawn from a battery. The current then divides and passes through two resistors in parallel. One of the resistors has a value of 64 Ω and the current through it is 0.20 A. What is the value of the other resistor?

a. 8.00 Ω
b. 16.0 Ω
c. 24.0 Ω
d. 32.0 Ω

12. What is the maximum number of 100-W light bulbs you can connect in parallel in a 120-volt home circuit without tripping the 20-amp circuit breaker?

a. 11
b. 17
c. 23
d. 29

Chapter 19

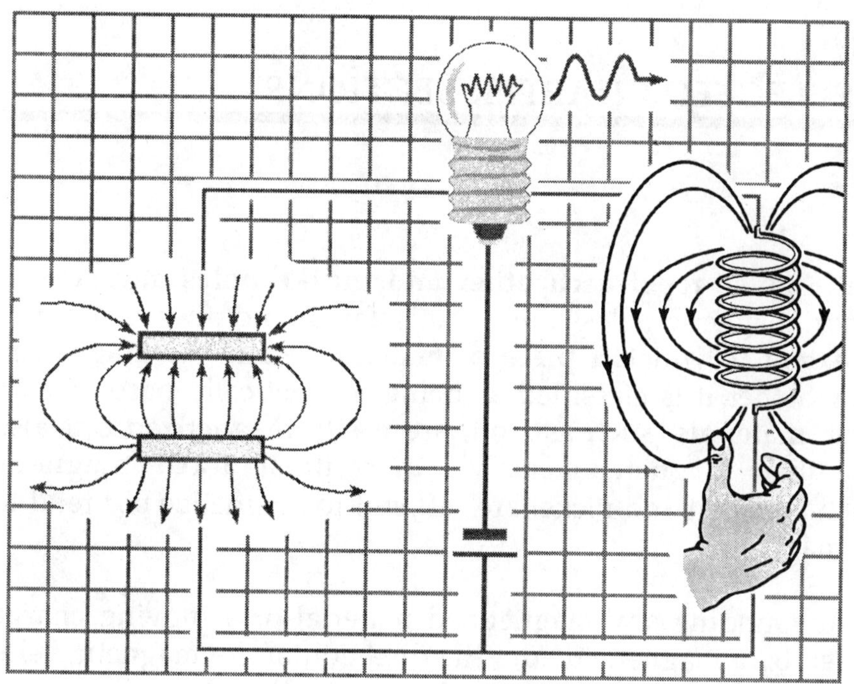

MAGNETISM

Chapter 19

MAGNETISM

As we investigate magnetism in this chapter, you will find that the subject cannot be divorced from electricity. For example, magnetic fields affect moving charges and moving charges produce magnetic fields. The ultimate source of all magnetic fields is electric current, whether it be the current in a wire or the current produced by the motion of charges within atoms or molecules.

NOTES FROM SELECTED CHAPTER SECTIONS

19.1 Magnets

Like magnetic poles repel each other and unlike poles attract each other. Further, magnetic poles cannot be isolated, and always occur in pairs. The extent to which a piece of material retains its magnetism depends on whether it is classified as being magnetically hard or soft. **Soft** magnetic materials, such as iron, are easily magnetized but also tend to lose their magnetism easily. In contrast, **hard** magnetic materials such as cobalt and nickel are difficult to magnetize but tend to retain their magnetism.

The region surrounding any magnetized material or a moving charge is characterized by a magnetic field. The direction of the magnetic field, **B**, at any location is the direction in which the north pole of a compass needle points at that location.

19.2 Magnetic Field of the Earth

The magnetic north pole corresponds to the south geographic pole, and the magnetic south pole corresponds to the north geographic pole. The angle between the direction of the earth's magnetic field and the horizontal is called the **dip angle.**

19.3 Magnetic Fields

Particles with charge q, moving with velocity **v** in a magnetic field **B**, experience a magnetic force **F**:

1. The magnetic force is proportional to the charge q and velocity **v** of the particle.
2. The magnitude and direction of the magnetic force depend on the velocity of the particle and on the magnitude and direction of the magnetic field.
3. When a charged particle moves in a direction **parallel** to the magnetic field vector, the magnetic force **F** on the charge is **zero**.
4. The magnetic force acts in a direction perpendicular to both **v** and **B**; that is, **F** is perpendicular to the plane formed by **v** and **B**.
5. The magnetic force on a positive charge is in the direction opposite to the force on a negative charge moving in the same direction.
6. If the velocity vector makes an angle θ with the magnetic field, the magnitude of the magnetic force is proportional to $\sin\theta$.

19.4 Magnetic Force on a Current-Carrying Conductor

The total magnetic force on a **closed** current loop in a **uniform** magnetic field is **zero.**

19.6 The Galvanometer and its Applications

The **galvanometer** is a device used in the construction of both ammeters and voltmeters. The basic operation of this instrument makes use of the fact that a **torque** acts on a current loop in the presence of a magnetic field.

In order to convert a galvanometer into an ammeter, a resistor (shunt) is placed in **parallel** with the galvanometer coil. For use as a voltmeter, a resistor is place in **series** with the galvanometer coil.

19.7 Motion of a Charged Particle in a Magnetic Field

When a charged particle moves in an external magnetic field, the work **done** by the magnetic force on the particle is **zero.** The magnetic force changes the direction of the velocity vector but does not change its magnitude.

When a charged particle enters an external magnetic field along a direction perpendicular to the field, the particle will move in a circular path in a plane perpendicular to the magnetic field.

19.8 Magnetic Field of a Long, Straight Wire and Ampère's Law

The direction of the magnetic field due to a current in a conductor is given by the right-hand rule:

> If the wire is grasped in the right hand with the thumb in the direction of the current, the fingers will wrap (or curl) in the direction of **B**.

Ampère's law is valid only for **steady** currents and is useful only in those cases where the current configuration has a **high degree** of **symmetry**.

19.9 Magnetic Force Between Two Parallel Conductors

Parallel conductors carrying currents in the **same direction attract** each other, whereas parallel conductors carrying currents in **opposite directions repel** each other.

The force between two parallel wires each carrying a current is used to define the ampere as follows:

> If two long, parallel wires 1.00 m apart carry the same current and the force per unit length on each wire is 2.00×10^{-7} N/m, then the current is defined to be 1.00 A.

EQUATIONS AND CONCEPTS

The magnitude of the magnetic field at a point in space is defined in terms of the magnetic force exerted on a moving electric charge. The magnetic force will be of maximum magnitude when the charge moves along a direction perpendicular to the direction of the magnetic field.

$$F_{max} = qvB \quad (19.4)$$

In general, the velocity vector may be directed along some direction other than 90.0° relative to the magnetic field. In this case, the magnetic force on the moving charge is less than its maximum value.

$$F = qvB\sin\theta \quad (19.1)$$

The SI unit of magnetic field intensity is the tesla (T).

$$1\ \mathrm{T} = 1\frac{\mathrm{N\cdot s}}{\mathrm{C\cdot m}}$$

Equation 19.1 can be written in a form which serves to define the magnitude of the magnetic field.

$$B \equiv \frac{F}{qv\sin\theta} \quad (19.2)$$

The cgs unit of magnetic field is the gauss (G).

$$1\ \mathrm{T} = 10^4\ \mathrm{G}$$

In order to determine the direction of the magnetic force, apply right-hand rule *A:* hold your open right hand with your fingers pointing in the direction of B and your thumb pointing in the direction of v. The magnetic force on a positive charge is then directed out of the palm of your hand. If the charge is negative, then the direction of the force is reversed.

Comment on the direction of the magnetic force on a moving charge.

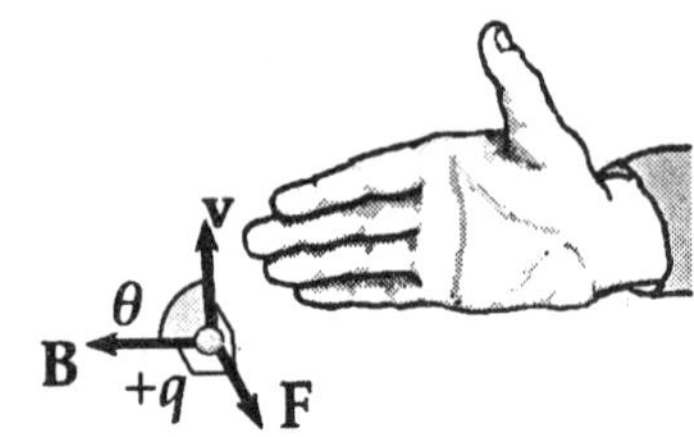

A magnetic force is exerted on a straight conductor placed in a magnetic field. The magnitude of the magnetic force on the conductor depends on the angle between the direction of the conductor and the direction of the field.

$$F = BI\ell \sin\theta \qquad (19.6)$$

The magnetic force will be maximum when the conductor is directed perpendicular to the magnetic field.

$$F_{max} = BI\ell \qquad (19.5)$$

In order to determine the direction of the magnetic force on a current-carrying conductor, use right-hand rule *A* with your thumb in the direction of conventional current.

Comment on the direction of the magnetic force on a current-carrying conductor.

When a closed conducting loop is placed in a uniform external magnetic field, a net torque will be exerted on the loop. The magnitude of the torque will depend on the angle between the direction of the magnetic field and the direction of the normal (or perpendicular) to the plane of the loop.

$$\tau = BIA\sin\theta \qquad (19.8)$$

The magnitude of the torque will be maximum when the magnetic field is parallel to the plane of the loop.

$$\tau_{\max} = BIA \qquad (19.7)$$

The direction of rotation of the loop is such that the normal to the plane of the loop turns into a direction parallel to the magnetic field.

Comment on the direction of rotation of the loop.

When a charged particle enters the region of a uniform magnetic field with the velocity of the particle initially perpendicular to the direction of the field, the particle will move in a circular path of constant radius. The circular path will have a radius which is proportional to the linear momentum of the moving charge and will be in a plane which is perpendicular to the direction of the magnetic field.

$$r = \frac{mv}{qB} \qquad (19.10)$$

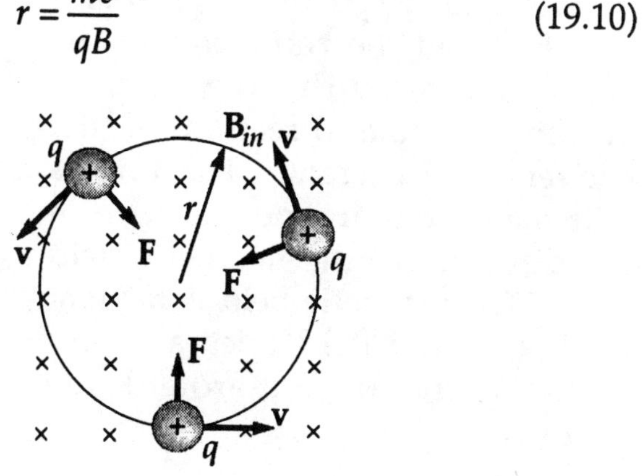

The period of revolution in the circular path is independent of the value of the radius.

$$T = \frac{2\pi r}{v} = \frac{2\pi m}{qB}$$

The magnetic field at a distance r from a long straight current-carrying conductor.

$$B = \frac{\mu_0 I}{2\pi r} \qquad (19.11)$$

The proportionality constant in Equation 19.11 is called the permeability of free space.

$$\mu_0 = 4\pi \times 10^{-7}\ \text{T}\cdot\text{m/A}$$

The magnetic field at the center of a current loop of radius R.

$$B = \frac{\mu_0 I}{2R}$$

The magnetic field inside a solenoid which has n turns of conductor per unit length.

$$B = \mu_0 n I \qquad (19.15)$$

The direction of the magnetic field due to current in a long wire is determined by using right-hand rule B: hold the conductor in the right hand with the thumb pointing in the direction of the conventional current. The fingers will then wrap around the wire in the direction of the magnetic field lines. The magnetic field is tangent to the circular field lines at every point in the region around the conductor.

Comment on the direction of the magnetic field due to different current geometries.

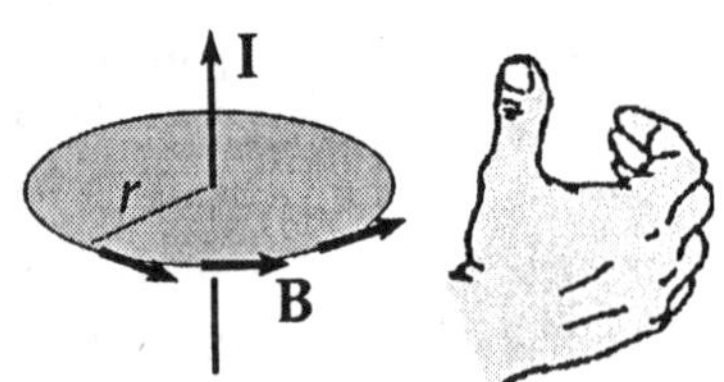

The direction of the magnetic field at the center of a current loop is perpendicular to the plane of the loop and directed in the sense given right-hand rule *B*.

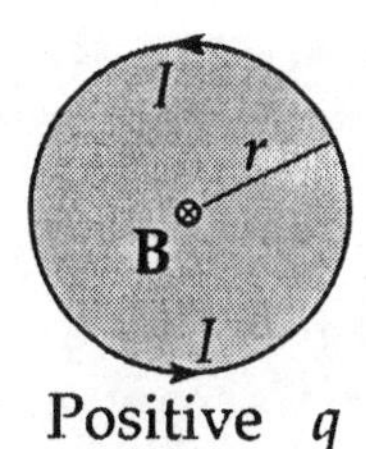

Positive q

Within a solenoid, the magnetic field is parallel to the axis of the solenoid and pointing in a sense determined by applying right-hand rule *B* to one of the coils.

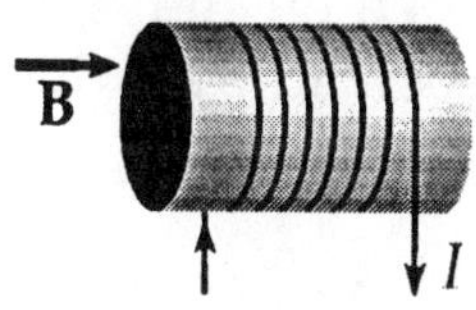

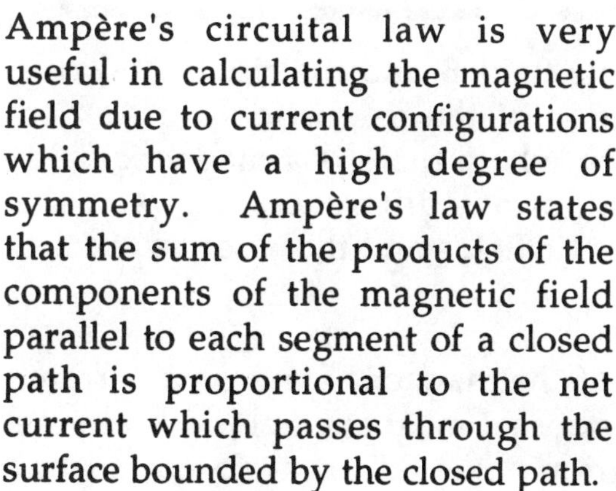
Ampère's circuital law is very useful in calculating the magnetic field due to current configurations which have a high degree of symmetry. Ampère's law states that the sum of the products of the components of the magnetic field parallel to each segment of a closed path is proportional to the net current which passes through the surface bounded by the closed path.

$$\sum B_{\parallel} \, \Delta \ell = \mu_0 I \qquad (19.13)$$

The magnitude of the magnetic force per unit length between parallel conductors is proportional to the product of the two currents and inversely proportional to the distance between the two conductors.

$$\frac{F_1}{\ell} = \frac{\mu_0 I_1 I_2}{2\pi d} \qquad (19.14)$$

Parallel currents in the same direction attract each other and parallel currents in opposite directions repel each other.

Comment on the direction of magnetic force between parallel conductors.

SUGGESTIONS, SKILLS, AND STRATEGIES

To remember the symbols for vectors that point away from you and towards you, think of a three-dimensional archery arrow as it turns to point first away from you (shown on the left), and then towards you, (shown on the right). The point on the arrow is represented by a dot; the feathers are represented by an 'x'.

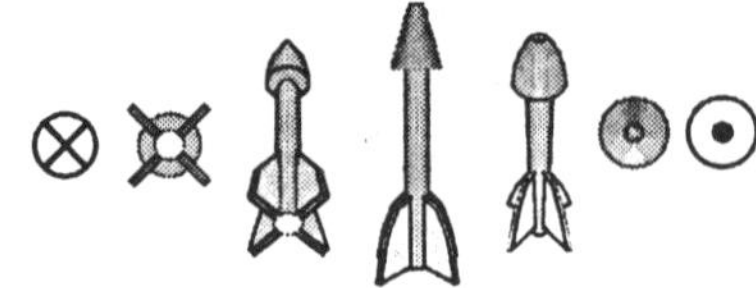

REVIEW CHECKLIST

▷ Use the defining equation for a magnetic field **B** and the right-hand rule to determine the magnitude and direction of the magnetic force exerted on an electric charge moving in a region where there is a magnetic field. You should understand clearly the important differences between the forces exerted on electric charges by electric fields and those forces exerted on moving charges by magnetic fields.

▷ Calculate the magnitude and direction of the magnetic force on a current-carrying conductor when placed in an external magnetic field. Determine the magnitude and direction of the torque exerted on a closed current loop in an external magnetic field.

▷ Describe how a moving coil galvanometer can be converted to either an ammeter or a voltmeter.

▷ Calculate the period and radius of the circular orbit of a charged particle moving in a uniform magnetic field.

▷ Calculate the magnitude and determine the direction of the magnetic field for the following cases: a point in the vicinity of a long, straight current-carrying conductor at the center of a current loop, and at interior points of a solenoid. Correctly apply the right-hand rule for these situations.

SOLUTIONS TO SELECTED END-OF-CHAPTER PROBLEMS

7. An electron is accelerated through 2400 V from rest and then enters a region where there is a uniform 1.70-T magnetic field. What are the (a) maximum and (b) minimum magnitudes of the magnetic force this charge can experience?

Solution When a particle with charge q is accelerated from rest through a potential difference ΔV, the gain of kinetic energy equals the loss in potential energy, or $\Delta KE = -\Delta PE = |q|\Delta V$.

For an electron that starts from rest, this becomes $\frac{1}{2}m_e v^2 - 0 = e\Delta V$

Given $\Delta V = 2400$ V, $v = \sqrt{\dfrac{2e\Delta V}{m_e}} = \sqrt{\dfrac{2\left(1.60\times10^{-19}\text{ C}\right)(2400\text{ V})}{9.11\times10^{-31}\text{ kg}}} = 2.90\times10^{7}$ m/s

A charged particle moving through a magnetic field experiences a force of magnitude $F = B|q|v\sin\theta$, where B is the magnetic field strength and θ is the angle between the direction of the magnetic field and the velocity of the particle. For an electron with speed $v = 2.90\times10^{7}$ m/s in a field of strength $B = 1.70$ T, this force is $F = Bev\sin\theta$, or

$$F = (1.70\text{ T})\left(1.60\times10^{-19}\text{ C}\right)\left(2.90\times10^{7}\text{ m/s}\right)\sin\theta = \left(7.89\times10^{-12}\text{ N}\right)\sin\theta$$

(a) The maximum force occurs when the particle moves perpendicular to the field and $\sin\theta = 1.00$. Then, $F = F_{max} = 7.89\times10^{-12}$ N ◊

(b) The minimum magnitude of the force occurs when the particle moves either parallel or anti-parallel to the field, so $\theta = 0°$ or $\theta = 180°$ and $\sin\theta = 0$

In that case, $F = F_{min} = 0$ ◊

11. A current, I = 15 A, is directed along the positive x axis and perpendicularly to a magnetic field. The conductor experiences a magnetic force per unit length of 0.12 N/m in the negative y direction. Calculate the magnitude and direction of the magnetic field in the region through which the current passes.

Solution

The magnetic force exerted on a straight conductor of length ℓ carrying a current I through a magnetic field of strength B is $F = BI\ell \sin\theta$. Here, θ is the angle between the directions of the current and the field. The direction of the force is given by the right-hand rule. Hold your right hand flat with the palm facing downward (the direction of the force on the conductor) and your thumb pointing in the direction of the current (i.e., to the right or in the $+x$-direction).

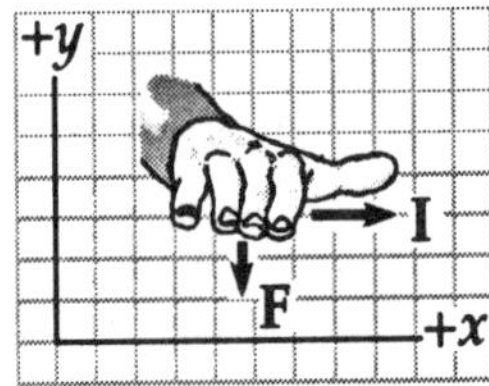

Then, your fingers are pointing in the direction of the field. This should be out of the page (toward you), or in the positive z direction. ◊

Since I is perpendicular to the field $(\sin\theta = 1.0)$, the magnitude of the field is

$$B = \frac{F/\ell}{I\sin\theta} = \frac{0.12\ \text{N/m}}{(15\ \text{A})(1.0)} = 8.0\times10^{-3}\ \text{T} \qquad ◊$$

17. An unusual message delivery system is pictured in Figure P19.17. A 15-cm length of conductor that is free to move is held in place between two thin conductors. When a 5.0-A current is directed as shown in the figure, the wire segment moves upward at a constant velocity. If the mass of the wire is 15 g, find the magnitude and direction of the minimum magnetic field that is required to move the wire. (The wire slides without friction on the two vertical conductors.)

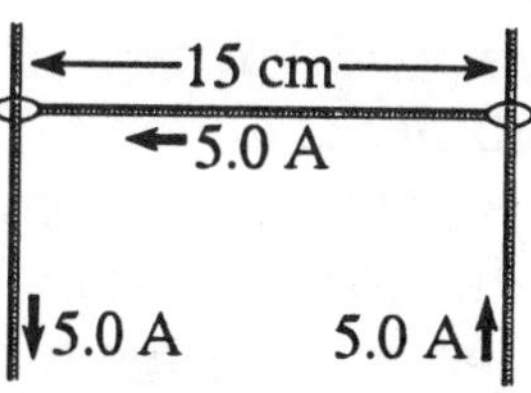

Figure P19.17

Solution For the wire to move upward at constant velocity, the net force acting on it must be zero. Thus, the magnetic force must be directed upward and have a magnitude equal to the weight of the wire, $F = w = mg$. In general, the magnitude of the magnetic force acting on a current carrying conductor is $F = BI\ell \sin\theta$, where B is the magnitude of the field, I is the current in the conductor, ℓ is the conductor's length, and θ is the angle between the directions of the current and the magnetic field. Therefore, to move the wire at constant velocity, the magnitude of the field needed is

$$B = \frac{mg}{I\ell \sin\theta}$$

If B is to be a minimum, it is necessary for $\sin\theta$ to have its maximum value (1.0). Thus,

$$B_{min} = \frac{mg}{I\ell} = \frac{\left(15\times10^{-3}\ \text{kg}\right)\left(9.8\ \text{m/s}^2\right)}{(5.0\ \text{A})\left(15\times10^{-2}\ \text{m}\right)} = 0.20\ \text{T} \qquad \lozenge$$

To find the direction of this minimum magnitude field, realize that if $\sin\theta = 1.00$, then $\theta = 90°$ and the field is perpendicular to the current. Using Figure P19.17 and the right-hand rule, hold your right hand flat with the palm upward (the required direction of the magnetic force), and the thumb pointing to the left (the direction of the current). You should find that your fingers are pointing out of the page toward you. This is the required direction of the magnetic field. ◊

21. A rectangular loop consists of 100 closely wrapped turns and has dimensions 0.40 m by 0.30 m. The loop is hinged along the y axis, and the plane of the coil makes an angle of 30.0° with the x axis (Fig. P19.21). What is the magnitude of the torque exerted on the loop by a uniform magnetic field of 0.80 T directed along the x axis, when the current in the windings has a value of 1.2 A in the direction shown? What is the expected direction of rotation of the loop?

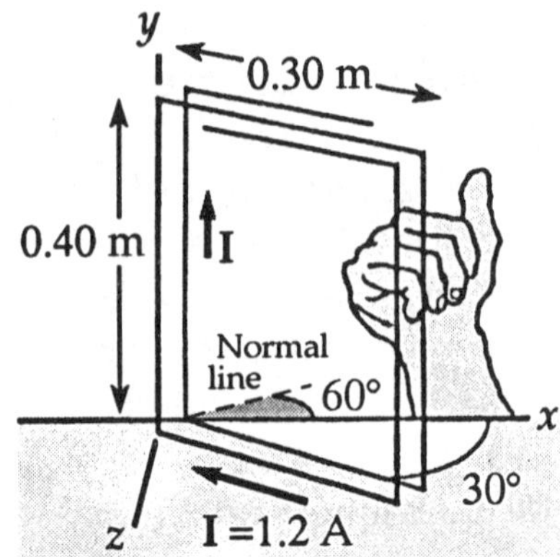

Figure P19.21

Solution

The torque that a magnetic field exerts on a current loop has a magnitude of $\tau = NBIA\sin\theta$, where N is the number of turns on the loop, B is the magnitude of the field, I is the current, A is the area enclosed by the loop, and θ is the angle between the line perpendicular to the plane of the loop (i.e., the normal line) and the direction of the field. Note from the figure that when the plane of the coil makes a 30° angle with the direction of the field (positive x direction), the angle between the normal line and field direction is $\theta = 60°$.

The area enclosed by this current loop is $A = (0.40 \text{ m})(0.30 \text{ m}) = 0.12 \text{ m}^2$. Thus, the magnitude of the torque exerted on this loop by a uniform magnetic field of 0.80 T is

$$\tau = (100)(0.80 \text{ T})(1.2 \text{ A})\left(0.12 \text{ m}^2\right)\sin(60°) = 10 \text{ N}\cdot\text{m} \qquad \Diamond$$

The torque exerted on a current loop by a magnetic field tends to rotate the loop so the line normal to the plane of the loop is parallel to the direction of the field ($\tau \to 0$ when $\theta = 0$).

Therefore, an observer located above the loop and looking in the negative y direction, will see the loop rotate in a clockwise direction. ◊

25. A galvanometer has a resistance of 50.0 Ω and deflects full-scale when the voltage across it is 50.0 mV. What is the magnitude of the shunt resistance needed to convert this galvanometer into an ammeter that reads 10.0 A at full-scale deflection?

Solution

To convert a galvanometer into an ammeter, a small resistance is connected in parallel with the galvanometer as shown in the sketch.

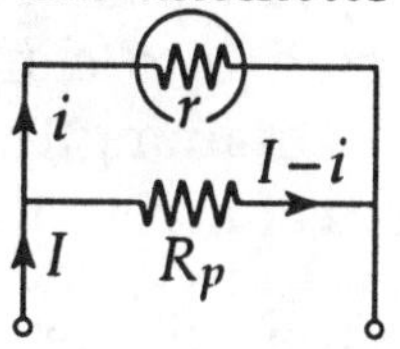

Galvanometer

If the galvanometer has a resistance $r = 50.0\ \Omega$ and a potential difference of 50.0 mV must exist across the galvanometer for full-scale deflection, the galvanometer current required for full-scale deflection is

$$i = \frac{\Delta V}{r} = \frac{50.0\ \text{mV}}{50.0\ \Omega} = 1.00\ \text{mA} = 1.00 \times 10^{-3}\ \text{A}$$

Therefore, when the total current is $I = 10.0\ \text{A}$ and full-scale deflection occurs, the current in the shunt resistor is $I - i = 10.0\ \text{A} - 0.001\ \text{A}$.

Since the galvanometer and the shunt resistor are connected in parallel, the same potential difference always exists across these two elements. Hence, $\Delta V = ir = (I - i)R_p$ and the magnitude of the required shunt resistance is

$$R_p = \frac{ir}{(I - i)} = \frac{(0.001\ \text{A})(50.0\ \Omega)}{10.0\ \text{A} - 0.001\ \text{A}} = 5.00 \times 10^{-3}\ \Omega$$ ◊

29. The galvanometer of Problem 28 is to be converted to a multirange voltmeter using the circuit shown in Figure P19.29. Find the values of R_1, R_2, and R_3 that will enable the meter to give the full-scale readings in the figure.

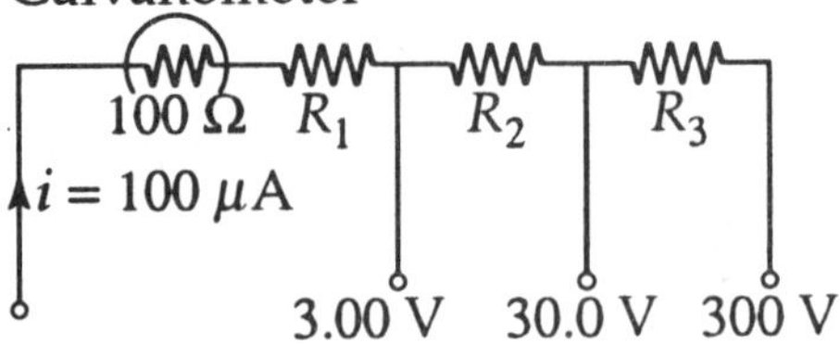

Figure P19.29

Solution The galvanometer of Problem 28 and shown in the figure has an internal resistance of $r = 100\ \Omega$ and deflects full-scale when a current of $i = 100\ \mu\text{A} = 1.00 \times 10^{-4}$ A flows through it. The galvanometer is to be converted into a multi-range voltmeter with the leftmost terminal and one of the remaining three terminals (chosen according to the desired range) used to connect across the potential difference being measured.

When the voltmeter is deflecting full-scale on the 3.00 V range, the current $i = 100\ \mu$A flows through the internal resistance (100 Ω) and through the added resistance R_1 before exiting at the 3 V terminal. The total potential drop between these two terminals is $i(r + R_1) = 3.00$ V and the value of R_1 is:

$$R_1 = \frac{3.00\text{ V}}{i} - r = \frac{3.00\text{ V}}{1.00 \times 10^{-4}\text{ A}} - 100\ \Omega = 2.99 \times 10^4\ \Omega \qquad \Diamond$$

When a 30.0 V potential difference produces a full-scale deflection, the current i passes through the series combination of r, R_1 and R_2 before exiting at the 30 V terminal. Therefore, $i(r + R_1 + R_2) = 30.0$ V and the needed value of R_2 is , or

$$R_2 = \frac{30.0\text{ V}}{i} - (r + R_1) = \frac{30.0\text{ V}}{1.00 \times 10^{-4}\text{ A}} - \left(100\ \Omega + 2.99 \times 10^4\ \Omega\right) = 2.70 \times 10^5\ \Omega \quad \Diamond$$

If full-scale deflection occurs when the voltmeter is connected (using the leftmost and the rightmost terminals) across a 300-V potential difference, current i passes through all four resistances connected in series. Thus,

$$i(r + R_1 + R_2 + R_3) = 300\text{ V} \quad \text{or} \quad R_3 = \frac{300\text{ V}}{i} - (r + R_1 + R_2)$$

The required value for the remaining resistance is then

$$R_3 = \frac{300\text{ V}}{1.00\times10^{-4}\text{ A}} - \left(100\ \Omega + 2.99\times10^{4}\ \Omega + 2.70\times10^{5}\ \Omega\right) = 2.70\times10^{6}\ \Omega \quad \Diamond$$

---

33. A singly charged positive ion has a mass of 2.50×10^{-26} kg. After being accelerated through a potential difference of 250 V, the ion enters a magnetic field of 0.500 T, in a direction perpendicular to the field. Calculate the radius of the path of the ion in the field.

Solution Since the ion moves perpendicularly to the magnetic field, the magnitude of the magnetic force exerted on it is $F = qvB\sin\theta = qvB$, where q is the charge of the ion, v is its speed, and B is the magnitude of the field. A singly charged positive ion has a charge of $q = +e$. The kinetic energy of the ion is the same as the potential energy that it lost, or

$$\frac{1}{2}mv^2 = e\Delta V \quad \text{and} \quad v = \sqrt{\frac{2e\Delta V}{m}} = \sqrt{\frac{2\left(1.60\times10^{-19}\text{ C}\right)(250\text{ V})}{2.50\times10^{-26}\text{ kg}}} = 5.66\times10^{4}\text{ m/s}$$

Thus, the magnitude of the force exerted on the ion by a 0.500 T magnetic field is

$$F = \left(1.60\times10^{-19}\text{ C}\right)\left(5.66\times10^{4}\text{ m/s}\right)(0.500\text{ T}) = 4.53\times10^{-15}\text{ N}$$

This force supplies the centripetal force needed to hold the ion in its circular path. Therefore, $F = mv^2/r$ and the radius of the ion's path is

$$r = \frac{mv^2}{F} = \frac{\left(2.50\times10^{-26}\text{ kg}\right)\left(5.66\times10^{4}\text{ m/s}\right)^2}{4.53\times10^{-15}\text{ N}} = 1.77\times10^{-2}\text{ m} = 1.77\text{ cm} \quad \Diamond$$

---

39. The two wires in Figure P19.39 carry currents of 3.00 A and 5.00 A in the direction indicated. (a) Find the direction and magnitude of the magnetic field at a point midway between the wires. (b) Find the magnitude and direction of the magnetic field at point *P*, located 20.0 cm above the wire carrying the 5.00-A current.

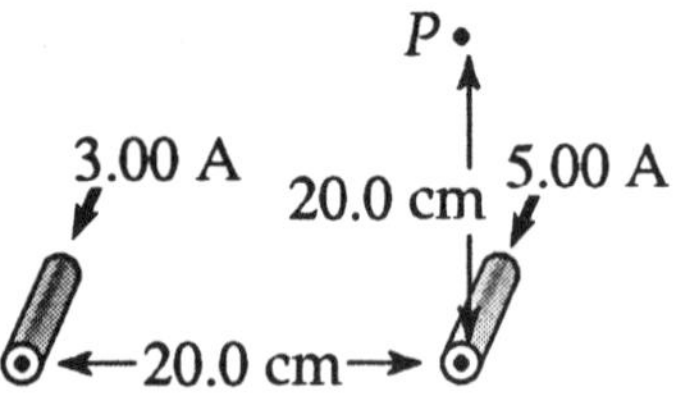

Figure P19.39

Solution (a) The point midway between the two wires is located at distances of $r_1 = r_2 = 10.0\text{ cm} = 0.100\text{ m}$ from each wire. At this location, the magnetic field produced by the leftmost wire has a magnitude of

$$B_1 = \frac{\mu_0 I_1}{2\pi r_1} = \frac{\left(4\pi\times10^{-7}\text{ T}\cdot\text{m/A}\right)(3.00\text{ A})}{2\pi(0.100\text{ m})} = 6.00\times10^{-6}\text{ T}$$

and is directed vertically upward. The magnetic field produced by the 5.00-A current in the rightmost wire is directed straight downward and has a magnitude of

$$B_2 = \frac{\mu_0 I_2}{2\pi r_2} = \frac{\left(4\pi\times10^{-7}\text{ T}\cdot\text{m/A}\right)(5.00\text{ A})}{2\pi(0.100\text{ m})} = 1.00\times10^{-5}\text{ T}$$

The resultant magnetic field at this location is then

$$B_{\text{net}} = B_2 - B_1 = 4.00\times10^{-6}\text{ T downward.}$$ ◊

(b) The distance from the leftmost wire to point *P* is $r_1 = (0.200\text{ m})\sqrt{2} = 0.283\text{ m}$. At point *P* the magnetic field produced by the leftmost wire is perpendicular to the line connecting this wire and point *P* (it is directed at 45° above the horizontal as shown in the sketch), and has a magnitude of

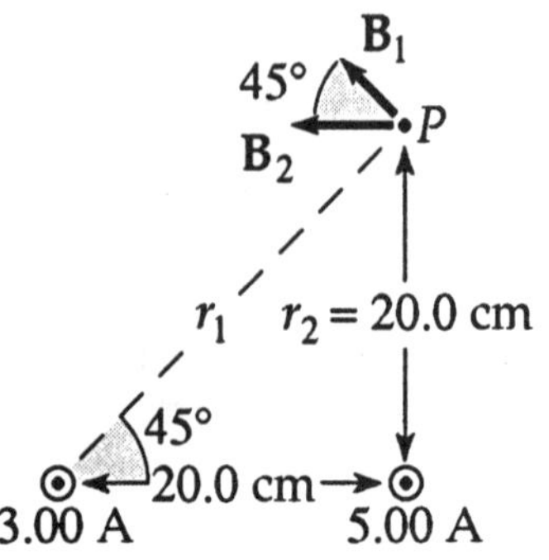

$$B_1 = \frac{\mu_0 I_1}{2\pi r_1} = \frac{\left(4\pi\times10^{-7}\ \text{T}\cdot\text{m/A}\right)(3.00\ \text{A})}{2\pi(0.283\ \text{m})} = 2.12\times10^{-6}\ \text{T}$$

The magnetic field produced by the rightmost wire at point P is horizontal and toward the left. Its magnitude is

$$B_2 = \frac{\mu_0 I_2}{2\pi r_2} = \frac{\left(4\pi\times10^{-7}\ \text{T}\cdot\text{m/A}\right)(5.00\ \text{A})}{2\pi(0.200\ \text{m})} = 5.00\times10^{-6}\ \text{T}$$

The resultant field at P is the vector sum of B_1 and B_2. The y component of the resultant field is

$$B_y = +B_1 \sin 45° + 0 = \left(2.12\times10^{-6}\ \text{T}\right)\frac{\sqrt{2}}{2} = 1.50\times10^{-6}\ \text{T}$$

and the x component is $B_x = B_{1x} + B_{2x} = -B_1 \cos 45° - B_2$, or

$$B_x = -\left(2.12\times10^{-6}\ \text{T}\right)\frac{\sqrt{2}}{2} - 5.00\times10^{-6}\ \text{T} = -6.50\times10^{-6}\ \text{T}$$

Thus, the magnitude of the resultant field is:

$$B = \sqrt{B_x^{\,2} + B_y^{\,2}} = \sqrt{\left(-6.50\times10^{-6}\ \text{T}\right)^2 + \left(1.50\times10^{-6}\ \text{T}\right)^2} = 6.67\times10^{-6}\ \text{T} \quad \lozenge$$

This resultant field at point P is directed at an angle of

$$\theta = \arctan\left(\frac{|B_x|}{B_y}\right) = \arctan\left(\frac{6.50\times10^{-6}\ \text{T}}{1.50\times10^{-6}\ \text{T}}\right) = 77.0° \quad \text{to the left of the vertical} \ \lozenge$$

---

43. A wire with a weight per unit length of 0.080 N/m is suspended directly above a second wire. The top wire carries a current of 30.0 A and the bottom wire carries a current of 60.0 A. Find the distance of separation between the wires so that the top wire will be held in place by magnetic repulsion.

Solution In the sketch at the right, it is assumed that the current in the lower wire (wire 2) is coming out of the page toward the reader. Then, the magnetic field lines due to the current in wire 2 are circular and centered this wire. At the location of wire 1, the magnetic field due to the current in wire 2 is horizontal and toward the left with a magnitude of $B_2 = \mu_0 I_2 / 2\pi d$. Here, $I_2 = 60.0$ A is the current in wire 2 and d is the distance separating the two wires.

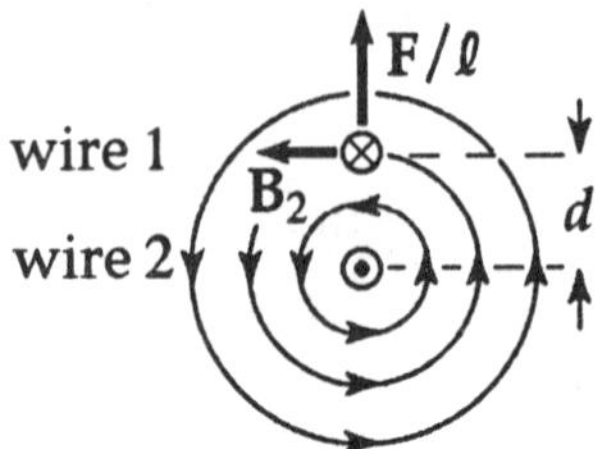

If wire 1 is to be suspended in the magnetic field B_2, the magnetic force per unit length, F/ℓ, exerted on it by the field must be directed upward and its magnitude must equal the weight per unit length of wire 1 (i.e. $F/\ell = 0.080$ N/m. Note that if this force is to be directed upward as needed, the current in wire 1 must flow into the page as indicted in the sketch. The magnetic force exerted on a conductor of length ℓ carrying current I_1 perpendicularly to a field of magnitude B_2 is $F = B_2 I_1 \ell$. Thus, the magnetic force per unit length acting on wire 1 is

$$F/\ell = B_2 I_1 = \left(\frac{\mu_0 I_2}{2\pi d}\right) I_1$$

Therefore, if wire 1 is to be suspended, it is necessary that $\mu_0 I_2 I_1 / 2\pi d = 0.080$ N/m or the distance between the wires must be

$$d = \frac{\mu_0 I_2 I_1}{2\pi(0.080 \text{ N/m})} = \frac{\left(4\pi \times 10^{-7} \text{ N/A}^2\right)(60.0 \text{ A})(30.0 \text{ A})}{2\pi(0.080 \text{ N/m})} = 4.50 \times 10^{-3} \text{ m} \quad \Diamond$$

47. An electron moves at a speed of 1.0×10^4 m/s in a circular path of radius 2.0 cm inside a solenoid. The magnetic field of the solenoid is perpendicular to the plane of the electron's path. Find (a) the strength of the magnetic field inside the solenoid and (b) the current in the solenoid if it has 25 turns per centimeter.

Solution

(a) The electron is always moving perpendicular to the direction of the magnetic field. Therefore, the magnetic force exerted on it has a magnitude of $F = evB$. This force is directed toward the center of the path and produces the centripetal acceleration needed to keep the electron moving in a circle. Thus, $mv^2/r = F = evB$ and the required magnetic field strength is:

$$B = \frac{mv}{er} = \frac{(9.11\times10^{-31}\text{ kg})(1.0\times10^{4}\text{ m/s})}{(1.60\times10^{-19}\text{ C})(2.0\times10^{-2}\text{ m})} = 2.8\times10^{-6}\text{ T} \qquad \Diamond$$

(b) The magnetic field inside a solenoid is $B = \mu_0 nI$ where $n = N/\ell$ is the number of turns per unit length on the solenoid and I is the current in the windings. In this case, $B = 2.8\times10^{-6}$ T as found in part (a), and

$$n = \frac{25}{1.0\text{ cm}} = \frac{25}{0.010\text{ m}} = 2500\text{ turns per meter}$$

The current must then be

$$I = \frac{B}{\mu_0 n} = \frac{2.8\times10^{-6}\text{ T}}{(4\pi\times10^{-7}\text{ T}\cdot\text{m/A})(2500\text{ m}^{-1})} = 9.1\times10^{-4}\text{ A} = 0.91\text{ mA} \qquad \Diamond$$

51. Two species of singly charged positive ions of masses 20.0×10^{-27} kg and 23.4×10^{-27} kg enter a magnetic field at the same location with a speed of 1.00×10^5 m/s. If the strength of the field is 0.200 T, and they move perpendicularly to the field, find their distance of separation after they complete one half of their circular path.

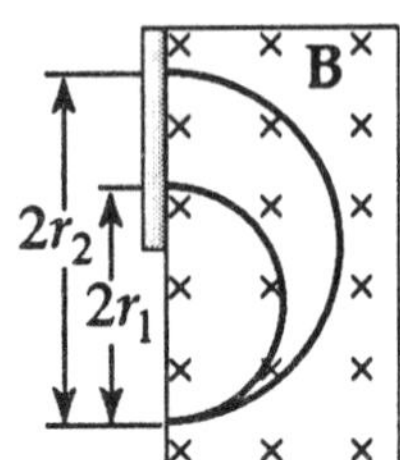

Solution When a charged particle moves perpendicularly to a magnetic field, a magnetic force that is always perpendicular to the velocity of the particle acts on it. This force, of magnitude $F = qvB$, produces a centripetal acceleration and causes the particle to follow a circular path. The force required to make a particle with mass m and speed v to move in a circle of radius r is given by $F = ma_c = mv^2/r$.

Thus, $qvB = mv^2/r$, and the radius of the path is $r = mv/qB$. Both species of ions are positive and singly charged $(q = +e)$, and both move at a speed of $v = 1.00 \times 10^5$ m/s. The radii of the paths of the two types of ions are:

$$r_1 = \frac{m_1 v}{qB} = \frac{\left(20.0 \times 10^{-27}\ \text{kg}\right)\left(1.00 \times 10^5\ \text{m/s}\right)}{\left(1.60 \times 10^{-19}\ \text{C}\right)(0.200\ \text{T})} = 6.25 \times 10^{-2}\ \text{m} = 6.25\ \text{cm}$$

and $$r_2 = \frac{m_2 v}{qB} = \frac{\left(23.4 \times 10^{-27}\ \text{kg}\right)\left(1.00 \times 10^5\ \text{m/s}\right)}{\left(1.60 \times 10^{-19}\ \text{C}\right)(0.200\ \text{T})} = 7.31 \times 10^{-2}\ \text{m} = 7.31\ \text{cm}$$

After completing one half of their circular paths, the two ions are separated by a distance equal to the difference in the diameters of their paths (see the sketch). Thus,

$$\Delta d = 2(r_2 - r_1) = 2(7.31\ \text{cm} - 6.25\ \text{cm}) = 2.12\ \text{cm}$$ ◊

57. A straight wire of mass 10.0 g and length 5.0 cm is suspended from two identical springs that, in turn, form a closed circuit (Fig. P19.57). The springs stretch a distance of 0.50 cm under the weight of the wire. The circuit has a total resistance of 12 Ω. When a magnetic field is turned on, directed out of the page (indicated by the dots in Fig. P19.57), the springs are observed to stretch an additional 0.30 cm. What is the strength of the magnetic field? (The upper portion of the circuit is fixed.)

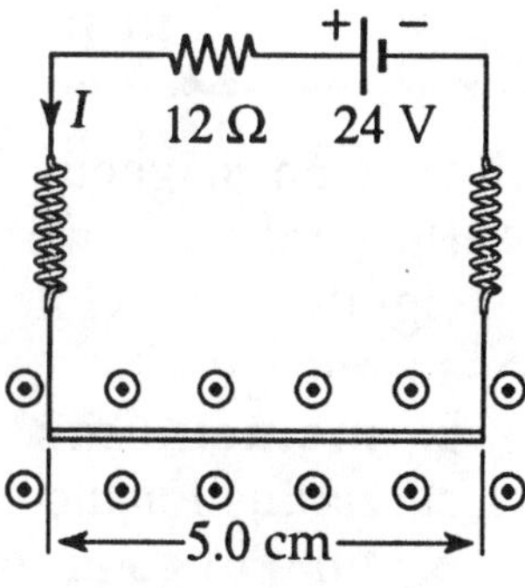

Figure P19.57

Solution The weight of the wire forming the bottom of the current loop is

$$w = mg = \left(10\times 10^{-3}\text{ kg}\right)\left(9.8\text{ m/s}^2\right) = 0.098\text{ N}$$

This force acting alone stretches the two identical springs a distance of 0.50 cm. Thus $mg = 2F_S = 2(kx)$ and the force constant of each spring is

$$k = \frac{mg}{2x} = \frac{0.098\text{ N}}{2\left(0.50\times 10^{-2}\text{ m}\right)} = 9.8\text{ N/m}$$

When the magnetic field is turned on, the magnetic force exerted on the wire is directed downward and adds to the weight of the wire. These combined forces stretch the springs an additional 0.30 cm (for a total elongation of the springs of 0.80 cm). The sum of the tensions in the springs must equal the magnetic force plus the weight, or $2F_S = 2(kx) = F + mg$ and the magnitude of the magnetic force is

$$F = 2(kx) - mg = 2(9.8\text{ N/m})\left(0.80\times 10^{-2}\text{ m}\right) - 0.098\text{ N} = 5.9\times 10^{-2}\text{ N}$$

The current in the circuit is $I = V/R = 24\text{ V}/12\ \Omega = 2.0\text{ A}$, and the magnetic force exerted on the wire of length $\ell = 5.0$ cm at the bottom of the current loop is $F = BI\ell$. The strength of the magnetic field must then be

$$B = \frac{F}{I\ell} = \frac{5.9\times 10^{-2}\text{ N}}{(2.0\text{ A})\left(5.0\times 10^{-2}\text{ m}\right)} = 0.59\text{ T}$$ ◊

CHAPTER SELF-QUIZ

1. When a magnetic field causes a charged particle to move in a circular path, the only quantity listed below which the magnetic force changes significantly while the particle goes around in a circle is the particle's
   a. energy
   b. momentum
   c. radius for the circle
   d. time to go around the circle once

2. A 10.0-ohm, 25.0-mA galvanometer is to be converted into a voltmeter which reads 20.0 V at full-scale deflection. What resistance should be placed in series with the galvanometer coil?
   a. 810 ohms
   b. 790 ohms
   c. 500 ohms
   d. 450 ohms

3. A 10.0-ohm, 15.0-mA galvanometer is converted into an ammeter by placing a 1.50×10^{-2} ohm shunt in parallel with the galvanometer coil. What is the maximum scale reading on the galvanometer?
   a. 25.0 A
   b. 15.0 A
   c. 10.0 A
   d. 5.00 A

4. A circular loop carrying a current of 2.00 A is oriented in a magnetic field of 3.50 T. The loop has an area of 0.12 m^2 and is mounted on an axis, perpendicular to the magnetic field, which allows the loop to rotate. What is the torque on the loop when its plane is oriented at a 37.0° angle to the field?
   a. 46.0 N·m
   b. 0.67 N·m
   c. 0.51 N·m
   d. 0.10 N·m

5. A current-carrying wire of length 0.300 m is positioned perpendicular to a uniform magnetic field. If the current is 5.00 A and it is determined that there is a resultant force of 2.25 N on the wire due to the interaction of the current and field, what is the magnetic field strength?
   a. 0.60 T
   b. 1.50 T
   c. 1.85×10^3 T
   d. 6.70×10^3 T

6. A proton, mass 1.67×10^{-27} kg and charge $+1.60 \times 10^{-19}$ C, moves in a circular orbit perpendicular to a uniform magnetic field of 0.75 T. Find the time for the proton to make one complete circular orbit.
   a. 4.30×10^{-8} s
   b. 8.70×10^{-8} s
   c. 4.90×10^{-7} s
   d. 9.80×10^{-7} s

7. An electron moves through a region of crossed electric and magnetic fields. The electric field **E** = 1000 V/m and is directed straight down. The magnetic field **B** = 0.40 T and is directed to the left. For what velocity **v** of the electron into the paper (perpendicular to the plane) will the electric force exactly cancel the magnetic force?
   a. 2500 m/s
   b. 4000 m/s
   c. 5000 m/s
   d. 8000 m/s

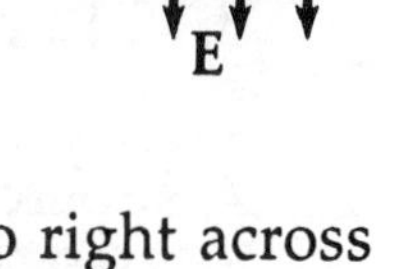

8. A proton is released such that its initial velocity is from left to right across this page. The proton's path, however, is deflected in a direction toward the bottom edge of the page due to the presence of a uniform magnetic field. What is the direction of this field?
   a. out of the page
   b. into the page
   c. from bottom edge to top edge of the page
   d. from right to left across the page

9. If a proton is released at the equator and falls toward the Earth under the influence of gravity, the magnetic force on the proton will be toward the
   a. north
   b. south
   c. east
   d. west

10. Two parallel cables of a high-voltage transmission line carry equal and opposite currents of 1500 A. The distance between the cables is 3.00 m. What is the magnetic force acting on a 40.0-meter length of each cable?
    a. 6 N
    b. 60 N
    c. 600 N
    d. 6000 N

11. A square loop (L = 0.200 m) consists of 50 closely-wrapped turns which each carry a current of 0.500 A. The loop is oriented as shown in a uniform magnetic field of 0.40 T directed in the positive y direction. What is the magnitude of the torque on the loop?
    a. 0.21 N·m
    b. 0.20 N·m
    c. 0.35 N·m
    d. 0.12 N·m

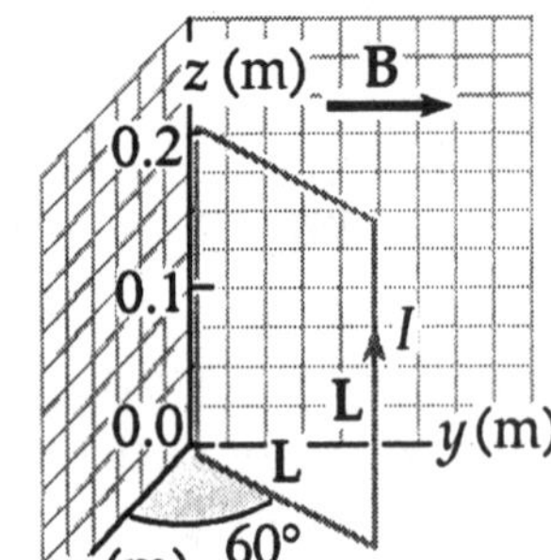

12. A superconducting solenoid is to be designed to generate a magnetic field of 10.0 T. If the solenoid winding has 2000 turns/meter, what is the required current? ($\mu_0 = 4\pi \times 10^{-7}$ A·m/T)
    a. 1000 A
    b. 1990 A
    c. 3980 A
    d. 5000 A

Chapter 20

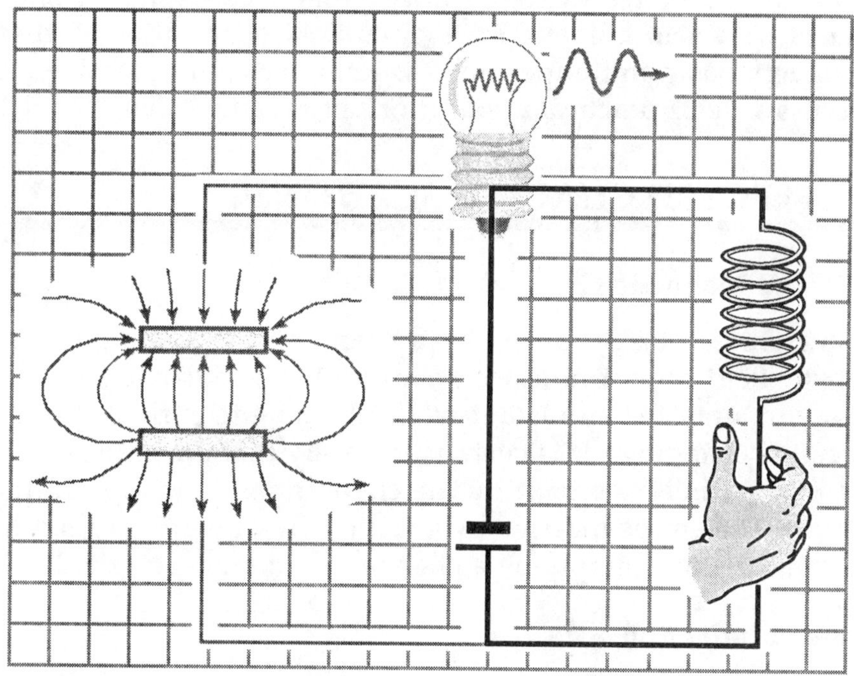

INDUCED VOLTAGES AND INDUCTANCE

Chapter 20

INDUCED VOLTAGES AND INDUCTANCE

Oersted (1819) discovered that a magnetic compass experiences a force in the vicinity of an electric current; this was the first evidence of a link between electricity and magnetism. Because nature is often symmetric, the discovery that electric currents produce magnetic fields led scientists to suspect that magnetic fields could produce electric currents. Indeed, experiments conducted by Michael Faraday in England and, independently, Joseph Henry in the United States in 1831 showed that a changing magnetic field could induce an electric current in a circuit. The results of these experiments led to a very basic and important law known as Faraday's law. In this chapter we discuss several practical applications of Faraday's law.

NOTES FROM SELECTED CHAPTER SECTIONS

20.1 Induced EMF and Magnetic Flux

An electric current can be produced by a changing magnetic field. (A steady magnetic field cannot produce a current.) The current produced in a circuit occurs only for an instant while the magnetic field through a nearby circuit is changing. It is customary to say that **an induced emf is produced in a secondary circuit by the changing magnetic field in a primary circuit.** The emf is induced by a change in a quantity called the magnetic flux rather than simply by a change in the magnetic field.

20.2 Faraday's Law of Induction

The emf induced in a circuit is proportional to the time rate of change of magnetic flux through the circuit. The polarity of the induced emf is such that it tends to produce a current that will create a magnetic flux to oppose the **change in flux** through the circuit.

20.3 Motional emf

A potential difference will be maintained across a conductor moving in a magnetic field as long as the direction of motion through the field is not parallel to the field direction. If the motion is reversed, the polarity of the potential difference will also be reversed.

20.5 Generators

Generators and motors are important practical devices that operate on the principle of electromagnetic induction. In its simplest form, the **alternating current** (ac) **generator** consists of a wire loop rotated in a magnetic field by some external means; the ends of a loop rotating in a magnetic field are connected to slip rings that rotate with the loop. Connections to the external circuit are made by stationary brushes in contact with the slip rings, and the induced emf varies sinusoidally with time with a frequency of 60 Hz (in the US and Canada.)

In a **direct current** (dc) **generator** the contacts to the rotating loop are made by a split ring, or commutator. In this design, the output voltage always has the same polarity and the current is a pulsating direct current.

Motors are devices that convert electrical energy to mechanical energy. Essentially, **a motor is a generator run in reverse.** Instead of a current being generated by a rotating loop, a current is supplied to the loop by a source of emf, and the magnetic force on the current-carrying loop causes it to rotate.

20.7 Self-Inductance

The self-induced emf is always proportional to the **time rate of change** of current in the circuit. The **inductance** of a device (an inductor) depends on its **geometry.**

20.8 *RL* Circuits

If a resistor and an inductor are connected in series to a battery, the current in the circuit will reach an **equilibrium** value ($\mathcal{E} / R$) after a time which is long compared to the **time constant** of the circuit, τ.

20.9 Energy Stored in a Magnetic Field

In an *RL* circuit, the rate at which energy is supplied by the battery equals the sum of the rate at which heat is dissipated in the resistor and the rate at which energy is stored in the inductor. The **energy density** is proportional to the **square of the magnetic field.**

EQUATIONS AND CONCEPTS

The total magnetic flux through a plane area, A, placed in a uniform magnetic field depends on the angle between the direction of the magnetic field and the direction perpendicular to the surface area.

$$\Phi \equiv B_{\perp} A = BA\cos\theta \qquad (20.1)$$

$$\Phi_{max} = BA$$

The maximum flux through the area occurs when the magnetic field is perpendicular to the plane of the surface area. When the magnetic field is parallel to the plane of the surface area, the flux through the area is zero. The unit of magnetic flux is the weber, Wb.

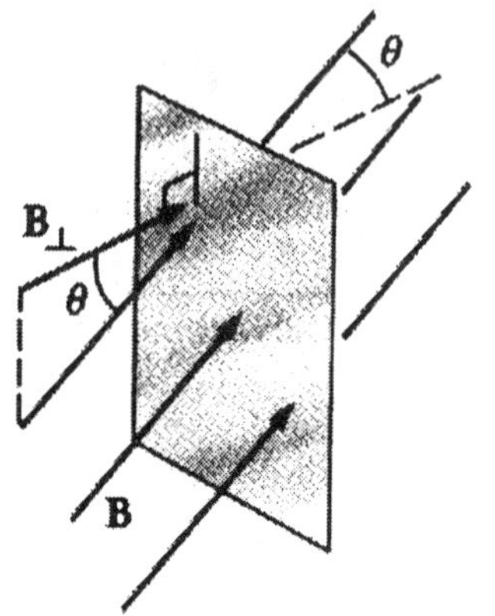

Faraday's law of induction states that the average emf induced in a circuit is proportional to the rate of change of magnetic flux through the circuit. The minus sign is included to indicate the polarity of the induced emf, which can be found by use of Lenz's law.

$$\mathcal{E} = -N\frac{\Delta\Phi}{\Delta t} \qquad (20.2)$$

Lenz's law states that the polarity of the induced emf (and the direction of the associated current in a closed circuit) produces a current whose magnetic field opposes the change in the flux through the loop. That is, the induced current tends to maintain the original flux through the circuit.

Comment on Lenz's law.

A "motional" emf is induced in a conductor of length, ℓ, moving with speed, v, perpendicular to a magnetic field.

$$\mathcal{E} = \frac{\Delta\Phi}{\Delta t} = B\ell v \qquad (20.4)$$

If the moving conductor is part of a complete circuit of resistance, R, a current will be induced in the circuit.

$$I = \frac{\mathcal{E}}{R} = \frac{B\ell v}{R} \qquad (20.5)$$

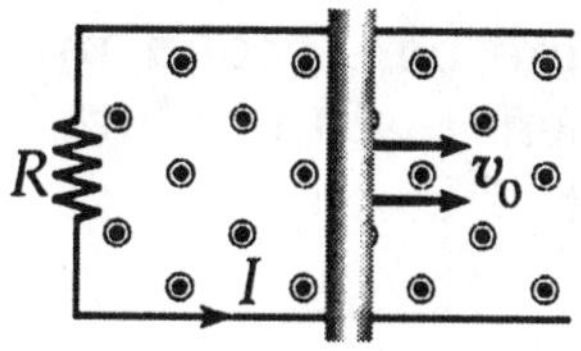

When a conducting loop of N turns and cross-sectional area, A, rotates with a constant angular velocity in a magnetic field, the emf induced in the loop will vary sinusoidally in time. For a given loop, the maximum value of the induced emf will be proportional to the angular velocity of the loop.

$$\mathcal{E} = NBA\omega \sin \omega t \qquad (20.7)$$

$$\mathcal{E}_{\text{max}} = NBA\omega \qquad (20.8)$$

A coil, solenoid, toroid, coaxial cable, or other conducting device is characterized by a parameter called its inductance, L. A change of current in the circuit will result in a self-induced emf which will be proportional to the rate of change of the current.

$$\mathcal{E} \equiv -L\frac{\Delta I}{\Delta t} \qquad (20.9)$$

The inductance of a given device, for example a coil, depends on its physical makeup: diameter, number of turns, type of material on which the wire is wound, and other geometric parameters. A circuit element which has a large inductance is called an inductor. The SI unit of inductance is the henry, H. A rate of change of current of 1 ampere per second in an inductor of 1 henry will produce a self-induced emf of 1 volt.

Comment on inductance

$$1\,\mathrm{H} = 1\frac{\mathrm{V \cdot s}}{\mathrm{A}}$$

The inductance of a coil can be calculated if the magnetic flux through the coil is known for a given current.

$$L = \frac{N\Phi}{I} \quad (20.10)$$

If the switch in a series circuit which contains a battery, resistor, and inductor is closed at time $t = 0$, the current in the circuit will increase in a characteristic fashion toward a maximum value of $I = \mathcal{E}/R$. The time required for the current to reach 63 percent of its final value is known as the time constant of the circuit.

$$\tau \equiv \frac{L}{R} \quad (20.15)$$

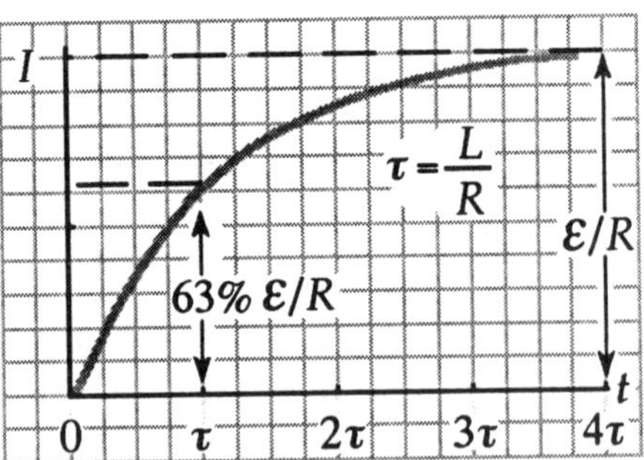

The equation for the current in a series RL circuit as a function of time is

$$I = \frac{\mathcal{E}}{R}\left(1 - e^{-Rt/L}\right)$$

The energy stored in the magnetic field of an inductor is proportional to the square of the current; and the energy stored in the electric field of a charged capacitor is proportional to the square of the potential difference between the plates.

$$PE_L = \frac{1}{2}LI^2 \qquad (20.16)$$

$$PE_C = \frac{1}{2}CV^2$$

REVIEW CHECKLIST

▷ Calculate the emf (or current) induced in a circuit when the magnetic flux through the circuit is changing in time. The variation in flux might be due to a change in (i) the area of the circuit, (ii) the magnitude of the magnetic field, (iii) the direction of the magnetic field, or (iv) the orientation/location of the circuit in the magnetic field.

▷ Apply Lenz's law to determine the direction of an induced emf or current. You should also understand that Lenz's law is a consequence of the law of conservation of energy.

▷ Calculate the emf induced between the ends of a conducting bar as it moves through a region where there is a constant magnetic field (motional emf).

▷ Define self-inductance, L, of a circuit in terms of appropriate circuit parameters. Calculate the total magnetic energy stored in a magnetic field if you are given the values of the inductance of the device with which the field is associated and the current in the circuit.

▷ Qualitatively describe the manner in which the instantaneous value of the current in an RL circuit changes while the current is either increasing or decreasing with time.

SOLUTIONS TO SELECTED END-OF-CHAPTER PROBLEMS

7. A cube of edge length $\ell = 2.5$ cm is positioned as shown in Figure P20.7. There is a uniform magnetic field throughout the region with components of $B_x = +5.0$ T, $B_y = +4.0$ T, and $B_z = +3.0$ T. (a) Calculate the flux through the shaded face of the cube. (b) What is the total flux emerging from the volume enclosed by the cube (i.e., total flux through all six faces)?

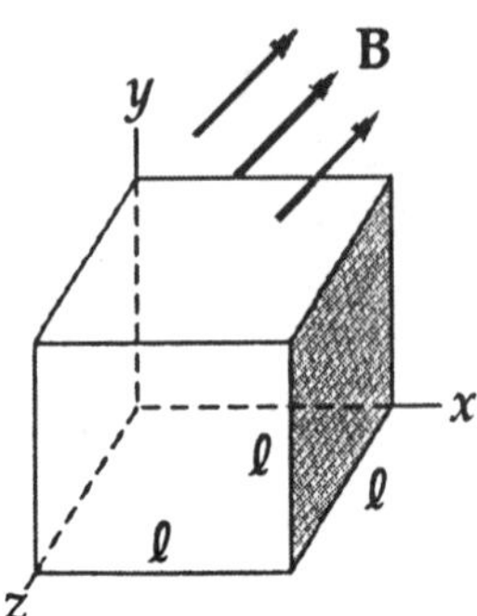

Figure P20.7

Solution (a) In a uniform magnetic field, the flux through a plane area is $\Phi = B_\perp A = BA\cos\theta$. Here, θ is the angle between the magnetic field **B** and the normal (perpendicular) to the area A, and $B_\perp = B\cos\theta$ is the component of **B** perpendicular to the area. In the figure, the area of the shaded face is

$$A = \ell^2 = \left(2.5\times10^{-2}\text{ m}\right)^2 = 6.3\times10^{-4}\text{ m}^2$$

and the component of the field perpendicular to this face is $B_\perp = B_x = +5.0$ T. Thus, the flux through the shaded face of the cube is

$$\Phi = B_x A = (5.0\text{ T})\left(6.3\times10^{-4}\text{ m}^2\right) = 3.1\times10^{-3}\text{ T}\cdot\text{m}^2 \qquad \Diamond$$

(b) When a surface is closed (i.e., completely surrounds or encloses some volume), the normal line to that surface at any point is considered to point outward, away from the enclosed volume. Thus, the normal to the shaded face of the cube is in the positive x direction and $B_\perp = B_x$ for that face. For the opposite side of the cube (the face in the $x = 0$ plane), the normal points in the negative x direction and $B_\perp = -B_x$. The flux through this face of the cube is $\Phi' = -B_x\ell^2$, which is the negative of the flux found for the shaded face. The net flux through these two opposite faces of the cube is then $\Phi_{net} = \Phi + \Phi' = 0$. In the same manner, it is found that the net flux through any pair of opposite faces of the cube is zero. Thus, the total flux emerging from the enclosed volume (i.e., the total flux through all six faces) is zero. ◊

11. The plane of a rectangular coil, 5.0 cm by 8.0 cm, is perpendicular to the direction of a magnetic field, **B**. If the coil has 75 turns and a total resistance of 8.0 Ω, at what rate must the magnitude of **B** change to induce a current of 0.10 A in the windings of the coil?

Solution To produce a current of $I = 0.10$ A in the windings of a coil with a resistance of 8.0 Ω, the magnitude of the induced emf in the coil must be

$$|\mathcal{E}| = IR = (0.10\text{ A})(8.0\ \Omega) = 0.80\text{ V}$$

But, the magnitude of the induced emf is given by $|\mathcal{E}| = N\,\Delta\Phi/\Delta t$ where N is the number of turns on the coil and Φ is the flux through each turn. Since the plane of the coil is perpendicular to the magnetic field, the flux through each turn on the coil is $\Phi = BA$. The area enclosed by a turn on the coil has a constant value of $A = (5.0\text{ cm})(8.0\text{ cm}) = 40\text{ cm}^2 = 4.0\times10^{-3}\text{ m}^2$, so any change in flux is due to a change in field strength, $\Delta\Phi = (\Delta B)A$. Therefore, the magnitude of the induced emf becomes $|\mathcal{E}| = NA(\Delta B/\Delta t)$ and the rate of change in the field must be

$$(\Delta B/\Delta t) = \frac{|\mathcal{E}|}{NA} = \frac{0.80\text{ V}}{(75)(4.0\times10^{-3}\text{ m}^2)} = 2.7\text{ T/s} \qquad \Diamond$$

---

15. A 300-turn solenoid with a length of 20 cm and a radius of 1.5 cm carries a current of 2.0 A. A second coil of four turns is wrapped tightly about this solenoid so that it can be considered to have the same radius as the solenoid. Find (a) the change in the magnetic flux through the coil and (b) the magnitude of the average induced emf in the coil when the current in the solenoid increases to 5.0 A in a period of 0.90 s.

Solution (a) By Section 19.11 in the text, the magnetic field inside the solenoid is parallel to the axis and has a magnitude of $B = \mu_0 nI$. Here, I is the current in its windings, and the number of turns per unit length of the solenoid is $n = N/\ell = (300)/(0.20\text{ m}) = 1.5\times10^3\text{ m}^{-1}$. As the current increases from 2.0 A to 5.0 A, the change in the magnetic field strength is

$$\Delta B = \mu_0 n(\Delta I) = (4\pi\times10^{-7}\text{ T}\cdot\text{m/A})(1.5\times10^3\text{ m}^{-1})(5.0\text{ A} - 2.0\text{ A}) = 5.7\times10^{-3}\text{ T}$$

The change in flux through the second coil is $\Delta\Phi = (\Delta B)A = (\Delta B)\pi r^2$, or

$$\Delta\Phi = \left(5.7\times10^{-3}\ \text{T}\right)\pi\left(1.5\times10^{-2}\ \text{m}\right)^2 = 4.0\times10^{-6}\ \text{T}\cdot\text{m}^2 \qquad \Diamond$$

(b) The magnitude of the average induced emf in the coil is given by $|\mathcal{E}| = N\,\Delta\Phi/\Delta t$, where N is the number of turns on the coil and Δt is the time during which the change in flux occurs. Therefore,

$$|\mathcal{E}| = (4)\left(\frac{4.0\times10^{-6}\ \text{T}\cdot\text{m}^2}{0.90\ \text{s}}\right) = 1.8\times10^{-5}\ \text{V} = 18\ \mu\text{V} \qquad \Diamond$$

---

19. A helicopter has blades of length 3.0 m, rotating at 2.0 rev/s about a central hub. If the vertical component of the earth's magnetic field is 5.0×10^{-5} T, what is the emf induced between the blade tip and the central hub?

Solution The time for one of the helicopter's blades to complete a full revolution is

$$\Delta t = \frac{1}{\text{frequency}} = \frac{1}{2.0\ \text{rev/s}} = 0.50\ \text{s}$$

During this time, the blade sweeps out a circular area of $A = \pi\ell^2 = \pi(3.0\ \text{m})^2 = 9.0\pi\ \text{m}^2$. The number of magnetic field lines (i.e., the flux) the blade cuts through in this period is

$$\Delta\Phi = B_\perp A = \left(5.0\times10^{-5}\ \text{T}\right)\left(9.0\pi\ \text{m}^2\right) = 1.4\times10^{-3}\ \text{T}\cdot\text{m}^2$$

Thus, the emf induced between the blade tip and the central hub is

$$|\mathcal{E}| = \frac{\Delta\Phi}{\Delta t} = \frac{1.4\times10^{-3}\ \text{T}\cdot\text{m}^2}{0.50\ \text{s}} = 2.8\times10^{-3}\ \text{V} = 2.8\ \text{mV} \qquad \Diamond$$

---

25. Find the direction of the current through the resistor in Figure P20.25, (a) at the instant the switch is closed, (b) after the switch has been closed for several minutes, and (c) at the instant the switch is opened.

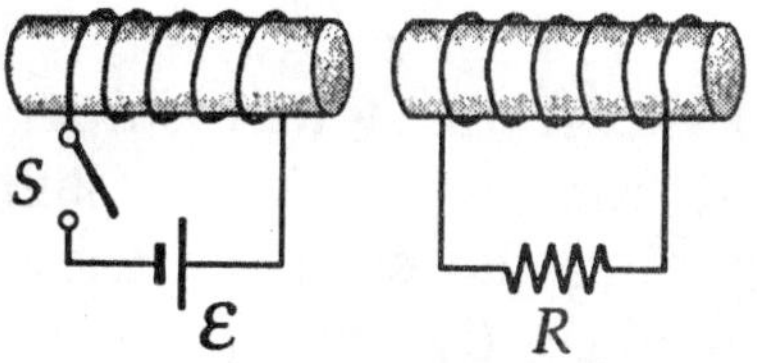

Figure P20.25

Solution (a) When the switch is closed, the battery produces a counterclockwise current (as viewed from the right end) around the solenoid on the left. Note that if you grip this solenoid with the fingers of your right hand going around in the direction of the current, your thumb points toward the right. Thus, when the switch is closed, the right end of this solenoid behaves like the north pole of a bar magnet. Having an increasing current in this solenoid (as is the case for a short period after the switch is closed) increases the flux, directed from left to right, through the second solenoid. This is equivalent to moving a north pole toward the left end of this solenoid. According to Lenz's law, an induced current flows in a direction that makes the left end behave like a north pole (hence, opposing the approaching "north pole"). Grip the second solenoid with the thumb of the right hand pointing toward the left end and notice that your fingers go clockwise around the solenoid (as viewed from the right end). This is the direction the induced current must flow so the left end will behave like a north pole. Hence, the induced current flows from left to right through the resistance R. ◊

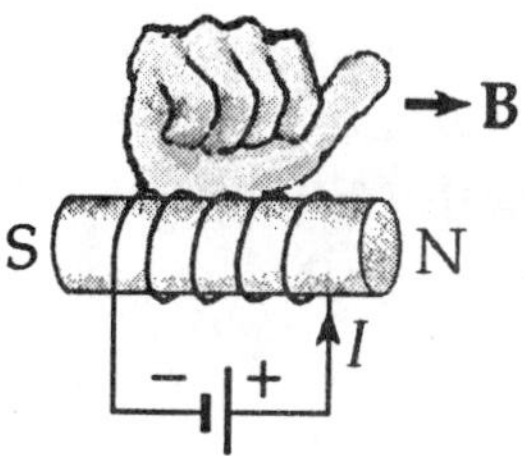

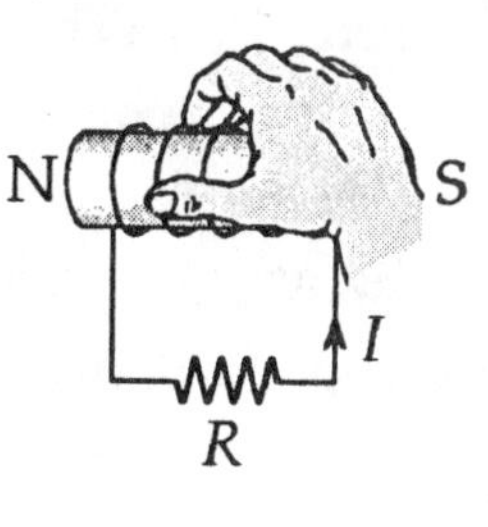

(b) After the switch has been closed for several minutes, the current in the first solenoid is constant. Thus, the magnetic field generated by this current is not varying in time and the flux through the second solenoid is constant. There is no induced emf, and hence no current through the resistor, when the flux is constant. ◊

(c) When the switch is opened, the current stops flowing in the first solenoid, and stops producing a magnetic flux through the second solenoid. Therefore, opening the switch removes a north pole from the left end of the second solenoid. By Lenz's law, while the flux decreases, a current is induced in the second solenoid to make the left end behave as a south pole and oppose the departure of the "north pole". Grip the second solenoid with the thumb of your right hand pointing toward the right end (the north pole). Your fingers curl around the solenoid in a counterclockwise direction (as viewed from the right end). Thus, the induced current flows counterclockwise around this solenoid, and right to left through R. ◊

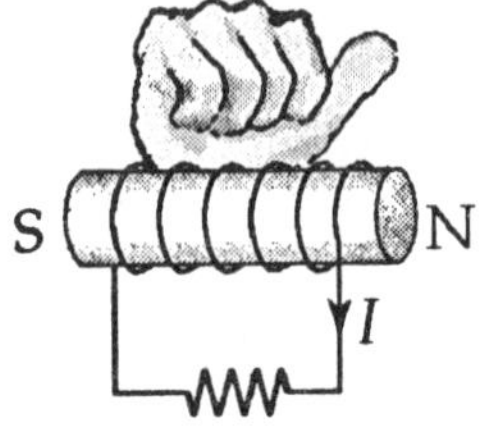

---

31. A loop of area 0.10 m^2 is rotating at 60 rev/s with its axis of rotation perpendicular to a 0.20-T magnetic field. (a) If there are 1000 turns on the loop, what is the maximum voltage induced in the loop? (b) When the maximum induced voltage occurs, what is the orientation of the loop with respect to the magnetic field?

Solution (a) When a loop rotates with its axis perpendicular to a magnetic field, the maximum value of the induced emf in the coil is $\mathcal{E}_{\max} = NBA\omega$ where N is the number of turns on the loop, B is the magnetic field strength, A is the area enclosed by the loop, and ω is the angular frequency (in radians per second). For the case described, $\omega = (60 \text{ rev/s})(2\pi \text{ rad/rev}) = (120\pi \text{ rad/s})$.

Thus, $\mathcal{E}_{\max} = (1000)(0.20 \text{ T})(0.10 \text{ m}^2)(120\pi \text{ rad/s}) = 7.5\times10^3 \text{ V}$ ◊

(b) The magnetic flux through the loop at any instant is $\Phi = BA\cos\theta$ where θ is the angle between the normal to the plane of the loop and the magnetic field direction. As the loop makes one complete revolution, θ varies from 0 to 2π. Thus,

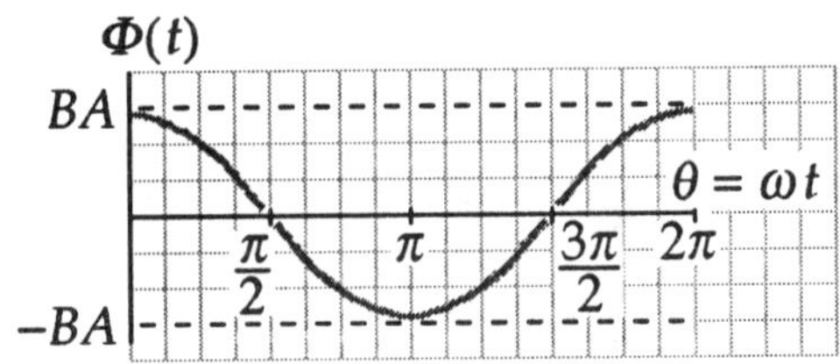

the flux Φ varies as a cosine function as shown in the sketch. Observe that the slope of the line tangent to this curve (i.e., the rate of change of the flux, $\Delta\Phi/\Delta t$) has maximum magnitude at $\theta = \pi/2$ and $\theta = 3\pi/2$. Hence, the magnitude of the induced emf $|\mathcal{E}| = N|\Delta\Phi/\Delta t|$ is a maximum when the line normal to the plane of the loop is also perpendicular to the magnetic field. This means that the plane of the loop is parallel to the field at an instant when the induced emf has maximum magnitude. ◊

---

37. An emf of 24.0 mV is induced in a 500-turn coil when the current is changing at a rate of 10.0 A/s. What is the magnetic flux through each turn of the coil at an instant when the current is 4.00 A?

Solution When the current in a coil changes, the emf induced in that coil is $\mathcal{E} = -L(\Delta I/\Delta t)$, where L is the self-inductance of the coil. If $|\mathcal{E}| = 24.0\ \text{mV}$ when the rate of change of the current is $\Delta I/\Delta t = 10.0$ A/s, the self inductance is

$$L = \frac{|\mathcal{E}|}{|\Delta I/\Delta t|} = \frac{24.0 \times 10^{-3}\ \text{V}}{10.0\ \text{A/s}} = 2.4 \times 10^{-3}\ \Omega \cdot \text{s} = 2.4\ \text{mH}$$

The induced emf can also be expressed in terms of the rate of change of the flux as $\mathcal{E} = -N(\Delta\Phi/\Delta t)$. Equating the two expressions for the induced emf gives $L(\Delta I/\Delta t) = N(\Delta\Phi/\Delta t)$. Thus, the self-inductance can be written as

$$L = N\left(\frac{\Delta\Phi}{\Delta I}\right) \qquad \text{and} \qquad L = \frac{N\Phi}{I}$$

At an instant when the current in the windings of this coil is $I = 4.00$ A, the flux through each turn on the coil is

$$\Phi = \frac{LI}{N} = \frac{(2.4 \times 10^{-3}\ \text{H})(4.00\ \text{A})}{500} = 1.92 \times 10^{-5}\ \text{T} \cdot \text{m}^2$$ ◊

---

41. A 25-mH inductor, an 8.0-Ω resistor, and a 6.0-V battery are connected in series. The switch is closed at t = 0. Find the voltage drop across the resistor (a) at t = 0 and (b) after one time constant has passed. Also, find the voltage drop across the inductor (c) at t = 0 and (d) after one time constant has elapsed.

$R = 8.0\ \Omega$ $L = 25$ mH

S $\Delta V = 6.0$ V

+ −

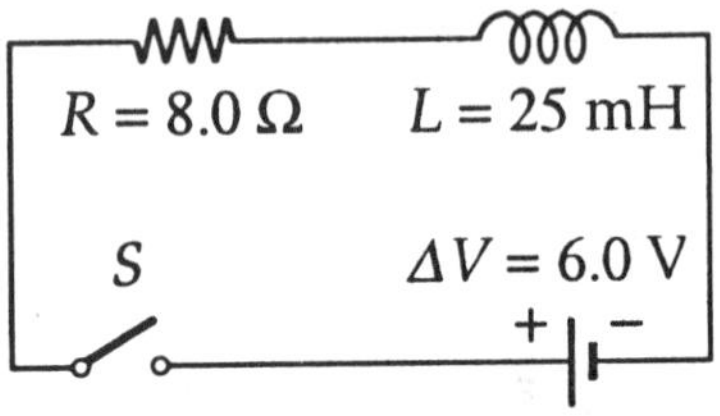

Solution (a) At t = 0, the switch is closed and the current will begin to increase. At this instant, before any time has elapsed, the current is still zero. Thus, the voltage across the resistor at this time is $\Delta V_R = IR = 0$. ◊

(b) After one time constant has passed, the current will have increased to 63.2% of its final value. For this circuit, the final current will be $I_{max} = \dfrac{\Delta V}{R} = \dfrac{6.0\text{ V}}{8.0\ \Omega} = 0.75$ A

Thus, at $t = \tau$, $I = 0.632 I_{max} = 0.632(0.75\text{ A}) = 0.47$ A

and $\Delta V_R = (0.47\text{ A})(8.0\ \Omega) = 3.8$ V ◊

When the switch is closed, Kirchoff's loop rule may be applied to this circuit to find that $-\Delta V_R - \Delta V_L + \Delta V = 0$, or $\Delta V_L = \Delta V - \Delta V_R$ at **all times.** Here, ΔV_R is the voltage drop across the resistor and ΔV_L is that across the inductor.

(c) At t = 0, it was found that $\Delta V_R = 0$ [see part (a) above]. Therefore at this time,

$$\Delta V_L = 6.0\text{ V} - 0 = 6.0\text{ V}$$ ◊

(d) After one time constant has elapsed, $\Delta V_R = 3.8$ V [see part (b) above]. Thus, at $t = \tau$ the voltage drop across the inductor is

$$\Delta V_L = \Delta V - \Delta V_R = 6.0\text{ V} - 3.8\text{ V} = 2.2\text{ V}$$ ◊

45. A 24-V battery is connected in series with a resistor and an inductor, where $R = 8.0\ \Omega$ and $L = 4.0$ H. Find the energy stored in the inductor (a) when the current reaches its maximum value and (b) one time constant after the switch is closed.

Solution If a coil has self-inductance L, an energy input is required to establish a current in it. The work required to overcome the opposing induced emf and establish the current is $PE_L = \frac{1}{2}LI^2$. This is labeled potential energy because it is useful to think of it as stored in the magnetic field surrounding the inductor when current I flows. The inductor releases this energy attempting to prevent the current from decreasing when the power source is removed or turned down.

(a) After the circuit has been intact for a very long time (as $t \to \infty$), the current reaches its final value of $I_{max} = \Delta V/R = 24\ \text{V}/8.0\ \Omega = 3.0\ \text{A}$. The stored energy at this time is

$$PE_L = \frac{1}{2}LI_{max}^2 = \frac{1}{2}(4.0\ \text{H})(3.0\ \text{A})^2 = 18\ \text{J} \qquad \Diamond$$

(b) When one time constant has elapsed since the circuit was completed (i.e., at $t = \tau$) the current is 63.2% of the final value. Since the final current is $I_{max} = 3.0$ A, the current at $t = \tau$ is $I = 0.632(3.0\ \text{A}) = 1.9\ \text{A}$, and the stored energy is

$$PE_L = \frac{1}{2}LI^2 = \frac{1}{2}(4.0\ \text{H})(1.9\ \text{A})^2 = 7.2\ \text{J} \qquad \Diamond$$

---

51. An 820-turn wire coil of resistance 24.0 Ω is placed on top of a 12 500-turn, 7.00-cm-long solenoid, as in Figure P20.51. Both coil and solenoid have cross-sectional areas of $1.00 \times 10^{-4}\ \text{m}^2$. (a) How long does it take the solenoid current to reach 0.632 times its maximum value? (b) Determine the average back emf caused by the self-inductance of the solenoid during this interval. (c) Determine the average rate of change in magnetic flux through each turn of the coil during this interval. (d) Find the magnitude of the average induced current in the coil.

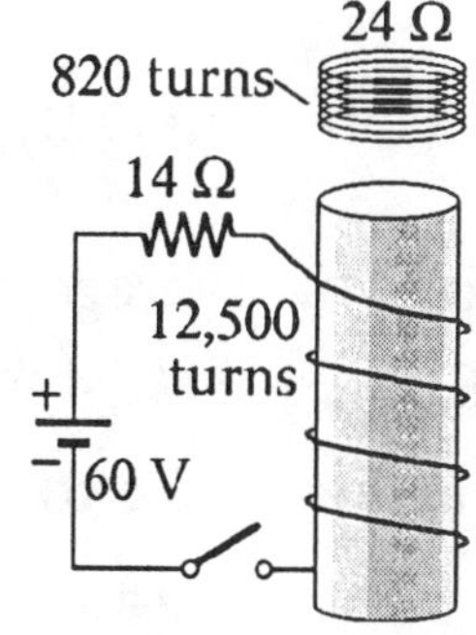

Figure P20.51

Solution (a) The self-inductance of a solenoid is $L = \mu_0 N^2 A / \ell$ where N is the number of turns, A is the cross-sectional area, and ℓ is the length of the solenoid. For the solenoid in this circuit,

$$L = \frac{\left(4\pi \times 10^{-7}\ \text{T}\cdot\text{m/A}\right)\left(1.25\times 10^{4}\right)^2\left(1.00\times 10^{-4}\ \text{m}^2\right)}{7.00\times 10^{-2}\ \text{m}} = 0.280\ \text{H}$$

The current in an RL circuit rises to 63.2% of the final value in a time

$$t = \tau \equiv \frac{L}{R} = \frac{0.280\ \text{H}}{14.0\ \Omega} = 2.00\times 10^{-2}\ \text{s}$$ ◊

(b) At $t = \tau$, the current in the solenoid is $I = 0.632 I_{max} = (0.632)\Delta V/R$, or $I = (0.632)(60.0\ \text{V}/14.0\ \Omega) = 2.71\ \text{A}$. The average rate of change in the current since the switch was closed is

$$\left(\frac{\Delta I}{\Delta t}\right)_{avg} = \frac{I-0}{\tau - 0} = \frac{2.71\ \text{A}}{2.00\times 10^{-2}\ \text{s}} = 135\ \text{A/s}$$

Hence, the average induced emf opposing the increase in current (i.e., back emf) during this time is

$$(\mathcal{E})_{avg} = L(\Delta I/\Delta t)_{avg} = (0.280\ \text{H})(135\ \text{A/s}) = 37.8\ \text{V}$$ ◊

(c) The average back emf is also related to the change of flux by $\mathcal{E}_{avg} = N(\Delta\Phi/\Delta t)_{avg}$. The average rate at which the flux in the solenoid and each turn of the 820-turn coil changes during this interval is therefore

$$(\Delta\Phi/\Delta t)_{avg} = \frac{\mathcal{E}_{avg}}{N} = \frac{37.8\ \text{V}}{1.25\times 10^{4}} = 3.02\times 10^{-3}\ \text{V}$$ ◊

(d) The average induced emf in the 820-turn coil between $t = 0$ and $t = \tau$ is $\mathcal{E}_{coil,\,avg} = N_{coil}(\Delta\Phi/\Delta t)_{avg} = (820)\left(3.02\times 10^{-3}\ \text{V}\right) = 2.48\ \text{V}$, and the average current in the coil is

$$I_{coil,\,avg} = \mathcal{E}_{coil,\,avg}/R_{coil} = 2.48\ \text{V}/24.0\ \Omega = 0.103\ \text{A}$$ ◊

55. A horizontal wire is free to slide on the vertical rails of a conducting frame, as in Figure P20.55. The wire has mass m and length ℓ, and the resistance of the circuit is R. If a uniform magnetic field is directed perpendicularly to the frame, what is the terminal speed of the wire as it falls under the force of gravity? (Neglect friction.)

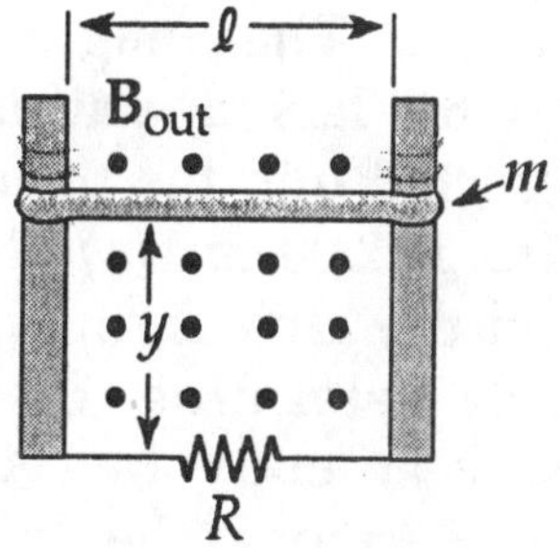

Figure P20.55

Solution Consider the rectangular conducting path formed by the wire, the two rails and the resistance line at the bottom. The area enclosed by this loop is $A = \ell y$ and the flux through this area is $\Phi = BA = B\ell y$. As the wire falls, the flux through the area decreases at a rate

$$\frac{\Delta\Phi}{\Delta t} = \frac{\Delta(B\ell y)}{\Delta t} = B\ell\left(\frac{\Delta y}{\Delta t}\right) = B\ell v$$

where v is the speed of the falling wire. This changing flux produces an induced emf of magnitude $|\mathcal{E}| = |\Delta\Phi/\Delta t| = B\ell v$ in the loop. An induced current $I = |\mathcal{E}|/R = B\ell v/R$ will therefore flow around the conducting path. According to Lenz's law, the flux produced by the current is directed out of the page through the area enclosed by the loop, opposing the flux decrease which generates the emf. Thus, the current must flow counterclockwise around the loop, or right to left through the falling wire.

The wire now carries a current from right to left through a magnetic field directed out of the page. According to the right-hand rule, the magnetic field then exerts on the wire a vertically upward force of magnitude

$$F = BI\ell = B\left(\frac{B\ell v}{R}\right)\ell = \frac{B^2\ell^2 v}{R}$$

The net **downward** force acting on the wire is $F_{net} = mg - F = mg - B^2\ell^2 v/R$. Notice that as the wire gains speed, the net downward force decreases. When the net force becomes zero, the wire has attained its terminal speed, $v = v_t$. Thus, $B^2\ell^2 v_t/R = mg$, the terminal speed of the wire is: $v_t = \dfrac{mgR}{B^2\ell^2}$ ◊

61. The magnetic field shown in Figure P20.61 has a uniform magnitude of 25.0 mT directed into the paper. The initial diameter of the kink is 2.00 cm. (a) The wire is quickly pulled taut, and the kink shrinks to a diameter of zero in 50.0 ms. Determine the average voltage induced between endpoints *A* and *B*. Include the polarity. (b) Suppose the kink is undisturbed, but the magnetic field increases to 100 mT in 4.00×10^{-3} s. Determine the average voltage across terminals *A* and *B*, including polarity, during this period.

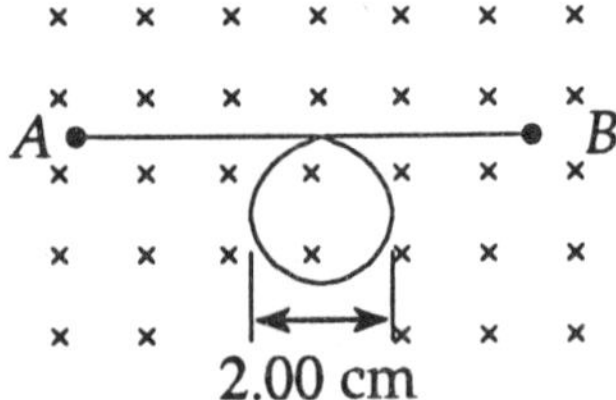

Figure P20.61

Solution Initially, the magnetic field strength is $B_i = 25.0 \times 10^{-3}$ T and the area enclosed by the loop in the wire is $A_i = \pi d_i^2/4$, or

$$A_i = \pi\left(2.00 \times 10^{-2} \text{ m}\right)^2/4 = 3.14 \times 10^{-4} \text{ m}^2$$

Thus, the initial flux, directed into the page through the loop in the wire is

$$\Phi_i = B_i A_i = \left(25.0 \times 10^{-3} \text{ T}\right)\left(3.14 \times 10^{-4} \text{ m}^2\right) = 7.85 \times 10^{-6} \text{ T} \cdot \text{m}^2$$

(a) When the loop shrinks to zero diameter, the enclosed area and the flux through that area both go to zero. The magnitude of the change in flux through the loop is then

$$|\Delta\Phi| = \left|\Phi_f - \Phi_i\right| = 7.85 \times 10^{-6} \text{ T} \cdot \text{m}^2$$

If this change in flux occurs in a time interval of $\Delta t = 50.0 \times 10^{-3}$ s, the magnitude of the average induced emf during this period is

$$\mathcal{E}_{\text{avg}} = \frac{|\Delta\Phi|}{\Delta t} = \frac{7.85 \times 10^{-6} \text{ T}}{50.0 \times 10^{-3} \text{ s}} = 1.57 \times 10^{-4} \text{ V} = 0.157 \text{ mV} \qquad \lozenge$$

An induced current will attempt to flow in a way that produces flux directed into the page through the enclosed area. This would oppose the decrease in the original flux as required by Lenz's law. Thus, the current would flow clockwise around the loop, or the induced emf in the wire must attempt to move charges from A toward B. Since the conductor terminates at point B, positive charges will accumulate there, making point B positive relative to point A. ◊

(b) If the magnetic field increases to $B_f = 100\times10^{-3}$ T while the area enclosed by the loop remains constant, the change in flux directed into the page is

$$\Delta\Phi = \left(B_f - B_i\right)A_i \quad \text{or} \quad \Delta\Phi = \left(+75.0\times10^{-3}\ \text{T}\right)\left(3.14\times10^{-4}\ \text{m}^2\right) = +2.36\times10^{-5}\ \text{T}\cdot\text{m}^2$$

This change occurs in $\Delta t = 4.00\times10^{-3}$ s, and the average magnitude of the induced emf is

$$\mathcal{E}_{avg} = \frac{|\Delta\Phi|}{\Delta t} = \frac{2.36\times10^{-5}\ \text{T}\cdot\text{m}^2}{4.00\times10^{-3}\ \text{s}} \quad \text{or} \quad \mathcal{E}_{avg} = 5.89\times10^{-3}\ \text{V} = 5.89\ \text{mV} \quad ◊$$

An induced current will attempt to flow from B to A (or counterclockwise around the loop). A current flowing in this manner would produce flux directed out of the page through the enclosed area, and hence oppose the increased flux into the page due to the increasing field strength. Since the conductor terminates at point A, positive charges accumulate there, making point A positive relative to B. ◊

CHAPTER SELF-QUIZ

1. In a circuit made up of inductor, resistance, ammeter, battery, and switch in series, at which one of the following time intervals after the switch is closed will the rate of current increase be greatest?
   a. zero
   b. one time constant
   c. reciprocal of one time constant
   d. ten time constants

2. A 0.20-m wire is moved perpendicular to a 0.50-T magnetic field at a speed of 1.50 m/s. What emf is induced across the ends of the wire?
   a. 2.25 V
   b. 1.00 V
   c. 0.60 V
   d. 0.15 V

3. A coil with a self-inductance of 0.75 mH experiences a constant increase in current from zero to 5.00 A in 0.125 s. What is the induced emf during this interval?
   a. 0.045 V
   b. 0.030 V
   c. 0.470 V
   d. 0.019 V

4. What is the self-inductance in a coil which experiences a 1.50-V induced emf when the current is changing at a rate of 55 A/s?
   a. 83 mH
   b. 45 mH
   c. 37 mH
   d. 27 mH

5. An inductor, battery, resistor, ammeter, and switch are connected in series. If the switch, initially open, is now closed, what is the current's final value?
   a. zero
   b. battery voltage divided by inductance
   c. battery voltage times inductance
   d. battery voltage divided by resistance

6. A coil with a self-inductance of 1.50 mH will show what time rate of change of current when inducing an emf of 0.30 V?
   a. 2.00×10^2 A/s
   b. 0.45×10^{-4} A/s
   c. 5.00×10^{-2} A/s
   d. 0.30 A/s

7. The self-inductance of a solenoid increases under which of the following conditions?
   a. solenoid length is increased
   b. cross-sectional area is decreased
   c. number of coils per unit length is decreased
   d. number of coils is increased

8. A 12.0-V battery is connected in series with a switch, resistor, and coil. If the circuit's time constant is 4.00×10^{-4} s and the final steady current after the switch is closed becomes 2.00 A, what is the value of the inductance?
   a. 1.20 mH
   b. 2.40 mH
   c. 9.60 mH
   d. 48.0 mH

9. A bar magnet is falling through a loop of wire with constant velocity with the north pole entering first. As the north pole enters the wire, the induced current as viewed from above will be:
   a. clockwise
   b. counterclockwise
   c. zero
   d. along the length of the magnet

10. A uniform 1.50-T magnetic field passes perpendicularly through the plane of a wire loop 0.300 m^2 in area. What flux passes through the loop?
    a. 5.00 Wb
    b. 0.45 Wb
    c. 0.25 Wb
    d. 0.135 Wb

11. An *RL*-series circuit has the following components: 2.50-mH coil, 0.500-ohm resistor, 6.00-V battery, ammeter, and switch. What is the time constant of this circuit?
    a. 12.5×10^{-3} s
    b. 5.00×10^{-3} s
    c. 2.50×10^{-2} s
    d. 200 s

12. Electricity may be generated by rotating a loop of wire between the poles of a magnet. The induced current is greatest when:
    a. the plane of the loop is parallel to the magnetic field
    b. the plane of the loop is perpendicular to the magnetic field
    c. the magnetic flux through the loop is a maximum
    d. the plane of the loop makes an angle of 45.0° with the magnetic field

Chapter 21

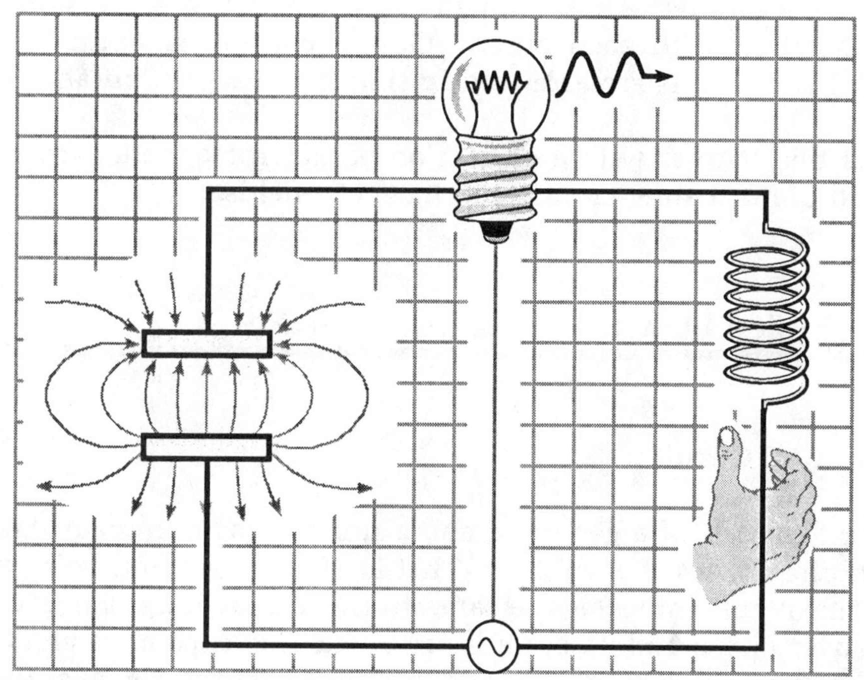

AC CIRCUITS AND ELECTROMAGNETIC WAVES

Chapter 21

ALTERNATING CURRENT CIRCUITS AND ELECTROMAGNETIC WAVES

It is important to understand the basic principles of alternating current (ac) circuits because they are so much a part of our everyday life. We begin our study of ac circuits by examining the characteristics of a circuit containing a source of emf and a single circuit element: a resistor, a capacitor, or an inductor. Then we examine what happens when these elements are connected in combination with each other. Our discussion is limited to situations in which the elements are arranged in simple series configurations.

We conclude this chapter with a discussion of electromagnetic waves, which are composed of fluctuating electric and magnetic fields.

NOTES FROM SELECTED CHAPTER SECTIONS

21.1 Resistors in an AC Circuit

If an ac circuit consists of a generator and a resistor, the **current in the circuit is in phase with the voltage.** That is, the current and voltage reach their maximum values at the same time. The **average** value of the current over **one complete cycle** is zero. The **rms** current refers to **root mean square,** which simply means the square root of the average value of the square of the current.

21.2 Capacitors in an AC Circuit

When an alternating voltage is applied across a capacitor, the voltage reaches its maximum value one quarter of a cycle after the current reaches its maximum value. In this situation, it is common to say that the **voltage always lags the current** through a capacitor by 90.0°.

21.3 Inductors in an AC Circuit

When a sinusoidal voltage is applied across an inductor, the voltage reaches its maximum value one quarter of an oscillation period before the current reaches its maximum value. In this situation, we say that the **voltage always leads the current** by 90.0°.

21.4 The *RLC* Series Circuit

The ac current at all points in a series ac circuit has the same amplitude and phase. Therefore, the voltage across each element will have **different** amplitudes and different phases relative to the common current. The voltage across the resistor is in phase with the current, the voltage across the inductor leads the current by 90.0°, and the voltage across the capacitor lags behind the current by 90.0°.

21.5 Power in an AC Circuit

The average power delivered by the generator in an AC circuit is dissipated as heat in the resistor. There is **no power** loss in an ideal inductor or capacitor.

21.6 Resonance in a Series *RLC* Circuit

The current in a series *RLC* circuit reaches its peak value when the frequency of the generator equals f_0; that is, when the "driving" frequency matches the **resonance frequency.**

21.7 The Transformer

A transformer is a device designed to raise or lower an ac voltage and current without causing an appreciable change in the product $I\Delta V$. In its simplest form, it consists of a primary coil of N_1 turns and a secondary coil of N_2 turns, both wound on a common soft iron core. In an ideal transformer, the power delivered by the generator must equal the power dissipated in the load.

21.8 Maxwell's Predictions and
21.9 Hertz's Discoveries

Electromagnetic waves are generated by accelerating electric charges. The radiated waves consist of oscillating electric and magnetic fields, which are **at right angles to each other** and also **at right angles to the direction of wave propagation.**

The fundamental laws governing the behavior of electric and magnetic fields are Maxwell's equations. In this unified theory of electromagnetism, Maxwell showed that electromagnetic waves are a natural consequence of these fundamental laws. The theory he developed is based upon the following four pieces of information:

1. Electric fields originate on positive charges and terminate on negative charges. The electric field due to a point charge can be determined at a location by applying Coulomb's force law to a test charge placed at that location.

2. Magnetic field lines always form closed loops; that is, they do not begin or end anywhere.

3. A varying magnetic field induces an emf and hence an electric field. This is a statement of Faraday's law (Chapter 20).

4. Magnetic fields are generated by moving charges (or currents), as summarized in Ampère's law (Chapter 19).

21.11 Properties of Electromagnetic Waves

Following is a summary of the properties of electromagnetic waves:

1. The solutions of Maxwell's third and fourth equations are wavelike, where both **E** and **B** satisfy the same wave equation.

2. Electromagnetic waves travel through empty space with the speed of light, $c = 1/\sqrt{\epsilon_0 \mu_0}$.

3. The electric and magnetic field components of plane electromagnetic waves are perpendicular to each other and also perpendicular to the direction of wave propagation. The latter property can be summarized by saying that electromagnetic waves are transverse waves.

4. The relative magnitudes of **E** and **B** in empty space are related by $E/B = c$.

5. Electromagnetic waves obey the principle of superposition.

EQUATIONS AND CONCEPTS

The output of an ac generator is sinusoidal where Δv is the instantaneous voltage and ΔV_m is the maximum voltage.

$$\Delta v = \Delta V_m \sin 2\pi f t \qquad (21.1)$$

These two equations relate the rms values of current and voltage to the maximum values of these quantities.

$$I = \frac{I_m}{\sqrt{2}} = 0.707\, I_m \qquad (21.2)$$

$$\Delta V = \frac{\Delta V_m}{\sqrt{2}} = 0.707\, \Delta V_m \qquad (21.3)$$

The rms voltage across a resistor is related to the rms current in the resistor by Ohm's law.

$$\Delta V_R = IR \qquad (21.4)$$

The impeding effect of a capacitor to the current in an ac circuit is expressed in terms of a factor called the **capacitive reactance.**	$X_C \equiv \dfrac{1}{2\pi f C}$	(21.5)
	$\Delta V_C = I X_C$	(21.6)
The effective resistance of a coil in an ac circuit is measured by a quantity called the inductive reactance.	$X_L \equiv 2\pi f L$	(21.8)
	$\Delta V_L = I X_L$	(21.9)
The instantaneous voltage across a resistor is in phase with the current.	$i = I_m \sin 2\pi f t$	
	$\Delta v_R = \Delta V_{R,\max} \sin 2\pi f t$	
The instantaneous voltage across an inductor leads the current by a quarter cycle.	$\Delta v_L = \Delta V_{L,\max} \cos 2\pi f t$	
The instantaneous voltage across the capacitor lags the current by a quarter cycle.	$\Delta v_C = -\Delta V_{C,\max} \cos 2\pi f t$	
The net instantaneous voltage across all three circuit elements is the sum of the instantaneous voltages across the separate elements; the net voltage is "out-of-step" with the instantaneous current by an amount called the phase angle, ϕ.	$\Delta v = \Delta V_{\max} \sin(2\pi f t + \phi)$	

The rms voltage across the combination of resistor, inductor, and capacitor can be determined in terms of the rms voltage values across the individual components. The circuit rms voltage can also be calculated in terms of the common circuit current and the values of resistance, inductive reactance, and capacitive reactance.

$$\Delta V = \sqrt{\Delta V_R{}^2 + (\Delta V_L - \Delta V_C)^2} \quad (21.10)$$

$$\Delta V = I\sqrt{R^2 + (X_L - X_C)^2} \quad (21.12)$$

By defining a quantity, Z, called the impedance, it is possible to relate the rms voltage and rms current in the form of a generalized Ohm's law.

$$Z \equiv \sqrt{R^2 + (X_L - X_C)^2} \quad (21.13)$$

$$\Delta V = IZ \quad (21.14)$$

The phase angle between rms voltage and rms current in the circuit can be determined from the impedance triangle or from the voltage triangle.

$$\tan\phi = \frac{X_L - X_C}{R} \quad (21.15)$$

$$\tan\phi = \frac{\Delta V_L - \Delta V_C}{\Delta V_R} \quad (21.11)$$

The only element in an ac circuit which dissipates energy is the resistor. The average power dissipated in an ac circuit can be expressed in terms of the power factor of the circuit, $\cos\phi$.

$$P_{avg} = I^2 R \quad (21.16)$$

$$P_{avg} = I\,\Delta V \cos\phi \quad (21.17)$$

When the frequency of an ac circuit is equal to the resonance frequency f_0, the impedance of the circuit becomes equal to the circuit resistance and the current becomes maximum.

$$f_0 = \frac{1}{2\pi\sqrt{LC}} \tag{21.19}$$

The primary voltage and current values in a transformer are related to the values of those quantities in the secondary in terms of the ratio of the number of turns in the primary and secondary coils. In a step-up transformer, N_2 is greater than N_1.

$$\Delta V_2 = \frac{N_2}{N_1}\Delta V_1 \tag{21.22}$$

$$I_1\Delta V_1 = I_2\Delta V_2 \tag{21.23}$$

The speed of an electromagnetic wave is related to the permeability and permittivity of the medium through which it travels. Electromagnetic waves travel with the speed of light.

$$c = \frac{1}{\sqrt{\mu_0 \epsilon_0}} \tag{21.24}$$

Permeability constant of free space. $\mu_0 = 4\pi \times 10^{-7}\ \mathrm{T \cdot m / A}$

Permittivity constant of free space. $\epsilon_0 = 8.85 \times 10^{-12}\ \mathrm{C / N \cdot m^2}$

Accurate value for the speed of light in a vacuum. $c = 2.99792 \times 10^{8}\ \mathrm{m / s}$

The ratio of the electric to the magnetic field in an electromagnetic wave is constant and equal to the speed of light.

$$\frac{E}{B} = c \qquad (21.26)$$

Electromagnetic waves carry energy as they travel through space. The rate of flow of energy, or power per unit area, transported perpendicular to a surface can be expressed in several alternate forms involving the maximum values of the electric and magnetic fields. The power per unit area given by these equations is the average power per unit area. Also, it can be shown that the energy carried by an electromagnetic wave is shared equally by the electric and magnetic fields.

$$\frac{P_{avg}}{A} = \frac{E_m B_m}{2\mu_0} \qquad (21.27)$$

$$\frac{P_{avg}}{A} = \frac{E_m{}^2}{2\mu_0 c} = \frac{c}{2\mu_0} B_m{}^2 \qquad (21.28)$$

where

$\frac{P_{avg}}{A} =$ Average power per unit area

The product of frequency and wavelength of an electromagnetic wave propagating in vacuum is constant and equal to c.

$$c = f\lambda \qquad (21.31)$$

Electromagnetic waves transport linear momentum, p, as well as energy, U. When an electromagnetic wave is incident upon a surface, momentum is transferred to the surface, and pressure is exerted on the surface.

$$p = \frac{U}{c} \quad \text{(Total absorption)} \qquad (21.29)$$

$$p = \frac{2U}{c} \quad \text{(Total reflection)} \qquad (21.30)$$

SUGGESTIONS, SKILLS, AND STRATEGIES

ALTERNATING CURRENT PROBLEMS

1. First analyze alternating current circuits is to calculate as many of the unknown quantities such as X_L and X_C as possible. When calculating X_C, express capacitance in farads, rather than, say, microfarads.
2. Apply the equation $\Delta V = IZ$ to the portion of the circuit that is of interest. That is, if you want to know the voltage drop across the combination of an inductor and a resistor, the equation reduces to $\Delta V = I\sqrt{R^2 + X_L^2}$.

REVIEW CHECKLIST

▷ Apply the formulas that give the reactance values in an ac circuit as a function of (i) capacitance, (ii) inductance, and (iii) frequency. Interpret the meaning of the terms **phase angle** and **power factor** in an ac circuit.

▷ Given an *RLC* series circuit in which values of resistance, inductance, capacitance, and the characteristics of the generator (source of emf) are known, calculate: (i) the instantaneous and rms voltage drop across each component, (ii) the instantaneous and rms current in the circuit, (iii) the phase angle by which the current leads or lags the voltage, (iv) the power expended in the circuit, and (v) the resonance frequency of the circuit.

▷ Understand the manner in which step-up and step-down transformers are used in the process of transmitting electrical power over large distances; and make calculations of primary to secondary voltage and current ratios for an ideal transformer.

▷ Describe the contribution by James Clerk Maxwell, properly relating the significance of the information available to him, to the theoretical understanding of the nature of electromagnetic radiation. Summarize the properties of electromagnetic waves.

▷ Relate the relative orientation of magnetic field, electric field, and direction of propagation in the corresponding electromagnetic wave. Justify the statement that electromagnetic waves carry both energy and momentum.

SOLUTIONS TO SELECTED END-OF-CHAPTER PROBLEMS

5. An ac power supply produces a peak voltage of $\Delta V_m = 100$ V. This power supply is connected to a 24-Ω resistor, and the current and resistor voltage are measured with an ideal ac ammeter and an ideal ac voltmeter, as shown in Figure P21.5. What does each meter read? Recall that an ideal ammeter has zero resistance and an ideal voltmeter has infinite resistance.

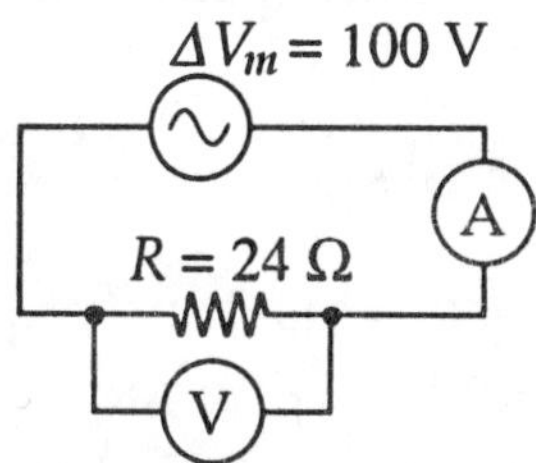

Figure P21.5

Solution First, realize that ac ammeters and voltmeters are designed to measure rms values rather than peak values. Since the ideal ammeter has zero resistance, the voltage maintained by the power supply, with a peak value of $\Delta V_m = 100$ V, exists across the resistor. The reading on the voltmeter is therefore

$$\Delta V = \Delta V_{\text{rms}} = \frac{\Delta V_m}{\sqrt{2}} = \frac{100\text{ V}}{\sqrt{2}} = 70.7\text{ V} \qquad \Diamond$$

Since the ideal voltmeter has infinite resistance, the resistance of the parallel combination of resistor and voltmeter is just the resistance of the resistor. From Ohm's law, the rms current through the circuit, and hence the reading of the ammeter, is

$$I = \frac{\Delta V}{R} = \frac{70.7\text{ V}}{24\ \Omega} = 3.0\text{ A} \qquad \Diamond$$

---

9. What value of capacitor must be inserted in a 60-Hz circuit in series with a generator of 170 V maximum output voltage to produce an rms current output of 0.75 A?

Solution The ac voltage across a capacitor and the ac current in a branch containing a capacitor obey a relation similar to Ohm's law, namely

$\Delta V_C = IX_C$. Here, one may use either peak values for the voltage and current or rms values for both quantities. However, one should never mix peak and rms values. The quantity X_c is the capacitive reactance and has a value of

$$X_c = \frac{1}{\omega C} = \frac{1}{2\pi f C}$$

where f is the frequency of the ac voltage and current. In this circuit, $\omega = 2\pi f = 120\pi$ rad/s, the rms voltage across the capacitor is $\Delta V_c = 170 \text{ V}/\sqrt{2} = 120$ V, and the rms current is 0.75 A.

Thus, the capacitive reactance is $$X_c = \frac{\Delta V_C}{I} = \frac{120 \text{ V}}{0.75 \text{ A}} = 1.6 \times 10^2 \ \Omega$$

and the capacitance is $$C = \frac{1}{\omega X_c} = \frac{1}{(120\pi \text{ rad/s})(1.6 \times 10^2 \ \Omega)} = 17 \ \mu\text{F}$$ ◊

---

13. An inductor has a 54.0-Ω reactance at 60.0 Hz. What will be the **peak** current if this inductor is connected to a 50.0-Hz source that produces a 100-V rms voltage?

Solution The inductive reactance of this inductor at 60.0 Hz is $X_L = \omega L = 2\pi(60.0 \text{ Hz})L = 54.0 \ \Omega$. Therefore, the self-inductance of the inductor is

$$L = \frac{X_L}{\omega} = \frac{54.0 \ \Omega}{2\pi(60.0 \text{ Hz})} = 0.143 \text{ H}$$

At a frequency of 50.0 Hz, the reactance of this inductor will be $X_L = \omega L = 2\pi(50.0 \text{ Hz})(0.143 \text{ H}) = 45.0 \ \Omega$. Hence, at 50.0 Hz, the rms current when a 100-V rms voltage exists across the inductor is

$$I = \frac{\Delta V_L}{X_L} = \frac{100 \text{ V}}{45.0 \ \Omega} = 2.22 \text{ A}$$

The **peak** current is then given by $$I_m = I\sqrt{2} = (2.22 \text{ A})\sqrt{2} = 3.14 \text{ A}$$ ◊

---

19. A resistor ($R = 900\ \Omega$), a capacitor ($C = 0.25\ \mu$F), and an inductor ($L = 2.5$ H) are connected in series across a 240-Hz ac source for which $\Delta V_m = 140\ \text{V}$. Calculate the (a) impedance of the circuit, (b) peak current delivered by the source, and (c) phase angle between the current and voltage. (d) Is the current leading or lagging behind the voltage?

Solution The capacitive reactance in this series circuit is

$$X_c = \frac{1}{\omega C} = \frac{1}{2\pi f C} = \frac{1}{2\pi(240\ \text{Hz})\left(0.25\times 10^{-6}\ \text{F}\right)} = 2.7\times 10^3\ \Omega$$

and the inductive reactance is

$$X_L = \omega L = 2\pi f L = 2\pi(240\ \text{Hz})(2.5\ \text{H}) = 3.8\ \Omega\times 10^3\ \Omega$$

(a) The total impedance of the series RLC circuit is $Z = \sqrt{R^2 + (X_L - X_c)^2}$, or

$$Z = \sqrt{(900\ \Omega)^2 + \left(3.8\ \Omega\times 10^3\ \Omega - 2.7\times 10^3\ \Omega\right)^2} = 1.4\times 10^3\ \Omega$$ ◊

(b) The peak current delivered to the circuit by the source is

$$I_m = \frac{\Delta V_m}{Z} = \frac{140\ \text{V}}{1.4\times 10^3\ \Omega} = 0.10\ \text{A}$$ ◊

(c) The phase of the voltage relative to the current is given by the phase angle

$$\phi = \arctan\left(\frac{X_L - X_c}{R}\right) = \arctan\left(\frac{+1.1\times 10^3\ \Omega}{900\ \Omega}\right) = \arctan(1.22) = 51^\circ$$ ◊

(d) Since $X_L > X_c$ for this circuit, the phase angle ϕ is positive, and the voltage leads the current (or the current lags behind the voltage). ◊

27. An inductor and a resistor are connected in series. When connected to a 60-Hz, 90-V source, the voltage drop across the resistor is found to be 50 V and the power dissipated in the circuit is 14 W. Find (a) the value of the resistance and (b) the value of the inductance.

Solution

(a) The average power dissipated in an ac circuit is $P = I(\Delta V)\cos\phi$. From the phasor diagram (shown in the sketch), it is clear that the voltage across the resistor is $\Delta V_R = \Delta V\cos\phi$. Thus, the power can be written as $P = I(\Delta V_R)$ and the rms current in the circuit is

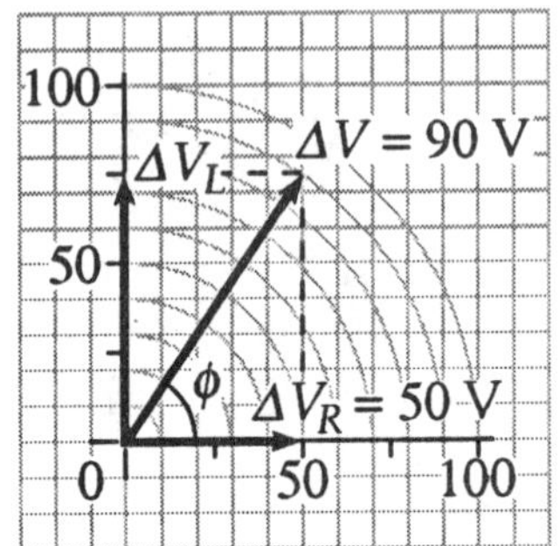

Phasor diagram for an *RL* circuit

$$I = \frac{P}{\Delta V_R} = \frac{14\text{ W}}{50\text{ V}} = 0.28\text{ A}$$

The resistance of the resistor in the circuit is then given by

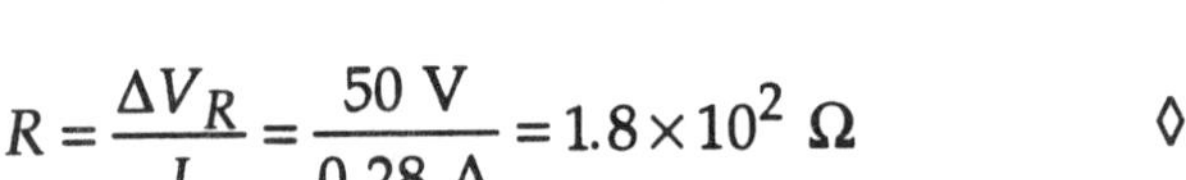

$$R = \frac{\Delta V_R}{I} = \frac{50\text{ V}}{0.28\text{ A}} = 1.8\times10^2\ \Omega \qquad \Diamond$$

(b) Observe from the phasor diagram that $(\Delta V)^2 = (\Delta V_R)^2 + (\Delta V_L)^2$ Therefore, the voltage across the inductor is

$$\Delta V_L = \sqrt{(\Delta V)^2 - (\Delta V_R)^2} = \sqrt{(90\text{ V})^2 - (50\text{ V})^2} \quad \text{or} \quad \Delta V_L = 75\text{ V}$$

The inductive reactance, $X_L = \Delta V_L/I$, is then $X_L = 75\text{ V}/0.28\text{ A} = 2.7\times10^2\ \Omega$ and the self-inductance in the circuit is

$$L = \frac{X_L}{\omega} = \frac{2.7\times10^2\ \Omega}{2\pi(60.0\text{ Hz})} = 0.71\text{ H} \qquad \Diamond$$

31. The AM band extends from approximately 500 kHz to 1600 kHz. If a 2.0-μH inductor is used in a tuning circuit for a radio, what are the extremes that a capacitor must reach in order to cover the complete band of frequencies?

Solution

A radio is tuned by adjusting the resonance frequency of a series *RLC* circuit to match the broadcast frequency of the desired station. Resonance is achieved when the capacitive and inductive reactances are equal, or $\omega_0 L = 1/(\omega_0 C)$. The angular frequency at resonance is therefore $\omega_0 = 2\pi f_0 = 1/\sqrt{LC}$, and the tuned broadcast frequency (in Hertz) is

$$f_0 = \frac{1}{2\pi\sqrt{LC}}$$

For a fixed inductance, $L = 2.0\ \mu\text{H}$, the capacitance needed in the circuit to tune to a frequency f_0 is

$$C = \frac{1}{(4\pi^2 L)f_0^{\,2}} = \frac{1}{4\pi^2(2.0\times10^{-6}\,\text{H})f_0^{\,2}}$$

To reach the lowest frequency in the AM band, the needed capacitance is

$$C_{max} = \frac{1}{4\pi^2(2.0\times10^{-6}\,\text{H})(500\times10^3\ \text{Hz})^2} = 5.1\times10^{-8}\ \text{F} = 51\ \text{nF}$$

The capacitance required to tune the highest frequency in the AM band is

$$C_{min} = \frac{1}{4\pi^2(2.0\times10^{-6}\,\text{H})(1600\times10^3\ \text{Hz})^2} = 4.9\times10^{-9}\ \text{F} = 4.9\ \text{nF}$$

Thus, to reach the full range of frequencies in the AM radio band, the capacitance must be adjustable from 4.9 nF to 51 nF. ◊

37. A transformer on a pole near a factory steps the voltage down from 3600 V to 120 V. The transformer is to deliver 1000 kW to the factory at 90% efficiency. Find (a) the power delivered to the primary, (b) the current in the primary, and (c) the current in the secondary.

Solution

(a) The efficiency of the transformer is: $\text{eff} = \dfrac{P_{\text{output}}}{P_{\text{input}}} = 0.90$

Since the factory requires a delivered power (i.e., power output from the transformer) of 1000 kW, the required power input to the primary of the transformer is

$$P_{\text{input}} = \frac{P_{\text{output}}}{0.90} = \frac{1000 \text{ kW}}{0.90} = 1.1 \times 10^3 \text{ kW} \qquad \Diamond$$

(b) In terms of the primary voltage and current, the input power for the transformer is $P_{\text{input}} = (\Delta V_p) I_p$. Therefore, the required primary current is

$$I_p = \frac{P_{\text{input}}}{\Delta V_p} = \frac{1.1 \times 10^3 \text{ kW}}{\Delta V_p} = \frac{1.1 \times 10^6 \text{ W}}{3600 \text{ V}} = 3.1 \times 10^2 \text{ A} \qquad \Diamond$$

(c) The power output from a transformer is the product of the secondary voltage and current, or $P_{\text{output}} = (\Delta V_s) I_s$. Hence, the secondary current for this transformer is

$$I_s = \frac{P_{\text{output}}}{\Delta V_s} = \frac{1000 \text{ kW}}{\Delta V_s} = \frac{1.0 \times 10^6 \text{ W}}{120 \text{ V}} = 8.3 \times 10^3 \text{ A} \qquad \Diamond$$

41. A particular electromagnetic wave traveling in vacuum has a maximum magnetic field intensity of 1.5×10^{-7} T. Find (a) the electric field intensity and (b) the average power per unit area associated with the wave.

Solution (a) In an electromagnetic wave, the ratio of the intensity of the associated electric and magnetic fields is equal to the speed of light, or $E/B = \left(E\sqrt{2}/B\sqrt{2}\right) = E_m/B_m = c$. Thus, the peak intensity of the electric field in this wave is

$$E_m = cB_m = \left(3.00 \times 10^8 \text{ m/s}\right)\left(1.5 \times 10^{-7} \text{ T}\right) = 45 \text{ V/m} = 45 \text{ N/C} \qquad \Diamond$$

(b) The intensity of the wave is identical to the average power per unit area transported by an electromagnetic wave. Therefore, the intensity is

$$\text{Intensity} = \frac{\text{power}}{\text{area}} = \frac{E_m B_m}{2\mu_0} = \frac{(45 \text{ N/C})\left(1.5 \times 10^{-7} \text{ T}\right)}{2\left(4\pi \times 10^{-7} \text{ N/A}^2\right)} = 2.7 \text{ W/m}^2 \qquad \Diamond$$

47. An important news announcement is transmitted by radio waves to people who are 100 km away, sitting next to their radios, and by sound waves to people sitting across the newsroom, 3.0 m from the newscaster. Who receives the news first? Explain. Take the speed of sound in air to be 343 m/s.

Solution The time for the radio wave to travel 100 km (at the speed of light) to the people by their radio is:

$$t_1 = \frac{100 \text{ km}}{c} = \frac{100 \times 10^3 \text{ m}}{3.00 \times 10^8 \text{ m/s}} = 3.33 \times 10^{-4} \text{ s} = 0.333 \text{ ms}$$

The time required for the sound waves to travel 3.0 m (at the speed of sound) across the newsroom is

$$t_2 = \frac{3.0 \text{ m}}{v_s} = \frac{3.0 \text{ m}}{343 \text{ m/s}} = 8.7 \times 10^{-3} \text{ s} = 8.7 \text{ ms}$$

Therefore, the listeners by the radio 100 km away hear the news before the people located 3.0 m from the newscaster. The difference in the reception times is

$$\Delta t = t_2 - t_1 = 8.7 \text{ ms} - 0.333 \text{ ms} = 8.4 \text{ ms} = 8.4 \times 10^{-3} \text{ s} \qquad \Diamond$$

53. A series *RLC* circuit has a resonance frequency of $2000/\pi$ Hz. When it is operating at a frequency of $\omega > \omega_0$, $X_L = 12\ \Omega$ and $X_C = 8.0\ \Omega$. Calculate the values of *L* and *C* for the circuit.

Solution At the resonance frequency, the capacitive and inductive reactances are equal, or $X_L = X_C$. Thus, at resonance,

$$2\pi f_0 L = \frac{1}{2\pi f_0 C}$$

Solving for the product of the capacitance and the inductance, and recognizing that $f_0 = 2000/\pi$ Hz, we obtain

$$LC = \frac{1}{(2\pi f_0)^2} \quad \text{or} \quad LC = \frac{1}{4\pi^2(2000/\pi\ \text{Hz})^2} = 6.25\times10^{-8}\ \text{s}^2 \qquad \textbf{[Equation 1]}$$

At any frequency, the product of the inductive and capacitive reactances is

$$X_L X_C = (2\pi f L)\left(\frac{1}{2\pi f C}\right) = \frac{L}{C}$$

Therefore, if $X_L = 12\ \Omega$ and $X_C = 8.0\ \Omega$ at some frequency greater than f_0, the ratio of the inductance to the capacitance in this circuit is

$$\frac{L}{C} = X_L X_C = (12\ \Omega)(8.0\ \Omega) = 96\ \Omega^2 \qquad \textbf{[Equation 2]}$$

From Equation 2, $L = (96\ \Omega^2)C$ and substituting this result into Equation 1 gives $(96\ \Omega^2)C^2 = 6.25\times10^{-8}\ \text{s}^2$. The capacitance in this circuit is then

$$C = \sqrt{\frac{6.25\times10^{-8}\ \text{s}^2}{96\ \Omega^2}} = 2.6\times10^{-5}\ \text{F} = 26\ \mu\text{F} \qquad \lozenge$$

Then, Equation 2 gives

$$L = (96\ \Omega^2)C = (96\ \Omega^2)(2.6\times10^{-5}\ \text{F}) = 2.5\times10^{-3}\ \text{H} = 2.5\ \text{mH} \qquad \lozenge$$

55. Two connections allow contact with two circuit elements in series inside a box, but it is not known whether the circuit elements are *R*, *L*, or *C*. In an attempt to find what is inside the box, you make some measurements, with the following results. When a 3.0-V dc power supply is connected across the terminals, there is a direct current of 300 mA in the circuit. When a 3.0-V, 60-Hz source is connected, the current becomes 200 mA. (a) What are the two elements in the box? (b) What are their values of *R*, *L*, or *C*?

Solution Since a steady dc current can flow through the series circuit inside the box, neither of the circuit elements can be a capacitor. Only a transitory dc current can flow in a branch containing a capacitor (as the capacitor is charging or discharging). Also, since the current is finite when a dc power supply is used, at least one of the elements must be a resistor. The total resistance within the box is

$$R = \frac{\Delta V_{dc}}{I_{dc}} = \frac{3.0 \text{ V}}{0.30 \text{ A}} = 10 \ \Omega$$

When a 60-Hz ac source is used, the circuit is found to have an impedance of

$$Z = \frac{\Delta V_{ac}}{I_{ac}} = \frac{3.0 \text{ V}}{0.20 \text{ A}} = 15 \ \Omega$$

Note that this is greater than the total resistance found above. Thus, the circuit has a non-zero reactance. Since a capacitor has been ruled out as a possible circuit element, one must conclude that the second element is an inductor. The inductive reactance is found from

$$Z = \sqrt{R^2 + X_L^2}, \text{ or } X_L = \sqrt{Z^2 - R^2} = \sqrt{(15 \ \Omega)^2 - (10 \ \Omega)^2} = 11 \ \Omega$$

Hence, the inductance in the circuit is

$$L = \frac{X_L}{2\pi f} = \frac{11 \ \Omega}{2\pi (60 \text{ Hz})} = 0.030 \text{ H} = 30 \text{ mH}$$

The conclusion is that the series combination within the box must consist of a 10 Ω resistance and a 30 mH inductance. ◊

62. A transmission line with a resistance per unit length of $4.5 \times 10^{-4}\ \Omega/\text{m}$ is to be used to transmit 5000 kW of power over a distance of 400 miles (6.44×10^5 m). The output voltage of the generator is 4500 V. (a) What is the line loss if a transformer is used to step up the voltage to 500 kV? (b) What fraction of the input power is lost to the line under these circumstances? (c) What difficulties would be encountered in an attempt to transmit the 5000 kW of power at the generator voltage of 4500 V?

Solution From the resistance per unit length, we find the total resistance of the transmission line, $R_{\text{line}} = \left(4.5 \times 10^{-4}\ \Omega/\text{m}\right)\left(6.44 \times 10^5\ \text{m}\right) = 2.9 \times 10^2\ \Omega$

(a) If the power is transmitted at a voltage of 500 kV, the current in the transmission line is $I = \text{Power}/\Delta V = 5000 \times 10^3\ \text{W}/500 \times 10^3\ \text{V} = 10\ \text{A}$, and the power loss in the line is

$$P_{\text{loss}} = I^2 R_{\text{line}} = (10\ \text{A})^2\left(2.9 \times 10^2\ \Omega\right) = 2.9 \times 10^4\ \text{W} = 29\ \text{kW} \qquad \Diamond$$

(b) To get the fractional power loss, we divide the loss by the delivered power. $P_{\text{loss}}/P_{\text{delivered}} = 29\ \text{kW}/5000\ \text{kW} = 0.0058$. Thus, when the power is transmitted at a voltage of 500 kV, the percentage lost as heat in the transmission line is 0.58%. ◊

(c) It is impossible to deliver 5000 kW of power to the customer through this transmission line with an input voltage of only 4500 V. With an input voltage of 4500 V, the power input from the generator is $P_{\text{input}} = (\Delta V)I = (4500\ \text{V})I$. The maximum current that can exist in this line with a 4500-V input occurs when the transmission line is shorted out at the customer's end, and the minimum resistance of the circuit is $R_{\min} = R_{\text{line}} = 290\ \Omega$. Hence, $I_{\max} = \Delta V/R_{\min} = 4500\ \text{V}/290\ \Omega = 15.5\ \text{A}$, and the maximum power input to this transmission line with an input voltage of 4500 V is

$$\left(P_{\text{input}}\right)_{\max} = (4500\ \text{V})(15.5\ \text{A}) = 6.98 \times 10^4\ \text{W} = 69.8\ \text{kW}$$

This is far less than the power the customer wants delivered (5000 kW) and it is all dissipated in the transmission line. ◊

CHAPTER SELF-QUIZ

1. When a series *RLC* circuit is in resonance, the maximum voltage across
   a. the resistor and inductance must be equal
   b. the resistor and capacitor must be equal
   c. the inductance and capacitor must be equal
   d. none of them must be equal

2. Find the resonant frequency for a series *RLC* circuit where $R = 10.0\ \Omega$, $C = 5.00\ \mu F$, and $L = 2.00$ mH.
   a. 998 Hz
   b. 1.59 kHz
   c. 2.45 kHz
   d. 11.3 kHz

3. A radio tuning circuit has a coil with an inductance of 5.00 mH. What must be the capacitance if the set is to be tuned to 980 kHz?
   a. 5.27×10^{-6} microfarads
   b. 4.71×10^{-3} microfarads
   c. 1.89×10^{-3} microfarads
   d. 5.14 microfarads

4. An ac voltage source is connected to a capacitor. The charge on the capacitor will be a maximum when
   a. the voltage and current in the circuit are both a maximum
   b. the voltage and current are both zero
   c. the voltage is a maximum and the current is zero
   d. the current is a maximum and the voltage is zero

5. The units of frequency times capacitance ($F \cdot s^{-1}$) is the same as the units for
   a. current (A)
   b. energy (J)
   c. resistance (R)
   d. conductivity (R^{-1})

6. When an ac voltage is connected to an inductance coil, the energy stored in the magnetic field of the coil will be a maximum when
   a. the voltage across the coil is a maximum
   b. the voltage across the coil is zero
   c. the product of voltage times current is a maximum
   d. there is never any energy stored in such a magnetic field

7. A resistor, inductor, and capacitor are connected in series, with effective rms voltages of 65.0 V, 140 V, and 80.0 V, respectively. What is the power factor in the circuit?
   a. 0.68
   b. 0.74
   c. 0.87
   d. 0.93

8. An *LC* circuit is set up to produce electromagnetic waves. If the inductor has a value of 2.00 mH and the capacitor a value of 1.26×10^{-12} F, what is the resonant frequency of the circuit?
   a. 3 MHz
   b. 30 MHz
   c. 300 MHz
   d. 3000 MHz

9. At a distance of 10.0 km from a radio transmitter, the amplitude of the E-field is 0.20 volts/meter. What is the total power emitted by the radio transmitter?
   a. 10.0 kW
   b. 67.0 kW
   c. 140 kW
   d. 245 kW

10. What is the self-inductance of a coil which has an inductive reactance of 35.0 ohms at an angular frequency of 500 rad/s?
   a. 14.3 mH
   b. 11.0 mH
   c. 70.0 mH
   d. 71.5 mH

11. What is the capacitance of a capacitor which has a capacitive reactance of 85.0 ohms at an angular frequency of 500 rad/s?
   a. 1.85 μF
   b. 3.70 μF
   c. 9.00 μF
   d. 23.5 μF

12. A 100-kW radio station emits EM waves in all directions from an antenna on top of a mountain. What is the intensity in Watts/m^2 of the signal at a distance of 10.0 km?
   a. 8.00×10^{-5} W/m^2
   b. 8.00×10^{-6} W/m^2
   c. 8.00×10^{-4} W/m^2
   d. 0.80 W/m^2

Chapter 22

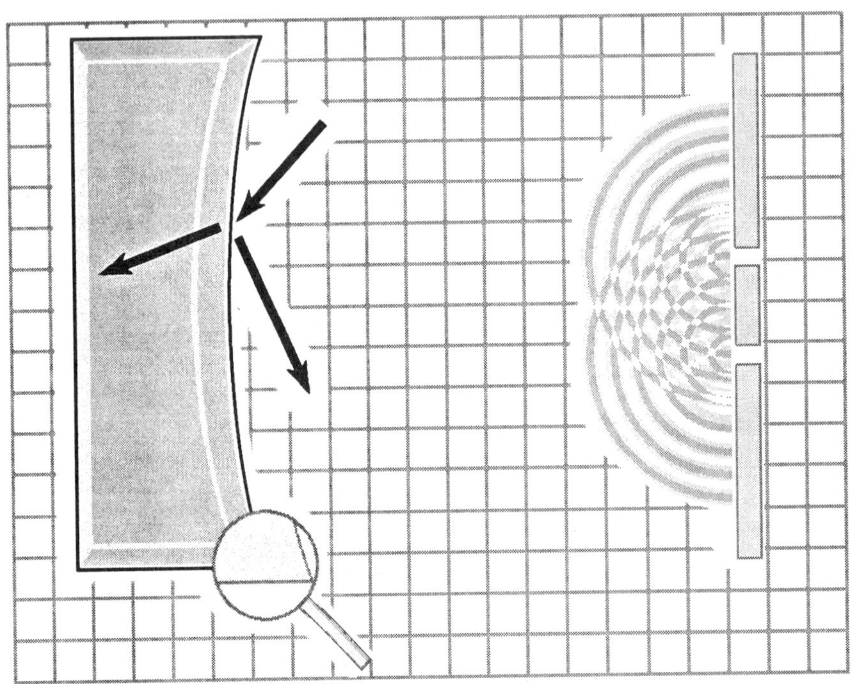

REFLECTION AND REFRACTION OF LIGHT

Chapter 22

REFLECTION AND REFRACTION OF LIGHT

The chief architect of the particle theory of light was Newton. With this theory he provided simple explanations of some known experimental facts concerning the nature of light, namely the laws of reflection and refraction.

In 1678 a Dutch physicist and astronomer, Christian Huygens (1629-1695), showed that a wave theory of light could also explain the laws of reflection and refraction. The wave theory did not receive immediate acceptance for several reasons. All the waves known at the time (sound, water, and so on) traveled through some sort of medium, but light from the Sun could travel to Earth through empty space. Furthermore, it was argued that if light were some form of wave, it would bend around obstacles; hence, we should be able to see around corners. It is now known that light does indeed bend around the edges of objects. This phenomenon, known as **diffraction,** is not easy to observe because light waves have such short wavelengths. For more than a century most scientists rejected the wave theory and adhered to Newton's particle theory. This was, for the most part, due to Newton's great reputation as a scientist.

NOTES FROM SELECTED CHAPTER SECTIONS

22.1 The Nature of Light

Until the beginning of the 19th century, light was considered to be a stream of particles, emitted by a light source. The first clear demonstration of the wave nature of light was provided in 1801 by Thomas Young, who showed that under appropriate conditions, light exhibits interference behavior.

An important development concerning the theory of light was the prediction that light was a form of high-frequency electromagnetic wave. On the other hand, the photoelectric effect could only be explained on the basis that light is composed of "corpuscles" or discontinuous quanta of energy. In view of these developments, light must be regarded as having a **dual nature**. That is, **in some cases light acts as a wave and in others it acts as a particle.**

22.2 Measurements of the Speed of Light

The first successful **estimate** of the speed of light was made in 1675 by the Danish astronomer Roemer and involved astronomical observations of one of the moons of Jupiter. Roemer's experiment is important historically because it demonstrated that light does have a **finite speed** and established a **rough estimate** ($\sim 2.1\times 10^8$ m/s) of the magnitude of that speed.

The first successful method of **measuring** the speed of light using purely earthbound techniques was developed in 1849 by Fizeau. Using a rotating, toothed wheel, Fizeau arrived at a value of $c = 3.1\times 10^8$ m/s. Similar measurements made by subsequent investigators have resulted in more accurate values for c, approximately 2.9977×10^8 m/s. The speed of light has been determined with such high accuracy that it is now used to define the SI unit of length, the meter.

22.4 Reflection and Refraction

A line drawn perpendicular to a surface at the point where an incident ray strikes the surface is called the **normal line**. Angles of reflection and refraction are measured relative to the normal.

When an incident ray undergoes partial reflection and partial refraction, the incident, reflected and refracted rays are all **in the same plane.**

The **path of a light ray through a refracting surface is reversible**.

As light travels from one medium into another, the **frequency does not change.**

22.6 Dispersion and Prisms

An important property of the index of refraction (n) is that its value in anything but vacuum depends on the wavelength of light. This phenomenon is called **dispersion**. Since n is a function of the wavelength, when a light beam is incident on the surface of a refracting material, **different wavelengths are bent at different angles.** The index of refraction of a given material decreases with increasing wavelength. This means that blue light bends more than red light, when passing into a refracting material.

22.8 Huygens' Principle

Every point on a given wave front can be considered as a point source for a **secondary wavelet.** At some later time, the new position of the wave front is determined by the surface tangent to the set of secondary wavelets.

22.9 Total Internal Reflection

Total internal reflection is possible only when light rays traveling in one medium are incident on an interface bounding a second medium of **lesser** index of refraction than the first. Total internal reflection of light occurs at angles of incidence $\theta \geq \theta_c$ (See Equation 22.9), where $n_1 > n_2$.

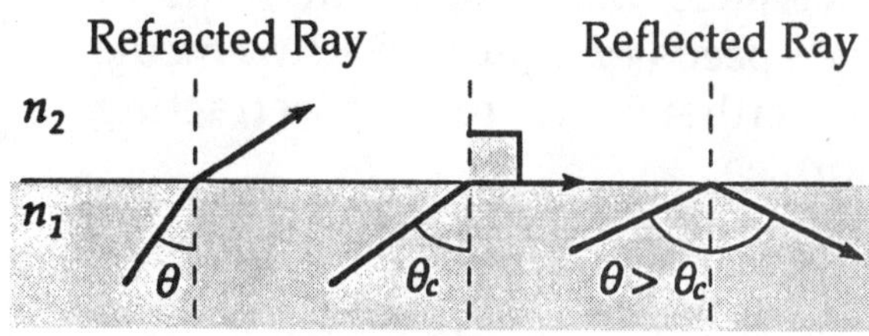

Diagram of total internal reflection

EQUATIONS AND CONCEPTS

The energy of a photon is proportional to the frequency of the associated electromagnetic wave.

$$E = hf \qquad (22.1)$$

The constant of proportionality in Equation 22.1 is known as Planck's constant.

$$h = 6.63 \times 10^{-34}\ \text{J} \cdot \text{s}$$

The law of reflection states that the angle of incidence (the angle measured between the incident ray and the normal line) equals the angle of reflection (the angle measured between the reflected ray and the normal line).

$$\theta_1' = \theta_1 \qquad (22.2)$$

This is one form of the statement of Snell's law. The angle of refraction (measured relative to the normal line) depends on the angle of incidence, and also on the ratio of the speeds of light in the two media on either side of the refracting surface.

$$\frac{\sin\theta_2}{\sin\theta_1} = \frac{v_2}{v_1} = \text{constant} \qquad (22.3)$$

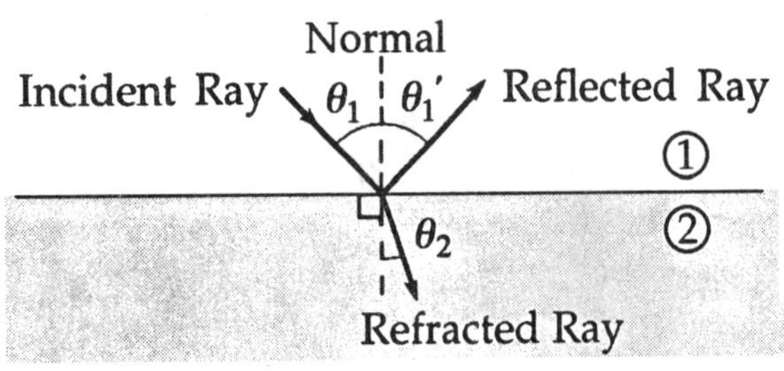

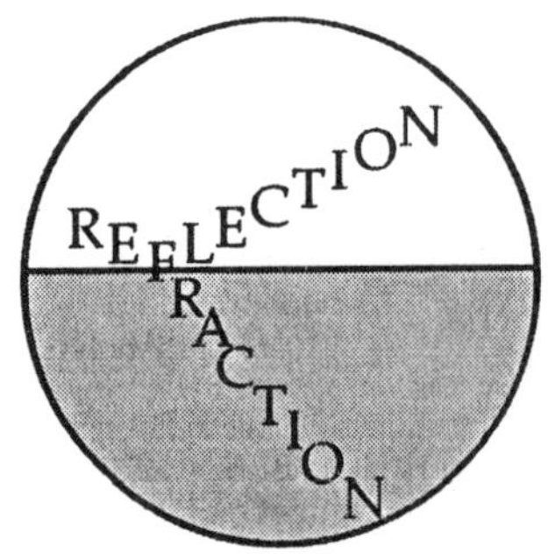

The most widely used and most practical form of Snell's law involves a parameter called the index of refraction. The index of refraction is defined in Equations 22.4 and 22.7.

$$n_1 \sin\theta_1 = n_2 \sin\theta_2 \quad (22.8)$$

Each particular transparent medium is characterized by a particular index of refraction, which equals the ratio of the speed of light in vacuum to the speed of light in the medium.

$$n = \frac{c}{v} \quad (22.4)$$

The frequency of any wave is characteristic of its source. Thus, as light travels from one medium into another of different index of refraction, the frequency remains constant but the wavelength changes. The index of refraction of a given medium can be expressed as the ratio of the wavelength of light in vacuum to the wavelength in that medium.

$$n = \frac{\lambda_0}{\lambda_n} \quad (22.7)$$

For angles of incidence equal to or greater than the critical angle, the incident ray will be totally internally reflected back into the first medium.

$$\sin\theta_c = \frac{n_2}{n_1} \quad (22.9)$$

$$n_1 > n_2$$

Total internal reflection is possible only when a light ray is directed from a medium of high index of refraction into a medium of lower index of refraction.

Comment on internal reflection

REVIEW CHECKLIST

▷ Describe the various experimental results which support the view of the dual nature of light (including Young's experiment and the photoelectric effect).

▷ Describe the methods used by Roemer and Fizeau for the measurement of c.

▷ Understand the conditions under which total internal reflection can occur in a medium and determine the critical angle for a given pair of adjacent media.

▷ Describe the process of dispersion of a beam of white light as it passes through a prism.

▷ Describe the conditions under which internal reflection of a light ray is possible; and calculate the critical angle for internal reflection at a boundary between two optical media of known indices of refraction. Describe the application of internal reflection to fiber optics techniques.

SOLUTIONS TO SELECTED END-OF-CHAPTER PROBLEMS

3. Albert A. Michelson very carefully measured the speed of light using an alternative version of the technique developed by Fizeau. Figure P22.3 shows the approach he used. Light was reflected from one face of a rotating eight-sided mirror toward a stationary mirror 35.0 km away. At certain rates of rotation, the returning beam of light was directed toward the eye of an observer as shown. (a) What minimum angular speed

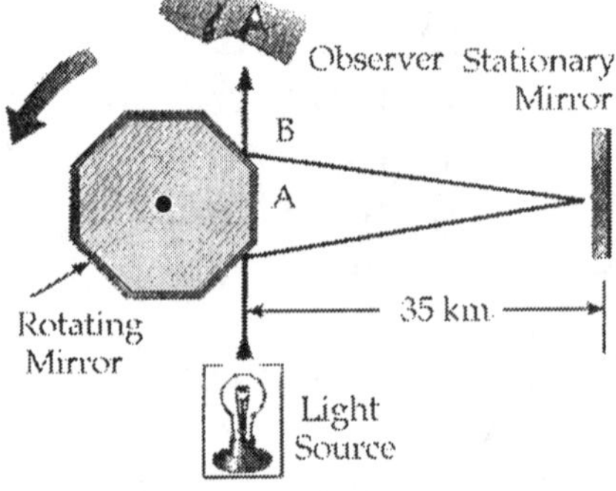

Figure P22.3

must the rotating mirror have in order that side A will have rotated to position B, causing the light to be reflected to the eye? (b) What is the next highest angular speed that will enable the source of light to be seen?

Solution

(a) For light to reach the eye of the observer, the rotating mirror must be in the orientation shown. The time required for light to travel from the rotating mirror to the stationary mirror and back is

$$\Delta t = \frac{2(\text{distance to stationary mirror})}{\text{speed of light}} = \frac{2(35.0 \times 10^3 \text{ m})}{3.00 \times 10^8 \text{ m/s}} = 2.33 \times 10^{-4} \text{ s}$$

If side A of the mirror is to be in position B when the light returns, the smallest acceptable rotation of the mirror in this elapsed time is one-eight of a revolution. This requires that the rotating mirror have an angular speed of

$$\omega_1 = \frac{1/8 \text{ rev}}{\Delta t} = \frac{0.125 \text{ rev}}{2.33 \times 10^{-4} \text{ s}} = 536 \text{ rev/s} \quad \lozenge$$

(b) The next larger rotation the rotating mirror could make in the elapsed time and be in proper orientation to reflect the returning light to the observer's eye is two-eighths (or one-quarter) revolution. In this mode, the angular speed of the mirror would be

$$\omega_2 = \frac{2/8 \text{ rev}}{\Delta t} = \frac{0.250 \text{ rev}}{2.33 \times 10^{-4} \text{ s}} = 1.07 \times 10^3 \text{ rev/s} \quad \lozenge$$

Note: Further consideration of this situation should reveal that an angular speed equal to any integral multiple of that found in part (a) will allow the light to reach the observer's eye.

---

5. The angle between the two mirrors in Figure P22.5 is a right angle. The beam of light in the vertical plane P strikes mirror 1 as shown. (a) Determine the distance the reflected light beam travels before striking mirror 2. (b) In what direction does the light beam travel after being reflected from mirror 2?

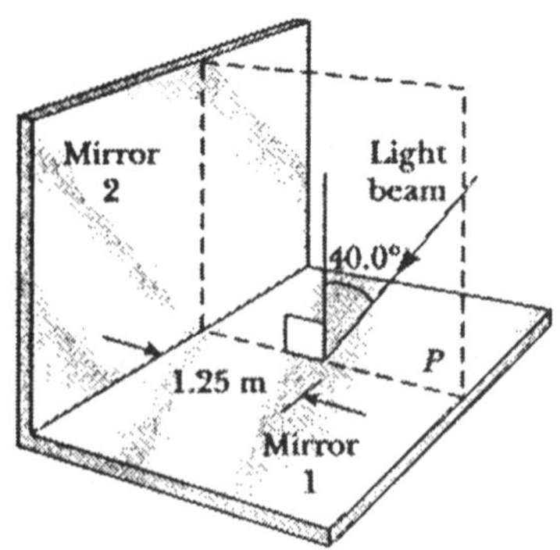

Figure P22.5

Solution (a) The incident light beam, the normal to the surface at the point of reflection, and the reflected beam all lie in the same plane at each reflection. Thus, the beam of light shown remains in the vertical plane P as it reflects from each mirror.

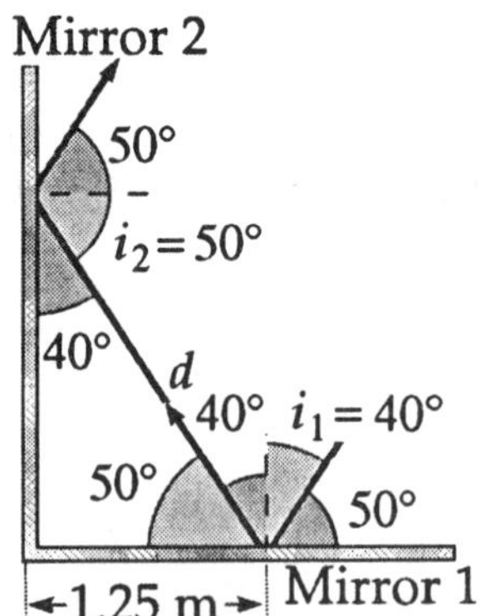

The second diagram shows the two reflections in this plane. The angle of reflection equals the original angle of incidence, $i_1 = 40°$, at the first reflection. Thus, it travels at 50° above the horizontal after leaving the first mirror.

The distance d the reflected beam travels before striking mirror 2 is the hypotenuse of a right triangle formed by the reflected beam and the two mirrors. Its magnitude is found from

$$\frac{1.25\text{ m}}{d} = \cos 50° \qquad \text{or} \qquad d = \frac{1.25\text{ m}}{\cos 50°} = 1.94\text{ m}$$ ◊

(b) The angle of incidence at the second mirror is $i_2 = 50°$. Hence, the reflected beam leaves this mirror at 50° above the horizontal, or 40° from the vertical as shown.

Observe that this is parallel to the original incident beam. ◊

11. A ray of light is incident on the surface of a block of clear ice at an angle of 40.0° with the normal. Part of the light is reflected and part is refracted. Find the angle between the reflected and refracted light.

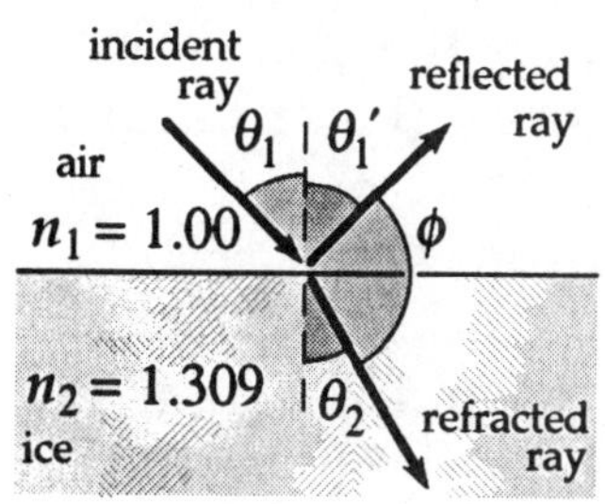

Solution

Consistent with the law of reflection, the angle of reflection is $\theta_1' = \theta_1 = 40.0°$. The angle of refraction is found from Snell's law:

$$n_2 \sin\theta_2 = n_1 \sin\theta_1$$

Thus,

$$\sin\theta_2 = \left(\frac{n_1}{n_2}\right)\sin\theta_1 = \left(\frac{1.00}{1.309}\right)\sin 40.0°$$

Solving for θ_2,

$$\sin\theta_2 = 0.491 \text{ and } \theta_2 = 29.4°$$

From the diagram, observe that

$$\theta_1' + \phi + \theta_2 = 180.0°$$

Thus, we can solve for the angle between the reflected and refracted rays:

$$\phi = 180.0° - \theta_1' - \theta_2$$

or

$$\phi = 180.0° - 40.0° - 29.4° = 110.6° \quad \Diamond$$

17. When the light ray in Problem 16 passes through the glass block, it is shifted laterally by a distance *d* (Fig. P22.16). Find the value of *d*.

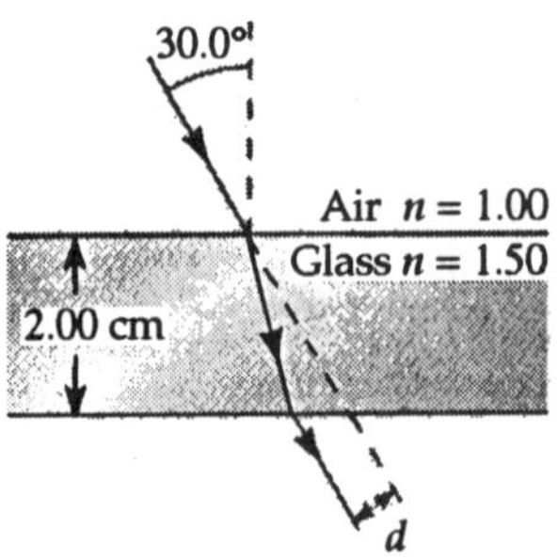

Figure P22.16

Solution

The second diagram gives a magnified view of the path the light ray follows through the glass. As the ray goes from air into glass at point *a*, Snell's law gives the angle of refraction as

$$\sin\theta_2 = \frac{n_1 \sin\theta_1}{n_2} = \frac{(1.00)\sin 30.0^\circ}{1.50} = 0.333 \quad \text{or} \quad \theta_2 = 19.5^\circ$$

Observe from the triangle *abc* that $\cos\theta_2 = \overline{ab}/\overline{ac}$. Thus, the ray travels a distance in the glass of

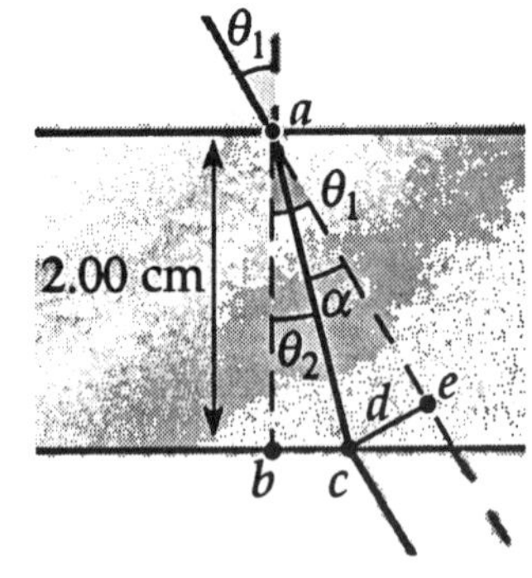

$$\overline{ac} = \frac{\overline{ab}}{\cos\theta_2} = \frac{2.00 \text{ cm}}{\cos 19.5^\circ} = 2.12 \text{ cm}$$

Also observe that angle α in the triangle *ace* is

$$\alpha = \theta_1 - \theta_2 = 30.0^\circ - 19.5^\circ = 10.5^\circ$$

Then, using triangle *ace*, the lateral shift *d* of the ray as it passes through the 2.00-cm thick glass block is seen to be $d = (\overline{ac})\sin\alpha$, or

$$d = (2.12 \text{ cm})\sin(10.5^\circ) = 0.388 \text{ cm} = 3.88 \text{ mm}$$ ◊

---

23. A cylindrical tank with an open top has a diameter of 3.00 m and is completely filled with water. When the setting Sun reaches an angle of 28.0° above the horizon, sunlight ceases to illuminate the bottom of the tank. How deep is the tank?

Solution The diagram shows the path followed by the last ray of sunlight to reach the bottom of the tank. The direction of the incident ray is 28.0° above the horizontal. Thus, the angle of incidence, the angle the ray makes with the normal to the surface of the water at the top of the tank, is $\theta_1 = 62.0°$. From Snell's law,

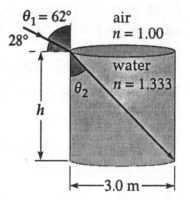

$$\sin\theta_2 = \frac{n_1 \sin\theta_1}{n_2} = \frac{1.00(\sin 62°)}{1.333} \quad \text{so} \quad \theta_2 = 41.5°$$

Now, consider the right triangle formed by the side of the tank, the diameter of the base, and the refracted ray. Observe that $\tan\theta_2 = (3.00\text{ m})/h$. Hence, the depth of the tank is $h = 3.00\text{ m}/\tan 41.5° = 3.39\text{ m}$ ◊

29. The index of refraction for violet light in silica flint glass is 1.66, and that for red light is 1.62. What is the angular dispersion of visible light passing through an equilateral prism of apex angle 60.0° if the angle of incidence is 50.0°? (See Fig. P22.29.)

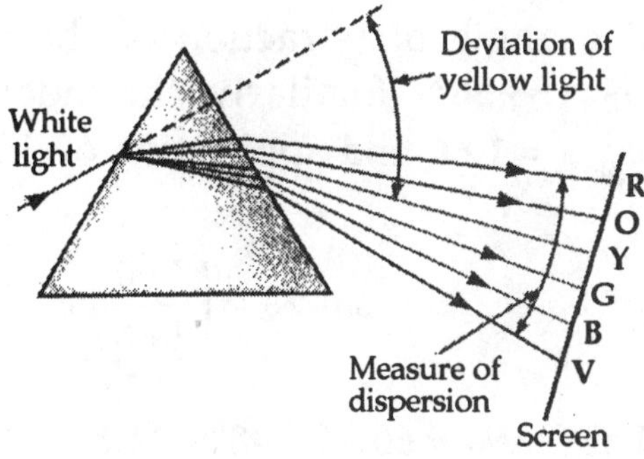

Figure P22.29

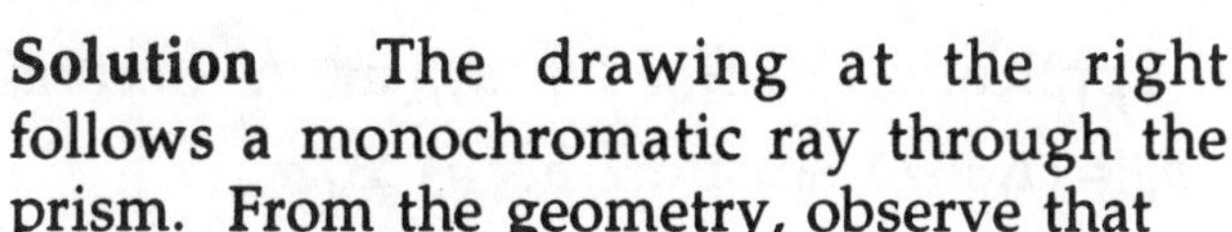

Solution The drawing at the right follows a monochromatic ray through the prism. From the geometry, observe that

$$\beta = 90.0° - \alpha \quad \text{and} \quad \gamma = 180.0° - 60.0° - \beta.$$

Combining these equations gives

$$\gamma = 180.0° - 60.0° - 90.0° + \alpha = 30.0° + \alpha$$

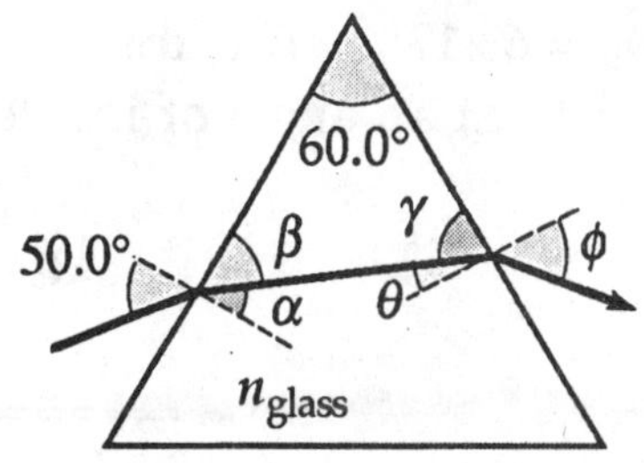

Also, observe that $\theta = 90.0° - \gamma = 90.0° - 30.0° - \alpha$

or $\theta = 60.0° - \alpha$

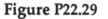

Note that in the previous equation, α is the angle of refraction at the first surface, and θ is the angle of incidence at the second surface.

For red light, the index of refraction of the glass is $n_{glass} = 1.62$. Applying Snell's law to the refraction of red light at the first surface,

$$\sin\alpha_R = \left(\frac{n_{air}}{n_{glass}}\right)\sin 50.0° = \left(\frac{1.00}{1.62}\right)\sin 50.0° = 0.473 \quad \text{or} \quad \alpha_R = 28.22°$$

Thus, $\theta_R = 60° - \alpha_R = 60° - 28.22° = 31.78°$ is the angle of incidence at the second surface. Then, applying Snell's law to the refraction at the second surface gives

$$\sin\phi_R = \left(\frac{n_{glass}}{n_{air}}\right)\sin\theta_R = \left(\frac{1.62}{1.00}\right)\sin 31.8° = 0.853$$

The angle of refraction of the red light at the second surface is therefore $\phi_R = 58.56°$. Similarly, the index of refraction of the glass for violet light is $n_{glass} = 1.66$ and application of Snell's law at the first surface gives

$$\sin\alpha_V = \left(\frac{1.00}{1.66}\right)\sin 50.0° = 0.461 \quad \text{or} \quad \alpha_V = 27.48°$$

Then, $\theta_V = 60° - 27.48° = 32.52°$, and application of Snell's law at the second surface for the violet light gives $\sin\phi_V = (1.66/1.00)\sin 32.52° = 0.892$.

The angle of refraction of the violet light at the second surface is then $\phi_V = 63.17°$. The total angular dispersion of visible light incident on this prism at an angle of incidence of 50.0° is then

$$\Delta\phi = \phi_V - \phi_R = 63.17° - 58.56° = 4.61°$$ ◊

31. A beam of light is incident from air on the surface of a liquid. If the angle of incidence is 30.0° and the angle of refraction is 22.0°, find the critical angle for the liquid when surrounded by air.

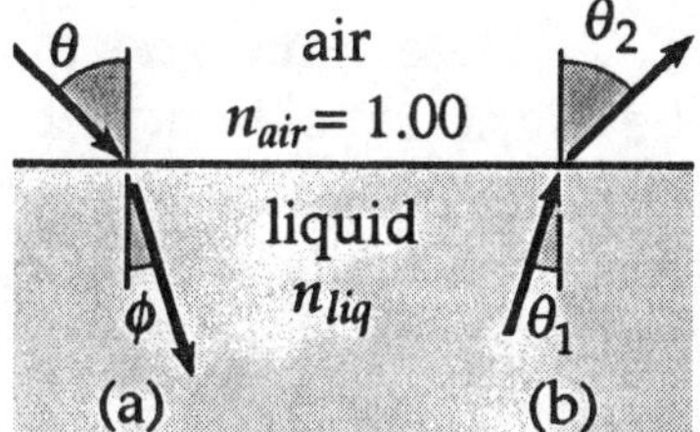

Solution

When light goes from one material into another having a higher index of refraction, it bends toward the normal line as shown in part (a) of the sketch. It is given that when $\theta = 30.0°$, the angle of refraction in the liquid is $\phi = 22.0°$. Thus, from Snell's law, the index of refraction of the liquid is:

$$n_{\text{liq}} = \frac{n_{\text{air}} \sin\theta}{\sin\phi} = \frac{(1.00)\sin 30.0°}{\sin 22.0°} = 1.33$$

When light goes from the liquid into the air as shown in part (b), it is going from one material into a second that has a smaller index of refraction. Under these conditions, the refraction is away from the normal $(\theta_2 > \theta_1)$. If the angle of incidence is equal to the critical angle, θ_c, the angle of refraction is $\theta_2 = 90.0°$ (i.e., the refracted ray goes parallel to the surface and never actually enters the second medium).

For any angle of incidence $\theta_1 > \theta_c$, the light is totally internally reflected. Snell's law gives the critical angle as

$$\sin\theta_c = \frac{n_2 \sin 90.0°}{n_1} = \frac{n_2}{n_1}$$

In this case, where the first medium is the liquid and the second medium is air, the critical angle is found as

$$\sin\theta_c = \frac{n_{\text{air}}}{n_{\text{liq}}} = \frac{1.00}{1.33} = 0.749 \quad \text{and} \quad \theta_c = \arcsin(0.749) = 48.5° \qquad \Diamond$$

35. Determine the maximum angle, θ, for which the light rays incident on the end of the pipe in Figure P22.35 are subject to total internal reflection along the walls of the pipe. Assume that the pipe has an index of refraction of 1.36 and that the outside medium is air.

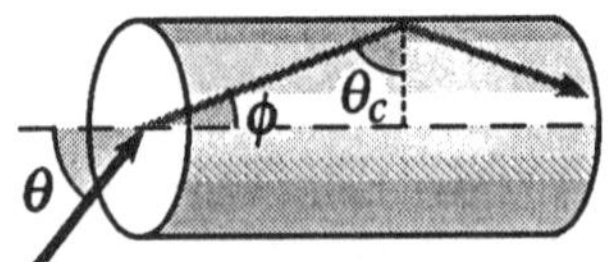

Figure P22.35

Solution If the light ray is incident on the end of the pipe at the maximum angle θ which permits total internal reflection to occur at the wall of the pipe, the angle of incidence when the refracted ray strikes the wall must be the critical angle, θ_c, as shown in the sketch above. From the right triangle formed by the two normal lines (shown as dashed lines) and the refracted ray, observe that the angle of refraction at the end of the pipe is given by $\phi = 90° - \theta_c$. Application of Snell's law to the refraction at the end of the pipe then gives:

$$\sin\theta = \left(\frac{n_{\text{pipe}}}{n_{\text{air}}}\right)\sin\phi = \left(\frac{1.36}{1.00}\right)\sin(90° - \theta_c)$$

Using the trigonometric identity $\sin(90° - \theta_c) = \cos\theta_c$, this equation for the maximum angle of incidence reduces to:

$$\sin\theta = 1.36\cos\theta_c \qquad \textbf{[Equation 1]}$$

When light, traveling in a medium with index of refraction n_1, strikes the boundary between that medium and a second medium of index $n_2 > n_1$, the critical angle is given by the equation $\sin\theta_c = n_2/n_1$. As light inside the pipe approaches the wall,

$$\sin\theta_c = \frac{1.00}{1.36} = 0.735 \qquad \text{and the critical angle is} \qquad \theta_c = 47.3°$$

Equation 1 then gives $\sin\theta = 1.36\cos 47.3° = 0.922$. Hence, the maximum angle of incidence at the end of the pipe that allows total internal reflection to occur at the wall of the pipe is $\theta = \arcsin(0.922) = 67.2°$ ◊

42. A layer of ice, having parallel sides, floats on water. If light is incident on the upper surface of the ice at an angle of incidence of 30.0°, what is the angle of refraction in the water?

Solution

First, we use Snell's law at the upper surface of the ice. If the angle of incidence as the light goes from the air into the ice is $\theta_1 = 30.0°$, then

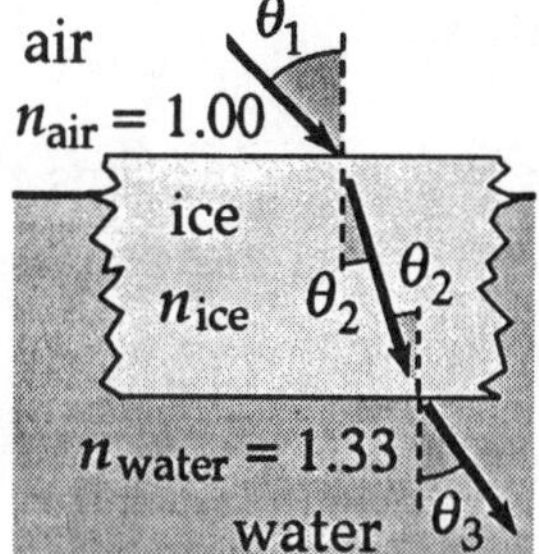

$$n_{ice} \sin\theta_2 = n_{air} \sin 30.0° = (1.00)(0.500)$$

or $$n_{ice} \sin\theta_2 = 0.500 \qquad \textbf{[Equation 1]}$$

Observe that since the layer of ice has parallel sides, the normal lines (shown as dashed lines) are also parallel. Thus, the angle of incidence at the second surface is equal to the angle of refraction at the first surface (alternate interior angles).

Therefore, application of Snell's law to the refraction at the second surface gives

$$n_{ice} \sin\theta_2 = n_{water} \sin\theta_3 \qquad \text{or} \qquad \sin\theta_3 = \frac{n_{ice}\sin\theta_2}{n_{water}}$$

Using Equation 1, this becomes: $$\sin\theta_3 = \frac{0.500}{n_{water}} = \frac{0.500}{1.33} = 0.375$$

The angle of refraction in the water is then $$\theta_3 = \arcsin(0.375) = 22.0° \quad \Diamond$$

Note that the numeric value of the index of refraction of the ice was never used in this solution. Hence, a layer of any transparent material with parallel sides could be substituted for the ice, and the final result would be unchanged.

47. A narrow beam of light is incident from air onto a glass surface with index of refraction 1.56. Find the angle of incidence for which the corresponding angle of refraction is one-half the angle of incidence. (**Hint:** You might want to use the trigonometric identity $\sin 2\theta = 2\sin\theta\cos\theta$.

Solution

If, as the light crosses the boundary from air to glass, the angle of incidence is θ_1 and the angle of refraction is θ_2, Snell's law is

$$n_{glass}\sin\theta_2 = n_{air}\sin\theta_1$$

When the angle of refraction is one-half the angle of incidence (i.e., $\theta_1 = 2\theta_2$), this becomes

$$n_{glass}\sin\theta_2 = n_{air}\sin(2\theta_2)$$

Using the suggested identity gives $\quad n_{glass}\sin\theta_2 = n_{air}2\sin\theta_2\cos\theta_2$

which reduces to $\quad \cos\theta_2 = \frac{1}{2}\left(\frac{n_{glass}}{n_{air}}\right)$

The index of refraction for glass and air are: $\quad n_{glass} = 1.56 \qquad n_{air} = 1.00$

Thus we can solve for the angle of refraction: $\quad \cos\theta_2 = \frac{1}{2}\left(\frac{1.56}{1.00}\right) = 0.780$

$$\theta_2 = 38.7°$$

The desired angle of incidence is then $\quad \theta_1 = 2\theta_2 = 77.5°$ ◊

51. Two light pulses are emitted simultaneously from a source. Both pulses travel to a detector, but one first passes through 6.20 m of ice. Determine the difference in the pulses' times of arrival at the detector.

Solution

The only difference in the paths of the two light pulses is that one travels through 6.20 m of ice while the other travels through 6.20 m of air. The time for one pulse to travel this distance in air is

$$t_1 = \frac{d}{c} = \frac{6.20\text{ m}}{c}$$

The index of refraction of a material is defined as $n = c/v$, where v is the speed of light in that material.

Thus, the speed of light in ice is given by $v_{\text{ice}} = c/n_{\text{ice}}$ and the time required for the second pulse to travel through 6.20 m of ice is

$$t_2 = \frac{6.20\text{ m}}{v_{\text{ice}}} = n_{\text{ice}}\left(\frac{6.20\text{ m}}{c}\right).$$

Since the two pulses start simultaneously, the difference in the arrival times at the detector is

$$\Delta t = t_2 - t_1 = \left(n_{\text{ice}} - 1\right)\left(\frac{6.20\text{ m}}{c}\right)$$

The index of refraction of ice is $n_{\text{ice}} = 1.309$, so this becomes

$$\Delta t = (1.309 - 1)\left(\frac{6.20\text{ m}}{3.00\times 10^{8}\text{ m/s}}\right) = 6.39\times 10^{-9}\text{ s} = 6.39\text{ ns} \qquad \Diamond$$

55. A laser beam strikes one end of a slab of material, as in Figure P22.55. The index of refraction of the slab is 1.48. Determine the number of internal reflections of the beam before it emerges from the opposite end of the slab.

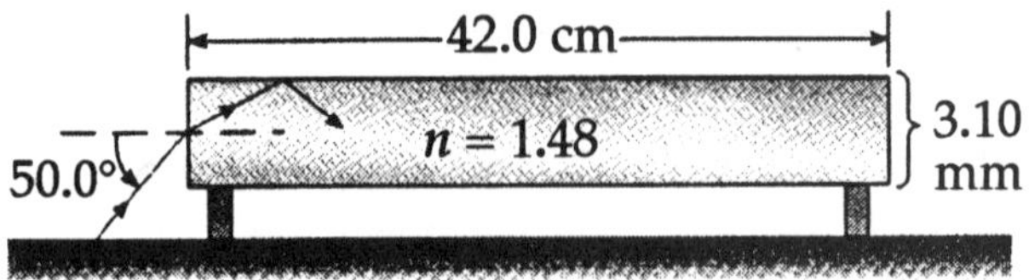

Figure P22.55

Solution As the beam enters the end of the slab, the angle of refraction is found from Snell's law:

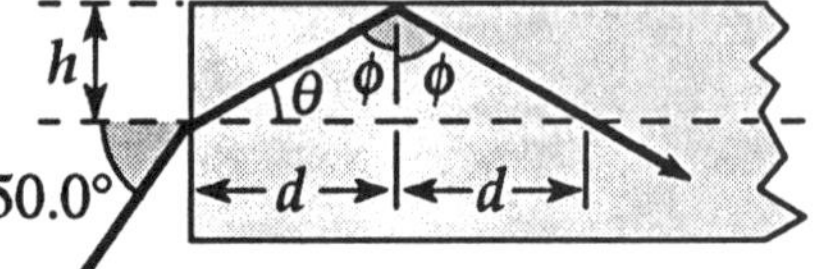

$$\sin\theta = \frac{n_{air}}{n_{slab}}\sin 50.0° = \left(\frac{1.00}{1.48}\right)\sin 50.0° = 0.518 \qquad \text{and} \qquad \theta = 31.2°$$

Note that the two normal lines are perpendicular to each other. Using the right triangle having these lines as two of its sides, observe that the angle of incidence at the top of the slab is $\phi = 90.0° - \theta = 58.8°$. The critical angle as the light tries to go from the slab back into air is

$$\theta_c = \arcsin\left(\frac{n_{air}}{n_{slab}}\right) = \arcsin\left(\frac{1.00}{1.48}\right) = 42.5°$$

Since $\phi > \theta_c$, total internal reflection will indeed occur at the top and bottom of the slab. The distance the beam travels down the length of the slab for each reflection is $2d$, where d is the base of the right triangle shown in the sketch. Given that the altitude of the triangle, h, is one half the thickness of the slab,

$$d = \frac{h}{\tan\theta} = \frac{(3.10 \text{ mm})/2}{\tan\theta} \quad \text{and} \quad 2d = \frac{3.10\times10^{-1}\text{ cm}}{\tan 31.2°} = 5.12\times10^{-1}\text{ cm}$$

The number of internal reflections made before reaching the opposite end of the slab is then

$$N = \frac{\text{length of slab}}{2d} = \frac{42.0 \text{ cm}}{5.12\times10^{-1}\text{ cm}} = 82.0$$ ◊

CHAPTER SELF-QUIZ

1. One phenomenon that demonstrates the particle nature of light is
   a. the photoelectric effect
   b. diffraction effects
   c. interference effects.
   d. the prediction by Maxwell's electromagnetic theory

2. Light in air enters a diamond (n = 2.42) at an angle of incidence of 30.0°. What is the angle of refraction inside the diamond?
   a. 11.9°
   b. 20.0°
   c. 23.8°
   d. 27.6°

3. An underwater scuba diver sees the Sun at an apparent angle of 45.0° from the vertical. How far is the Sun above the horizon? (n_{water} = 1.333)
   a. 9.90°
   b. 13.8°
   c. 19.5°
   d. 27.0°

4. A beam of light is incident upon a flat piece of glass (n = 1.50) at an angle of incidence of 30.0°. Part of the beam is transmitted and part is reflected. What is the angle between the reflected and transmitted rays?
   a. 79.5°
   b. 90.0°
   c. 130.5°
   d. 140°

5. Dispersion occurs when
   a. some materials bend light more than other materials
   b. a material slows down some wavelengths more than others
   c. a material changes some frequencies more than others
   d. light has different speeds in different materials

6. If the velocity of light through an unknown liquid is measured at 2.40×10^8 m/s, what is the index of refraction of this liquid? ($c = 3.00 \times 10^8$ m/s)
   a. 1.80
   b. 1.25
   c. 1.33
   d. 0.80

7. If the critical angle for internal reflection inside a certain transparent material is found to be 48.0°, what is the index of refraction of the material? (Air is outside the material).
   a. 1.35
   b. 1.48
   c. 1.49
   d. 0.743

8. Which of the following describes what will happen to a light ray incident on an air-to-glass boundary at less than the critical angle?
   a. total reflection
   b. total transmission
   c. partial reflection, partial transmission
   d. partial reflection, total transmission

9. A beam of light in air is incident at an angle of 35.0° to the surface of a rectangular block of clear plastic (n = 1.49). The light beam first passes through the block and re-emerges from the opposite side into air at what angle to the normal to that surface?
   a. 42.0°
   b. 23.0°
   c. 35.0°
   d. 59.0°

10. A beam of light in air is incident on the surface of a rectangular block of clear plastic (n = 1.49). If the velocity of the beam before it enters the plastic is 3.00×10^8 m/s, what is its velocity after emerging from the block?
    a. 3.00×10^8 m/s
    b. 1.93×10^8 m/s
    c. 2.01×10^8 m/s
    d. 1.35×10^8 m/s

11. A ray of light in air is incident on an air-to-glass boundary at an angle of 30.0° with the normal. If the index of refraction of the glass is 1.65, what is the angle of the refracted ray within the glass with respect to the normal?
    a. 56.1°
    b. 46.3°
    c. 30.0°
    d. 17.6°

12. A light ray in air is incident on an air-to-glass boundary at an angle of 45.0° and is refracted in the glass of 30.0° with the normal. What is the index of refraction of the glass?
    a. 2.13
    b. 1.74
    c. 1.23
    d. 1.41

Chapter 23

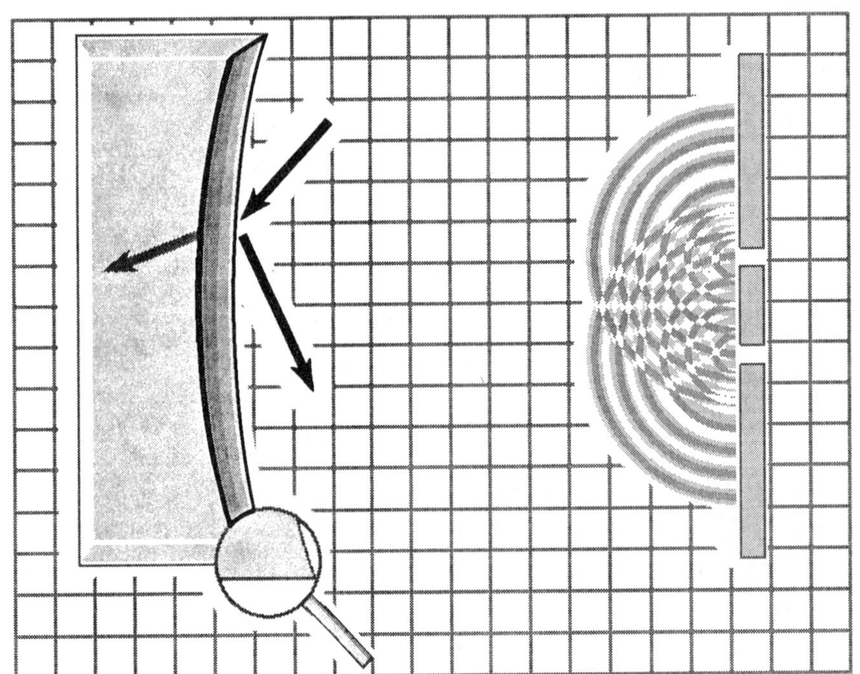

MIRRORS AND LENSES

Chapter 23

MIRRORS AND LENSES

This chapter is concerned with the formation of images when plane and spherical waves fall on plane and spherical surfaces. Images can be formed by either reflection or refraction. Mirrors and lenses form images in both ways. In our study of mirrors and lenses, we continue to use the ray approximation and to assume that light travels in straight lines (in other words, we ignore diffraction). This chapter completes our study of geometric optics. In Chapters 25 and 26 we shall examine the construction and properties of some optical instruments that use mirrors and lenses.

NOTES FROM SELECTED CHAPTER SECTIONS

In equations and diagrams, the **object distance,** p, is the distance from the object to the mirror (or lens). The **image distance,** q, is the distance from the mirror (or lens) to the location of the image.

23.1 Flat Mirrors

The image formed by a flat mirror has the following properties:

1. The image is as far behind the mirror as the object is in front.

2. The image is unmagnified, virtual, and upright. (By upright, we mean that, if the object arrow points upward, so does the image arrow.)

3. The image has right-left reversal.

23.2 Images Formed by Spherical Mirrors

A spherical mirror is a reflecting surface which has the shape of a segment of a sphere. If the inner surface is reflecting the mirror is **concave**; if the outer surface of the sphere is reflecting, then the mirror is **convex.**

Real images are formed at a point when **reflected light actually passes through the point.**

Virtual images are formed at a point when light rays **appear to diverge from the point.**

The point of intersection of any two of the following rays in a ray diagram for mirrors locates the image:

1. The first ray is drawn from the top of the object parallel to the optical axis and is reflected back through the focal point, *F*.
2. The second ray is drawn from the top of the object to the vertex of the mirror and is reflected with the angle of incidence equal to the angle of reflection.
3. The third ray is drawn from the top of the object through the center of curvature, *C*, which is reflected back on itself.

23.6 Thin Lenses

The following three rays form the ray diagram for a thin lens:

1. The first ray is drawn **parallel** to the optic axis. After being refracted by the lens, this ray passes **through** (or appears to come from) one of the focal points.

2. The second ray is drawn **through the center** of the lens. This ray continues in a **straight line.**

3. The third ray is drawn through the **focal point** F, and emerges from the lens **parallel to the optic axis.**

23.7 Lens Aberrations

Aberrations are responsible for the formation of imperfect images by lenses and mirrors. Spherical aberration is due to the variation in focal points for parallel incident rays that strike the lens at various distances from the optical axis. Chromatic aberration arises from the fact that light of different wavelengths focuses at different points when refracted by a lens.

EQUATIONS AND CONCEPTS

Lateral magnification is defined as the ratio of image height to the object height.

$$M = \frac{h'}{h} \tag{23.1}$$

Images formed by plane mirrors have the following properties: (i) the image is as far behind the mirror as the object is in front; (ii) the image is the same size as the object ($M = 1$), virtual, and upright; and (iii) the image has right-left reversal.

Comment on reflection by plane mirrors

Lateral magnification of spherical mirrors can also be expressed as the negative of the ratio of the image distance to the object distance.

$$M = \frac{h'}{h} = -\frac{q}{p} \tag{23.2}$$

The mirror equation is used to determine the location of an image formed by reflection of paraxial rays from a spherical surface. The focal point of a spherical mirror is located midway between the center of curvature and the vertex of the mirror.

$$\frac{1}{p} + \frac{1}{q} = \frac{1}{f} \tag{23.6}$$

$$f = \frac{R}{2} \tag{23.5}$$

A magnified image of an object can be formed by a single spherical refracting surface of radius, R, which separates two media whose indices of refraction are n_1 and n_2.

$$\frac{n_1}{p} + \frac{n_2}{q} = \frac{n_2 - n_1}{R} \tag{23.7}$$

$$M = \frac{h'}{h} = -\frac{n_1 q}{n_2 p} \tag{23.8}$$

A special case is that of a virtual image formed by a plane refracting surface (when the radius, R, is infinite).

$$q = -\frac{n_2}{n_1} p \tag{23.9}$$

The equations for magnification and image location for thin lenses are the same as the corresponding equations for spherical mirrors.

$$M = \frac{h'}{h} = -\frac{q}{p} \qquad (23.10)$$

$$\frac{1}{p} + \frac{1}{q} = \frac{1}{f} \qquad (23.11)$$

A major portion of this chapter is devoted to the development and presentation of equations which can be used to determine the location and nature of images formed by various optical components acting either singly or in combination. It is essential that these equations be used with the correct algebraic sign associated with each quantity involved. You must understand clearly the sign conventions for mirrors, refracting surfaces and lenses.

Comment on sign conventions

The lens makers' equation can be used to calculate the focal length of a thin lens in terms of the radii of curvature of the front and back surfaces and the index of refraction of the lens material.

$$\frac{1}{f} = (n-1)\left(\frac{1}{R_1} - \frac{1}{R_2}\right) \qquad (23.12)$$

SUGGESTIONS, SKILLS, AND STRATEGIES

A major portion of this chapter is devoted to the development and presentation of equations which can be used to determine the location and nature of images formed by various optical components acting either singly or in combination. It is essential that these equations be used with the correct algebraic sign associated with each quantity involved. You must understand clearly the sign conventions for mirrors, refracting surfaces, and lenses. The following discussion represents a review of these sign conventions.

SIGN CONVENTIONS FOR MIRRORS

Equation: $\frac{1}{p}+\frac{1}{q}=\frac{1}{f}=\frac{2}{R}$ $\qquad M=\frac{h'}{h}=-\frac{q}{p}$

The front side of the mirror is the region on which light rays are incident and reflected.

p is + if the object is in front of the mirror (real object).
p is – if the object is in back of the mirror (virtual object).

q is + if the image is in front of the mirror (real image).
q is – if the image is in back of the mirror (virtual image).

Both f and R are + if the center of curvature is in front of the mirror (concave mirror).

Both f and R are – if the center of curvature is in back of the mirror (convex mirror).

If M is positive, the image is upright.
If M is negative, the image is inverted.

You should check the sign conventions as stated against the situations described in Figure 23.1.

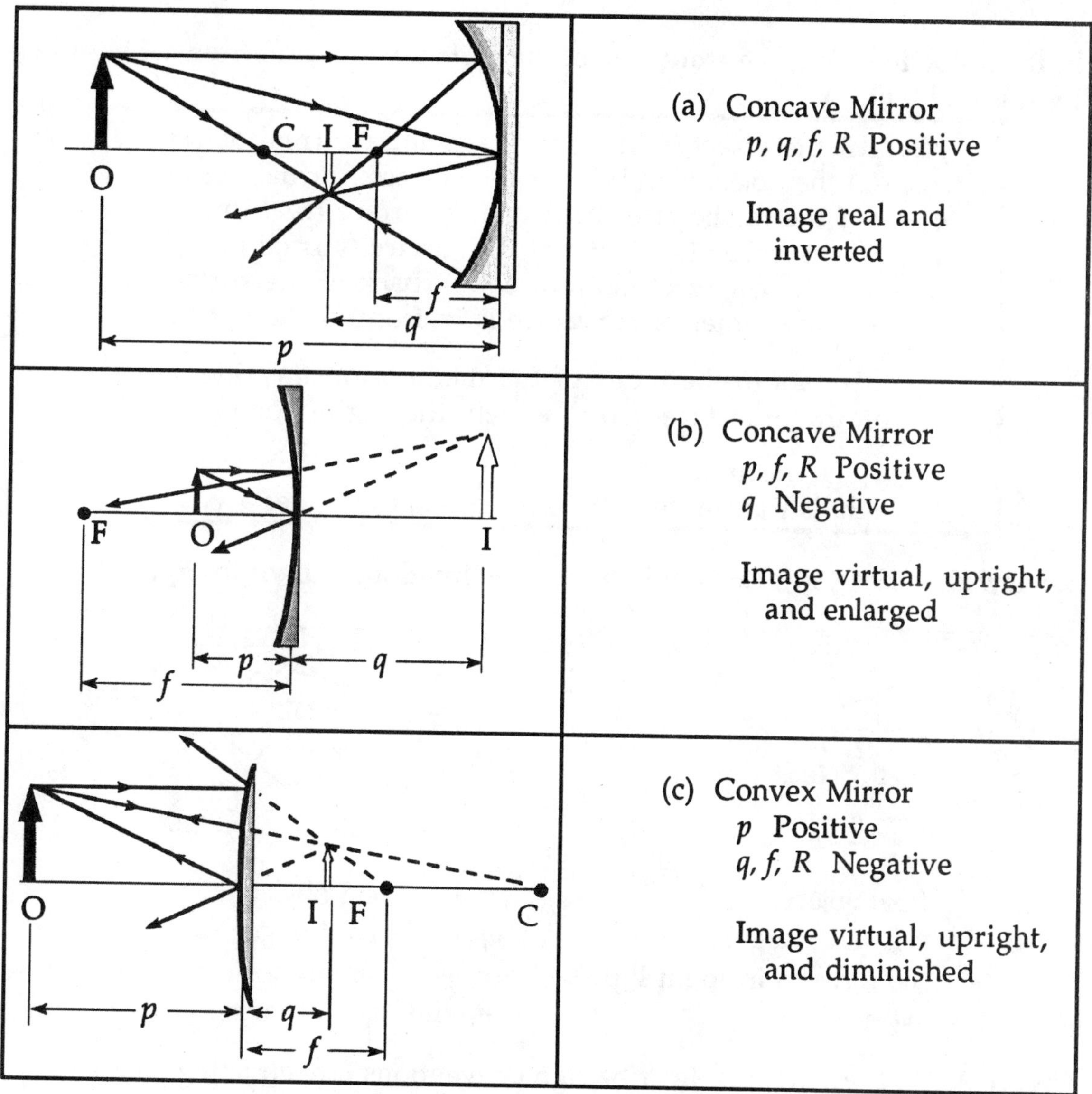

Figure 23.1 Figures describing sign conventions for mirrors.

SIGN CONVENTIONS FOR REFRACTING SURFACES

Equations: $$\frac{n_1}{p} + \frac{n_2}{q} = \frac{n_2 - n_1}{R} \qquad M = \frac{h'}{h} = -\frac{n_1 q}{n_2 p}$$

In the following table, the **front** side of the surface is the side **from which the light is incident.**

p is + if the object is in front of the surface (real object). p is – if the object is in back of the surface (virtual object). q is + if the image is in back of the surface (real image). q is – if the image is in front of the surface (virtual image). R is + if the center of curvature is in back of the surface. R is – if the center of curvature is in front of the surface. n_1 refers to the index of the medium on the side of the interface from which the light comes. n_2 is the index of the medium into which the light is transmitted after refraction at the interface.

Review the above sign conventions for the situations shown in Figure 23.2.

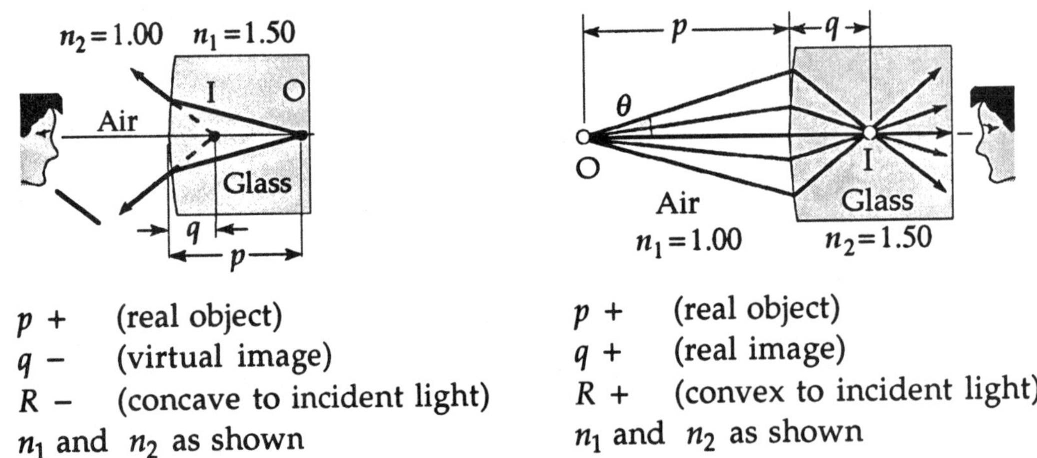

Figure 23.2 These figures describe sign conventions for refracting surfaces. In the first case, the object appears closer than it actually is. In the second case, an object beyond the glass appears to be in the glass.

SIGN CONVENTIONS FOR THIN LENSES

Equations: $\frac{1}{p}+\frac{1}{q}=\frac{1}{f}=(n-1)\left(\frac{1}{R_1}-\frac{1}{R_2}\right)$ $\qquad M=\frac{h'}{h}=-\frac{q}{p}$

In the following table, the **front** of the lens is the **side from which the light is incident.**

p is + if the object is in front of the lens. p is − if the object is in back of the lens.
q is + if the image is in back of the lens. q is − if the image is in front of the lens.
R_1 and R_2 are + if the center of curvature is in back of the lens.
R_1 and R_2 are − if the center of curvature is in front of the lens.

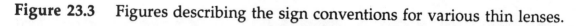

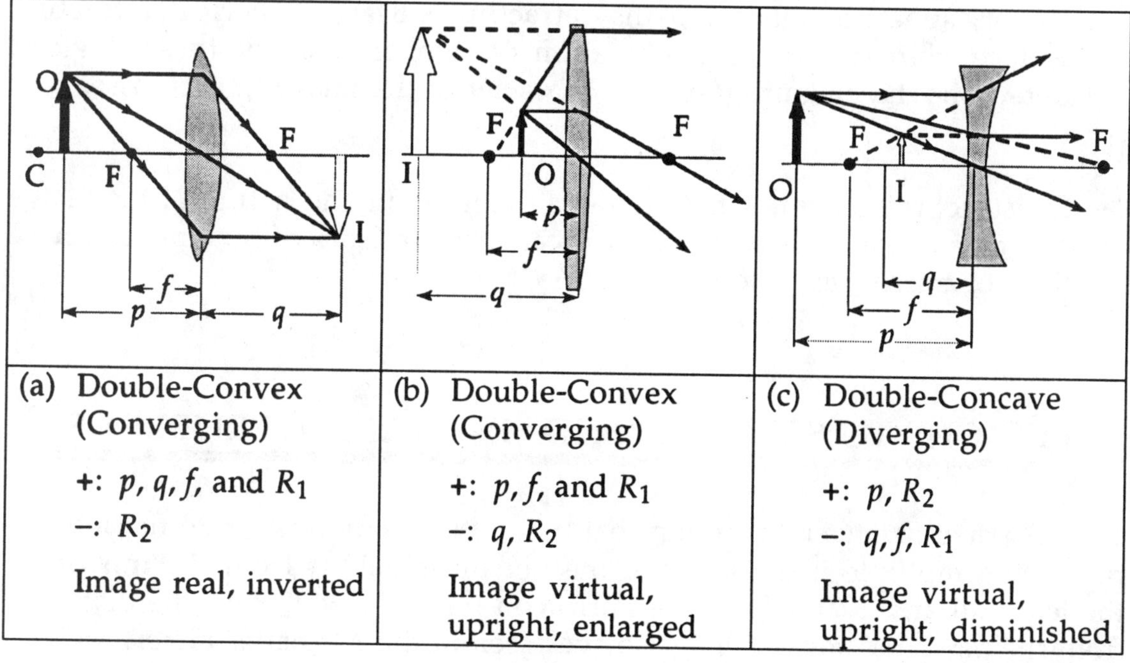

Figure 23.3 Figures describing the sign conventions for various thin lenses.

Chapter 23

REVIEW CHECKLIST

▷ Identify the following properties which characterize an image formed by a lens or mirror system with respect to an object: position, magnification, orientation (i.e. inverted, upright or right-left reversal) and whether real or virtual.

▷ Understand the relationship of the algebraic signs associated with calculated quantities to the nature of the image and object: real or virtual, upright or inverted.

▷ Calculate the location of the image of a specified object as formed by a flat mirror, spherical mirror, plane refracting surface, spherical refracting surface, thin lens, or a combination of two or more of these devices. Determine the magnification and character of the image in each case.

▷ Construct ray diagrams to determine the location and nature of the image of a given object when the geometrical characteristics of the optical device (lens or mirror) are known.

SOLUTIONS TO SELECTED END-OF-CHAPTER PROBLEMS

1. A person walks into a room that has, on opposite walls, two flat mirrors producing multiple images. When the person is 5.00 ft from the mirror on the left wall and 10.0 ft from the mirror on the right wall, find the distances from the person to the first three images seen in the left-hand mirror.

Solution The image formed by a flat mirror is an upright, virtual image located as far behind the mirror as the object is in front of the mirror. The object for the first image formed by the left-hand mirror is the person located 5.00 ft in front of that mirror. This first image (I_{1L}) is located 5.00 ft behind the mirror, or 10.0 ft from the person. ◊

Images formed by the right hand mirror are located in front of the left-hand mirror and also serve as objects for the left-hand mirror. The person is 10.0 ft in front of the right-hand mirror, so that mirror forms an image (I_{1R}) located 10.0 ft behind it. Since the mirrors are 15.0 ft apart, image I_{1R} is 25.0 ft in front of the left-hand mirror. The left-hand mirror then forms a virtual image (I_{2L}) 25.0 ft behind it. This second image formed by the left-hand mirror is 30.0 ft from the person. ◊

Image I_{1L} is located $5.0\text{ ft} + 15.0\text{ ft} = 20.0\text{ ft}$ in front of the right-hand mirror and serves as a real object for that mirror. The right-hand mirror then forms an image (I_{2R}) 20.0 ft behind it. This image is $20.0\text{ ft} + 15.0\text{ ft} = 35.0\text{ ft}$ in front of the left-hand mirror. With I_{2R} serving as the object, the left-hand mirror then forms an image (I_{3L})

35.0 ft behind it and 40.0 ft from the person. ◊

The process whose beginning is described above continues, with each mirror forming an infinite number of images. Each image formed by a mirror is located farther behind the mirror than the previous image, and it serves as an object for the opposite mirror.

---

5. A concave spherical mirror has a radius of curvature of 20.0 cm. Locate the images for object distances of (a) 40.0 cm, (b) 20.0 cm, and (c) 10.0 cm. In each case, state whether the image is real or virtual and upright or inverted, and find the magnification.

Solution The radius of curvature of a concave mirror is positive. Thus, $R = +20.0$ cm, and the focal length of this mirror is $f = R/2 = +10.0$ cm. The location of the images and the magnifications may be found from the mirror equation:

$$\frac{1}{p} + \frac{1}{q} = \frac{1}{f} \quad \text{yields an image distance} \quad q = \frac{pf}{p-f}$$

$$\text{and a magnification} \quad M = \frac{\text{image height}}{\text{object height}} = -\frac{q}{p}$$

Here, p is the object distance, q is the image distance, and f is the focal length.

(a) If $p = +40.0$ cm then $$q = \frac{(+40.0 \text{ cm})(+10.0 \text{ cm})}{40.0 \text{ cm} - 10.0 \text{ cm}} = +13.3 \text{ cm}$$

Since $q > 0$, the image is real and located 13.3 cm in front of the mirror. The magnification is $M = -q/p = -13.3 \text{ cm}/40.0 \text{ cm} = -0.333$, so the image is inverted $(M < 0)$ and its size is one-third the size of the object $(|M| = 0.333)$. ◊

(b) When $p = +20.0$ cm $$q = \frac{(+20.0 \text{ cm})(+10.0 \text{ cm})}{20.0 \text{ cm} - 10.0 \text{ cm}} = +20.0 \text{ cm}$$

The magnification is then $M = -q/p = -20.0 \text{ cm}/20.0 \text{ cm} = -1.00$. In this case, the image is real $(q > 0)$, inverted $(M < 0)$, the same size as the object $(|M| = 1.00)$, and located 20.0 cm in front of the mirror. ◊

(c) When $p = +10.0$ cm $$q = \frac{(+10.0 \text{ cm})(+10.0 \text{ cm})}{10.0 \text{ cm} - 10.0 \text{ cm}} = \infty$$

Thus, no image is formed. When the object is located at the focal point of a concave mirror, the rays leaving the mirror are parallel to each other. ◊

13. A concave makeup mirror is designed so that a person 25 cm in front of it sees an upright image magnified by a factor of two. What is the radius of curvature of the mirror?

Solution The object distance for this mirror is $p = +25\text{ cm}$. Since the image is upright, the magnification is positive. Because the size of the image is twice the size of the object, the magnitude of the magnification is $|M| = 2.0$.

From the magnification equation $M = -\dfrac{q}{p} = -\dfrac{q}{25\text{ cm}} = +2.0$, $\quad q = -50\text{ cm}$.

The mirror equation $\dfrac{1}{f} = \dfrac{1}{p} + \dfrac{1}{q}$

yields a focal length of $f = \dfrac{pq}{p+q} = \dfrac{(+25\text{ cm})(-50\text{ cm})}{25\text{ cm} - 50\text{ cm}} = +50\text{ cm}$

Since the focal length is positive, the mirror must be concave with a radius of curvature given by: $R = 2f = 2(+50\text{ cm}) = +1.0 \times 10^2\text{ cm} = 1.0\text{ m}$. ◊

---

19. A spherical mirror is to be used to form an image, five times as tall as an object, on a screen positioned 5.0 m from the mirror. (a) Describe the type of mirror required. (b) Where should the mirror be positioned relative to the object?

Solution Assuming the object is located in front of the mirror, this is a real object and the object distance is positive $(p > 0)$. Since the image is formed on a screen, it is real and the image distance is positive $(q > 0)$. With both p and q positive, the magnification given by $M = h'/h = -q/p$ must be negative. The magnitude of the magnification is 5.0 since the image is five times as tall as the object, Hence, $M = -q/p = -5.0$

It is given that the screen is 5.0 m from the mirror. Thus, the image distance is $q = +5.0\ \text{m}$. The object distance is then

$$p = -\frac{q}{M} = -\frac{(+5.0\ \text{m})}{(-5.0)} = +1.0\ \text{m}$$

Substituting $f = R/2$ into the mirror equation yields the mirror radius R:

$$\frac{1}{p}+\frac{1}{q}=\frac{2}{R} \quad \text{becomes} \quad \frac{2}{R}=\frac{1}{1.0\ \text{m}}+\frac{1}{5.0\ \text{m}} \quad \text{so that} \quad R = +1.67\ \text{m}$$

(a) Since $R > 0$, a concave mirror having a radius of curvature of $R = 1.67\ \text{m}$ is required. ◊

(b) As found above, the object distance must be $p = +1.0\ \text{m}$. Therefore, the object should be positioned 1.0 m in front of the mirror. ◊

---

23. A paperweight is made of a solid glass hemisphere of index of refraction 1.50. The radius of the circular cross section is 4.0 cm. The hemisphere is placed on its flat surface with the center directly over a 2.5-mm-long line drawn on a sheet of paper. What length of line is seen by someone looking vertically down on the hemisphere?

Solution

The center of curvature of the convex surface of the hemisphere is on the side of the surface from which the light is coming (i.e., in front of the surface). Thus, the radius of curvature is negative by the sign convention adopted in Section 23.4 of the text, and $R = -4.0\ \text{cm}$. A 2.5-mm-long line located at the center of curvature serves as a real object for the spherical refracting surface.

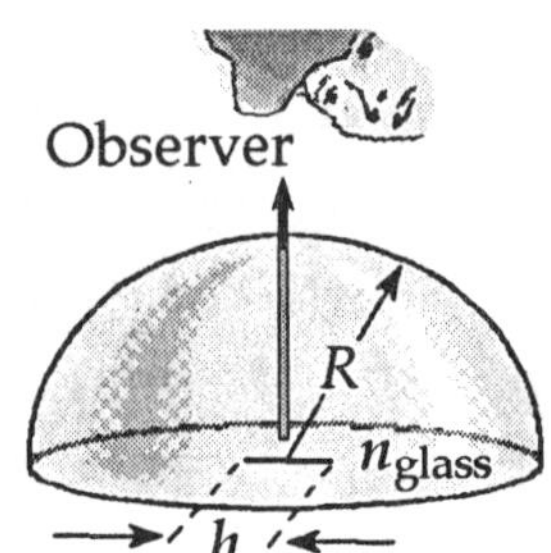

Thus, the object distance is $p = +4.0\ \text{cm}$. The index of refraction on the side of the surface the light is coming from is $n_1 = n_{glass} = 1.50$, and that on the side the light is going toward is $n_2 = n_{air} = 1.00$. The location of the image is found from

$$\frac{n_1}{p} + \frac{n_2}{q} = \frac{n_2 - n_1}{R} \quad \text{which becomes} \quad \frac{1.00}{q} = \frac{1.00 - 1.50}{(-4.0\ \text{cm})} - \frac{1.50}{(+4.0\ \text{cm})}$$

Solving gives $q = -4.0\ \text{cm}$, so the image is virtual and located at the same position as the object.

The magnification when light is refracted by a single spherical surface is given by

$$M = \frac{\text{image size}}{\text{object size}} = \frac{h'}{h} = -\frac{n_1 q}{n_2 p}$$

The image size formed by the hemispherical paper weight is:

$$h' = Mh = \left(-\frac{n_1 q}{n_2 p}\right)h = -\frac{(1.50)(-4.0\ \text{cm})}{(1.00)(+4.0\ \text{cm})}(2.5\ \text{mm}) = +3.8\ \text{mm}$$

Thus, the image is upright, virtual, located at the same position as the object, and is 3.8 mm long. ◊

---

31. A diverging lens (n = 1.50) is shaped like that in Figure 23.24c. The radius of the first surface is 15.0 cm, and that of the second surface is 10.0 cm. (a) Find the focal length of the lens. Determine the positions of the images for object distances of (b) infinity, (c) $3|f|$, (d) $|f|$, and (e) $|f|/2$.

Solution (a) The lens in Figure 23.24c is a bi-concave lens with the radii of curvature and index of refraction given in the sketch at the right. The center of curvature of the first surface is in front of the lens while the center of curvature of the second surface is in back of the lens (see the Suggestions, Skills, and Strategies section for clarification of the terms "front" and "back").

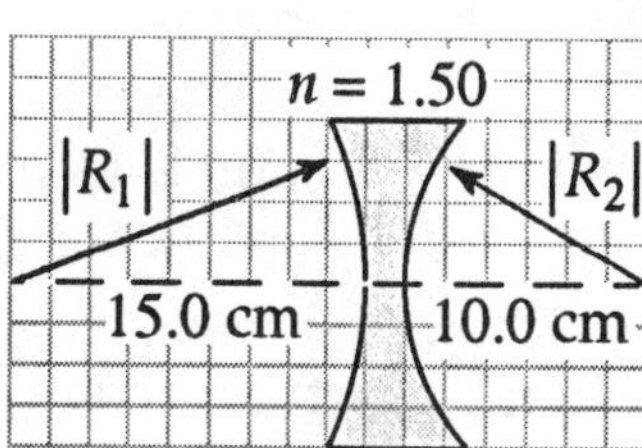

Thus, using the sign convention for thin lenses from Table 23.3 in the text, $R_1 < 0$ and $R_2 > 0$, so $R_1 = -15.0$ cm and $R_2 = +10.0$ cm. Since $n = 1.50$, the Lens-maker's equation becomes:

$$\frac{1}{f} = (n-1)\left(\frac{1}{R_1} - \frac{1}{R_2}\right) = (1.50 - 1.00)\left(\frac{1}{-15.0 \text{ cm}} - \frac{1}{10.0 \text{ cm}}\right)$$

Solving for the focal length, we get $f = -12$ cm ◊

(b) The thin lens equation is $\frac{1}{p} + \frac{1}{q} = \frac{1}{f}$, Thus, if $p \approx \infty$, the image distance is $q = f = -12$ cm. This is a virtual image 12 cm in front of the lens. ◊

(c) The thin lens equation may be solved for the image distance:

From $\frac{1}{p} + \frac{1}{q} = \frac{1}{f}$, the image distance is $q = \frac{pf}{p-f}$

Given that $p = 3|f| = +36$ cm, $q = \frac{(+36 \text{ cm})(-12 \text{ cm})}{36 \text{ cm} - (-12 \text{ cm})} = -9.0$ cm

Therefore a virtual image is located 9.0 cm in front of the lens. ◊

(d) Given that $p = |f| = +12$ cm, $q = \frac{(+12 \text{ cm})(-12 \text{ cm})}{12 \text{ cm} - (-12 \text{ cm})} = -6.0$ cm

Thus, a virtual image is formed 6.0 cm in front of the lens. ◊

(e) When $p = |f|/2 = +6.0$ cm, $q = \frac{(+6.0 \text{ cm})(-12 \text{ cm})}{6.0 \text{ cm} - (-12 \text{ cm})} = -4.0$ cm

Thus, a virtual image is formed 4.0 cm in front of the lens in this case. ◊

39. A diverging lens is to be used to produce a virtual image one-third as tall as the object. Where should the object be placed?

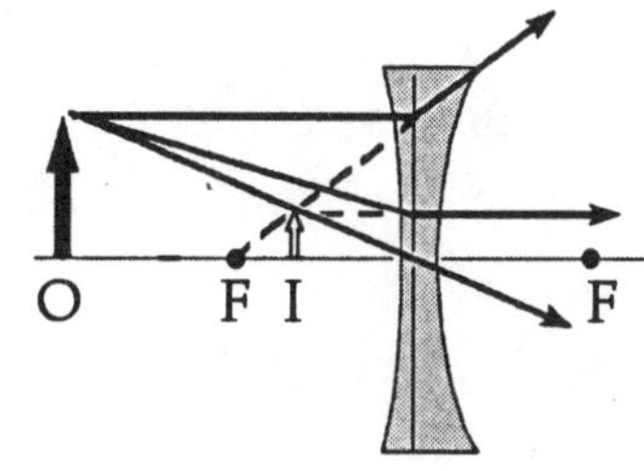

Solution A diverging lens always forms an upright, virtual, image of real objects as shown in the ray diagram at the right. Since the image will be upright, the magnification is positive. If the size of the image is one-third the size of the object, the magnitude of the magnification is $|M| = 1/3$. We can then use the magnifiction to solve for the image distance:

$$M = \frac{\text{image height}}{\text{object height}} = -\frac{q}{p} = +\frac{1}{3}$$

Thus, $q = -\frac{p}{3}$ and the thin lens equation, $\frac{1}{p} + \frac{1}{q} = \frac{1}{f}$, then becomes $\frac{1}{p} - \frac{3}{p} = \frac{1}{f}$

This simplifies to $-2/p = 1/f$ or $p = -2f$. The focal length of a diverging lens is negative, so $-f = |f|$, and the required object distance is seen to be $p = 2|f|$. Therefore, if the virtual image is to be one-third as tall as the object, the object should be placed in front of the lens at a distance equal to twice the magnitude of the focal length. ◊

---

45. A microscope slide is placed in front of a converging lens with a focal length of 2.44 cm. The lens forms an image of the slide 12.9 cm from the slide. How far is the lens from the slide if the image is (a) real? (b) virtual?

Solution (a) For a converging lens, the focal length is positive. Thus, $f = +2.44$ cm. If the image formed by the converging lens is real, it is located on the opposite side of the lens from the real object and the distance between image and object is $p + q = 12.9$ cm as shown in Figure A. The thin lens equation then becomes

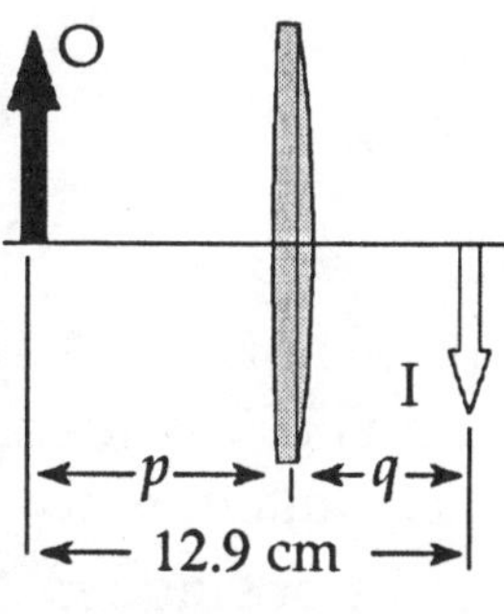

Figure A

$$\frac{1}{p} + \frac{1}{12.9\text{ cm} - p} = \frac{1}{2.44\text{ cm}}$$

Finding a common denominator and simplifying, this gives

$$(2.44 \text{ cm})(12.9 \text{ cm} - p + p) = p(12.9 \text{ cm} - p) \quad \text{or} \quad p^2 - (12.9 \text{ cm})p + 31.5 \text{ cm}^2 = 0$$

The quadratic formula then yields two positive solutions, $p = 3.27$ cm and $p = 9.63$ cm. Both are valid object distances for the situation described. ◊

If $p = 3.27$ cm, then the image distance and the magnification are

$$q = 12.9 \text{ cm} - 3.27 \text{ cm} = 9.63 \text{ cm} \qquad \text{and} \qquad M = -\frac{q}{p} = -\frac{9.63 \text{ cm}}{3.27 \text{ cm}} = -2.94$$

or the real image is inverted and 2.94 times the size of the object.

If the other solution is used, $p = 9.63$ cm and $q = 3.27$ cm (i.e., the values of the image and object distances are interchanged from the above case). In this case,

$$M = -\frac{q}{p} = -\frac{3.27 \text{ cm}}{9.63 \text{ cm}} = -0.340$$

so the real image is inverted and its size is 34.0% of that of the object.

(b) If the real object is located inside the focal point, a converging lens forms an upright, magnified, virtual image as shown in Figure B. In this case, the image distance is negative and has a magnitude of $|q| = p + 12.9$ cm. Thus, the thin lens equation becomes

$$\frac{1}{p} - \frac{1}{p + 12.9 \text{ cm}} = \frac{1}{2.44 \text{ cm}}$$

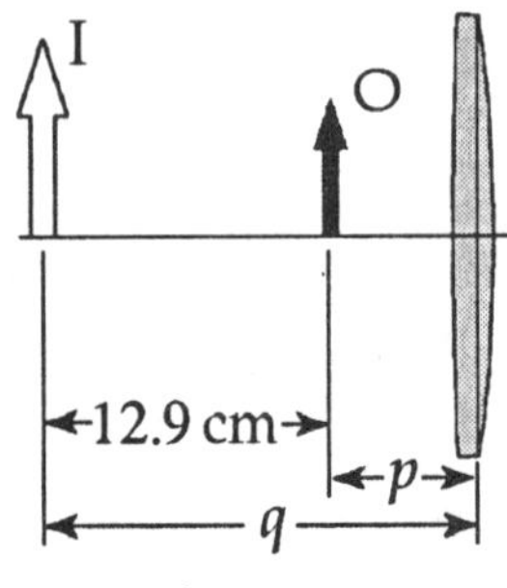

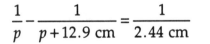

Figure B

which reduces to $p^2 + (12.9 \text{ cm})p - 31.5 \text{ cm}^2 = 0$. This quadratic equation has solutions of $p = +2.10$ cm and $p = -15.0$ cm. Since the microscope slide serves as a real object for the lens, the negative solution must be rejected. Thus, if a virtual image is to be located 12.9 cm from the object, the microscope slide must be positioned 2.10 cm in front of the lens. ◊

49. Lens L_1 in Figure P23.49 has a focal length of 15.0 cm and is located a fixed distance in front of the film plane of a camera. Lens L_2 has a focal length of 13.3 cm, and the distance, d, that it is from the film plane can be varied from 5.00 cm to 10.0 cm. Determine the range of distances for which objects can be focused on the film.

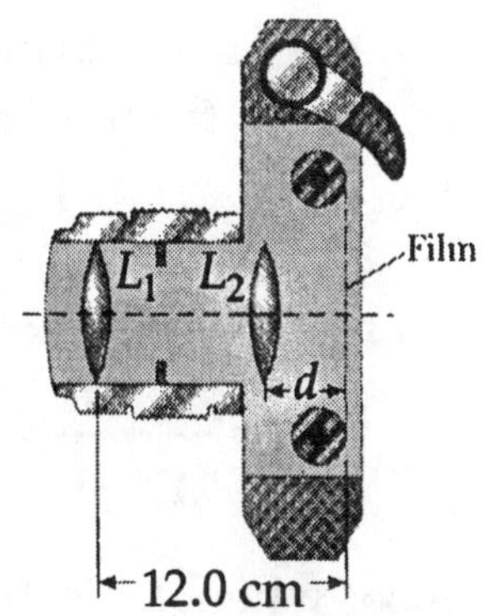

Figure P23.49

Solution The image formed by lens L_1 serves as the object for lens L_2. The image formed by L_2 is focused on the film. The two lenses shown are both converging lenses, so both have positive focal lengths. The focal length of L_1 is $f_1 = +15.0$ cm, and that of L_2 is $f_2 = +13.3$ cm. If $d = 5.00$ cm, the image distance for L_2 is $q_2 = +5.00$ cm, and the distance between the lenses is $D = 12.0\text{ cm} - d = 7.00$ cm. The object distance for L_2 is found from the thin lens equation as

$$\frac{1}{p_2} = \frac{1}{f_2} - \frac{1}{q_2} = \frac{1}{13.3\text{ cm}} - \frac{1}{5.00\text{ cm}}$$

which gives $p_2 = -8.01$ cm. Thus, the image formed by L_1 serves as a virtual object for L_2, and is located 8.01 cm to the right of L_2. This means that the image distance for L_1 is $q_1 = D + |p_2| = 7.00\text{ cm} + 8.01\text{ cm} = +15.01\text{ cm}$. The thin lens equation then gives the object distance for L_1 as

$$\frac{1}{p_1} = \frac{1}{f_1} - \frac{1}{q_1} = \frac{1}{15.0\text{ cm}} - \frac{1}{15.01\text{ cm}} \quad \text{or} \quad p_1 = 2.25 \times 10^4\text{ cm} = 225\text{ m}$$

Thus, when L_2 is 5.00 cm from the film plane, the camera is focused on an object 225 m in front of it. ◊

If $d = 10.0$ cm, then $q_2 = +10.0$ cm and the separation of the two lenses is $D = 12.00\text{ cm} - d = 2.00$ cm. In this case, the object distance for L_2 is found from

$$\frac{1}{p_2} = \frac{1}{f_2} - \frac{1}{q_2} = \frac{1}{13.3\text{ cm}} - \frac{1}{10.0\text{ cm}} \quad \text{or} \quad p_2 = -40.3\text{ cm}$$

Thus, the image formed by lens L_1 is located 40.3 cm to the right of the second lens, giving for the first lens (L_1) an image distance of:

$$q_1 = D + |p_2| = 2.00\text{ cm} + 40.3\text{ cm} = +42.3\text{ cm}$$

and an object distance of

$$\frac{1}{p_1} = \frac{1}{f_1} - \frac{1}{q_1} = \frac{1}{15.0\text{ cm}} - \frac{1}{42.3\text{ cm}} \qquad \text{or} \qquad p_1 = 23.2\text{ cm} = 0.232\text{ m}$$

Hence, when L_2 is 10.0 cm from the film plane, the camera is focused on an object 0.232 m in front of it. ◊

Since the distance from the film plane to lens L_2 can vary continuously from 5.00 cm to 10.0 cm, the range of distances for objects that the camera can focus on is from 0.232 m to 225 m. ◊

---

51. An object placed 10.0 cm from a concave spherical mirror produces a real image 8.00 cm from the mirror. If the object is moved to a new position 20.0 cm from the mirror, what is the position of the image? Is the final image real or virtual?

Solution When the object distance for this mirror is $p = +10.0\text{ cm}$, the image distance is found to be $q = +8.00\text{ cm}$ (q is positive because the image is real). The mirror equation gives the radius of curvature as follows:

$$\frac{2}{R} = \frac{1}{p} + \frac{1}{q} = \frac{1}{10.0\text{ cm}} + \frac{1}{8.00\text{ cm}} \qquad \text{which becomes} \qquad R = 8.89\text{ cm}$$

If the object is moved to a position 20.0 cm from the mirror, then $p = +20.0\text{ cm}$ and the mirror equation gives the new image distance as

$$\frac{1}{q} = \frac{2}{R} - \frac{1}{p} = \frac{2}{8.89\text{ cm}} - \frac{1}{20.0\text{ cm}} \qquad \text{or} \qquad q = +5.71\text{ cm}$$

Thus, the new image is located 5.71 cm in front of the mirror, and since $q > 0$, it is seen to be a real image. ◊

---

55. The lens and mirror in Figure P23.55 are separated by 1.00 m and have focal lengths of +80.0 cm and –50.0 cm, respectively. If an object is placed 1.00 m to the left of the lens as shown, locate the final image formed by light that has gone through the lens twice. State whether the image is upright or inverted, and determine the overall magnification.

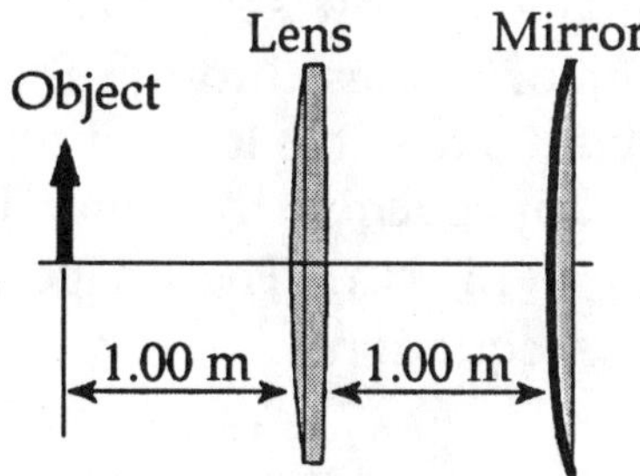

Figure P23.55

Solution

The focal length of the converging lens is $f_L = +80.0$ cm, and that of the convex mirror is $f_M = -50.0$ cm. The original object is located to the left of the lens with an object distance of $p_1 = +100$ cm, so the thin lens equation gives the position of the first image (formed by the lens) as

$$\frac{1}{q_1} = \frac{1}{f_L} - \frac{1}{p_1} = \frac{1}{80.0 \text{ cm}} - \frac{1}{100 \text{ cm}} \quad \text{or} \quad q_1 = +400 \text{ cm}$$

Thus, this image (I_1) is located 400 cm to the right of the lens, or 300 cm to the right of the mirror. Image I_1 then serves as a virtual object for the mirror (virtual because it is located behind the mirror).

The object distance for the mirror is $p_2 = -300$ cm (negative because it is a virtual object). The mirror equation, gives the image distance for the mirror:

$$\frac{1}{p} + \frac{1}{q} = \frac{2}{R} = \frac{1}{f_M} \quad \text{or} \quad \frac{1}{q_2} = \frac{1}{f_M} - \frac{1}{p_2} = \frac{1}{(-50.0 \text{ cm})} - \frac{1}{(-300 \text{ cm})}$$

Solving this equation, we find that $q_2 = -60.0$ cm. Hence, the second image (I_2) is a virtual image located 60.0 cm behind the mirror, or 160 cm to the right of the lens.

Image I_2 now serves as an object for the lens as the light reflected from the mirror passes through the lens going from right to left. This will be a real object since it is located on the side of the lens that the light is coming from as it approaches the lens for this passage. Thus, the object distance is $p_3 = +160$ cm. The image distance for the final image is found using the thin lens equation:

$$\frac{1}{q_3} = \frac{1}{f_L} - \frac{1}{p_3} = \frac{1}{80.0 \text{ cm}} - \frac{1}{160 \text{ cm}} \quad \text{so that} \quad q_3 = +160 \text{ cm}$$

Thus, the final image (I_3) is a real image located 160 cm to the left of the lens (i.e., on the side of the lens the reflected light was going toward as it passed through the lens). ◊

The magnifications associated with the formation of each of these images are found from $M = -q/p$:

$$M_1 = -\frac{400 \text{ cm}}{100 \text{ cm}} = -4.00 \qquad M_2 = -\frac{(-60.0 \text{ cm})}{(-300 \text{ cm})} = -0.200 \qquad M_3 = -\frac{160 \text{ cm}}{160 \text{ cm}} = -1.00$$

The overall magnification (i.e., ratio of the size of the final image to the size of the original object) is then given by

$$M_{\text{overall}} = M_1 \cdot M_2 \cdot M_3 = (-4.00)(-0.200)(-1.00) = -0.800 \qquad ◊$$

Since the overall magnification is negative $(M_{\text{overall}} < 0)$, the final image is inverted relative to the original object. ◊

---

61. Object O_1 is 15.0 cm to the left of a converging lens of 10.0-cm focal length. A second lens is positioned 10.0 cm to the right of the first lens and is observed to form a final image at the position of the original object, O_1. (a) What is the focal length of the second lens? (b) What is the overall magnification of this system? (c) What is the nature (i.e., real or virtual, upright or inverted) of the final image?

Solution The converging lens forms an image I_1 of the object O_1. The object distance for this process is $p_1 = +15.0$ cm (positive since the object is real) and the focal length of this lens is $f_1 = +10.0$ cm (positive since this is a converging lens). The image distance for this image is given by

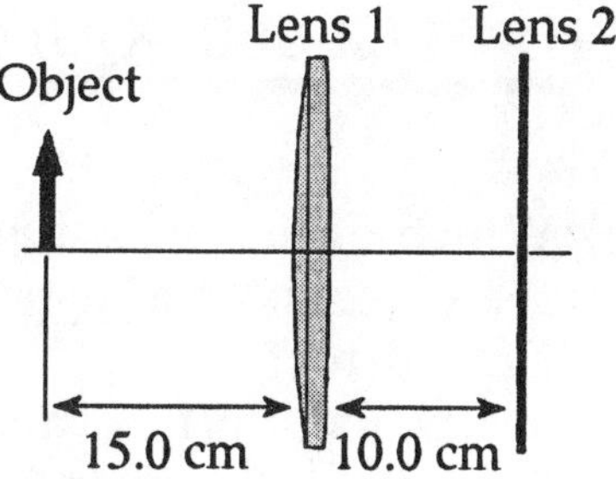

$$\frac{1}{q_1} = \frac{1}{f_1} - \frac{1}{p_1} = \frac{1}{10.0\text{ cm}} - \frac{1}{15.0\text{ cm}} \qquad \text{or} \qquad q_1 = +30.0\text{ cm}$$

Thus, the image I_1 is located 30.0 cm to the right of the first lens or 20.0 cm to the right of the second lens. Image I_1 serves as a virtual object for the second lens (virtual since the light is going through the lens from left to right but this object is located to the right of the lens). Thus, the object distance for the second lens is $p_2 = -20.0$ cm. It is given that image I_2, formed by the second lens, is located at the position of the original object O_1. Thus, this is a virtual image with an image distance of $q_2 = -25.0$ cm.

(a) From the thin lens equation, we obtain the focal length of the second lens:

$$\frac{1}{f_2} = \frac{1}{p_2} + \frac{1}{q_2} = \frac{1}{(-20.0\text{ cm})} + \frac{1}{(-25.0\text{ cm})} \qquad \text{so that} \qquad f_2 = -11.1\text{ cm}$$ ◊

(b) The magnification produced by the first and the second lens are

$$M_1 = -\frac{q_1}{p_1} = -\frac{30.0\text{ cm}}{15.0\text{ cm}} = -2.00 \qquad M_2 = -\frac{q_2}{p_2} = -\frac{(-25.0\text{ cm})}{(-20.0\text{ cm})} = -1.25$$

The overall magnification produced by this system of lenses is

$$M_{overall} = M_1 \cdot M_2 = (-2.00)(-1.25) = +2.50$$ ◊

(c) Since the final image distance, q_2, is negative, the final image is virtual. The overall magnification is positive, so the final image is upright relative to the original object. The magnitude of the overall magnification is 2.50. Thus, the size of the final image is 2.50 times that of the original object. In summary, the final image is virtual, upright, and enlarged by a factor of 2.50. ◊

CHAPTER SELF-QUIZ

1. Which of the following best describes the image of a concave mirror when the object is at a distance greater than twice the focal point distance from the mirror?
   a. virtual, upright; and magnification greater than one
   b. real, inverted; and magnification less than one
   c. virtual, upright; and magnification less than one
   d. real, inverted; and magnification greater than one

2. If a man's face is 30.0 cm in front of a concave shaving mirror creating an upright image 1.50 times as large as the object, what is the mirror's focal length?
   a. 12.0 cm
   b. 20.0 cm
   c. 70.0 cm
   d. 90.0 cm

3. An object is placed at a distance of 40.0 cm from a thin lens along the axis. If a virtual image forms at a distance of 50.0 cm from the lens, on the same side as the object, what is the focal length of the lens?
   a. 22.0 cm
   b. 45.0 cm
   c. 90.0 cm
   d. 200 cm

4. A glass block, for which $n = 1.48$, has a bubble blemish located 6.40 cm from one surface. At what distance from that surface does the image of the blemish appear to the outside observer?
   a. 3.20 cm
   b. 4.32 cm
   c. 7.64 cm
   d. 9.52 cm

5. Which of the following effects is the result of the fact that the index of refraction of glass will vary with wavelength?
   a. spherical aberration
   b. mirages
   c. chromatic aberration
   d. light scattering

6. Two thin lenses of focal lengths 15.0 and 20.0 cm, respectively, are placed in contact in an orientation so that their optic axes coincide. What is the focal length of the two in combination?
   a. 8.57 cm
   b. 17.5 cm
   c. 35.0 cm
   d. 60.0 cm

7. For a diverging lens with one flat surface, the radius of curvature for the curved surface is 10.0 cm. What must the index of refraction be so that the focal length is –10.0 cm?
   a. 1.50
   b. 2.00
   c. 3.00
   d. 0.50

8. If atmospheric refraction did not occur, how would the apparent time of sunrise and sunset be changed?
   a. both would be later
   b. both would be earlier
   c. sunrise would be later and sunset earlier
   d. sunrise would be earlier and sunset later

9. Parallel rays of light that hit a concave mirror will come together
   a. at the center of curvature
   b. at the focal point
   c. at a point half way to the focal point
   d. at infinity

10. A magnifying glass has a convex lens of focal length 15.0 cm. At what distance from a postage stamp should you hold this lens to get a magnification of +2?
    a. 10.0 cm
    b. 7.50 cm
    c. 5.00 cm
    d. 2.50 cm

11. When the reflection of an object is seen in a concave mirror, the image will
    a. always be real
    b. always be virtual
    c. may be either real or virtual
    d. will always be magnified

12. An object is 15.0 cm from the surface of a spherical Christmas tree ornament that is 5.00 cm in diameter. What is the magnification of the image?
    a. –0.055
    b. +0.033
    c. +0.077
    d. +0.154

Chapter 24

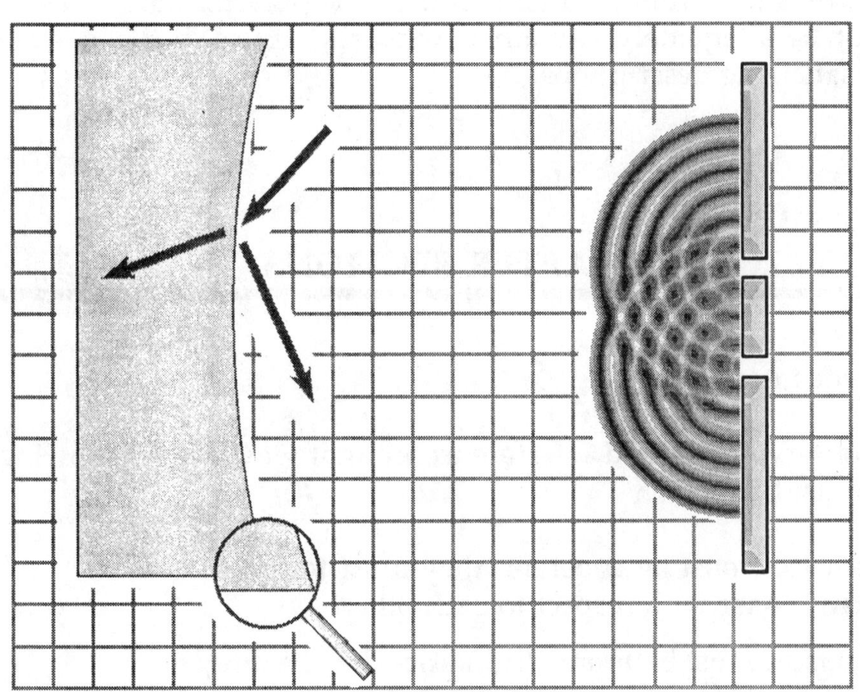

WAVE OPTICS

Chapter 24

WAVE OPTICS

Our discussion of light has thus far been concerned with what happens when light passes through a lens or reflects from a mirror. Because explanations of such phenomena rely on a geometric analysis of light rays, that part of optics is often called geometric optics. We now expand our study of light into an area called **wave optics.** The three primary topics we examine in this chapter are interference, diffraction, and polarization. These phenomena cannot be adequately explained with ray optics, but the wave theory leads us to satisfying descriptions.

NOTES FROM SELECTED CHAPTER SECTIONS

24.1 Conditions for Interference

In order to observe **sustained** interference in light waves, the following conditions must be met:

1. The sources must be coherent; they must maintain a **constant phase** with respect to each other.
2. The sources must be **monochromatic**—be of a **single wavelength.**
3. The **superposition principle** must apply.

24.2 Young's Double-Slit Experiment

A schematic diagram illustrating the geometry used in Young's double-slit experiment is shown in the figure below. The two slits S_1 and S_2 serve as coherent monochromatic sources. The **path difference** $\delta = r_2 - r_1 = d\sin\theta$.

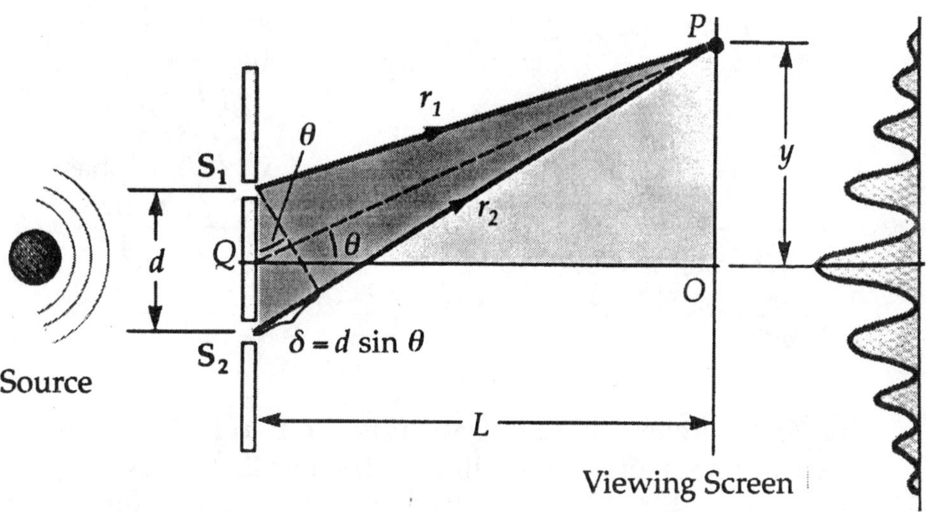

24.3 Change of Phase due to Reflection

An electromagnetic wave undergoes a **phase change of 180°** upon reflection from a medium that is **optically more dense** (with a higher index of refraction) than the one in which it was traveling. There is also a 180° phase change upon reflection from a **conducting surface.**

24.4 Interference in Thin Films

Interference effects in thin films depend on the difference in length of path traveled by the interfering waves as well as any phase changes which may occur due to reflection. It can be shown by the Lloyd's mirror experiment that when light is reflected from a surface of index of refraction greater than the index of the medium in which it is initially traveling, the reflected ray undergoes a 180° (or π radians) phase change. In analyzing interference effects, this can be considered equivalent to the gain or loss of a half wavelength in path difference. Therefore, there are two different cases to consider: (a) a film surrounded by a common medium and (b) a thin film located between two different media. These cases are illustrated in the figures on the next page.

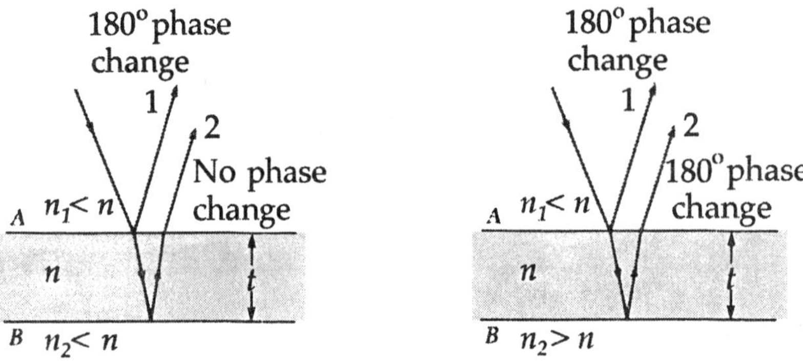

(a) Interference of light resulting from reflections at two surfaces of a thin film of thickness t, index of refraction n. (b) A thin film between two different media.

In case (a), where a phase change occurs at the top surface, the reflected rays will be in phase (constructive interference) if the thickness of the film is an odd number of quarter wavelengths—so that the path lengths will differ by an odd number of half wavelengths.

In case (b), the phase changes due to reflection at **both** the top and bottom surfaces are offsetting and, therefore, constructive interference for the reflected rays will occur when the film thickness is an integer number of half wavelengths—and, therefore, the path difference will be a whole number of wavelengths.

24.5 Diffraction
24.6 Single Slit Diffraction

The divergence of light from its initial line of travel when waves pass through small openings, around obstacles, or by sharp edges, is called **diffraction**.

The diffraction pattern produced by a narrow slit consists of a broad, intense central band, the **central maximum**, flanked by a series of narrower and less intense secondary bands (called **secondary maxima**) alternating with a series of dark bands, or **minima**.

One type of diffraction, called **Fraunhofer diffraction**, occurs when the rays reaching the observing screen are approximately parallel.

In the case of single slit diffraction, **each portion of the slit acts as a source of waves**; and **light from one portion of the slit can interfere with light from another portion**. The resultant intensity on the screen depends on the direction of the angle θ between the perpendicular to the plane of the slit and the direction to a point on the screen.

24.7 Polarization

The electric and magnetic vectors associated with an electromagnetic wave are at right angles to each other and are perpendicular to the direction of wave propagation. The electric field vector of an **unpolarized** light wave vibrates in a plane perpendicular to the direction of propagation; and all directions of vibration in the plane are equally probable.

In a linearly polarized wave the electric field vibrates in the same direction at all times at a particular point. It is possible to obtain a linearly polarized wave from an unpolarized wave by removing from the unpolarized wave all components except those whose electric field vectors oscillate in a single plane which contains the direction of propagation. Three important processes for producing polarization are: (1) selective absorption, (2) reflection, and (3) scattering.

EQUATIONS AND CONCEPTS

In the arrangement used for the Young's double-slit experiment, two slits separated by a distance, d, serve as monochromatic coherent sources. The light intensity at any point on the screen is the resultant of light reaching the screen from both slits. Also, light from the two slits reaching any point on the screen (except the center) travel unequal path lengths. This difference in length of path is called the path difference.

$$\delta = r_2 - r_1 = d\sin\theta \quad (24.1)$$

Bright fringes (constructive interference) will appear at points on the screen for which the path difference is equal to an integral multiple of the wavelength. The positions of bright fringes can also be located by calculating their vertical distance from the center of the screen (y). In each case, the number m is called the order number of the fringe. The central bright fringe ($\theta = 0$, $m = 0$) is called the zeroth-order maximum.

$$\delta = d\sin\theta = m\lambda \quad (24.2)$$

$$m = 0,\ \pm1,\ \pm2,\ \ldots$$

$$y_{\text{bright}} = \left(\frac{\lambda L}{d}\right)m \quad (24.5)$$

Dark fringes (destructive interference) will appear at points on the screen which correspond to path differences of an odd multiple of half wavelengths. For these points of destructive interference, waves which leave the two slits in phase arrive at the screen 180° out of phase.

$$\delta = d\sin\theta = \left(m + \frac{1}{2}\right)\lambda \qquad (24.3)$$

$m = 0, \pm 1, \pm 2, \ldots$

$$y_{\text{dark}} = \left(\frac{\lambda L}{d}\right)\left(m + \frac{1}{2}\right) \qquad (24.6)$$

The wavelength of light in a medium, λ_n, is less than the wavelength in free space, λ.

$$\lambda_n = \frac{\lambda}{n} \qquad (24.7)$$

When parallel light rays are incident on a slit of width, a, a diffraction pattern is formed on a screen in front of the slit. The condition for destructive interference (dark band) at a given point on the screen can be stated in terms of the angle θ (direction from the center of the slit to the point on the screen).

$$\sin\theta = m\frac{\lambda}{a} \quad \text{(Destructive)} \qquad (24.11)$$

$m = \pm 1, \pm 2, \pm 3, \ldots$

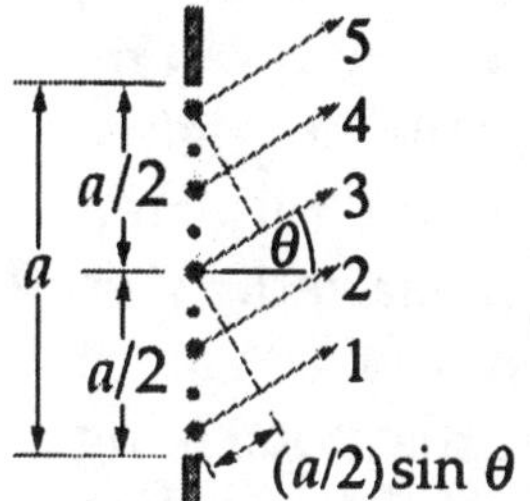

There is a broad central bright fringe at the center of the screen. There are secondary bright fringes (of progressively diminished intensity) located approximately midway between adjacent dark fringes.

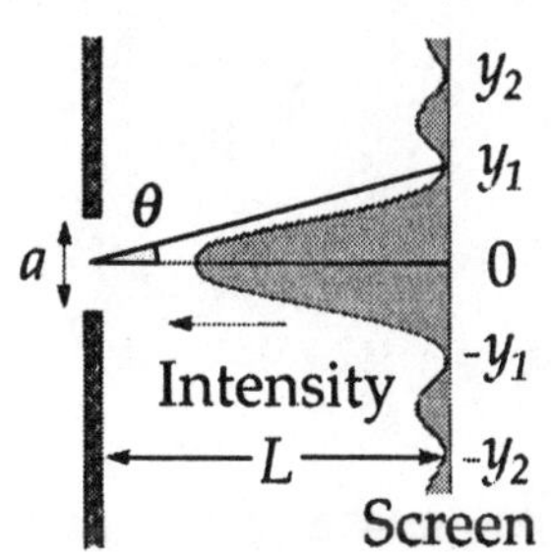

$\sin\theta = 2\lambda/a$
$\sin\theta = \lambda/a$
$\sin\theta = 0$
$\sin\theta = -\lambda/a$
$\sin\theta = -2\lambda/a$

When a wave with initial intensity I_0 passes through any two polarizing materials whose transmission axes are at an angle of θ to each other, the intensity of the transmitted (polarized) light varies according to Malus's law. From this expression, note that the transmitted intensity is a maximum when the transmission axes are parallel ($\theta = 0°$ or $180°$). When the transmission axes are perpendicular to each other, the light is completely absorbed by the analyzer, and the transmitted intensity is zero.

$$I = I_0 \cos^2 \theta \qquad (24.12)$$

Brewster's law gives the value of the polarizing angle for a surface of index of refraction, n. The polarizing angle is the angle of incidence for which the reflected beam is completely polarized. It is also true that when the angle of incidence is equal to the polarizing angle, the reflected ray is perpendicular to the transmitted ray.

$$n = \tan \theta_p \qquad (24.13)$$

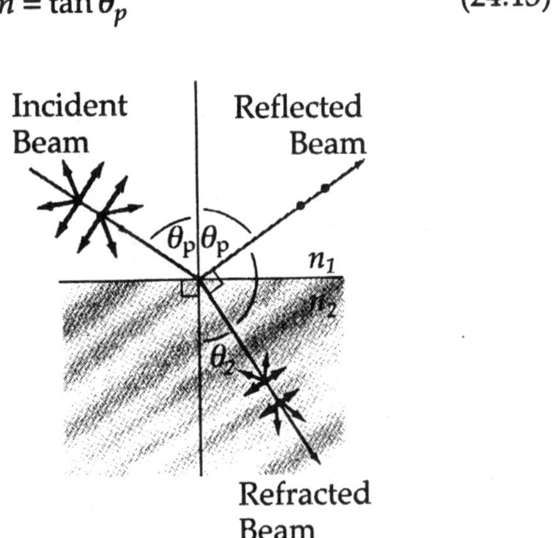

SUGGESTIONS, SKILLS, AND STRATEGIES

THIN FILM INTERFERENCE PROBLEMS

1. Identify the thin film from which interference effects are being observed.

2. The type of interference that occurs in a specific problem is determined by the phase relationship between that portion of the wave reflected at the upper surface of the film and that portion reflected at the lower surface of the film.

3. Phase differences between the two portions of the wave occur because of differences in the distance traveled by the two portions and by phase changes occurring upon reflection.

   The wave reflected from the lower surface of the film has to travel a distance equal to twice the thickness of the film before it returns to the upper surface of the film where it interferes with that portion of the wave reflected at the upper surface. When **distances alone are considered**, if this extra distance is equal to an integral multiple of $\frac{1}{2}\lambda$, the interference will be constructive; if the extra distance equals $\frac{1}{4}\lambda$, $\frac{3}{4}\lambda$, and so forth, destructive interference will occur.

   Reflections may change the results above which are based solely on the distance traveled. When a wave traveling in a particular medium reflects off a surface having a higher index of refraction than the one it is in, a 180° phase shift occurs. This has the same effect as if the wave lost $\frac{1}{2}\lambda$. These losses must be considered in addition to those losses that occur because of the extra distance one wave travels over another.

4. When distance and phase changes upon reflection are both taken into account, the interference will be constructive if the waves are out of phase by an integral number of wavelengths. Destructive interference will occur when the phase difference is an odd number of half-wavelengths and so forth.

REVIEW CHECKLIST

▷ Describe Young's double-slit experiment to demonstrate the wave nature of light. Account for the phase difference between light waves from the two sources as they arrive at a given point on the screen. State the conditions for constructive and destructive interference in terms of each of the following: path difference, phase difference, distance from center of screen, and angle subtended by the observation point at the source midpoint.

▷ Account for the conditions of constructive and destructive interference in thin films considering both path difference and any expected phase changes due to reflection.

▷ Describe Fraunhofer diffraction produced by a single slit. Determine the positions of the maxima and minima in a single-slit diffraction pattern.

▷ Describe qualitatively the polarization of light by selective absorption, reflection, scattering, and double refraction. Also, make appropriate calculations using Brewster's law.

SOLUTIONS TO SELECTED END-OF-CHAPTER PROBLEMS

5. Light of wavelength 575 nm falls on a double slit, and the first bright fringe is seen at an angle of 16.5°. Find the distance between the two slits.

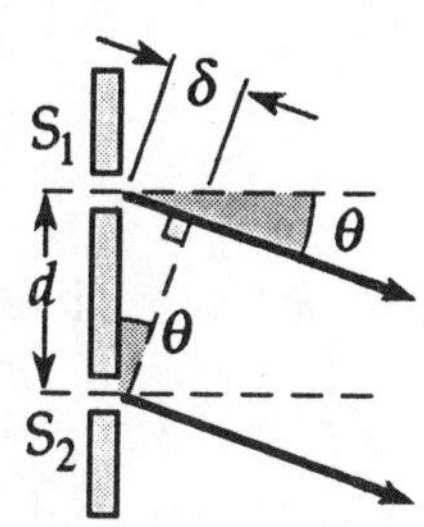

Solution When light passing through a double slit is viewed at an angle θ from the line perpendicular to the plane of the slits, the path difference (i.e., the additional distance a ray from one of the slits must travel) is seen in the sketch to be $\delta = d\sin\theta$ where d is the distance between the slits. If light from the two slits is to interfere constructively and a bright fringe observed, the path difference be $\delta = m\lambda$ where m is any integer and λ is the wavelength of the light.

Thus, the angles at which bright fringes will be found are given by the equation $\delta = d\sin\theta = m\lambda$. For the first-order bright fringe (i.e., the first bright fringe found on either side of the central maximum), $m = 1$. Therefore, if the first-order bright fringes are seen at $\theta = 16.5°$ when the wavelength of the light is $\lambda = 575$ nm, the distance between the slits is:

$$d = \frac{m\lambda}{\sin\theta} = \frac{(1)(575 \text{ nm})}{\sin 16.5°} = 2.02 \times 10^3 \text{ nm} = 2.02\ \mu\text{m} \qquad \Diamond$$

11. The waves from a radio station can reach a home receiver by two different paths. One is a straight-line path from the transmitter to the home, a distance of 30.0 km. The second path is by reflection from a storm cloud. Assume that this reflection takes place at a point midway between receiver and transmitter. If the wavelength broadcast by the radio station is 400 m, find the minimum height of the storm cloud that will produce destructive interference between the direct and reflected beams. (Assume no phase changes on reflection.)

Solution As shown in the sketch, the difference in the lengths of the two paths taken by the radio waves from the transmitter to the receiver is $\delta = 2d - 30.0\text{ km}$. If the direct and reflected beams are to interfere destructively, the path difference must be

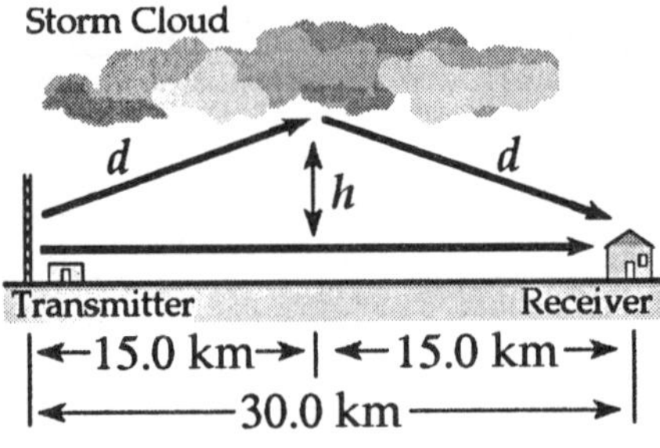

$$\delta = \left(m + \frac{1}{2}\right)\lambda$$

where m is an integer value.

For the minimum cloud height, h, and hence minimum path difference, one must require that $m = 0$. Thus, if $\lambda = 400\text{ m}$, the condition for the minimum cloud height which produces destructive interference becomes

$$\delta = 2d - 30.0 \times 10^3\text{ m} = \frac{\lambda}{2} = 200\text{ m} \qquad \text{or} \qquad d = 15.1 \times 10^3\text{ m}$$

From the sketch, it is seen that d is the hypotenuse of a right triangle having the cloud height h as its altitude. Thus, from the Pythagorean theorem,

$$h = \sqrt{d^2 - \left(15.0 \times 10^3\text{ m}\right)^2} = \sqrt{\left(15.1 \times 10^3\text{ m}\right)^2 - \left(15.0 \times 10^3\text{ m}\right)^2}$$

or the minimum cloud height is $h = 1.73 \times 10^3\text{ m} = 1.73\text{ km}$. ◊

13. Suppose the film shown in Figure 24.7 has an index of refraction of 1.36 and is surrounded by air on both sides. Find the minimum thickness, other than zero, that will produce constructive interference in the reflected light when the film is illuminated by light of wavelength 500 nm.

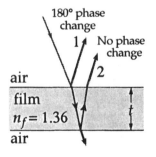

Figure 24.7

Solution When light reflects as it attempts to go from one material into another having a higher index of refraction (as at the upper surface of this film), it undergoes a 180° phase change. This phase change is equivalent to changing the path by one-half wavelength.

If the light reflects from a boundary when attempting to go from a material into one of lower index of refraction (as at the lower surface of this film), no phase change occurs. In this case the net difference, due to reflections, in the two reflected rays is one-half wavelength. The total difference between the reflected rays (labeled 1 and 2 in the figure) is

$$\delta_{\text{total}} = (\text{difference in path lengths}) + (\text{difference due to reflections})$$

assuming a normal angle of incidence, this becomes $\delta_{\text{total}} = (2t) + \lambda_f/2$, where t is the thickness of the film and λ_f is the wavelength of the light **in the film.** The wavelength of the light in the film is $\lambda_f = \lambda/n_f$, where λ is the wavelength of the light in a vacuum and n_f is the index of refraction of the film. Thus,

$$\delta_{\text{total}} = 2t + \frac{\lambda}{2n_f}$$

If the two reflected beams are to be in phase (and hence produce constructive interference), the total path difference must be an integral number of wavelengths of the light **in the film.** Hence, the necessary condition for constructive interference in the reflected light is

$$\delta_{\text{total}} = 2t + \frac{\lambda}{2n_f} = m\left(\frac{\lambda}{n_f}\right) \quad \text{or} \quad t = \left(m - \frac{1}{2}\right)\left(\frac{\lambda}{2n_f}\right) \quad (m = 1,\ 2,\ 3,\ \ldots)$$

For the minimum acceptable film thickness, $m = 1$, and $t_{\min} = \dfrac{\lambda}{4n_f}$

If the film has a index of refraction of $n_f = 1.36$ and the film is illuminated at normal incidence with 500 nm light, the desired film thickness is

$$t_{\min} = \frac{500\text{ nm}}{4(1.36)} = 91.9\text{ nm}$$ ◊

16. A transparent oil of index of refraction 1.29 spills on the surface of water (index of refraction 1.33), producing a maximum of reflection with normally incident orange light (wavelength 600 nm in air). Assuming the maximum occurs in the first order, determine the thickness of the oil slick.

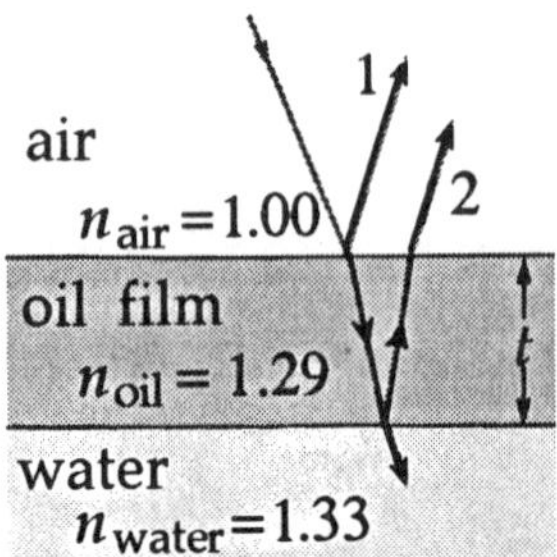

Solution As in the solution of Problem 24.11, the total difference in the two reflected rays has two possible contributions. These are 1) a difference in the actual distances traveled (path lengths), and 2) a difference (if any) due to the nature of the reflections undergone. For normal incidence, reflected ray 2 travels a distance $2t$ farther than reflected ray 1.

Since $n_{oil} > n_{air}$, ray 1 undergoes a 180° phase change upon reflection from the upper surface of the oil. Because $n_{water} > n_{oil}$, reflected ray 2 also undergoes a 180° phase change upon its reflection (at the lower surface of the oil). Note that the two reflected rays undergo identical phase changes upon reflection; thus, reflections produce zero net difference in the two reflected rays, and the total difference in the two rays is:

$$\delta_{total} = (\text{difference in path lengths}) + (\text{difference due to reflections})$$

or $$\delta_{total} = (2t) + (0) = 2t$$

For a maximum in the intensity of the reflected light, this total difference must be an integral multiple of the wavelength of the light in the oil $(\lambda_{oil} = \lambda/n_{oil})$,

or $$\delta_{total} = 2t = m\lambda_{oil} = m\left(\frac{\lambda}{n_{oil}}\right) \quad \text{where} \quad (m = 1, 2, 3, \ldots).$$

Note that $m = 0$ has been disallowed since that would correspond to zero thickness or a non-existent oil film. Assuming that the maximum in the intensity of the reflected light is due to first-order reflections, then $m = 1$ and the thickness of the oil film is:

$$t = (1)\left(\frac{\lambda}{2n_{oil}}\right) = \frac{600 \text{ nm}}{2(1.29)} = 233 \text{ nm}$$ ◊

24. A plano-convex lens (see Problem 21) with radius of curvature $R = 3.0\text{ m}$ is in contact with a flat plate of glass. A light source and the observer's eye are both close to the normal, as shown in Figure 24.8. The radius of the 50th bright Newton's ring is found to be 9.8 mm. What is the wavelength of the light produced by the source?

Solution

From the Pythagorean theorem and the right triangle shown in the sketch:

$$R^2 = (R-t)^2 + r^2 \quad \text{or} \quad (R-t) = \sqrt{R^2 - r^2}$$

The thickness of the air wedge at distance r from the axis of the lens is $t = R - \sqrt{R^2 - r^2}$

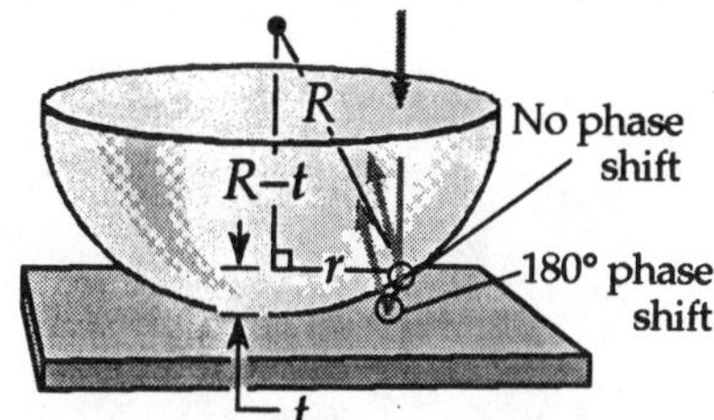

Figure not to scale: angles and distances are exaggerated to improve readability

If the radius of the lens is $R = 3.0\text{ m}$, the thickness of the air wedge at a distance $r = 9.8\text{ mm} = 9.8 \times 10^{-3}\text{ m}$ from the axis is:

$$t = 3.0\text{ m} - \sqrt{(3.0\text{ m})^2 - \left(9.8 \times 10^{-3}\text{ m}\right)^2} = 1.6 \times 10^{-5}\text{ m}.$$

The index of refraction of glass is greater than that of air $\left(n_{\text{glass}} > n_{\text{air}}\right)$. Thus, the light does not undergo a phase shift when it reflects at the upper surface of the air wedge but does undergo a shift when reflecting at the lower surface of the wedge. The total difference in the two reflected rays consists of a path difference of $2t$ due to the additional distance one ray travels through the air wedge plus a one-half wavelength difference due to the nature of the two reflections; in equation form,

$$\delta_{\text{total}} = 2t + \frac{\lambda_{\text{air}}}{2}$$

Note that $n_{\text{air}} \approx 1.0$ so that $\lambda_{\text{air}} = \lambda / n_{\text{air}} \approx \lambda$ where λ is the wavelength of the light in a vacuum.

For constructive interference at this point, this difference must be a integral number of wavelengths, or

$$\delta_{total} = 2t + \frac{\lambda}{2} = m\lambda \quad \text{and} \quad \left(m - \frac{1}{2}\right)\lambda = 2t \quad \text{where} \quad (m = 1, 2, 3, \ldots)$$

It is given that the 50th bright ring ($m = 50$) occurs where $r = 9.8$ mm and hence $t = 1.6 \times 10^{-5}$ m (see above). The wavelength of the light must then be

$$\lambda = \frac{t}{(m - 1/2)} = \frac{2\left(1.6 \times 10^{-5}\ \text{m}\right)}{49.5} = 6.5 \times 10^{-7}\ \text{m} = 6.5 \times 10^{2}\ \text{nm}$$ ◊

29. Microwaves of wavelength 5.00 cm enter a long, narrow window in a building that is otherwise essentially opaque to the microwaves. If the window is 36.0 cm wide, what is the distance from the central maximum to the first-order minimum along a wall 6.50 m from the window?

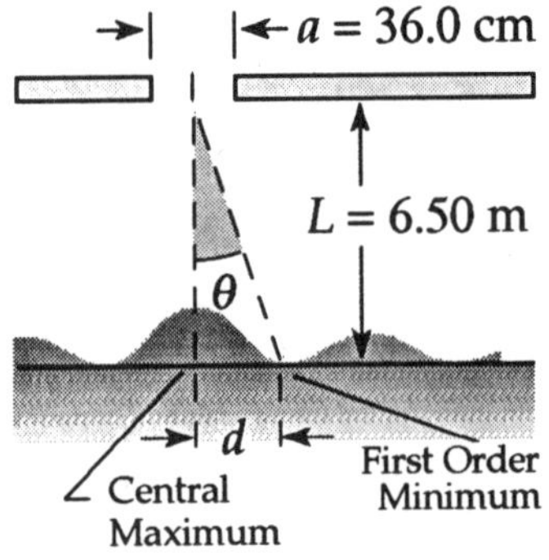

Solution In single-slit diffraction, the general condition for destructive interference is:

$$\sin\theta = m\frac{\lambda}{a} \qquad m = \pm 1, \pm 2, \pm 3, \ldots$$

where a is the width of the slit, λ is the wavelength of the waves passing through the slit, and θ is the angle the viewing direction makes with the perpendicular bisector of the slit as shown in the sketch above.

In this case a long, narrow window of width $a = 36.0$ cm acts as a slit diffracting microwaves of wavelength $\lambda = 5.00$ cm. The condition for a first-order ($m = \pm 1$) minimum is then

$$\sin\theta = (\pm 1)\frac{(5.00\ \text{cm})}{(36.0\ \text{cm})} = \pm 0.139$$

Therefore, the first-order minima are located at $\theta = \arcsin(\pm 0.139) = \pm 7.98°$. Along the opposite wall located at $L = 6.50$ m from the window, the distance from the central maximum to a first-order minimum is

$$d = L\tan\theta = (6.50 \text{ m})\tan(7.98°) = 0.912 \text{ m} = 91.2 \text{ cm} \qquad \lozenge$$

---

31. A slit of width 0.50 mm is illuminated with light of wavelength 500 nm, and a screen is placed 120 cm in front of the slit. Find the widths of the first and second maxima on each side of the central maximum.

Solution

When light passes through a single slit, minima or dark fringes appear in the diffraction pattern wherever $\sin\theta = m\lambda/a$. Here, a is the width of the slit, λ is the wavelength of the light, θ is as shown in the diagram, and m is any non-zero integer.

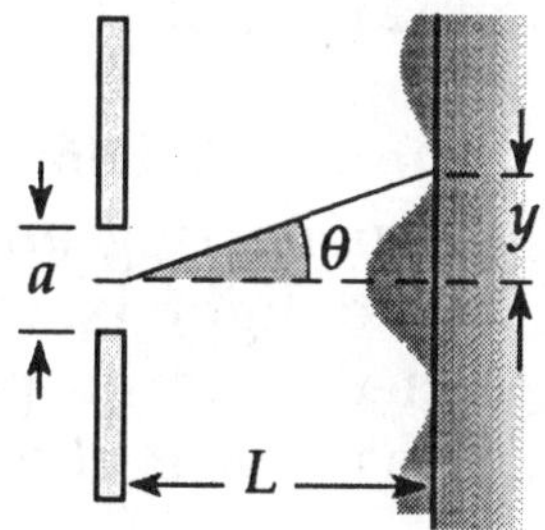

If the width of the slit is very large in comparison to the wavelength of the light $(\text{i.e., } \lambda/a << 1)$, then $\sin\theta$ and hence θ itself is very small for small m.

From the diagram, it is seen $y = L\tan\theta$. For small very angles, $\tan\theta \approx \sin\theta$ so the locations of the dark fringes on the screen are given by

$$y_m \approx L\sin\theta = mL\frac{\lambda}{a}$$

The first maximum beyond the central maximum extends from the first $(m = 1)$ minimum to the second $(m = 2)$ minimum. Its width is then

$$\Delta y = y_2 - y_1 \approx 2L(\lambda/a) - L(\lambda/a) = \lambda L/a$$

The second maximum extends from the second minimum to the third ($m = 3$) minimum and has a width of

$$\Delta y = y_3 - y_2 \approx 3L(\lambda/a) - 2L(\lambda/a) = \lambda L/a$$

Thus, the widths of both the first and second maxima on either side of the central maximum are given by $\Delta y \approx \lambda L/a$:

$$\Delta y \approx \frac{(500 \text{ nm})(120 \text{ cm})}{0.50 \text{ mm}} = \frac{(500 \times 10^{-9} \text{ m})(1.20 \text{ m})}{0.50 \times 10^{-3} \text{ m}} = 1.2 \times 10^{-3} \text{ m} = 1.2 \text{ mm}$$ ◊

35. At what angle above the horizon is the Sun if light from it is completely polarized upon reflection from water?

Solution When light is incident on a surface at the polarizing angle (or Brewster's angle), the reflected light is completely polarized and the refracted beam is perpendicular to the reflected beam as shown in the sketch.

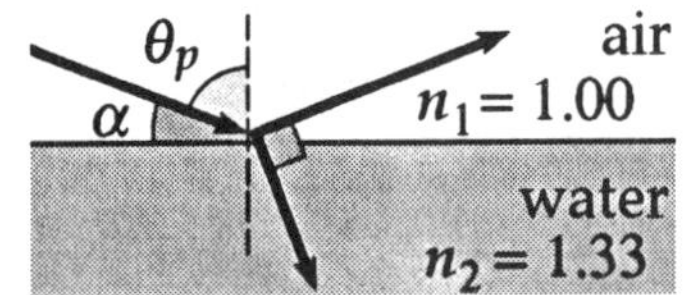

The polarizing angle is given by $\tan\theta_p = n_2/n_1$, where n_1 is the index of refraction of the medium the incident light travels through and n_2 is the index of refraction of the second medium. Thus, if sunlight incident in air reflects from a water surface, the polarizing angle is

$$\theta_p = \arctan\left(\frac{1.33}{1.00}\right) = 53.1°$$

Therefore, if the reflected sunlight is to be completely polarized, the angle of the Sun above the horizon (angle between the incident beam and the horizontal in the above sketch) must be

$$\alpha = 90.0° - \theta_p = 90.0° - 53.1° = 36.9°$$ ◊

39. Light of intensity I_0 and polarized parallel to the transmission axis of a polarizer, is incident on an analyzer. (a) If the transmission axis of the analyzer makes an angle of 45° with the axis of the polarizer, what is the intensity of the transmitted light? (b) What should the angle between the transmission axes be to make $I = I_0/3$?

Solution Linearly polarized light includes an oscillating electric field oriented perpendicular to the direction the light is traveling. The amplitude of this electric field, E_0, has a one component, $E_{\|}$, oriented parallel to the transmission axis of the analyzer shown in the sketch and one component, $E_{\perp}$, that is perpendicular to this axis.

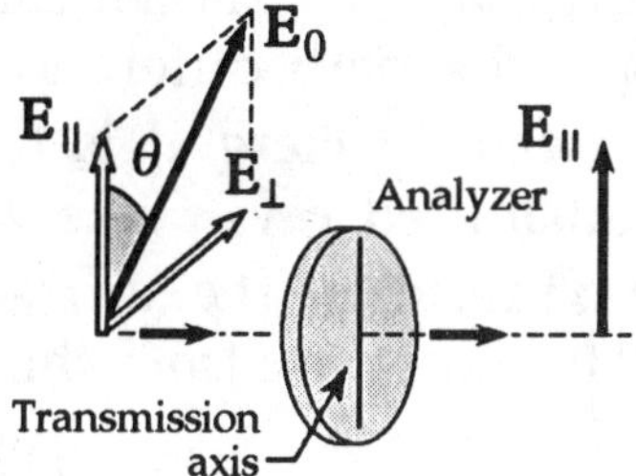

As the light passes through the analyzer, the component $E_{\perp}$ is absorbed while the component parallel to the axis $E_{\|} = E_0 \cos\theta$ is transmitted. Here, θ is the angle between the plane of polarization of the incident light and the transmission axis of the analyzer.

Since the intensity of a light wave is proportional to the square of the amplitude of the associated electric field, the ratio of the intensity of the transmitted light to that of the incident light is

$$\frac{I}{I_0} = \frac{(E_{\|})^2}{(E_0)^2} = \cos^2\theta$$

Thus, when polarized light of intensity I_0 is incident on an analyzer with the plane of polarization at an angle θ to the transmission axis, the intensity of the transmitted light is $I = I_0 \cos^2\theta$. This is known as **Malus's law.**

(a) If $\theta = 45°$, Malus's law gives the intensity of the transmitted light as

$$I = I_0 \cos^2 45° = I_0\left(\frac{\sqrt{2}}{2}\right)^2 = I_0/2 \qquad \Diamond$$

(b) If $\frac{I}{I_0} = \frac{1}{3}$, then $\cos\theta = \sqrt{I/I_0} = \sqrt{1/3} = 0.577$, and $\theta = 54.7°$ ◊

43. Light of wavelength 546 nm (the intense green line from a mercury source) produces a Young's interference pattern in which the second minimum from the central maximum is along a direction that makes an angle of 18.0 min of arc with the axis through the central maximum. What is the distance between the parallel slits?

Solution As shown earlier in the figure included in the solution of Problem 24.5, the path difference for light passing through a pair of parallel slits is $\delta = d\sin\theta$. Here, d is the distance between the slits and θ is the angle the transmitted rays make with the normal to the plane of the slits. The central maximum in the diffraction pattern occurs at $\theta = 0°$. For a direction at 18.0 min of arc from the axis through the central maximum,

$$\theta = 18.0 \text{ min}\left(\frac{1.00 \text{ degree}}{60.0 \text{ min}}\right) = 0.300°$$

If the light coming through the two slits is to interfere destructively (i.e., produce a minimum in intensity or a dark fringe), the path difference must be an odd number of half-wavelengths, or

$$\delta = d\sin\theta = (m + 1/2)\lambda \qquad \text{with} \qquad (m = 0, 1, 2, 3\ldots)$$

For the second minimum on either side of the central maximum, $m = 1$ and $\delta = 3\lambda/2$. If this second minimum is found at an angle of $\theta = 18.0 \text{ min of arc} = 0.300°$ when the wavelength is $\lambda = 546$ nm, the spacing between the slits must be

$$d = \frac{\delta}{\sin\theta} = \frac{3\left(546\times10^{-9}\text{ m}\right)}{2(\sin 0.300°)} = 1.56\times10^{-4}\text{ m} = 0.156\text{ mm} \qquad \Diamond$$

49. A pair of slits, separated by 0.150 mm, are illuminated by light having a wavelength of λ = 643 nm. An interference pattern is observed on a screen 140 cm from the slits. Consider a point on the screen located at y = 1.80 cm from the central maximum of this pattern. (a) What is the path difference, δ, for the two slits at this y location? (b) Express this path difference in terms of the wavelength. (c) Will the interference correspond to a maximum, a minimum, or an intermediate condition?

Solution (a) As shown in the solution to Problem 24.5, the path difference for rays from a pair of parallel slits is $\delta = d\sin\theta$ when those rays travel at angle θ from the direction normal to the plane of the slits. As seen in the sketch, the value of $\sin\theta$ at the screen location $y = 1.80$ cm is

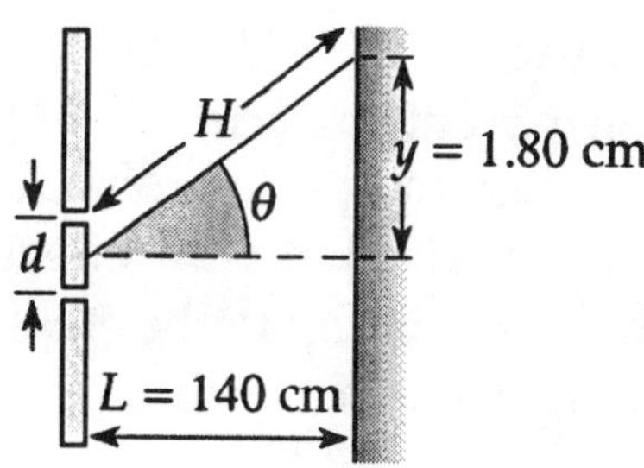

$$\sin\theta = \frac{y}{H} = \frac{1.80\text{ cm}}{\sqrt{(140\text{ cm})^2 + (1.80\text{ cm})^2}} = 1.29\times10^{-2}$$

Thus, if $d = 0.150$ mm, the path difference for rays striking this point on the screen is

$$\delta = d\sin\theta = \left(1.50\times10^{-4}\text{ m}\right)\left(1.29\times10^{-2}\right) = 1.93\times10^{-6}\text{ m} = 1.93\ \mu\text{m} \qquad \Diamond$$

(b) If the wavelength of the light passing through the slits is $\lambda = 643$ nm, then

$$\delta = 1.93\times10^{-6}\text{ m}\left(\frac{\lambda}{643\times10^{-9}\text{ m}}\right) = 3.00\lambda \qquad \Diamond$$

(c) Since the path difference is an integral number of wavelengths, a maximum (bright fringe) is at this location on the screen. ◊

---

55. (a) If light is incident at an angle of θ from a medium of index n_1 on a medium of index n_2 so that the angle between the reflected ray and refracted ray is β, show that

$$\tan\theta = \frac{n_2\sin\beta}{n_1 - n_2\cos\beta}$$

Hint: Use the following identity: $\sin(A+B) = \sin A\cos B + \cos A\sin B$. (b) Show that the foregoing equation for $\tan\theta$ reduces to Brewster's law when $\beta = 90°$, $n_1 = 1$, and $n_2 = n$.

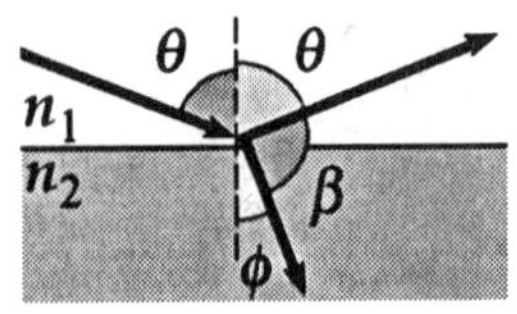

Solution (a) If light is incident on the surface of a transparent material at angle θ from the normal, part is reflected and part is refracted. The reflected light obeys the law of reflection and reflects at angle θ from the normal. The refracted light obeys Snell's law and refracts at angle ϕ where $n_2 \sin\phi = n_1 \sin\theta$.

From the sketch, observe that $\theta + \beta + \phi = \pi$, or $\phi = \pi - (\theta + \beta)$ where β is the angle between the reflected and refracted rays. Thus, $\sin\phi = \sin[\pi - (\theta + \beta)]$. Applying the suggested identity gives

$$\begin{aligned}\sin\phi &= \sin\pi\cos[-(\theta+\beta)] + \cos\pi\sin[-(\theta+\beta)] \\ &= 0 + (-1)\sin[-(\theta+\beta)] \\ &= \sin(\theta+\beta)\end{aligned}$$

Applying the identity again, $\quad \sin\phi = \sin\theta\cos\beta + \cos\theta\sin\beta$

Therefore, Snell's law becomes $\quad n_2\sin\theta\cos\beta + n_2\cos\theta\sin\beta = n_1\sin\theta$

or

$$n_2\left(\frac{\sin\theta}{\cos\theta}\right)\cos\beta + n_2\sin\beta = n_1\left(\frac{\sin\theta}{\cos\theta}\right)$$

Since $\dfrac{\sin\theta}{\cos\theta} = \tan\theta$, $\quad n_2\sin\beta = (n_1 - n_2\cos\beta)\tan\theta$

or

$$\tan\theta = \frac{n_2\sin\beta}{n_1 - n_2\cos\beta} \qquad \Diamond$$

(b) If $\beta = 90°$ (i.e., the reflected and refracted rays are perpendicular to each other), then $\cos\beta = 0$ and $\sin\beta = 1$. In this case, the result from part (a) becomes $\tan\theta = n_2/n_1$ which is the general form of Brewster's law. If the first medium is air, then $n_1 = 1$ and this result reduces to $\tan\theta = n$ where $n = n_2$ is the index of refraction of the second medium. This is the form of Brewster's law given in the textbook. ◊

CHAPTER SELF-QUIZ

1. A Young's double slit has a slit separation of 4.00×10^{-5} m on which a monochromatic light beam is directed. The resultant bright fringes on a screen 1.20 m from the double slit are separated by 2.15×10^{-2} m. What is the wavelength of this beam? (1 nanometer = 10^{-9} m)
   a. 573 nm
   b. 454 nm
   c. 717 nm
   d. 667 nm

2. What is the minimum thickness of a soap bubble film (n = 1.46) on which light of wavelength 500 nm shines, assuming one observes constructive interference of the reflected light?
   a. 63.0 nm
   b. 85.6 nm
   c. 125 nm
   d. 172 nm

3. A silicon monoxide thin film (n = 1.45) of thickness 97.4 nm is applied to a camera lens made of glass (n = 1.58). This will result in a destructive interference for reflected light of what wavelength?
   a. 720 nm
   b. 616 nm
   c. 565 nm
   d. 493 nm

4. A beam of unpolarized light in air strikes a flat piece of glass at an angle of incidence of 57.33°. If the reflected beam is completely polarized, what is the index of refraction of the glass?
   a. 1.60
   b. 1.56
   c. 1.52
   d. 2.48

5. In a Young's double-slit interference apparatus, by what factor is the distance between adjacent light and dark fringes changed when the wavelength of the source is tripled?
   a. 1/9
   b. 1/3
   c. 1.0
   d. 3.0

6. The dark spot observed in the center of a Newton's rings pattern is attributed to which of the following?
   a. polarization of light when reflected
   b. polarization of light when refracted
   c. phase shift of light when reflected
   d. phase shift of light when refracted

7. Light of wavelength 610 nm is incident on a slit of width 0.20 mm and a diffraction pattern is produced on a screen that is 1.50 m from the slit. What is the width of the central bright fringe? (1 nm = 10^{-9} m)
   a. 0.68 cm
   b. 0.92 cm
   c. 1.22 cm
   d. 1.35 cm

8. When the sun is near one of the horizons, an observer looking at the sky directly overhead will view partially polarized light. This effect is due to which of the following processes?
   a. reflection
   b. double refraction
   c. selective absorption
   d. scattering

9. Waves from a radio station with a wavelength of 400 m arrive at a home receiver a distance 30.0 km away from the transmitter by two paths. One is a direct-line path and the second by reflection from a mountain directly behind the receiver. What is the minimum distance between the mountain and receiver so that destructive interference occurs at the location of the listener?
   a. 100 m
   b. 200 m
   c. 300 m
   d. 400 m

10. When light shines on a lens placed on a flat piece of glass, interference occurs which causes circular fringes called Newton's rings. The two beams that are interfering come from
    a. the top and bottom surface of the lens
    b. the top surface of the lens and the top surface of the piece of glass
    c. the bottom surface of the lens and the top surface of the piece of glass
    d. the top and bottom surface of the flat piece of glass

11. Polaroid material, made from thin sheets of oriented molecules, polarizes light by the process of
    a. selective absorption
    b. reflection
    c. double refraction
    d. scattering

12. A possible means for making an airplane radar-invisible is to coat the plane with an antireflective polymer. If radar waves have a wavelength of 3.00 cm and the index of refraction of the polymer is $n = 1.50$, how thick would the coating be?
    a. 1.00 mm
    b. 2.00 mm
    c. 5.00 mm
    d. 2.00 cm

Chapter 25

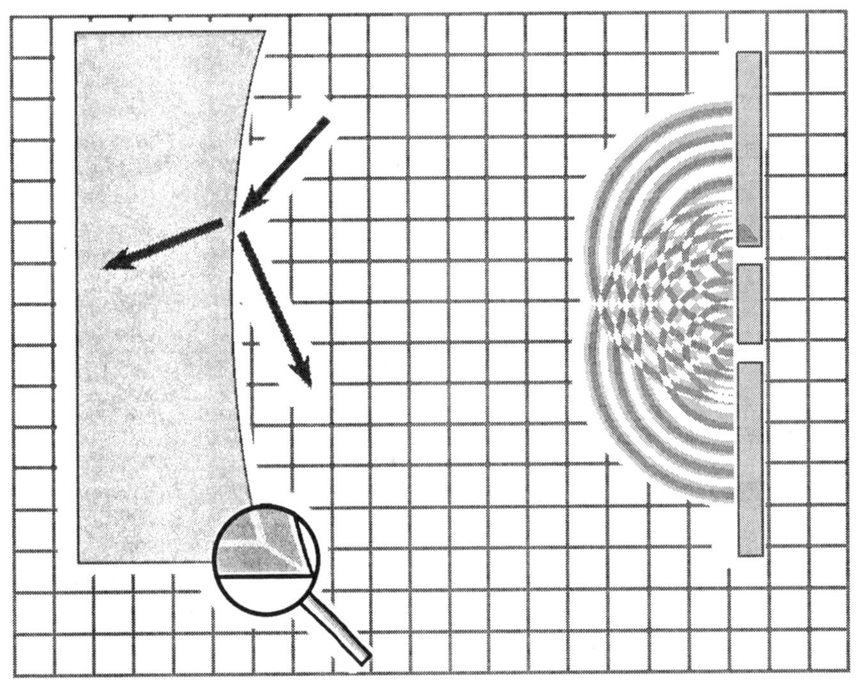

OPTICAL INSTRUMENTS

Chapter 25

OPTICAL INSTRUMENTS

We use devices made from lenses, mirrors, or other optical components every time we put on a pair of eyeglasses, take a photograph, look at the sky through a telescope, and so on. In this chapter we examine how these and other optical instruments work. For the most part, our analyses will involve the laws of reflection and refraction and the procedures of geometric optics. However, to explain certain phenomena, we must use the wave nature of light.

NOTES FROM SELECTED CHAPTER SECTIONS

25.2 The Eye

The **near point** represents the closest distance for which the lens will produce a sharp image on the retina. This distance usually increases with age and has an average value of around 25.0 cm.

Hyperopia (farsightedness) is a defect of the eye that results either when the eyeball is too short or when the ciliary muscle is unable to change the shape of the lens enough to form a properly focused image. Myopia (nearsightedness) is caused either when the eye is longer than normal or when the maximum focal length of the lens is insufficient to produce a clearly focused image on the retina.

The power of a lens in diopters equals the inverse of the focal length in meters.

25.4 The Compound Microscope

The overall magnification of a compound microscope of length L is equal to the product of the magnification produced by the objective of focal length f_0 and the magnification produced by the eyepiece of focal length f_e.

25.5 The Telescope

There are two fundamentally different types of telescopes, both designed to aid in viewing distant objects, such as the planets in our solar system. The two classifications are (1) the **refracting telescope,** which uses a combination of lenses to form an image, and (2) the **reflecting telescope,** which uses a curved mirror and a lens to form an image.

25.6 Resolution of Single-Slit and Circular Apertures

When the central maximum of one image falls on the first minimum of another image, the images are said to be just resolved. This limiting condition of resolution is known as Rayleigh's criterion.

25.7 The Michelson Interferometer

The **Michelson interferometer** is an optical instrument that has great scientific importance. The interferometer is an ingenious device that splits a light beam into two parts and then recombines them to form an interference pattern. The device is used for obtaining accurate length measurements.

25.8 The Diffraction Grating

A **diffraction grating** consists of a large number of equally spaced, parallel, identical slits.

A plane wave incident on a grating normal to the plane of the grating produces a pattern on a screen which is the result of the combined effects of interference and diffraction. Each slit produces diffraction, and the diffracted beams in turn interfere with one another to produce the intensity pattern.

EQUATIONS AND CONCEPTS

The brightness of the image on a film depends on the ratio of the focal length to the diameter of the camera lens. This ratio, called the f-number, is a measure of the light concentrating power of the lens, and determines what is called the speed of the lens. A fast lens has a small f-number — usually a short focal length and a large diameter.

$$f\text{-number} \equiv \frac{f}{D} \tag{25.1}$$

The power of a lens measured in diopters is equal to the inverse of the focal length measured in meters. The correct algebraic sign must be used with f. Optometrists and ophthalmologists usually prescribe lenses measured in diopters.

$$P = \frac{1}{f}$$

A simple magnifier increases the angular size of the object (the size of the angle subtended by the object at the eye). The angular magnification is the ratio of the angle subtended by the image formed by a convex lens to the angle subtended by the object when it is placed at the near point of the eye (25.0 cm) with no lens.

$$m \equiv \frac{\theta}{\theta_0} \tag{25.2}$$

This is an alternate form used to express the magnifying power or angular magnification of a simple lens when the image is formed at the near point of the eye.

$$m = 1 + \frac{25.0\text{ cm}}{f} \tag{25.5}$$

This is the expression for the magnification of a simple lens when the object is placed at the focal point of the lens. In this case the image will be formed at infinity and will allow the eye to focus in a more relaxed manner.

$$m = \frac{25.0\text{ cm}}{f} \tag{25.6}$$

A compound microscope contains an objective lens of short focal length f_0 and an eyepiece of focal length f_e. The two lenses are separated by a distance L. When an object is located just beyond the focal point of the objective, the two lenses in combination form an enlarged, virtual, and inverted image of overall magnification, M. The negative sign indicates the inverted nature of the image.

$$M = M_1 m_e = -\frac{L}{f_0}\left(\frac{25.0\text{ cm}}{f_e}\right) \tag{25.7}$$

The angular magnification of a telescope is equal to the ratio of the objective focal length to the eyepiece focal length. In this case, the two lenses are separated by a distance equal to the sum of their focal lengths. The negative sign indicates an inverted image.

$$m = -\frac{f_0}{f_e} \tag{25.8}$$

Rayleigh's criterion states the condition for the resolution of two images of closely spaced sources. For a slit the angular separation between sources (in radians) must be greater than the ratio of the wavelength to the slit width, a.

$$\theta_m \approx \frac{\lambda}{a} \tag{25.9}$$

(for a slit)

In the case of a circular aperture, the minimum angular separation which can be resolved depends on the diameter of the aperture, D.

$$\theta_m = 1.22\frac{\lambda}{D} \tag{25.10}$$

(for a circular aperture)

Rayleigh's criterion determines the limiting condition for resolution of adjacent sources. Under this criterion two sources are just resolved (seen as separate sources) when they are spaced so that the central maximum of the diffraction pattern of one source is located at the position of the first minimum of the diffraction pattern of the second source.

Comment on Rayleigh's criterion.

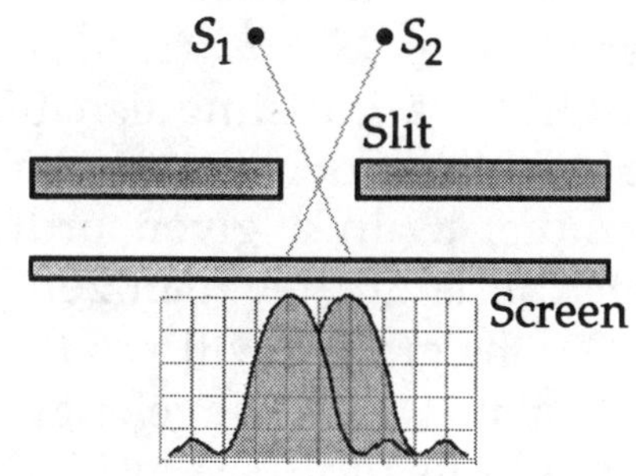

A diffraction grating of equally spaced parallel slits, separated by a distance, d, will produce an interference pattern in which there will be a series of maxima (bright lines) for each wavelength in the light source. The set of maxima due to wavelengths of all different values comprise a spectral order denoted by the number m.

$$d \sin\theta = m\lambda \qquad (25.11)$$

$m = 0, 1, 2 \ldots$

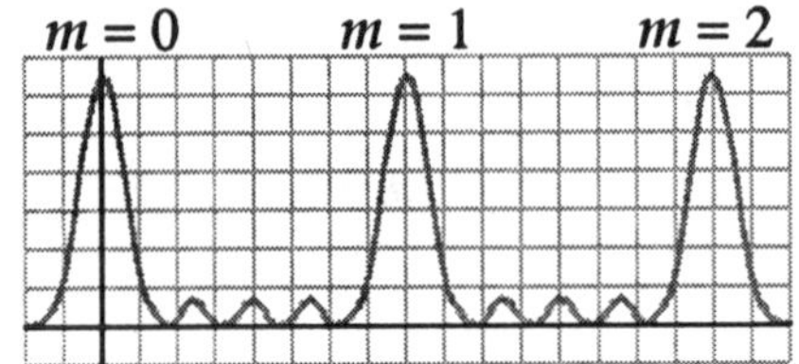

The resolving power of a diffraction grating increases as the number of lines illuminated increases; also, the resolving power is proportional to the order number in which the spectrum is observed.

$$R \equiv \frac{\lambda}{\lambda_2 - \lambda_1} = \frac{\lambda}{\Delta\lambda} \qquad (25.12)$$

$$R = Nm \qquad (25.13)$$

From Equation 25.12, it can be seen that a grating with a high resolving power can distinguish small differences between adjacent wavelengths. Equation 25.13 shows that in the zeroth order ($m = 0$), the resolution $R = 0$ (i.e. all wavelengths are indistinguishable for the zeroth order maximum). For example, if, in a given order (for which m is greater than zero), $R = 10{,}000$, the grating will produce a spectrum in which wavelengths differing in value by 1 part in 10,000 can be resolved.

Comment on resolving power of a grating.

REVIEW CHECKLIST

▷ Define the f-number of a camera lens and relate this criterion to shutter speed.

▷ Describe the geometry of the combination of lenses for each of several simple optical instruments: simple magnifier, compound microscope, reflecting telescope, and refracting telescope. Also, calculate the magnifying power for each instrument.

▷ Determine whether or not two sources under a given set of conditions are resolvable as defined by Rayleigh's criterion.

▷ Describe the technique employed in the Michelson interferometer for precise measurement of length based on known values for the wavelength of light.

▷ Determine the positions of the principal maxima in the interference pattern of a diffraction grating. Understand what is meant by the resolving power and the dispersion of a grating, and calculate the resolving power of a grating under specified conditions.

SOLUTIONS TO SELECTED END-OF-CHAPTER PROBLEMS

5. A camera is being used with the correct exposure at $f/4$ and a shutter speed of 1/32 s. In order to "stop" a fast-moving subject, the shutter speed is changed to 1/256 s. Find the new f-stop that should be used to maintain satisfactory exposure, assuming no change occurs in lighting conditions occurs.

Solution

If light of intensity I passes through an aperture of area A for a time Δt, the total energy transported through the opening is $E = IA(\Delta t)$.

The energy delivered to the film in the correct exposure (at $f/4$ with a shutter speed of $\Delta t = 1/32$ s) is $E_1 = IA_1(1/32\text{ s})$. If the shutter speed is changed to $\Delta t = 1/256$ s, the energy entering the camera during an exposure is $E_2 = IA_2(1/256\text{ s})$.

Note that since the lighting conditions have not changed, the intensity of the light is the same as before. Since the film is unchanged, the energy required for correct exposure is also unchanged. Thus, one must have $E_2 = E_1$, or $IA_2(1/256\text{ s}) = IA_1(1/32\text{ s})$ which reduces to $A_2 = 8A_1$. The aperture must be increased so its new area is eight times the original area. Assuming a circular opening with diameter d, this means that

$$\frac{\pi d_2^{\,2}}{4} = 8\frac{\pi d_1^{\,2}}{4}$$ or the new diameter of the aperture must be $d_2 = \sqrt{8}\, d_1$

The f-number is defined as $f\text{-number} = \dfrac{\text{focal length of lens}}{\text{diameter of opening}} = \dfrac{f}{d}$

Since the focal length of the camera lens is unchanged, the new f-number required is

$$(f-\text{number})_2 = \frac{f}{d_2} = \frac{f}{\sqrt{8}\, d_1} = \frac{1}{\sqrt{8}}\left(\frac{f}{d1}\right) = \frac{(f-\text{number})_1}{\sqrt{8}} = \frac{4}{\sqrt{8}} = 1.41$$

Thus, the new f-stop required to produce correct exposure at a shutter speed of $1/256$ s is $f/1.4$. ◊

13. A person is to be fitted with bifocals. She can see clearly when the object is between 30 cm and 1.5 m from the eye. (a) The upper portions of the bifocals should be designed to enable her to see distant objects clearly. What power should they have? (b) The lower portions of the bifocals should enable her to see objects comfortably at 25 cm. What power should they have (Fig. P25.13)?

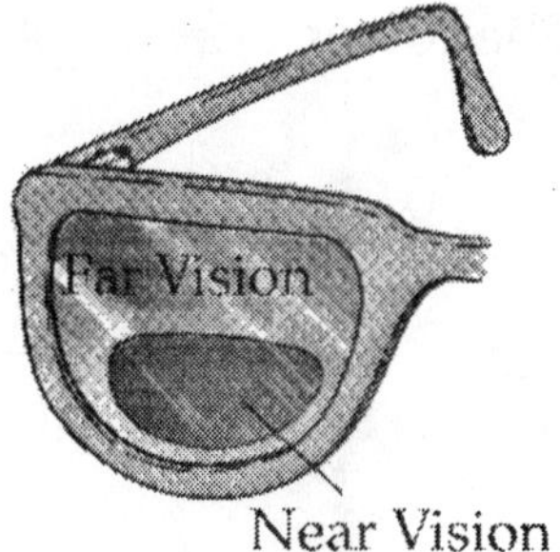

Figure P25.13

Solution

(a) To correct a nearsighted eye (fails to see distant objects clearly), the corrective lens should form an upright virtual image of the most distant objects (those having an object distance of $p \approx \infty$) at the far point of the eye. Since this person cannot focus on objects that are more than 1.5 m from the eye, the far point is 1.5 m and the required focal length of the corrective lens is found from the thin-lens equation as

$$\frac{1}{f} = \frac{1}{p} + \frac{1}{q} = \frac{1}{\infty} + \frac{1}{-1.5\text{ m}} \quad \text{or} \quad f = -1.5\text{ m}$$

Note that the image distance is negative since the lens must form a virtual image. Thus, the power of the upper portion of the bifocals should be

$$P_{\text{upper}} = \frac{1}{f} = \frac{1}{-1.5\text{ m}} = -0.67\text{ diopters} \qquad \Diamond$$

(b) The lower portion of the bifocals must allow the person to see objects comfortably at 25 cm from the eye. Since the nearest object the unaided eye can comfortable focus on is 30 cm from the eye, the corrective lens must form an upright virtual image 30 cm in front of the eye when the object is 25 cm from the eye. Using the thin lens equation, with $q = -30$ cm when $p = +25$ cm, gives the required focal length of the corrective lens as

$$\frac{1}{f} = \frac{1}{+25\text{ cm}} + \frac{1}{-30\text{ cm}} = \frac{+1}{150\text{ cm}} \quad \text{or} \quad f = +150\text{ cm} = +1.5\text{ m}$$

The lower portion of the bifocals should then have a power of

$$P_{\text{lower}} = \frac{1}{f} = \frac{1}{+1.5 \text{ m}} = +0.67 \text{ diopters} \qquad \Diamond$$

---

17. A biology student uses a simple magnifier to examine the structural features of the wing of an insect. The wing is held 3.50 cm in front of the lens, and the image is formed 25.0 cm from the eye. (a) What is the focal length of the lens? (b) What angular magnification is achieved?

Solution

(a) A simple magnifier consists of a single converging lens held close to the eye. The lens forms an upright, magnified, virtual image of an object located between the lens and its focal point. In this case, it is desired that the image distance be $q = -25.0$ cm when the object distance is $p = +3.50$ cm. The thin lens then gives the required focal length as

$$\frac{1}{f} = \frac{1}{+3.50 \text{ cm}} + \frac{1}{-25.0 \text{ cm}} \quad \text{or} \quad f = +4.07 \text{ cm} \qquad \Diamond$$

(b) When a simple magnifier forms its virtual image at the near point of the normal human eye (i.e., when $q = -25.0$ cm), the angular magnification produced by the magnifier is

$$M = 1 + \frac{25.0 \text{ cm}}{f} = 1 + \frac{25.0 \text{ cm}}{4.07 \text{ cm}} = 7.14 \qquad \Diamond$$

Note: The alternate expression for the angular magnification produced by a simple magnifier, $M = 25.0 \text{ cm}/f$, is valid only when the image is formed at $q \approx -\infty$ so the eye is most relaxed while viewing it. These are not the conditions present in this problem.

---

21. The length of a microscope tube is 15.0 cm. The focal length of the objective is 1.00 cm, and the focal length of the eyepiece is 2.50 cm. What is the magnification of the microscope, assuming it is adjusted so that the eye is relaxed?

Solution When the microscope is adjusted so the eye is relaxed, parallel rays enter the eye. Therefore, the object for the eyepiece must be located at its focal point $(p_e = f_e = 2.50\text{ cm})$, and the angular magnification produced by the eyepiece is

$$m_e = \frac{25.0\text{ cm}}{f_e} = \frac{25.0\text{ cm}}{2.50\text{ cm}} = +10.0$$

Since the object for the eyepiece is the image formed by the objective lens, the image distance for the objective lens is

$$q_o = L - p_e = 15.0\text{ cm} - 2.50\text{ cm} = 12.5\text{ cm}$$

where $L = 15.0\text{ cm}$ is the length of the microscope. Note that one of the approximations made in the textbook while deriving the overall magnification of the compound microscope, namely the assumption that $q_o \approx L$, is not valid in this case. Hence, the result of that derivation should not be used in this problem. The thin lens equation then gives the object distance for the objective lens as

$$p_o = \frac{q_o f_o}{q_o - f_o} = \frac{(12.5\text{ cm})(1.00\text{ cm})}{12.5\text{ cm} - 1.00\text{ cm}} = 1.09\text{ cm}$$

The lateral magnification produced by the objective lens and the overall magnification of the microscope are therefore:

$$M_o = -\frac{q_o}{p_o} = -\frac{12.5\text{ cm}}{1.09\text{ cm}} = -11.5 \quad \text{and} \quad M = M_o m_e = (-11.5)(+10.0) = -115 \quad \Diamond$$

The negative sign means that the image viewed by the eye is inverted relative to the original object.

25. The lenses of an astronomical telescope are 92 cm apart when adjusted for viewing a distant object with minimum eyestrain. The angular magnification produced by the telescope is 45. Compute the focal length of each lens.

Solution

An astronomical telescope views very distant objects so the object distance for the objective lens is $p_0 \approx \infty$, and the image is formed at its focal point $(q_0 = f_0)$.

If the telescope is adjusted for minimum eyestrain, the object for the eyepiece is located at its focal point $(p_f = f_e)$ and parallel rays enter the eye. Under these conditions, the length of the telescope (i.e. the distance between the objective and eyepiece lenses) is $L = f_0 + f_e$, and the angular magnification is $m = f_o / f_e$.

It is given that the lenses of the telescope are 92 cm apart. Hence,

$$L = f_0 + f_e = 92 \text{ cm} \qquad \textbf{[Equation 1]}$$

Also, the angular magnification produced is given as 45. Therefore,

$$m = \frac{f_o}{f_e} = 45, \text{ or } f_o = 45 f_e \qquad \textbf{[Equation 2]}$$

Substituting Equation 2 into Equation 1 gives $45f_e + f_e = 46f_e = 92$ cm

so the focal length of the eyepiece is $f_e = 2.0$ cm ◊

By Equation 2, the objective lens' focal length is:

$$f_0 = 45 f_e = 45(2.0 \text{ cm}) = 90 \text{ cm}$$ ◊

35. Suppose a 5.00-m-diameter telescope were constructed on the Moon, where the absence of atmospheric distortion would permit excellent viewing. If observations were made using 500-nm light, what minimum separation between two objects could just be resolved on Mars at closest approach (when Mars is 8.0×10^7 km from the Moon)?

Solution According to Rayleigh's criterion, the limiting angle of resolution for a circular aperture is $\theta_m = 1.22\lambda/d$ where d is the diameter of the aperture and λ is the wavelength of the light passing through it.

In the case of a large astronomical telescope, the objective mirror is the relevant aperture. Thus, when viewing with 500 nm light, the limiting angle of resolution for the described telescope is

$$\theta_m = 1.22\frac{\left(500 \times 10^{-9}\ \text{m}\right)}{5.00\ \text{m}} = 1.22 \times 10^{-7}\ \text{radians}$$

When this telescope is used to view the surface of Mars from a distance of

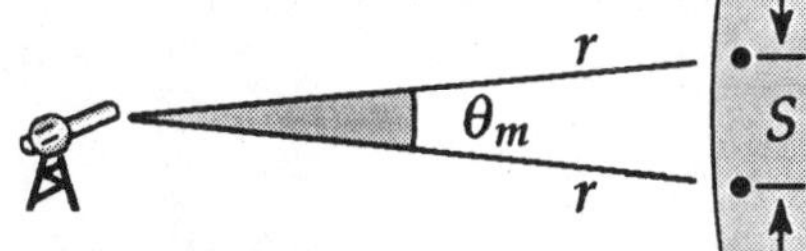

$$r = 8.0 \times 10^7\ \text{km}$$

the separation between two objects that can just be resolved is

$$S = r\theta_m = \left(8.0 \times 10^7\ \text{km}\right)\left(1.22 \times 10^{-7}\ \text{radians}\right) = 9.8\ \text{km}$$ ◊

39. The light path in one arm of a Michelson interferometer includes a transparent cell that is 5.00 cm long. How many fringe shifts would be observed if all the air were evacuated from the cell? The wavelength of the light source is 590 nm and the refractive index of air is 1.000 29. (See the **Hint** in Problem 38.)

Solution As the light travels down the arm of the interferometer and back again, it passes through the transparent cell twice. Thus, the total distance traveled in the cell is 2ℓ where ℓ is the length of the cell.

The number of wavelengths of the light that fit in this length when a vacuum exists in the cell is $N_{vac} = 2\ell/\lambda_{vac}$. When the cell was filled with air, the number of wavelength that would fit in the cell was $N_{air} = 2\ell/\lambda_{air}$,

or since $\lambda_{air} = \dfrac{\lambda}{n_{air}}$, $\qquad N_{air} = \dfrac{2\ell}{\lambda/n_{air}} = n_{air}\left(\dfrac{2\ell}{\lambda}\right)$

Thus, evacuating the air from the cell decreases the optical path length in this arm of the interferometer by $\delta = (\Delta N)\lambda = (N_{air} - N_{vac})\lambda = 2\ell(n_{air} - 1)$. A fringe shift (i.e., the changing of a fringe from light to dark or vise versa) occurs each time the optical path length is changed by one-half wavelength. Therefore, the number of fringe shifts that will be observed as the cell is evacuated is

$$\text{\# fringe shifts} = \frac{\delta}{\lambda/2} = \frac{4\ell(n_{air} - 1)}{\lambda}$$

If the cell has a length of ℓ = 5.00 cm, air has an index of refraction of $n_{air} = 1.000\,29$, and light of wavelength $\lambda = 590$ nm is used, the number of fringe shifts observed will be

$$\text{\# fringe shifts} = \frac{4\left(5.00 \times 10^{-2}\ \text{m}\right)\left(2.9 \times 10^{-4}\right)}{590 \times 10^{-9}\ \text{m}} = 98 \qquad \lozenge$$

43. Intense white light is incident on a diffraction grating which has 600 lines/mm. (a) What is the highest order in which the complete visible spectrum can be seen using this grating? (b) What is the angular separation between the violet edge (400 nm) and the red edge (700 nm) of the first order spectrum produced by this grating?

Solution (a) The longest wavelength in the visible spectrum is 700 nm. Since the longest wavelengths are diffracted the most (i.e., deviated through the largest angles), the highest order which yields a complete visible spectrum is the same as the highest order in which the 700 nm wavelength can be found. The grating equation is $m\lambda = d\sin\theta$ where m is the order number, d is the spacing between adjacent slits, and λ is the wavelength of the light deviated by the angle θ.

The spacing between adjacent slits on a grating with 600 lines/mm is

$$d = \frac{1.00\times10^{-3}\text{ m}}{600} = 1.67\times10^{-6}\text{ m}$$

Since the largest possible angle of deviation is $\theta = 90.0°$, the highest order in which the red edge of the visible spectrum can be seen is

$$m = \frac{d\sin\theta}{\lambda} = \frac{\left(1.67\times10^{-6}\text{ m}\right)\sin 90.0°}{700\times10^{-9}\text{ m}} = 2.38$$

The order number (m) must be an integer, so the highest order in which the complete visible spectrum can be seen is the second order $(m = 2)$. ◊

(b) In the first order, the violet edge $(\lambda = 400\text{ nm})$ of the spectrum is at

$$\theta = \arcsin\left(\frac{m\lambda}{d}\right) = \arcsin\left(\frac{(1)\left(400\times10^{-9}\text{ m}\right)}{1.67\times10^{-6}\text{ m}}\right) = \arcsin(0.240) = 13.9°$$

Similarly, the location of the red edge (700 nm) in the first order is

$$\theta = \arcsin\left(\frac{700\times10^{-9}\text{ m}}{1.67\times10^{-6}\text{ m}}\right) = \arcsin(0.420) = 24.8°$$

The angular separation between the violet and red edges of the first order spectrum produced by this grating is then $\Delta\theta = 24.8° - 13.9° = 10.9°$ ◊

49. The H_α line in hydrogen has a wavelength of 656.20 nm. This line differs in wavelength from the corresponding spectral line in deuterium (the heavy stable isotope of hydrogen) by 0.18 nm. (a) Determine the minimum number of lines a grating must have to resolve these two wavelengths in the first order. (b) Repeat part (a) for the second order.

Solution

(a) The difference in the wavelengths to be resolved is $\Delta\lambda = 0.18 \text{ nm}$. Thus, the required resolving power is

$$R = \frac{\lambda}{\Delta\lambda} = \frac{656.2 \text{ nm}}{0.18 \text{ nm}} = 3.65 \times 10^3$$

The resolving power of a diffraction grating is given by the product of the total number of slits on the grating and the order number of the image observed, i.e., $R = Nm$. Thus, the minimum number of slits required on the grating to produce the needed resolving power in the first order is

$$N = \frac{R}{m} = \frac{3.65 \times 10^3}{1} = 3.65 \times 10^3 \qquad \lozenge$$

(b) To resolve the same spectral lines in the second-order spectrum requires the same resolving power found above, $R = \lambda/\Delta\lambda = 3.65 \times 10^3$. The minimum number of slits required on the grating to yield this resolving power in the second order is

$$N = \frac{R}{m} = \frac{3.65 \times 10^3}{2} = 1.83 \times 10^3 \qquad \lozenge$$

---

51. The near point of an eye is 75.0 cm. (a) What should be the power of a corrective lens prescribed to enable the eye to see an object clearly at 25.0 cm? (b) If, using the corrective lens, the user can see an object clearly at 26.0 cm but not 25.0 cm, by how many diopters did the lens grinder miss the prescription?

Solution (a) Unaided, this far-sighted eye is unable to focus on objects that are less than 75.0 cm from it. To enable the eye to see an object located 25.0 cm in front of it, a corrective lens should form an upright virtual image 75.0 cm from the eye. Thus, the focal length of the lens should be such that the image distance is $q = -75.0\text{ cm}$ when the object distance is $p = +25.0\text{ cm}$. From the thin lens equation,

$$\frac{1}{f} = \frac{1}{p} + \frac{1}{q} = \frac{1}{25.0\text{ cm}} - \frac{1}{75.0\text{ cm}} = \frac{2}{75.0\text{ cm}} \quad \text{or} \quad f = +37.5\text{ cm} = +0.375\text{ m}$$

The power of the required lens is $P = \frac{1}{f} = \frac{1}{+0.375\text{ m}} = +2.67\text{ diopters}$ ◊

(b) If, when using the corrective lens, the user must position the object at a distance of $p = +26.0\text{ cm}$ in order to see it clearly (i.e., have the image formed at the near point or at $q = -75.0\text{ cm}$), the focal length of the lens is

$$f = \frac{pq}{p+q} = \frac{(+26.0\text{ cm})(-75.0\text{ cm})}{26.0\text{ cm} - 75.0\text{ cm}} = +39.8\text{ cm} = 0.398\text{ m}$$

The actual power of the corrective lens is then

$$P = \frac{1}{f} = \frac{1}{+0.398\text{ m}} = +2.51\text{ diopters}$$

compared to the required power of +2.67 diopters. The power of the actual lens is therefore 0.16 diopters less than what is needed. ◊

---

57. Sunlight is incident on a diffraction grating that has 2750 lines/cm. The second-order spectrum over the visible range (400–700 nm) is to be limited to 1.75 cm along a screen that is distance *L* from the grating. What is the required value of *L*?

Solution On a grating with 2750 lines/cm, the spacing between adjacent slits is

$$d = \frac{1.00\text{ cm}}{2750} = 3.64 \times 10^{-4}\text{ cm} = 3.64 \times 10^{-6}\text{ m} = 3.64 \times 10^{3}\text{ nm}$$

From the grating equation, the angular position of the violet edge (400 nm) of the second-order visible spectrum is

$$\theta_V = \arcsin\left(\frac{m\lambda}{d}\right) = \arcsin\left(\frac{2(400 \text{ nm})}{3.64\times 10^3 \text{ nm}}\right) = 12.7° = 0.222 \text{ rad}$$

The location of the red edge (700 nm) of the second-order visible spectrum is

$$\theta_R = \arcsin\left(\frac{m\lambda}{d}\right) = \arcsin\left(\frac{2(700 \text{ nm})}{3.64\times 10^3 \text{ nm}}\right) = 22.6° = 0.395 \text{ rad}$$

Thus, the total angular width of the second-order visible spectrum is

$$\Delta\theta = \theta_R - \theta_V = 0.395 \text{ rad} - 0.222 \text{ rad} = 0.173 \text{ rad}$$

If this spectrum is incident on a screen located distance L from the grating, the width of the spectrum on the screen is $S = L(\Delta\theta)$. Therefore, if the width on the screen is observed to be $S = 1.75$ cm, the distance from the grating to the screen is

$$L = \frac{S}{\Delta\theta} = \frac{1.75 \text{ cm}}{0.173 \text{ rad}} = 10.1 \text{ cm}$$ ◊

---

61. Light containing two different wavelengths passes through a diffraction grating with 1200 slits/cm. On a screen 15.0 cm from the grating, the third-order **maximum** of the shorter wavelength falls on top of the first-order **minimum** of the longer wavelength. If the neighboring maxima of the longer wavelength are 8.44 mm apart on the screen, what are the wavelengths in the light? (**Hint:** Use the small angle approximation.)

Solution The distance on the screen from the central maximum to the location of the mth order maximum for wavelength λ is $y_m = L\tan\theta$ where L is the distance from grating to screen and θ is the angular deviation as shown in the sketch.

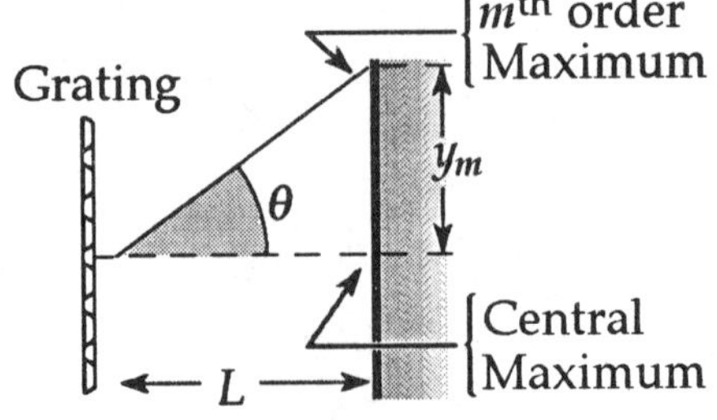

The angular deviation is given by the grating equation as $\sin\theta = m\lambda/d$, where d is the spacing between adjacent slits on the grating. Assuming that $d >> \lambda$, the angles of deviation will be small and $\sin\theta \approx \tan\theta$. Under these conditions, the screen locations of the maxima become $y_m \approx m(\lambda L/d)$.

The spacing between successive maxima for wavelength λ is therefore

$$\Delta y = y_{m+1} - y_m \approx (m+1)\left(\frac{\lambda L}{d}\right) - m\left(\frac{\lambda L}{d}\right) = \frac{\lambda L}{d}$$

It is given that for the longer wavelength passing through the grating,

$$\Delta y = 8.44 \text{ mm} = 8.44\times10^{-3} \text{ m} \quad \text{when} \quad d = 1 \text{ cm}/1200 = 8.33\times10^{-4} \text{ cm}$$

Thus, the longer wavelength passing through the grating ($L = 15.0$ cm) is

$$\lambda_1 = \frac{(\Delta y)d}{L} = \frac{\left(8.44\times10^{-3} \text{ m}\right)\left(8.33\times10^{-4} \text{ cm}\right)}{15.0 \text{ cm}} = 4.69\times10^{-7} \text{ m} = 469 \text{ nm} \quad \Diamond$$

The minima in intensity occur at angles of deviation for which the path difference for light passing through adjacent slits on the grating is

$$\delta = d\sin\theta = \left(m + \frac{1}{2}\right)\lambda \quad \text{with } m = 0, 1, 2, 3, \ldots$$

Hence, at the first-order **minimum** for the longer wavelength,

$$\sin\theta = \left(0 + \frac{1}{2}\right)\frac{\lambda_1}{d} = \frac{\lambda_1}{2d}$$

The angle at which the third-order **maximum** for the shorter wavelength λ_2 occurs is $\sin\theta = m\lambda/d = 3\lambda_2/d$. If this maximum coincides with the first-order minimum for λ_1, then

$$\frac{3\lambda_2}{d} = \frac{\lambda_1}{2d} \quad \text{or} \quad \lambda_2 = \frac{\lambda_1}{6} = \frac{469 \text{ nm}}{6} = 78.1 \text{ nm} \quad \Diamond$$

CHAPTER SELF-QUIZ

1. A camera lens with a focal length of 2.00 cm and an aperture opening diameter of 0.40 cm has what *f*-number?
   a. 0.20
   b. 0.80
   c. 5.00
   d. 10.0

2. A converging lens will be prescribed by the eye doctor to correct which of the following?
   a. nearsightedness
   b. glaucoma
   c. farsightedness
   d. astigmatism

3. What is the magnification of a refracting telescope with objective and eyepiece lenses of focal lengths 90.0 cm and 2.00 cm, respectively?
   a. 30
   b. 45
   c. 60
   d. 180

4. The Michelson interferometer is a device that may be used to measure which of the following?
   a. magnifying power of lenses
   b. light wavelength
   c. atomic masses
   d. electron charge

5. Doubling the *f*-number of a camera lens will change the light intensity admitted to the film by what factor?
   a. 0.25
   b. 0.50
   c. 2.00
   d. 4.00

6. Which eye defect is corrected by a lens having different curvatures in two perpendicular directions?
   a. myopia
   b. presbyopia
   c. hyperopia
   d astigmatism

7. Two thin lenses in combination, placed in contact with each other along a common axis, have the respective powers of 45.0 and –15.0 diopters. What is their combined power?
   a. 15.0 diopters
   b. 30.0 diopters
   c. 60.0 diopters
   d. –22.5 diopters

8. A compound microscope has objective and eyepiece lenses of focal lengths 0.80 and 4.00 cm, respectively. If the microscope length is 15.0 cm, what is the maximum magnification?
   a. 117.0
   b. 6.30
   c. 48.0
   d. 97.0

9. What resolving power must a diffraction grating have in order to distinguish wavelengths of 630.1 and 631.7 nm?
   a. 315
   b. 394
   c. 630.9
   d. 788

10. A camera uses a
    a. converging lens to form a real image
    b. converging lens to form an virtual image
    c. diverging lens to form a real image
    d. diverging lens to form an virtual image

11. A magnifier uses a
    a. converging lens to form a real image
    b. converging lens to form a virtual image
    c. diverging lens to form a virtual image
    d. diverging lens to form a real image

12. A 1.7-m-tall woman stands 5.0 m in front of a camera with a 5.0-cm focal length lens. What is the size of the image formed on film?
    a. 3.4 cm
    b. 2.6 cm
    c. 1.7 cm
    d. 0.85 cm

Chapter 26

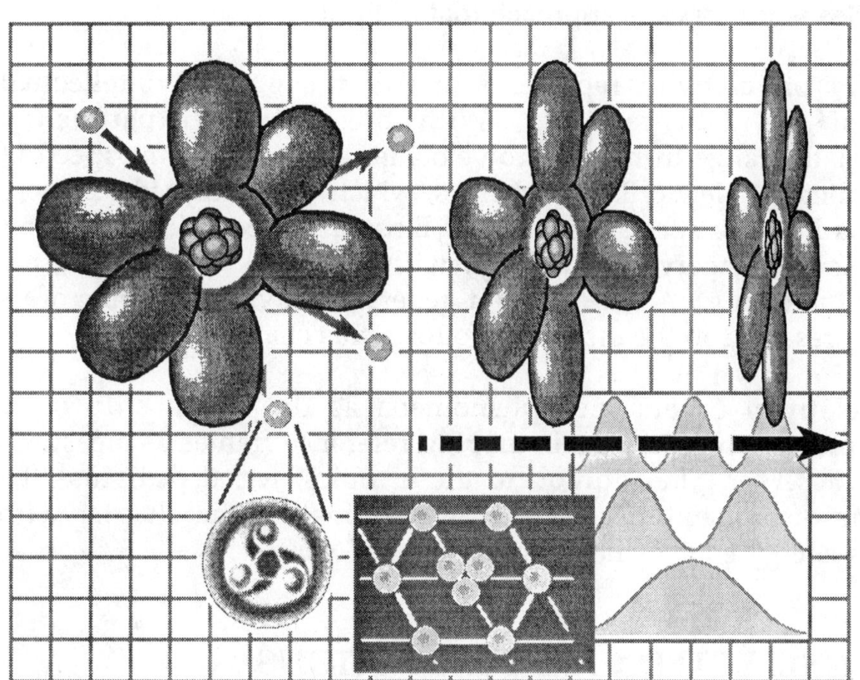

RELATIVITY

Chapter 26

RELATIVITY

Most of our everyday experiences and observations deal with objects that move at speeds much lower than the speed of light. Newtonian mechanics and the early ideas on space and time were formulated to describe the motion of such objects. As we saw in the chapters on mechanics, this formalism is very successful in describing a wide range of phenomena. Although Newtonian mechanics works very well at low speeds, it fails when applied to particles whose speeds approach that of light.

The theory of relativity represents one of the greatest intellectual achievements of the 20th century. With this theory, experimental observations over the range from $v = 0$ to velocities approaching the speed of light can be predicted. Newtonian mechanics, which was accepted for more than 200 years, is in fact a specialized case of Einstein's theory. This chapter introduces the special theory of relativity, with emphasis on some of the consequences of the theory. A discussion of general relativity and some of its consequences is presented in the essay that follows this chapter.

Special relativity covers such phenomena as the slowing down of clocks and the contraction of lengths in moving reference frames as measured by a stationary observer. In addition to these topics, we also discuss the relativistic forms of momentum and energy, terminating the chapter with the famous mass-energy equivalence formula, $E = mc^2$.

NOTES FROM SELECTED CHAPTER SECTIONS

26.1 Introduction

The special theory of relativity is based on two basic postulates:

1. The laws of physics are the same in all **inertial** reference systems.
2. The speed of light in vacuum is always measured to be 3.00×10^8 m/s, and the measured value is **independent** of the motion of the observer or of the motion of the source of light.

26.2 The Principle of Relativity

According to the principle of Newtonian relativity, the laws of mechanics are the same in all inertial frames of reference. Inertial frames of reference are those coordinate systems which are **at rest with respect to one another** or which move at constant velocity with respect to one another. **There is no preferred frame of reference for describing the laws of mechanics.**

26.4 The Michelson-Morley Experiment

This experiment was designed to detect the velocity of the Earth with respect to the **hypothetical ether.** The instrument used is called an interferometer and the measurement involves the observation of an interference pattern of two reflected light beams. The outcome of the experiment was **negative,** contradicting the ether hypothesis. Light is now understood to be an electromagnetic wave that requires no medium for its propagation.

26.5 Einstein's Principle of Relativity

Einstein's special theory of relativity is based upon two postulates:

1. The laws of physics are the same in every inertial frame of reference.

2. The speed of light has the same value as measured by all observers, independent of the motion of the light source or observer.

The second postulate is consistent with the negative results of the Michelson-Morley experiment which failed to detect the presence of an ether and suggested that the speed of light is the same in all inertial frames.

26.6 Consequences of Special Relativity

The concepts of length, time, and simultaneity are quite different in relativistic mechanics from what they are in Newtonian mechanics. **The distance between two points and the time interval between two events depend on the frame of reference in which they are measured.**

Two events that are simultaneous in one reference frame are, in general, not simultaneous in another frame which is moving with respect to the first.

The **proper time** is always the time **interval measured with a single clock at rest in the frame in which the events take place at the same position**. According to a stationary observer, a moving clock runs slower than an identical clock at rest. This effect is known as **time dilation.**

The **proper length** of an object is defined as the length of the object measured in the **reference frame in which the object is at rest.** The length of an object measured in a reference frame in which it is moving is always less than the proper length. This effect is known as **length contraction.** The contraction occurs only **along the direction of motion.**

26.7 Relativistic Momentum

To account for **relativistic effects,** it is necessary to modify the definition of momentum to satisfy the following conditions:

1. The relativistic momentum must be conserved in all collisions.

2. The relativistic momentum must approach the classical value, mv, as the quantity v/c approaches zero.

EQUATIONS AND CONCEPTS

As measured by a stationary observer, a moving clock runs slower than an identical stationary clock. This effect is known as time dilation. The time interval Δt_p is called the proper time and is always the time measured by an observer moving with the clock.

$$\Delta t = \frac{\Delta t_p}{\sqrt{1 - v^2 / c^2}} = \gamma \Delta t_p \qquad (26.8)$$

$$\gamma = \frac{1}{\sqrt{1 - v^2 / c^2}}$$

An observer moving with a speed v relative to an object will determine its length to be shorter than that measured by an observer at rest with respect to the object. The length measured in the reference frame in which the object is at rest is called the proper length. Note that length contraction is observed only along the direction of motion.

$$L = L_p \sqrt{1 - v^2 / c^2} \qquad (26.9)$$

This equation determines the momentum of a relativistic particle of mass m moving with a speed v. The equation satisfies the two essential requirements: (i) the relativistic momentum is conserved in all collisions, and (ii) the relativistic momentum approaches the classical value as the ratio v/c approaches zero.

$$p \equiv \frac{mv}{\sqrt{1 - v^2 / c^2}} = \gamma m v \qquad (26.10)$$

In these equations for relativistic energy and kinetic energy, mc^2 is the rest energy of the object and is independent of the object's speed. The term γmc^2 is dependent on the object's speed and is the total energy of the object. These equations show that mass is a form of energy.

$$KE = \gamma mc^2 - mc^2 \tag{26.12}$$

$$E = \gamma mc^2 = KE + mc^2 \tag{26.14}$$

$$E = \frac{mc^2}{\sqrt{1 - v^2 / c^2}} \tag{26.15}$$

In cases where the speed of an object is not known, it is useful to have an expression which relates the relativistic energy, E, to the relativistic momentum, p. When an object is at rest, its total energy is equal to its rest energy.

$$E^2 = p^2c^2 + (mc^2)^2 \tag{26.16}$$

This equation is an exact expression relating energy and momentum for photons which have a speed equal to c.

$$E = pc \tag{26.17}$$

Comment on basic postulates

Einstein's theory of special relativity is based upon two postulates: (i) the laws of physics are the same in all inertial frames of reference (an inertial frame of reference is a non-accelerated frame); and (ii) the speed of light has the same value as measured by all observers, independent of the relative motion between light source and observer. The second postulate is consistent with the negative results of the Michelson-Morley experiment which failed to detect the presence of an ether.

When dealing with subatomic particles, it is convenient to express their energy in electron volts (eV).

$$1\ \text{eV} = 1.60 \times 10^{-19}\ \text{J}$$

$$m_e c^2 = 0.511\ \text{MeV}$$

REVIEW CHECKLIST

- ▷ State Einstein's two postulates of the special theory of relativity.
- ▷ Understand the Michelson-Morley experiment, its objectives, results, and the significance of its outcome.
- ▷ Understand the idea of simultaneity, and the fact that simultaneity is not an absolute concept. That is, two events which are simultaneous in one reference frame are not simultaneous when viewed from a second frame moving with respect to the first.
- ▷ Make calculations using the equations for time dilation and length contraction.
- ▷ State the correct relativistic expressions for the momentum, kinetic energy, and total energy of a particle. Make calculations using these equations.

SOLUTIONS TO SELECTED END-OF-CHAPTER PROBLEMS

3. A deep-space probe moves away from Earth with a speed of $0.80c$. An antenna on the probe requires 3.0 s, probe time, to rotate through 1.0 rev. How much time is required for 1.0 rev according to an observer on Earth?

Solution The reference frame in which an observer would see the timed event (the rotation of the antenna) occur at a single position is the rest frame of the space probe. Thus, the proper time for the event is the probe time, $\Delta t_p = 3.0\text{ s}$. An observer on Earth is moving at speed $v = 0.80c$ relative to the event, and hence measures a dilated time interval given by

$$\Delta t = \gamma \Delta t_p \qquad \text{where} \qquad \gamma = \frac{1}{\sqrt{1-(v/c)^2}}$$

The time required for one revolution according to the observer on Earth is

$$\Delta t = \frac{\Delta t_p}{\sqrt{1-(v/c)^2}} = \frac{3.0\text{ s}}{\sqrt{1-(0.80)^2}} = 5.0\text{ s} \qquad \lozenge$$

---

7. A muon formed high in the Earth's atmosphere travels at speed $v = 0.99c$ for a distance of 4.6 km before it decays into an electron, a neutrino, and an anti-neutrino ($\mu^- = e^- + \nu + \bar{\nu}$). (a) How long does the muon live, as measured in its reference frame? (b) How far does the muon travel, as measured in its frame?

Solution (a) The observer on the Earth sees the muon travel 4.6 km at a speed of $v = 0.99c$ before it decays. Hence, this observer, who is in motion relative to the observed event (the life span of the muon), measures a dilated time interval of

$$\Delta t = \frac{L_p}{v} = \frac{4.6\text{ km}}{0.99c} = \frac{4.6\times10^3\text{ m}}{0.99\left(3.0\times10^8\text{ m/s}\right)} = 1.5\times10^{-5}\text{ s}$$

Note: The distance here is the proper length, L_p, because the observer on Earth is at rest with respect to the distance scale used to measure this length. The proper time interval (life span of the muon in its own rest frame) is then

$$\Delta t_p = \frac{\Delta t}{\gamma} = \Delta t\sqrt{1-(v/c)^2} = \left(1.6\times10^{-5}\text{ s}\right)\sqrt{1-(0.99)^2} = 2.2\times10^{-6}\text{ s} = 2.2\ \mu\text{s} \quad \lozenge$$

(b) The distance an observer in the rest frame of the muon, and hence in motion relative to the distance scale fixed on the Earth, will observe the scale move relative to the muon is

$$L = v(\Delta t_p) = 0.99(3.0\times10^8 \text{ m/s})(2.2\times10^{-6} \text{ s}) = 6.5\times10^2 \text{ m} = 0.65 \text{ km} \qquad \lozenge$$

Notice that this result could also have been obtained from the length contraction equation by realizing that the proper length is the 4.6 km measured by the observer on the Earth. The length contraction equation would give

$$L = L_p\sqrt{1 - v^2/c^2} = (4.6 \text{ km})\sqrt{1-(0.99)^2} = 0.65 \text{ km} \qquad \lozenge$$

---

13. An observer, moving at a speed of $0.995c$ relative to a rod (Figure P26.13), measures its length to be 2.00 m and sees its length to be oriented at 30.0° with respect to the direction of motion. (a) What is the proper length of the rod? (b) What is the orientation angle in a reference frame moving with the rod?

Solution (a) In the figure, the quantities

$$L_p,\ \theta_p,\ (L_p)_x,\ \text{and}\ (L_p)_y$$

are all proper quantities measured in the rest frame of the rod. The given data values, $L = 2.00$ m and $\theta = 30.0°$, are measured by an observer who is moving at a speed of $v = 0.995c$ relative to the rod.

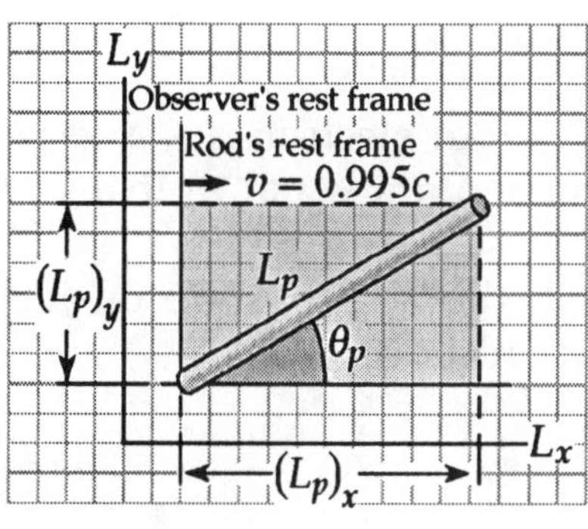

Figure P26.13

The components of the length of the rod, as measured by the moving observer, are then

$$L_x = L\cos\theta = (2.00 \text{ m})\cos 30.0° = 1.732 \text{ m}$$

and $$L_y = L\sin\theta = (2.00 \text{ m})\sin 30.0° = 1.00 \text{ m}$$

Component L_x is along the direction of motion and has undergone length contraction. The component L_y is perpendicular to the motion and is not contracted. Thus, the proper lengths of the components (measured in the rod's rest frame) are $\left(L_p\right)_y = L_y = 1.00\text{ m}$, and

$$\left(L_p\right)_x = \gamma L_x = \frac{L_x}{\sqrt{1 - v^2/c^2}} = \frac{1.732\text{ m}}{\sqrt{1-(0.995)^2}} = 17.34\text{ m}$$

Therefore, the proper length of the rod is

$$L_p = \sqrt{\left(L_p\right)_x^2 + \left(L_p\right)_y^2} = \sqrt{(17.34\text{ m})^2 + (1.00\text{ m})^2} = 17.37\text{ m} \qquad \lozenge$$

(b) The orientation angle in the rest frame of the rod is given by

$$\theta_p = \arctan\left[\frac{\left(L_p\right)_y}{\left(L_p\right)_x}\right] = \arctan\left(\frac{1.00\text{ m}}{17.3\text{ m}}\right) = 3.31° \qquad \lozenge$$

---

15. An electron has a speed v = 0.90c. At what speed will a proton have a momentum equal to that of the electron?

Solution A particle of mass m, moving at speed v, has a momentum of magnitude

$$p = \gamma m v = \frac{mv}{\sqrt{1-(v/c)^2}}$$

Thus, the momentum of an electron moving at $v = 0.90c$ has a magnitude of

$$p_e = \frac{\left(9.11\times10^{-31}\text{ kg}\right)(0.90c)}{\sqrt{1-(0.90)^2}} = \left(1.88\times10^{-30}\text{ kg}\right)c$$

If a proton is to have the same momentum, it is necessary that

$$p_p = \frac{m_p v}{\sqrt{1-(v/c)^2}} = p_e \quad \text{or} \quad p_p = \frac{\left(1.67\times10^{-27}\ \text{kg}\right)v}{\sqrt{1-(v/c)^2}} = \left(1.88\times10^{-30}\ \text{kg}\right)c$$

This may be rewritten as

$$\frac{v}{c} = \left(\frac{1.88\times10^{-30}\ \text{kg}}{1.67\times10^{-27}\ \text{kg}}\right)\sqrt{1-\left(\frac{v}{c}\right)^2} = \left(1.12\times10^{-3}\right)\sqrt{1-\left(\frac{v}{c}\right)^2}$$

Squaring both sides of this expression gives

$$\left(\frac{v}{c}\right)^2 = 1.3\times10^{-6} - 1.3\times10^{-6}\left(\frac{v}{c}\right)^2 \quad \text{or} \quad \left(\frac{v}{c}\right)^2 = \frac{1.3\times10^{-6}}{1-1.3\times10^{-6}} = 1.3\times10^{-6}$$

Hence, the proton must have a speed of

$$v = c\sqrt{1.3\times10^{-6}} = 1.1\times10^{-3}c = \left(1.1\times10^{-3}\right)\left(3.0\times10^{8}\ \text{m/s}\right) = 3.4\times10^{5}\ \text{m/s} \quad \Diamond$$

---

19. The nonrelativistic expression for the momentum of a particle, $p = mv$, can be used if $v \ll c$. For what speed does the use of this formula give an error in the momentum of (a) 1.00% and (b) 10.0% ?

Solution The percentage error incurred when using the nonrelativistic (or classical) expression for the momentum of a particle is

$$\%\ \text{error} = \left(\frac{p_{\text{relativistic}} - p_{\text{classical}}}{p_{\text{relativistic}}}\right)100\% = \left(\frac{\gamma mv - mv}{\gamma mv}\right)100\% = \left(1-\frac{1}{\gamma}\right)100\%$$

Since $\gamma = \dfrac{1}{\sqrt{1-(v/c)^2}}$ this becomes $\%\ \text{error} = \left(1-\sqrt{1-(v/c)^2}\right)100\%$

Note that for a given speed v, the percentage error incurred will be the same for all particles, regardless of the mass of the particle.

(a) If an error of 1.00 percent is to occur, $1-\sqrt{1-(v/c)^2} = 1.00\times10^{-2}$

or $1-(v/c)^2 = (1-0.0100)^2 = (0.990)^2$

Thus, for a 1.00 percent error, $v = c\sqrt{1-(0.990)^2} = 0.141c$ ◊

(b) When the error is 10.0 percent, $1-\sqrt{1-(v/c)^2} = 1.00\times10^{-1}$

and $1-(v/c)^2 = (1-0.100)^2 = (0.900)^2$

The speed is then $v = c\sqrt{1-(0.900)^2} = 0.436c$ ◊

---

23. A rocket moves with a velocity of 0.92*c* to the right with respect to a stationary observer A. An observer B moving relative to observer A finds that the rocket is moving with a velocity of 0.95*c* to the left. What is the velocity of observer B relative to observer A? (**Hint:** Consider observer B's velocity in the frame of reference of the rocket.)

Solution Choose the rest frame of observer A and the rest frame of the rocket as the reference frames under consideration. If the positive direction is toward the right, the velocity of the second frame (rocket) relative to the first (observer A) is $v_{rA} = +0.92c$. Observer B sees the rocket moving at $0.95c$ toward the left. Hence, the velocity of B **relative to the rocket** is $0.95c$ toward the right or $v_{Br} = +0.95c$.

The velocity addition equation then gives the velocity of observer B relative to A as

$$v_{BA} = \frac{v_{Br} + v_{rA}}{1 + v_{Br}\, v_{rA}/c^2} = \frac{0.95c + 0.92c}{1 + (+0.95c)(+0.92c)/c^2} = +0.998c$$

Therefore, observer B is moving toward the right with a speed of $0.998c$ relative to observer A. ◊

---

29. What is the speed of a particle whose kinetic energy is equal to its own rest energy?

Solution The total energy of a particle is $E = \gamma mc^2 = \gamma E_R$ where $E_R = mc^2$ is the rest energy of the particle. The total energy is also given by $E = KE + E_R$ where KE is the kinetic energy. If the kinetic energy of the particle equals its own rest energy, one has

$$E = E_R + E_R = 2E_R \qquad \text{and} \qquad E = \gamma E_R$$

Thus, $\gamma = \dfrac{1}{\sqrt{1-(v/c)^2}} = 2$, which gives $1 - \left(\dfrac{v}{c}\right)^2 = \dfrac{1}{4}$ and reduces to $\left(\dfrac{v}{c}\right)^2 = \dfrac{3}{4}$

Hence, the speed of a particle whose kinetic energy equals its own rest energy is

$$v = c\sqrt{\frac{3}{4}} = \frac{c\sqrt{3}}{2} = 0.866c$$ ◊

---

33. An unstable particle with a mass equal to 3.34×10^{-27} kg is initially at rest. The particle decays into two fragments that fly off with velocities of $0.987c$ and $-0.868c$. Find the masses of the fragments. (**Hint:** Conserve both mass-energy and momentum.)

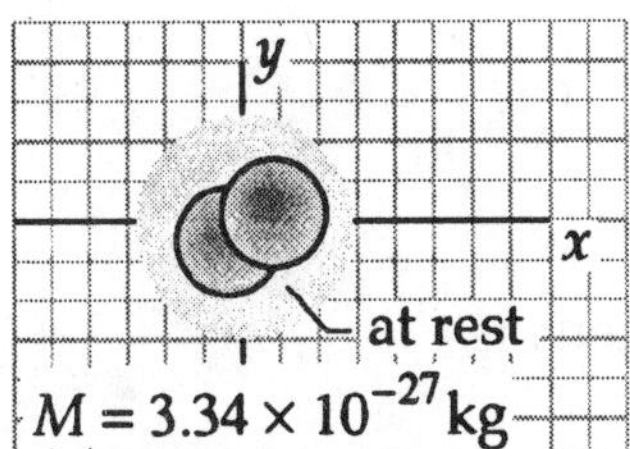

Before decay

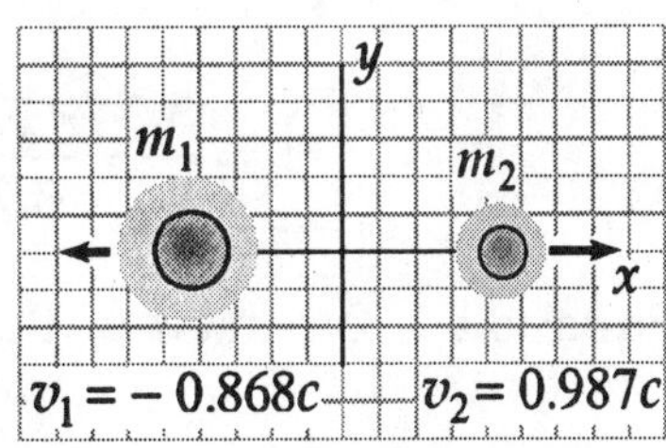

Immediately after decay

Solution To conserve momentum, the two fragments must travel in opposite directions along a line in space. Choose that line to be the x-axis and let the masses of the two fragments be m_1 and m_2 as shown in the figure above. Conservation of momentum then requires that

$$\gamma_1 m_1 v_1 + \gamma_2 m_2 v_2 = M(0) = 0 \qquad \text{or} \qquad m_2 = -\left(\frac{\gamma_1 v_1}{\gamma_2 v_2}\right) m_1$$

Using the observed speeds,

$$\gamma_1 v_1 = \frac{-0.868c}{\sqrt{1-(-0.868)^2}} = -1.75c \quad \text{and} \quad \gamma_2 v_2 = \frac{+0.987c}{\sqrt{1-(0.987)^2}} = +6.14c$$

Thus,

$$m_2 = -\left(\frac{\gamma_1 v_1}{\gamma_2 v_2}\right) m_1 = -\left(\frac{-1.75c}{+6.14c}\right) m_1 = 0.285 m_1 \qquad \textbf{[Equation 1]}$$

Conservation of energy, applied to the entire system, requires that

$$\gamma_1 m_1 c^2 + \gamma_2 m_2 c^2 = Mc^2 \qquad \text{or} \qquad \frac{m_2}{\sqrt{1-(0.868)^2}} + \frac{m_2}{\sqrt{1-(0.987)^2}} = M$$

which reduces to

$$2.01 m_1 + 6.22 m_2 = 3.34 \times 10^{-27} \text{ kg} \qquad \textbf{[Equation 2]}$$

Substituting Equation 1 into Equation 2 yields the mass of the larger fragment:

$$[2.01 + 6.22(0.285)] m_1 = 3.34 \times 10^{-27} \text{ kg}$$

or

$$m_1 = \frac{3.34 \times 10^{-27} \text{ kg}}{3.78} = 8.84 \times 10^{-28} \text{ kg} \qquad \Diamond$$

From Equation 1, the mass of the smaller fragment is

$$m_2 = 0.285\left(8.84 \times 10^{-28} \text{ kg}\right) = 2.52 \times 10^{-28} \text{ kg} \qquad \Diamond$$

37. What is the speed of a proton that has been accelerated from rest through a difference of potential of (a) 500 V and (b) 5.00×10^8 V?

Solution The rest energy of a proton is

$$E_R = m_p c^2 = \left(1.67\times10^{-27}\ \text{kg}\right)\left(3.00\times10^8\ \text{m/s}\right)^2 = 1.50\times10^{-10}\ \text{J}\left(\frac{1.00\ \text{MeV}}{1.60\times10^{-13}\ \text{J}}\right)$$

Thus, $E_R = 939\ \text{MeV}$. When the proton is accelerated from rest through a potential difference ΔV, the gain in kinetic energy equals the work done by the electric field or $KE = (\Delta V)e$. The total energy of the proton will then be

$$E = \gamma E_R = KE + E_R = (\Delta V)e + 939\ \text{MeV}.$$

(a) If $\Delta V = 500$ V, then $KE = (500\ \text{V})e = 500\ \text{eV} = 5.00\times10^{-4}\ \text{MeV}$ and the total energy is $E = 5.00\times10^{-4}\ \text{MeV} + 939\ \text{MeV} \approx 939\ \text{MeV} = E_R$. Since $E \approx E_R$, the particle is non-relativistic. Thus, $KE = \frac{1}{2}m_p v^2$ and the speed of the proton is

$$v = \sqrt{\frac{2KE}{m_p}} = \sqrt{\frac{2(500\ \text{eV})}{\left(1.67\times10^{-27}\ \text{kg}\right)}\left(\frac{1.60\times10^{-19}\ \text{J}}{1.00\ \text{ev}}\right)} = 3.10\times10^5\ \text{m/s} \qquad \lozenge$$

(b) If the accelerating voltage is $V = 5.00\times10^8$ V, the kinetic energy is $KE = \left(5.00\times10^8\ \text{V}\right)e = 5.00\times10^8\ \text{eV} = 500\ \text{MeV}$ and the total energy becomes $E = \gamma E_R = 500\ \text{MeV} + 939\ \text{MeV} = 1439\ \text{MeV}$. The speed of this relativistic proton may be found from

$$\frac{1}{\gamma^2} = 1 - (v/c)^2 = \left(\frac{E_R}{E}\right)^2 = \left(\frac{939\ \text{MeV}}{1439\ \text{MeV}}\right)^2 \quad \text{or} \quad \left(\frac{v}{c}\right)^2 = 1 - \left(\frac{939}{1439}\right)^2 = 0.574$$

Thus, $$v = c\sqrt{0.574} = 0.758c \qquad \lozenge$$

43. A certain quasar recedes from the Earth at $v = 0.870c$. A jet of material ejected from the quasar back toward the Earth moves at $0.550c$ relative to the quasar. Find the speed of the ejected material relative to the Earth.

Solution Consider the motion of the jet of material in two frames of reference, one fixed on Earth and one fixed on the quasar. Choosing the direction away from the Earth to be the positive direction, the velocity of the quasar relative to the Earth is $v_{qE} = +0.870c$. The velocity of the ejected material relative to the quasar is $v_{mq} = -0.550c$. The velocity of the ejected material relative to the Earth is then given by the velocity addition equation as

$$v_{mE} = \frac{v_{mq} + v_{qE}}{1 + v_{mq}v_{qE}/c^2} = \frac{(-0.550c) + 0.870c}{1 + (-0.550c)(0.870c)} = +0.614c$$

The ejected material moves **away from the Earth** at $0.614c$ ◊

---

46. A physics professor on Earth gives an exam to her students who are on a rocket ship traveling at speed of v with respect to Earth. The moment the ship passes the professor, she signals the start of the exam. If she wishes her students to have T_0 (rocket time) to complete the exam, show that she should wait a time of

$$T = T_0\sqrt{\frac{1 - v/c}{1 + v/c}} \qquad \text{(Earth time)}$$

before sending a light signal telling them to stop. (**Hint:** Remember that it takes some time for the second light signal to travel from the professor to the students.)

Solution The proper time for the exam is measured in the reference frame in which the students start and end the exam at the same position (i.e., in the rocket frame). The professor wants the proper time interval to have a value $\Delta t_p = T_0$. The dilated time of the exam, measured by the professor (moving at speed v relative to the rocket) is $\Delta t = \gamma\Delta t_p = \gamma T_0$.

This time is also given by $\Delta t = T + t$, where T is the time she waits before sending the signal and t is the time required for the signal to reach the students. Thus,

$$T + t = \gamma T_0 \qquad \text{[Equation 1]}$$

The distance (measured by the professor) the students have moved away from the Earth before the signal reaches them is $d = v(T+t)$. The time required for the light signal to travel this distance is $t = d/c$, so $d = ct = v(T+t)$.

Solving this for t gives the travel time of the signal as $t = \dfrac{(v/c)T}{1-(v/c)}$.

Substituting this expression for the travel time into Equation 1, and then finding a common denominator on the left side yields

$$T + \frac{(v/c)T}{1-(v/c)} = \gamma T_0 \quad \text{and} \quad \frac{T}{1-(v/c)} = \gamma T_0$$

Since $\gamma = \dfrac{1}{\sqrt{1-(v/c)^2}}$, this becomes $T = \left(\dfrac{1-(v/c)}{\sqrt{1-(v/c)^2}}\right) T_0 = T_0 \sqrt{\dfrac{[1-(v/c)]^2}{1-(v/c)^2}}$

The denominator may be factored as $1-(v/c)^2 = [1-(v/c)][1+(v/c)]$, so the time the professor should wait before sending the signal to stop the exam is

$$T = T_0 \sqrt{\frac{[1-(v/c)]^2}{[1-(v/c)][1+(v/c)]}} = T_0 \sqrt{\frac{1-(v/c)}{1+(v/c)}} \qquad \Diamond$$

50. (a) Show that a potential difference of 1.02×10^6 V would be sufficient to give an electron a speed equal to twice the speed of light if Newtonian mechanics remained valid at high speeds. (b) What speed would an electron actually acquire in falling through a potential difference of 1.02×10^6 V?

Solution When an electron undergoes a potential difference of $\Delta V = 1.02 \times 10^{6}$ V, starting from rest, it gains a kinetic energy of

$$KE = (\Delta V)e = \left(1.02 \times 10^{6}\ \text{V}\right)e = 1.02\ \text{MeV}$$

(a) Assuming that Newtonian mechanics remained valid at high speed, so $KE = \frac{1}{2}m_e v^2$, the speed of the electron would be

$$v = \sqrt{\frac{2KE}{m_e}} = \sqrt{\frac{2(1.02\ \text{MeV})}{9.11 \times 10^{-31}\ \text{kg}}\left(\frac{1.60 \times 10^{-13}\ \text{J}}{1.00\ \text{MeV}}\right)} = 5.99 \times 10^{8}\ \text{m/s}$$

or $$v = \left(5.99 \times 10^{8}\ \text{m/s}\right)\left(\frac{c}{3.00 \times 10^{8}\ \text{m/s}}\right) = 2.00c$$ ◊

(b) The rest energy of an electron is

$$E_R = m_e c^2 = \left(9.11 \times 10^{-31}\ \text{kg}\right)\left(3.00 \times 10^{8}\ \text{m/s}\right)^2 = 8.20 \times 10^{-14}\ \text{J}$$

or $$E_R = \left(8.20 \times 10^{-14}\ \text{J}\right)\left(\frac{1.00\ \text{MeV}}{1.60 \times 10^{-13}\ \text{J}}\right) = 0.511\ \text{MeV}$$

If the kinetic energy is $KE = 1.02\ \text{MeV}$, the total relativistic energy of the electron is $E = KE + E_R = 1.02\ \text{MeV} + 0.511\ \text{MeV} = 1.53\ \text{MeV} = 2.99E_R$. But this total energy is also given by $E = \gamma m_e c^2 = \gamma E_R$.

Thus, $\gamma E_R = 2.99E_R$, and $\gamma = \frac{1}{\sqrt{1-(v/c)^2}} = 2.99$ for this electron.

Solving for the speed of the electron then gives

$$\left(\frac{v}{c}\right)^2 = 1 - \frac{1}{\gamma^2} = 1 - \frac{1}{(2.99)^2} = 0.888 \quad \text{or} \quad v = c\sqrt{0.888} = 0.942c$$ ◊

CHAPTER SELF-QUIZ

1. The observed relativistic length of a super rocket moving by the observer at 0.800c will be what factor times that of the measured rocket length if it were at rest?
   a. 0.45
   b. 0.60
   c. 0.80
   d. 1.33

2. According to a postulate of Einstein, which of the following describes the nature of the laws of physics as one observes processes taking place in various inertial frames of reference?
   a. laws are same only on inertial frames with zero velocity
   b. laws are same only on inertial frames moving at low velocities
   c. laws are same only on inertial frames moving at near speed of light
   d. laws are same on all inertial frames

3. A super fast freight train has a 20.0 m-long box car when measured at rest. What box car length does the ground observer measure when the train is going by at a speed of 0.650c?
   a. 11.8 m
   b. 15.2 m
   c. 18.3 m
   d. 26.3 m

4. How fast would a rocket have to move past a ground observer if the latter were to observe a 1% length shrinkage in the rocket length? ($c = 3.00 \times 10^8$ m/s)
   a. 0.03×10^8 m/s
   b. 0.14×10^8 m/s
   c. 0.42×10^8 m/s
   d. 1.00×10^8 m/s

5. An unknown particle in an accelerator, moving at a speed of 2.00×10^8 m/s, has a measured relativistic energy of 1.80×10^{-9} J. What must its rest energy be?
   a. 5.85×10^{-10} J
   b. 7.29×10^{-10} J
   c. 9.09×10^{-10} J
   d. 1.34×10^{-9} J

6. When one ton of TNT is exploded, approximately 4.50×10^9 J of energy is released. How much mass would this represent in a mass-to-energy conversion? ($c = 3.00 \times 10^8$ m/s)
   a. 1.50 kg
   b. 5.00×10^{-8} kg
   c. 5.30×10^4 kg
   d. 1.70×10^{-3} kg

7. The astronaut whose heart rate on Earth is 65 per minute increases his velocity to $v = 0.950c$. What is his heart rate now as measured by an Earth observer?
   a. 20 per minute
   b. 30 per minute
   c. 52 per minute
   d. 81 per minute

8. I am stationary in a reference system but, if my reference system is **not** an inertial reference system, then, **relative to me,** a system that is an inertial reference system must
   a. remain at rest
   b. move with constant velocity
   c. be accelerating
   d. none of the above

9. The short lifetime of muons created in the upper atmosphere of the Earth would not allow them to reach the surface of the Earth unless their lifetime increased by time dilation. From the reference system of the muons, the muons can reach the surface of the Earth because
   a. time dilation increases their velocity
   b. time dilation increases their energy
   c. length contraction decreases the distance to the Earth
   d. the relativistic speed of the Earth toward them is added to their velocity

10. A spaceship of mass 10^6 kg is to be accelerated to 0.600c. How much energy does this require?
   a. 1.13×10^{22} J
   b. 2.25×10^{22} J
   c. 3.38×10^{22} J
   d. 4.43×10^{22} J

11. At what speed would a clock have to be moving in order to run at a rate that is one-half the rate of a clock at rest?
   a. $0.670c$
   b. $0.750c$
   c. $0.870c$
   d. $0.950c$

12. In a typical color television tube, the electrons are accelerated through a potential difference of 25,000 volts. What speed do the electrons have when they strike the screen? ($q_e = 1.60 \times 10^{-19}$ C, $m_e = 9.11 \times 10^{-31}$ kg, and $c = 3.00 \times 10^8$ m/s)
   a. $v = 0.150c$
   b. $v = 0.301c$
   c. $v = 0.450c$
   d. $v = 0.600c$

Chapter 27

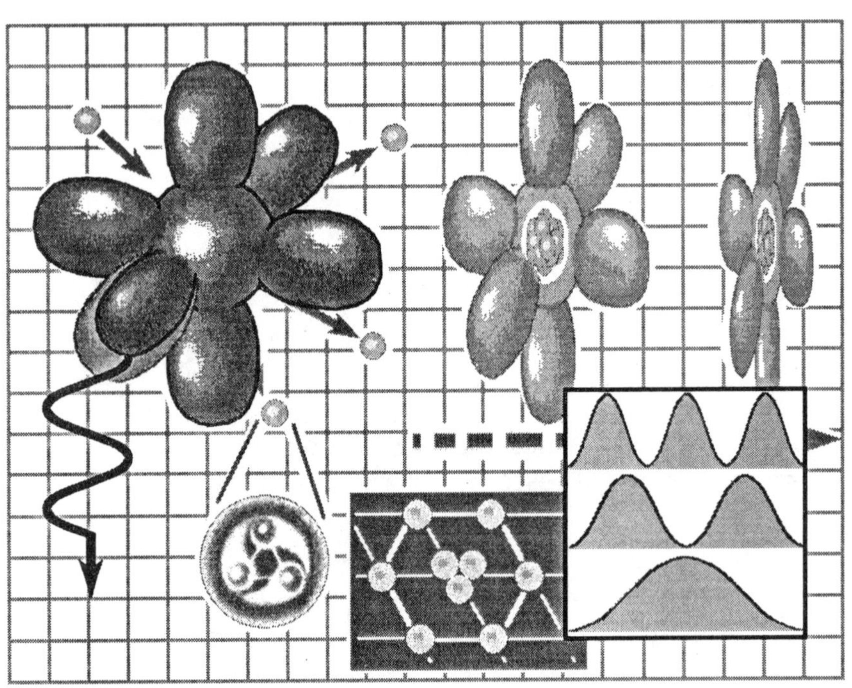

QUANTUM PHYSICS

Chapter 27

QUANTUM PHYSICS

In the previous chapter, we discussed why Newtonian mechanics must be replaced by Einstein's special theory of relativity when dealing with particles whose speeds are comparable to the speed of light. Although many problems were indeed resolved by the theory of relativity in the early part of the 20th century, many experimental and theoretical problems remained unsolved. Attempts to explain the behavior of matter on the atomic level with the laws of classical physics were totally unsuccessful. Various phenomena, such as blackbody radiation, the photoelectric effect, and the emission of sharp spectra lines by atoms in a gas discharge tube, could not be understood within the framework of classical physics. We shall describe these phenomena because of their importance in subsequent developments.

Another revolution took place in physics between 1900 and 1930. This was the era of a new and more general formulation called **quantum mechanics.** This new approach was highly successful in explaining the behavior of atoms, molecules, and nuclei. As with relativity, the quantum theory requires a modification of our ideas concerning the physical world.

An extensive study of quantum theory is certainly beyond the scope of this book. This chapter is simply an introduction to the underlying ideas of quantum theory and the wave-particle nature of matter. We also discuss some simple applications of quantum theory, including the photoelectric effect, the Compton effect, and x-rays.

NOTES FROM SELECTED CHAPTER SECTIONS

27.1 Blackbody Radiation and Planck's Hypothesis

A **black body** is an ideal body that absorbs all radiation incident on it. Any body at some temperature T emits thermal radiation which is characterized by the properties of the body and its temperature. The spectral distribution of blackbody radiation at various temperatures is sketched in the figure. As the temperature increases, the intensity of the radiation (area under the curve) increases, while the peak of the distribution shifts to lower wavelengths. Classical theories failed to explain blackbody radiation. An empirical formula, proposed by Max Planck, is consistent with this distribution at all wavelengths. Planck made two basic assumptions in the development of this result:

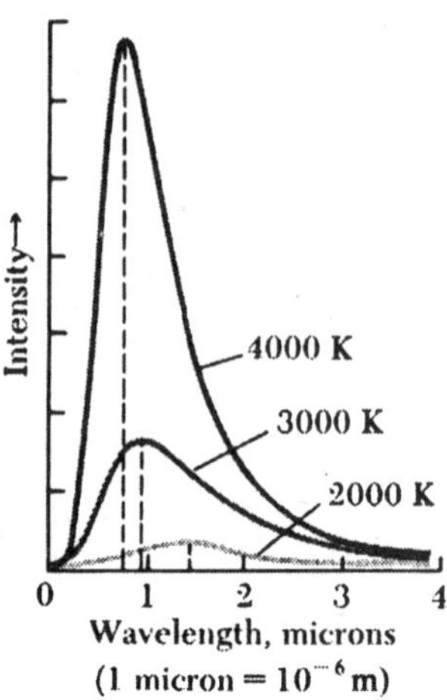

(1) The oscillators emitting the radiation could only have **discrete energies** given by $E_n = nhf$, where f is the oscillator frequency, n is a quantum number ($n = 1, 2, 3, \ldots$), and h is Planck's constant.

(2) These oscillators can emit or absorb energy in discrete units called quanta (or photons), where the energy of a light quantum obeys the relation $E = hf$.

Subsequent developments showed that the quantum concept was necessary in order to explain several phenomena at the atomic level, including the photoelectric effect, the Compton effect, and atomic spectra.

27.2 The Photoelectric Effect

When light is incident on certain metallic surfaces, electrons can be emitted from the surfaces. This is called the photoelectric effect, discovered by Hertz. One cannot explain many features of the photoelectric effect using classical concepts.

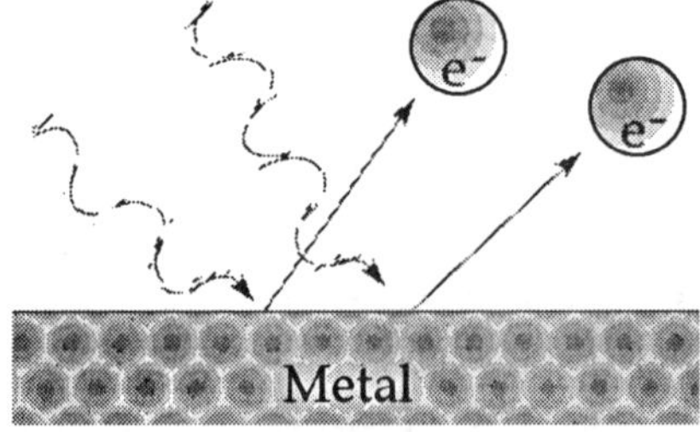

Chapter 27

In 1905, Einstein provided a successful explanation of the photoelectric effect by extending Planck's quantum concept to include electromagnetic fields.

Several features of the photoelectric effect could not be explained with classical physics or with the wave theory of light. However, each of these features can be explained and understood on the basis of the photon theory of light. These observations and their explanations include:

1. No electrons are emitted if the incident light frequency falls below some cutoff frequency, f_c, which is characteristic of the material being illuminated. For example, in the case of sodium, $f_c = 5.50 \times 10^{14}$ Hz. This is inconsistent with the wave theory, which predicts that the photoelectric effect should occur at any frequency, provided the light intensity is high enough.

   The fact that the photoelectric effect is not observed below a certain cutoff frequency follows from the fact that the energy of the photon must be greater than or equal to ϕ, the work function of the illuminated material. If the energy of the incoming photon is not equal to or greater than ϕ, the electrons will never be ejected from the surface, regardless of the intensity of the light.

2. If the light frequency exceeds the cutoff frequency, a photoelectric effect is observed and the number of photoelectrons emitted is proportional to the light intensity. However, the maximum kinetic energy of the photoelectrons is independent of light intensity, a fact that cannot be explained by the concepts of classical physics.

   The fact that K_{max} is independent of the light intensity can be understood with the following argument. If the light intensity is doubled, the number of photons is doubled, which doubles the number of photoelectrons emitted. However, their kinetic energy, which equals $hf - \phi$, depends only on the light frequency and the work function, not on the light intensity.

3. The maximum kinetic energy of the photoelectrons increases with increasing light frequency. The fact that K_{max} increases with increasing frequency is easily understood with Equation 27.6.

4. Electrons are emitted from the surface almost instantaneously (less than 10^{-9} s after the surface is illuminated), even at low light intensities. Classically, one would expect that the electrons would require some time to absorb the incident radiation before they acquire enough kinetic energy to escape from the metal.

   Finally, the fact that the electrons are emitted almost instantaneously is consistent with the particle theory of light, in which the incident energy appears in small packets and there is a one-to-one interaction between photons and electrons. This is in contrast to having the energy of the photons distributed uniformly over a large area.

27.4 X-Rays

X-Rays are a part of the electromagnetic spectrum, characterized by frequencies higher than those of ultraviolet radiation. They are produced when high-speed electrons are suddenly decelerated, for example, when a metal target is struck by electrons that have been accelerated through a potential difference.

The radiation emitted by an x-ray tube is characterized by a continuous broad wavelength spectrum (sometimes called **bremsstrahlung**) that depends on the voltage applied to the tube. As electrons undergo deceleration inside the target, they emit photons which give rise to the continuous spectrum.

A series of sharp, intense lines are superimposed on this spectrum, and depend on the nature of the target material. The sharp lines, or **characteristic x-rays**, represent radiation emitted by the target atoms as their electrons undergo rearrangements.

27.6 The Compton Effect

The Compton effect involves the scattering of an x-ray by an electron as shown in the following figure. The scattered x-ray undergoes a **change** in wavelength $\Delta\lambda$, called the Compton shift, which cannot be explained using classical concepts. By treating the x-ray as a photon (the quantum concept), the scattering process between the photon and electron predicts a shift in photon (x-ray) wavelength given by Equation 27.11, where θ is the angle between the incident and scattered x-ray and m is the mass of the electron. The formula is in excellent agreement with experimental results.

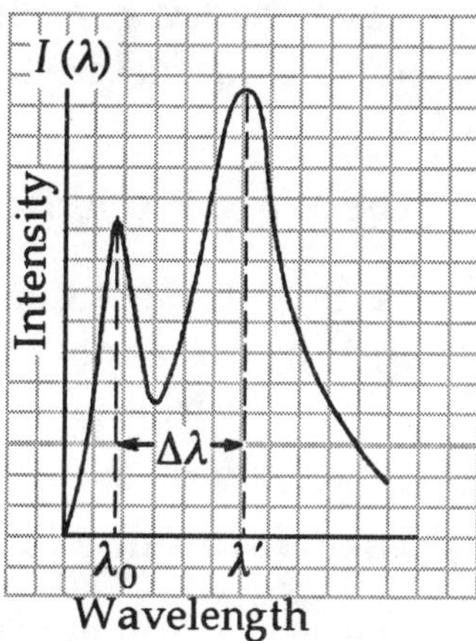

The figure at the right represents the data for Compton scattering at $\theta = 90.0°$ for scattering of x-rays from graphite. In this case, the Compton shift is $\Delta\lambda = 0.0236$ Å and λ_0 is the wavelength of the incident x-ray beam.

27.7 Pair Production and Annihilation

Pair production is a process in which a photon creates matter. In this process, an electron and a positron are simultaneously produced, while the photon disappears. The **minimum** energy that a photon must have to produce an electron-positron pair can be found using conservation of energy by equating the photon energy to the total rest energy of the pair. This then yields a minimum energy of 1.02 MeV.

Pair production cannot occur in a vacuum but can only take place in the presence of a massive particle such as an atomic nucleus. The massive particle must participate in the interaction in order that energy and momentum be conserved simultaneously.

Pair annihilation is a process in which an electron-positron pair produces two photons, moving in opposite directions, both with the same energy and magnitude of momentum.

27.8 Photons and Electromagnetic Waves

The results of some experiments are better described on the basis of the photon model of light; other experimental outcomes are better described in terms of the wave model. The photon theory and the wave theory complement each other—**light exhibits both wave and photon characteristics.**

27.9 The Wave Properties of Particles

de Broglie postulated that a particle in motion has wave properties and a corresponding wavelength inversely proportional to the particle's momentum.

27.10 The Wave Function

The wave function is a complex valued quantity, the absolute square of which gives the probability of finding a particle at a given point at some instant; and the wave function contains all the information that can be known about the particle.

The **Schrödinger equation** describes the manner in which matter waves change in time and space.

The probability of finding a certain value for a quantity (position, energy) is called the **expectation value** of the quantity.

27.11 The Uncertainty Principle

Quantum theory predicts that it is fundamentally impossible to make **simultaneous measurements** of a particle's position and velocity with infinite accuracy. This situation is described by equations 27.17 and 27.18.

EQUATIONS AND CONCEPTS

The Wien displacement law properly describes the distribution of wavelengths in the energy spectrum emitted by a blackbody radiator.

$$\lambda_{max}T = 0.2898\times10^{-2}\ \text{m}\cdot\text{K} \quad (27.1)$$

A **black body** is an ideal body that absorbs all radiation incident on it. Any body at some temperature T emits thermal radiation which is characterized by the properties of the body and its temperature. As the temperature increases, the intensity of the radiation (area under the curve) increases, while the peak of the distribution shifts to shorter wavelengths.

Comment on blackbody radiation

Vibrating molecules are characterized by discrete energy levels called quantum states. The positive integer n is called a quantum number.

$$E_n = nhf \quad (27.2)$$

$$h = 6.626\times10^{-34}\ \text{J}\cdot\text{s} \quad (27.3)$$

Molecules emit or absorb energy in discrete units of light energy called quanta. The energy of a quantum or photon corresponds to the energy difference between adjacent quantum states.

$$E = hf \quad (27.4)$$

Comment on the photoelectric effect

When light is incident on certain metallic surfaces, electrons can be emitted from the surfaces. This is the photoelectric effect, and was discovered by Hertz. Many of the features of the photoelectric effect cannot be explained through classical concepts. In 1905, Einstein successfully explained the effect by extending Planck's quantum concept to include electromagnetic fields. Einstein assumed that light consists of a stream of particles called **photons** whose energy is given by $E = hf$, where h is Planck's constant and f is their frequency.

The maximum kinetic energy of the ejected photoelectron also depends on the work function of the metal, which is typically a few eV. This model is in excellent agreement with experimental results, including the prediction of a cutoff (or threshold) wavelength above which no photoelectric effect is observed.

$$KE_{\text{max}} = hf - \phi \tag{27.6}$$

$$\lambda_c = \frac{hc}{\phi} \tag{27.7}$$

Comment on x-ray production

In x-ray production, a series of "braking collisions" results in energy losses by a decelerating electron. Each increment of energy lost by the electron appears as a quantum of radiation in the x-ray spectrum of the target.

If an electron is completely stopped in a single collision, the resulting photon will have the minimum wavelength possible for a given accelerating voltage.

$$\lambda_{\min} = \frac{hc}{eV} \qquad (27.9)$$

Bragg's law states the condition for constructive interference of x-rays diffracted by a crystal.

$$2d\sin\theta = m\lambda \qquad (27.10)$$

$$(m = 1, 2, 3, \ldots)$$

The Compton effect involves the scattering of an x-ray by an electron. The scattered x-ray undergoes a change in wavelength called the Compton shift, which cannot be explained using classical concepts. By treating the x-ray as a photon (the quantum concept), the scattering process between the photon and electron predicts a shift in photon (x-ray) wavelength, where θ is the angle between the incident and scattered x-ray and m_e is the mass of the electron. The formula is in excellent agreement with experimental results.

$$\Delta\lambda = \frac{h}{m_e c}(1 - \cos\theta) \qquad (27.11)$$

The quantity $h/(m_e c)$ = 0.00243 nm is called the Compton wavelength.

The minimum frequency that a photon must have to produce an electron-positron pair can be found from conservation of energy.

$$hf_{\min} = 2m_e c^2 \qquad (27.12)$$

The wavelength of a photon can be specified by its momentum.

$$p = \frac{h}{\lambda} \tag{27.14}$$

In 1924, de Broglie proposed that any particle of momentum $p = mv$ has wavelike properties, and a wavelength given by $\lambda = h/(mv)$ called the de Broglie wavelength. The waves associated with material particles are called matter waves, and have a frequency which obeys the Einstein relation, $E = hf$.

$$\lambda = \frac{h}{mv} \tag{27.15}$$

$$f = \frac{E}{h} \tag{27.16}$$

The uncertainty principle, proposed by Heisenberg, deals with the limited precision to which one can make simultaneous measurements of the position and velocity of a particle. If Δx and Δp_x are the uncertainties in the position and momentum, respectively, the product $\Delta x \Delta p_x$ is always greater than or equal to a number of the order of Planck's constant. In other words, it is physically impossible to measure simultaneously the exact position and exact momentum of a particle.

$$\Delta x\, \Delta p_x \geq \frac{h}{4\pi} \tag{27.17}$$

Another form of the uncertainty principle applies to the simultaneous measurement of energy and time.

$$\Delta E\, \Delta t \geq \frac{h}{4\pi} \tag{27.18}$$

REVIEW CHECKLIST

- ▷ Describe the formula for blackbody radiation proposed by Planck, and the assumption made in deriving this formula.

- ▷ Describe the Einstein model for the photoelectric effect, and the predictions of the fundamental photoelectric effect equation for the maximum kinetic energy of photoelectrons. Recognize that Einstein's model of the photoelectric effect involves the photon concept ($E = hf$), and the fact that the basic features of the photoelectric effect are consistent with this model.

- ▷ Describe the Compton effect (the scattering of x-rays by electrons) and be able to use the formula for the Compton shift (Eq. 27.11). Recognize that the Compton effect can only be explained using the photon concept.

- ▷ Discuss the wave properties of particles, the de Broglie wavelength concept, and the dual nature of both matter and light.

- ▷ Discuss the manner in which the uncertainty principle makes possible a better understanding of the dual wave-particle nature of light and matter.

SOLUTIONS TO SELECTED END-OF-CHAPTER PROBLEMS

7. An FM radio transmitter has a power output of 150 kW and operates at a frequency of 99.7 MHz. How many photons per second does the transmitter emit?

Solution The energy carried away by each photon emitted by the transmitter is $E_\gamma = hf$ where $h = 6.626\times10^{-34}$ J·s is Planck's constant and $f = 99.7\text{ MHz} = 99.7\times10^{6}$ Hz is the frequency of the radiation. Thus,

$$E_\gamma = \left(6.626\times10^{-34}\text{ J}\cdot\text{s}\right)\left(99.7\times10^{6}\text{ Hz}\right) = 6.61\times10^{-26}\text{ J}$$

The energy emitted each second by the transmitter is

$$E = P(\Delta t) = (150 \text{ kW})(1.00 \text{ s}) = 150 \text{ kJ} = 150\times 10^{3} \text{ J}$$

Hence, the number of photons emitted each second is

$$n = \frac{E}{E_{\gamma}} = \frac{150\times 10^{3} \text{ J}}{6.61\times 10^{-26} \text{ J}} = 2.27\times 10^{30} \qquad \lozenge$$

---

13. When light of wavelength 350 nm falls on a potassium surface, electrons are emitted that have a maximum kinetic energy of 1.31 eV. Find (a) the work function of potassium, (b) the cutoff wavelength, and (c) the frequency corresponding to the cutoff wavelength.

Solution (a) Einstein's photoelectric effect equation is $KE_{max} = E_{\gamma} - \phi$, where KE_{max} is the maximum kinetic energy of the emitted electrons, E_{γ} is the energy of the photons incident on the surface, and ϕ is the work function of the material of the surface. If the wavelength of the incident light is $\lambda = 350 \text{ nm}$,

$$E_{\gamma} = hf = \frac{hc}{\lambda} = \frac{\left(6.626\times 10^{-34} \text{ J}\cdot\text{s}\right)\left(3.00\times 10^{8} \text{ m/s}\right)}{350\times 10^{-9} \text{ m}} = 5.68\times 10^{-19} \text{ J}$$

or

$$E_{\gamma} = \left(5.68\times 10^{-19} \text{ J}\right)\left(\frac{1.00 \text{ eV}}{1.60\times 10^{-19} \text{ J}}\right) = 3.55 \text{ eV}$$

The observed maximum kinetic energy of the emitted electrons is $KE_{max} = 1.31 \text{ eV}$. Thus, the work function of the potassium surface is

$$\phi = E_{\gamma} - KE_{max} = 3.55 \text{ eV} - 1.31 \text{ eV} = 2.24 \text{ eV} \qquad \lozenge$$

(b) The cutoff wavelength is the wavelength of the lowest energy photons capable of freeing electrons from the surface. At the cutoff wavelength $(\lambda = \lambda_c)$, the freed electrons leave the surface with zero kinetic energy (i.e., $KE_{max} = 0$).

From the photoelectric effect equation,

$$0 = (E_\gamma)_{\text{min}} - \phi \quad \text{or} \quad (E_\gamma)_{\text{min}} = \frac{hc}{\lambda_c} = \phi \quad \text{and} \quad \lambda_c = \frac{hc}{\phi}$$

Thus, $$\lambda_c = \frac{(6.626\times10^{-34}\text{ J}\cdot\text{s})(3.00\times10^{8}\text{ m/s})}{2.24\text{ eV}(1.60\times10^{-19}\text{ J/1.00 eV})} = 5.55\times10^{-7}\text{ m} = 555\text{ nm} \quad \Diamond$$

(c) The frequency corresponding to the cutoff wavelength is

$$f_c = \frac{c}{\lambda_c} = \frac{3.00\times10^{8}\text{ m/s}}{5.55\times10^{-7}\text{ m}} = 5.41\times10^{14}\text{ Hz} \quad \Diamond$$

---

17. When light of wavelength 254 nm falls on cesium, the required stopping potential is 3.00 V. If light of wavelength 436 nm is used, the stopping potential is 0.900 V. Use this information to plot a graph like that shown in Figure 27.6, and from the graph determine the cutoff frequency for cesium and its work function.

Solution From the photoelectric equation, $KE_{\text{max}} = E_\gamma - \phi = hf - \phi$, observe that a linear relationship exists between KE_{max} and the frequency of the radiation. Thus, the graph of KE_{max} versus f is a straight line. The maximum kinetic energy of the electrons leaving a photosensitive surface may be expressed as $KE_{\text{max}} = V_s e$, where V_s is the retarding voltage required to stop these electrons (i.e., the stopping potential). Since $f = c/\lambda$ and $KE_{\text{max}} = V_s e$, the maximum kinetic energy is

$$KE_{\text{max}} = (3.00\text{ V})e = 3.00\text{ eV} \quad \text{when} \quad f = \frac{3.00\times10^{8}\text{ m/s}}{254\times10^{-9}\text{ m}} = 11.8\times10^{14}\text{ Hz}$$

and $$KE_{\text{max}} = (0.900\text{ V})e = 0.900\text{ eV} \quad \text{when} \quad f = \frac{3.00\times10^{8}\text{ m/s}}{436\times10^{-9}\text{ m}} = 6.88\times10^{14}\text{ Hz}$$

Therefore, your graph should be similar to the sketch at the below. At the cutoff frequency (the minimum frequency photon capable of freeing electrons from the surface), $KE_{max} = 0$. The cutoff frequency is equal to the horizontal intercept on the graph. Also, from the photoelectric effect equation, $KE_{max} = -\phi$ when $f = 0$. The vertical intercept of the graph is therefore equal to the negative of the work function. From your graph, you should find these values to be approximately

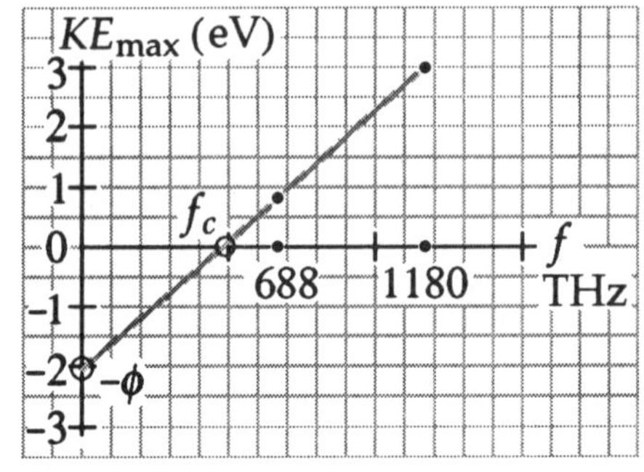

$$f_c = 4.77 \times 10^{14} \text{ Hz} \quad \text{and} \quad \phi = 2.03 \text{ ev} \quad \Diamond$$

---

21. What minimum accelerating voltage would be required to produce an x-ray with a wavelength of 0.0300 nm?

Solution The energy of an x-ray with a wavelength of $\lambda = 0.0300$ nm is

$$E_\gamma = hf = \frac{hc}{\lambda} = \frac{\left(6.626 \times 10^{-34} \text{ J} \cdot \text{s}\right)\left(3.00 \times 10^8 \text{ m/s}\right)}{0.0300 \times 10^{-9} \text{ m}}\left(\frac{1.00 \text{ eV}}{1.60 \times 10^{-19} \text{ J}}\right)$$

or $$E_\gamma = 4.14 \times 10^4 \text{ eV} = 41.4 \text{ keV}$$

In order to produce a 41.4 keV x-ray photon in a collision with a target, an electron must have a minimum kinetic energy of $KE_{min} = 41.4 \text{ keV}$. The kinetic energy an electron gains when accelerated through a voltage ΔV is $KE = (\Delta V)e$. Thus, the minimum accelerating voltage required to produce an x-ray with a wavelength of 0.0300 nm is

$$(\Delta V)_{min} = \frac{KE_{min}}{e} = \frac{41.4 \text{ keV}}{e} = 41.4 \text{ kV} = 4.14 \times 10^4 \text{ V} \quad \Diamond$$

---

25. X-rays of wavelength 0.140 nm are reflected from a certain crystal, and the first-order maximum occurs at an angle of 14.4°. What value does this give for the interplanar spacing of this crystal?

Solution Bragg's law summarizes the conditions necessary for constructive interference when x-rays are diffracted by a crystal. This law is

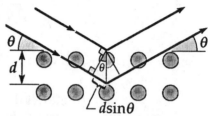

$$2d\sin\theta = m\lambda \quad (m = 1, 2, 3, \ldots)$$

where λ is the wavelength of the x-rays, θ is the angle the incident and reflected x-rays make with the planes (layers of atoms) in the crystal, and d is the spacing between successive planes as shown in the figure.

Thus, if the first-order $(m = 1)$ maximum occurs at $\theta = 14.4°$ when the wavelength of the x-rays is $\lambda = 0.140\text{ nm}$, the interplanar spacing for this crystal is

$$d = \frac{m\lambda}{2\sin\theta} = \frac{(1)(0.140\text{ nm})}{2\sin 14.4°} = 0.281\text{ nm} \qquad \Diamond$$

---

29. X-rays with an energy of 300 keV undergo Compton scattering from a target. If the scattered rays are deflected at 37° relative to the direction of the incident rays, find (a) the Compton shift at this angle, (b) the energy of the scattered x-ray, and (c) the kinetic energy of the recoiling electron.

Solution When a photon is scattered through an angle θ by a particle of mass m_e, the difference between the wavelengths of the scattered and incident photons (i.e., the Compton shift) is given by

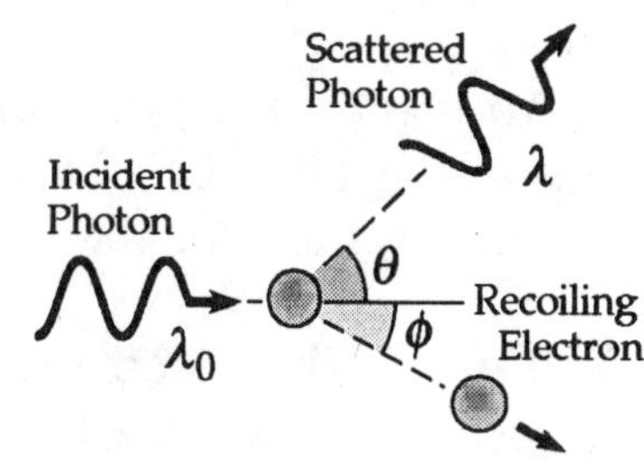

$$\Delta\lambda = \lambda - \lambda_0 = \frac{h}{m_e c}(1 - \cos\theta)$$

(a) If the scattering particle is a free electron, $m_e = 9.11 \times 10^{-31}$ kg. Thus, if $\theta = 37°$, the Compton shift is

$$\Delta\lambda = \frac{6.626 \times 10^{-34} \text{ J} \cdot \text{s}}{\left(9.11 \times 10^{-31} \text{ kg}\right)\left(3.00 \times 10^{8} \text{ m/s}\right)}(1 - \cos 37°) = 4.88 \times 10^{-13} \text{ m} \qquad \lozenge$$

(b) The wavelength of the incident 300-keV x-ray is

$$\lambda_0 = \frac{c}{f} = \frac{hc}{E_\gamma} = \frac{\left(6.626 \times 10^{-34} \text{ J} \cdot \text{s}\right)\left(3.00 \times 10^{8} \text{ m/s}\right)}{300 \text{ keV}\left(1.60 \times 10^{-16} \text{ J/1.00 keV}\right)} = 4.14 \times 10^{-12} \text{ m}$$

Hence, the wavelength of the scattered x-ray is

$$\lambda = \lambda_0 + \Delta\lambda = 4.14 \times 10^{-12} \text{ m} + 4.88 \times 10^{-13} \text{ m} = 4.63 \times 10^{-12} \text{ m}$$

Its energy is $E'_\gamma = \dfrac{hc}{\lambda} = \dfrac{\left(6.626 \times 10^{-34} \text{ J} \cdot \text{s}\right)\left(3.00 \times 10^{8} \text{ m/s}\right)}{4.63 \times 10^{-12} \text{ m}} = 4.29 \times 10^{-14} \text{ J}$

or $$E'_\gamma = 4.29 \times 10^{-14} \text{ J}\left(\frac{1.00 \text{ keV}}{1.60 \times 10^{-16} \text{ J}}\right) = 268 \text{ keV} \qquad \lozenge$$

(c) The kinetic energy of the recoiling electron equals the difference between the energy of the incident and scattered x-ray photons, or

$$KE = E_\gamma - E'_\gamma = 300 \text{ keV} - 268 \text{ keV} = 32 \text{ keV} \qquad \lozenge$$

35. An electron moving at a speed of $0.60c$ collides head on with a positron also moving at $0.60c$. Determine the energy and momentum of each photon produced in the process.

Solution The electron and the positron have the same mass, m_e. Since they both have a speed of $v = 0.6c$ before impact, the total energy of each particle is

$$E_{e^{\pm}} = \gamma m_e c^2 = \frac{m_e c^2}{\sqrt{1-(v/c)^2}} = \frac{\left(9.11\times10^{-31}\text{ kg}\right)\left(3.0\times10^{8}\text{ m/s}\right)^2}{\sqrt{1-(0.60)^2}}$$

or $$E_{e^{\pm}} = 1.02\times10^{-13}\text{ J} = \left(1.02\times10^{-13}\text{ J}\right)\left(\frac{1.0\text{ MeV}}{1.6\times10^{-13}\text{ J}}\right) = 0.64\text{ MeV}$$

The momentum of each particle is $p = \gamma m_e v = m_e v \big/ \sqrt{1-(v/c)^2}$

Since the two particles have the same mass and speed, their momenta have the same magnitudes but opposite directions before the head-on collision. Thus, the total momentum before collision is zero and, to conserve momentum, the total momentum after collision must also be zero. This can be true only if the momenta of the two photons produced in this pair annihilation have equal magnitudes and opposite directions. Since the energy and momentum of a photon are related by $E_\gamma = p_\gamma c$, two photons with equal momenta also have equal energies. Therefore to conserve energy, one must have $2E_\gamma = E_{e^-} + E_{e^+} = 2E_{e^{\pm}} = 2(0.64\text{ MeV})$. The energy of each photon produced in this process is then $E_\gamma = 0.64\text{ MeV}$ ◊

The momentum of each photon is then $p_\gamma = \dfrac{E_\gamma}{c} = 0.64\text{ MeV}/c$ ◊

---

38. Through what potential difference would an electron have to be accelerated from rest to give it a de Broglie wavelength of 1.0×10^{-10} m?

Solution The de Broglie wavelength of a particle is $\lambda = h/p$, where Planck's constant is $h = 6.626\times10^{-34}$ J·s and p is the linear momentum of the particle. Thus, if $\lambda = 1.0\times10^{-10}$ m, the electron must have a momentum of

$$p = \frac{h}{\lambda} = \frac{6.626\times10^{-34}\text{ J}\cdot\text{s}}{1.0\times10^{-10}\text{ m}} = 6.6\times10^{-24}\text{ J}\cdot\text{s/m}$$

The total relativistic energy of the electron is then found from $E^2 = (pc)^2 + E_R{}^2$ where the rest energy is $E_R = 0.511\ \text{MeV}$ for an electron, and for this electron,

$$pc = \left(6.6 \times 10^{-24}\ \text{J} \cdot \text{s/m}\right)\left(\frac{1.0\ \text{MeV}}{1.6 \times 10^{-13}\ \text{J}}\right)\left(3.0 \times 10^{8}\ \text{m/s}\right) = 0.012\ \text{MeV}$$

This yields $E = \sqrt{(0.012\ \text{MeV})^2 + (0.511\ \text{MeV})^2} \approx 0.511\ \text{MeV}$. Hence, $E \approx E_R$ and this electron is non-relativistic. The kinetic energy may then be computed by classical methods as

$$KE = \frac{1}{2}m_e v^2 = \frac{p^2}{2m_e} \quad \text{or} \quad KE = \frac{\left(6.6 \times 10^{-24}\ \text{J} \cdot \text{s/m}\right)^2}{2\left(9.11 \times 10^{-31}\ \text{kg}\right)} = 2.4 \times 10^{-17}\ \text{J}$$

The kinetic energy an electron gains when accelerated through a potential difference ΔV is $KE = (\Delta V)e$. Thus, the potential difference this electron must have been accelerated through is:

$$\Delta V = \frac{KE}{e} = \frac{2.4 \times 10^{-17}\ \text{J}}{1.6 \times 10^{-19}\ \text{C}} = 1.5 \times 10^{2}\ \text{V} \qquad \Diamond$$

41. De Broglie postulated that the relationship $\lambda = h/p$ is valid for relativistic particles. What is the de Broglie wavelength for a (relativistic) electron whose kinetic energy is 3.00 MeV?

Solution The kinetic energy of a particle with total relativistic energy E and rest energy E_R is $KE = E - E_R$, Thus, if $KE = 3.00\ \text{MeV}$ for an electron, the total energy is $E = KE + E_R = 3.00\ \text{MeV} + 0.511\ \text{MeV} = 3.51\ \text{MeV}$. The linear momentum of this relativistic electron may be found from $E^2 = (pc)^2 + E_R{}^2$. This yields

$$p = \frac{\sqrt{E^2 - E_R{}^2}}{c} = \frac{\sqrt{(3.51\ \text{MeV})^2 - (0.511\ \text{Mev})^2}}{c}$$

or $$p = \frac{3.47\ \text{MeV}}{c} = \frac{(3.47\ \text{MeV})}{(3.00 \times 10^{8}\ \text{m/s})}\left(\frac{1.60 \times 10^{-13}\ \text{J}}{1.00\ \text{MeV}}\right) = 1.85 \times 10^{-21}\ \text{J} \cdot \text{s/m}$$

The de Broglie wavelength of this relativistic electron is then

$$\lambda = \frac{h}{p} = \frac{6.626 \times 10^{-34}\ \text{J} \cdot \text{s}}{1.85 \times 10^{-21}\ \text{J} \cdot \text{s/m}} = 3.58 \times 10^{-13}\ \text{m}$$ ◊

47. Suppose optical radiation (λ = 5.00 × 10^{-7} m) is used to determine the position of an electron to within the wavelength of the light. What will be the resulting uncertainty in the electron's velocity?

Solution If the position is determined to within the wavelength of the light, the uncertainty in the position of the electron is $\Delta x = \lambda$. The uncertainty principle, $(\Delta x)(\Delta p_x) \geq h/4\pi$, places a lower limit on the uncertainty in the linear momentum as

$$(\Delta p_x) \geq \frac{h}{4\pi(\Delta x)} = \frac{h}{4\pi\lambda}$$

Assuming that the electron is non-relativistic, its linear momentum is $p_x = m_e v$. Thus, the uncertainty in momentum is $\Delta p_x = \Delta(m_e v) = m_e(\Delta v)$, and the uncertainty in the electron's velocity is

$$(\Delta v) = \frac{\Delta p_x}{m_e} = \frac{h/4\pi\lambda}{m_e} = \frac{h}{4\pi m_e \lambda}$$

With $\lambda = 5.00 \times 10^{-7}$ m and $m_e = 9.11 \times 10^{-31}$ kg, this gives

$$(\Delta v) = \frac{h}{4\pi m_e \lambda} = \frac{6.626 \times 10^{-34}\ \text{J} \cdot \text{s}}{4\pi(9.11 \times 10^{-31}\ \text{kg})(5.00 \times 10^{-7}\ \text{m})} = 116\ \text{m/s}$$ ◊

55. A light source of wavelength λ illuminates a metal and ejects photoelectrons with a maximum kinetic energy of 1.00 eV. A second light source of wavelength $\lambda/2$ ejects photoelectrons with a maximum kinetic energy of 4.00 eV. What is the work function of the metal?

Solution

The wavelengths of the light emitted by these two sources are both unknown. However, it is known that the wavelength of the light from the second source is one-half that from the first source. The energy of the photons emitted by the first source is $E_\gamma = hf = hc/\lambda$, and the energy of the photons from the second source is $E'_\gamma = hc/\lambda' = 2hc/\lambda = 2E_\gamma$. Since the maximum kinetic energy of the ejected photoelectrons when the first source is used is $KE_{max} = 1.00\text{ eV}$, the photoelectric effect equation $\left(KE_{max} = E_\gamma - \phi\right)$ gives:

$$1.00\text{ eV} = E_\gamma - \phi \qquad \textbf{[Equation 1]}$$

If the maximum kinetic energy of the photoelectrons is $KE_{max} = 4.00\text{ eV}$ when the second source $\left(\text{with } E'_\gamma = 2E_\gamma\right)$ is used, the photoelectric effect equation yields:

$$4.00\text{ eV} = 2E_\gamma - \phi \qquad \textbf{[Equation 2]}$$

From Equation 1, $E_\gamma = 1.00\text{ eV} + \phi$. Substituting this result into Equation 2 gives $4.00\text{ eV} = 2(1.00\text{ eV} + \phi) - \phi = 2.00\text{ eV} + \phi$, so the work function of the metal must be

$$\phi = 2.00\text{ eV} \qquad \lozenge$$

---

62. In a Compton scattering event, the scattered photon has an energy of 120.0 keV and the recoiling electron has a kinetic energy of 40.0 keV. Find (a) the wavelength of the incident photon, (b) the angle θ at which the photon is scattered, and (c) the recoil angle of the electron. (**Hint:** Conserve both mass-energy and relativistic momentum.)

Solution The figure at the right shows the situation just before and just after the Compton scattering event.

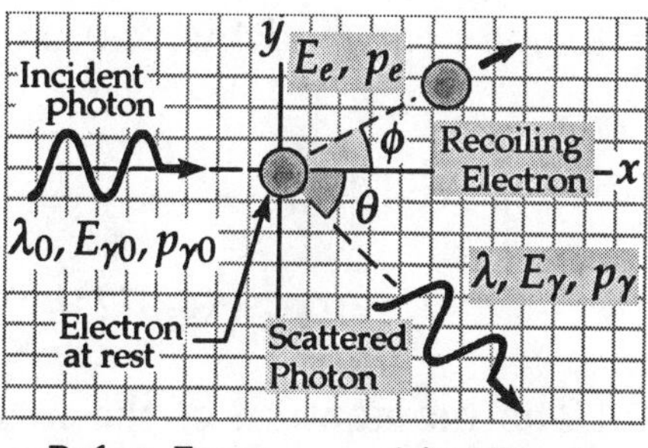

(a) Define $E_{\gamma o}$ as the energy of the original (incident) photon, E_γ as the energy of the scattered photon, E_R as the rest energy of an electron, and KE as the kinetic energy of the recoiling electron. Conserving mass-energy, the total energy before the event must equal that after the event:

$$E_{\gamma o} + E_R = E_\gamma + E_e = E_\gamma + (E_R + KE) \quad \text{which reduces to} \quad E_{\gamma o} = E_\gamma + KE$$

Since it is given that the energy of the scattered photon is $E_\gamma = 120 \text{ keV}$ and the kinetic energy of the recoiling electron is $KE = 40.0 \text{ keV}$, the energy of the incident photon must be $E_{\gamma o} = 120 \text{ keV} + 40.0 \text{ keV} = 160 \text{ keV}$, The wavelength of the incident photon is then:

$$\lambda_o = \frac{hc}{E_{\gamma o}} = \frac{(6.626 \times 10^{-34} \text{ J}\cdot\text{s})(3.00 \times 10^8 \text{ m/s})}{(160 \text{ keV})(1.60 \times 10^{-16} \text{ J/1.00 keV})} = 7.77 \times 10^{-12} \text{ m} \qquad \Diamond$$

(b) The wavelength of the scattered photon is

$$\lambda = \frac{hc}{E_\gamma} = \frac{(6.626 \times 10^{-34} \text{ J}\cdot\text{s})(3.00 \times 10^8 \text{ m/s})}{(120 \text{ keV})(1.60 \times 10^{-16} \text{ J/1.00 keV})} = 10.4 \times 10^{-12} \text{ m}$$

so the Compton shift that has occurred is

$$\Delta\lambda = \lambda - \lambda_o = 10.4 \times 10^{-12} \text{ m} - 7.77 \times 10^{-12} \text{ m} = 2.6 \times 10^{-12} \text{ m}$$

Taking m_e to be the mass of the electron, the angle θ at which the photon is scattered is found from the Compton shift formula, giving

$$\Delta\lambda = \frac{h}{m_e c}(1 - \cos\theta) \quad \text{or} \quad \cos\theta = 1 - \frac{(\Delta\lambda) m_e c}{h}$$

Thus, we can find the photon scattering angle θ:

$$\cos\theta = 1 - \frac{\left(2.6\times10^{-12}\text{ m}\right)\left(9.11\times10^{-31}\text{ kg}\right)\left(3.00\times10^{8}\text{ m/s}\right)}{6.626\times10^{-34}\text{ J}\cdot\text{s}} = -0.072$$

$$\theta = \arccos(-0.072) = 94°$$ ◊

(c) The momentum of the incident photon in the positive x direction is

$$p_{\gamma 0} = \frac{E_{\gamma 0}}{c} = 160\text{ keV/c}$$

The components of the total momentum before the event are then:

$$\left(p_{\text{before}}\right)_x = +160\text{ keV/c} \quad\text{and}\quad \left(p_{\text{before}}\right)_y = 0$$

The scattered photon has a momentum of magnitude $p_\gamma = E_\gamma / c = 120\text{ keV}/c$ with x and y components of

$$\left(p_\gamma\right)_x = p_\gamma\cos\theta = \left(120\text{ keV}/c\right)\cos 94° = -8.37\text{ keV}/c$$

and $$\left(p_\gamma\right)_y = -p_\gamma\sin\theta = -\left(120\text{ keV}/c\right)\sin 94° = -120\text{ keV}/c$$

Requiring that momentum be conserved in each direction, and taking p_e is the momentum of the recoiling electron and ϕ is its recoil angle,

$$\left(p_\gamma\right)_x + p_e\cos\phi = \left(p_{\text{before}}\right)_x \quad\text{and}\quad \left(p_\gamma\right)_y + p_e\sin\phi = \left(p_{\text{before}}\right)_y$$

Thus, $$p_e\sin\phi = \left(p_{\text{before}}\right)_y - \left(p_\gamma\right)_y = 120\text{ keV}/c$$

and $$p_e\cos\phi = \left(p_{\text{before}}\right)_x - \left(p_\gamma\right)_x = 160\text{ keV}/c - \left(-8.37\text{ keV}/c\right) = 169\text{ keV}/c$$

Dividing these two equations then gives

$$\frac{p_e\sin\phi}{p_e\cos\phi} = \tan\phi \quad\text{and}\quad \tan\phi = \frac{120\text{ keV}/c}{169\text{ keV}/c} = 0.710$$

Thus, the recoil angle of the electron is: $\phi = 35.4°$ ◊

CHAPTER SELF-QUIZ

1. According to the de Broglie hypothesis, which of the following statements is applicable to the wavelength of a moving particle?
   a. directly proportional to its energy
   b. directly proportional to its momentum
   c. inversely proportional to its energy
   d. inversely proportional to its momentum

2. As the temperature of a radiation-emitting black body becomes higher, what happens to the peak wavelength of the radiation?
   a. increases
   b. decreases
   c. remains constant
   d. is directly proportional to temperature

3. According to Einstein, what is true of the stopping potential for a photoelectric current as the wavelength of incident light becomes shorter?
   a. increases
   b. decreases
   c. remains constant
   d. stopping potential is directly proportional to wavelength

4. What is the wavelength of a monochromatic light beam where the photon energy is 2.00 eV? ($h = 6.63 \times 10^{-34}$ J·s, $c = 3.00 \times 10^8$ m/s, $1.00 \text{ nm} = 10^{-9}$ m, and $1.00 \text{ eV} = 1.60 \times 10^{-19}$ J)
   a. 414 nm
   b. 621 nm
   c. 746 nm
   d. 829 nm

5. What is the de Broglie wavelength for a proton ($m = 1.67 \times 10^{-27}$ kg) moving at a speed of 5.00×10^5 m/s? ($h = 6.63 \times 10^{-34}$ J·s)
   a. 1.10×10^{-12} m
   b. 0.42×10^{-12} m
   c. 1.80×10^{-12} m
   d. 0.79×10^{-12} m

6. The Compton experiment demonstrated which of the following when an x-ray photon collides with an electron?
   a. momentum is conserved
   b. energy is conserved
   c. momentum and energy are both conserved
   d. wavelength of scattered photon equals that of incident photon

7. Light of wavelength 480 nm is incident on a metallic surface with a resultant photoelectric stopping potential of 0.55 V. What is the maximum kinetic energy of the emitted electrons? ($h = 6.63 \times 10^{-34}$ J·s, $c = 3.00 \times 10^8$ m/s, 1.00 nm $= 10^{-9}$ m, and 1.00 eV = 1.60×10^{-19} J)
   a. 3.19 eV
   b. 0.55 eV
   c. 2.04 eV
   d. 2.59 eV

8. The Sun's surface temperature is 5800 K and the peak wavelength in its radiation is 500 nm. What is the surface temperature of a distant star where the peak wavelength is 475 nm?
   a. 5510 K
   b. 5626 K
   c. 6105 K
   d. 6350 K

9. According to Einstein, increasing the brightness of a beam of light without changing its color will increase the
   a. number of photons
   b. energy of each photon
   c. speed of the photons
   d. frequency of the photons

10. According to the principle of complementarity, everything acts in a given experiment as if it is either a wave or a particle. In which experiment is the wave aspect exhibited?
    a. the Davisson and Germer experiment
    b. the photoelectric effect
    c. pair production
    d. Compton scattering

11. How much energy (in eV) does a photon of red light (λ = 640 nm) have? ($h = 6.626 \times 10^{-34}$ J·s and 1.00 eV = 1.60×10^{-19} J)
    a. 3.26 eV
    b. 2.50 eV
    c. 1.94 eV
    d. 1.32 eV

12. An electron microscope operates with electrons of kinetic energy 40.0 keV. What is the wavelength of these electrons? ($h = 6.626 \times 10^{-34}$ J·s, 1.00 eV = 1.60×10^{-19} J, and $m_e = 9.11 \times 10^{-31}$ kg)
    a. 0.50×10^{-10} m
    b. 7.17×10^{-11} m
    c. 6.02×10^{-12} m
    d. 3.07×10^{-13} m

Chapter 28

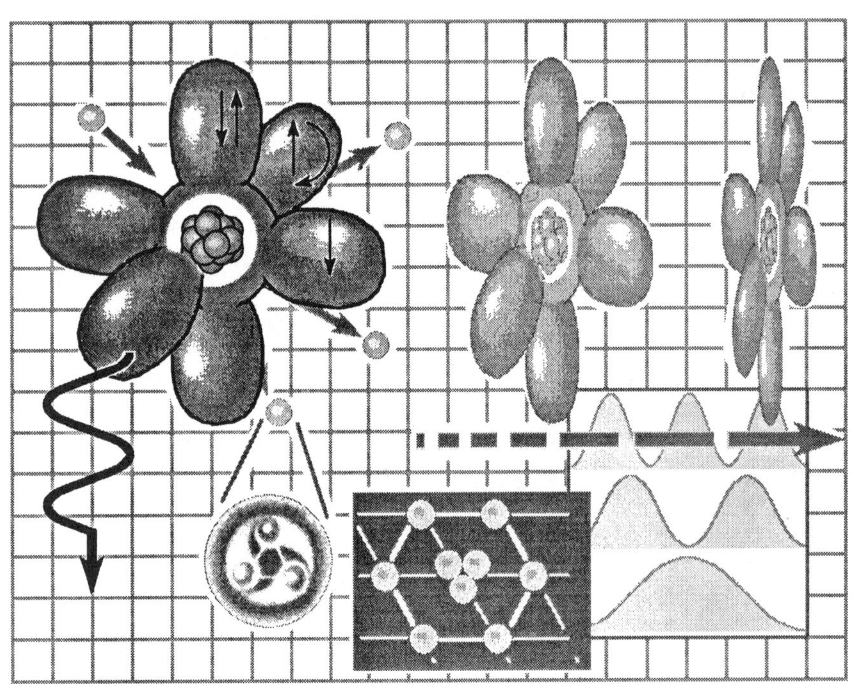

ATOMIC PHYSICS

Chapter 28

ATOMIC PHYSICS

A large portion of this chapter is concerned with the study of the hydrogen atom. Although the hydrogen atom is the simplest atomic system, it is especially important.

We first discuss the Bohr model of hydrogen, which helps us understand many features of hydrogen but fails to explain many finer details of atomic structure. Next we examine the hydrogen atom from the viewpoint of quantum mechanics and the quantum numbers used to characterize various atomic states. Additionally, we examine the physical significance of the quantum numbers and the effect of a magnetic field on certain quantum states. The Pauli exclusion principle is also presented. This physical principle is extremely important in understanding the properties of complex atoms and the arrangement of elements in the periodic table. Finally, we apply our knowledge of atomic structure to describe the mechanisms involved in the production of x-rays and the operation of a laser.

NOTES FROM SELECTED CHAPTER SECTIONS

28.1 Early Models of the Atom

One of the first indications that there was a need for modification of the Bohr theory became apparent when improved spectroscopic techniques were used to examine the spectral lines of hydrogen. It was found that many of the lines in the Balmer and other series were not single lines at all. Instead, each line was actually a group of lines spaced very close together. An additional difficulty arose when it was observed that, in some situations, certain single spectral lines were split into three closely spaced lines when the atoms were placed in a strong magnetic field.

28.3 The Bohr Theory of Hydrogen

The basic postulates of the Bohr model of the hydrogen atom are as follows:

1. The electron moves in circular orbits about the nucleus (the planetary model of the atom) under the influence of the Coulomb force of attraction between the electron and the positively charged nucleus.

2. The electron can exist only in very specific orbits; hence, the states are **quantized** (Planck's quantum hypothesis). The allowed orbits are those for which the angular momentum of the electron about the nucleus is an integral multiple of $\hbar = h/2\pi$, where h is Planck's constant.

3. When the electron is in one of its allowed orbits, it does not radiate energy; hence, the atom is stable. Such stable orbits are called stationary states.

4. The atom radiates energy only when the electron "jumps" from one allowed stationary orbit to another. This postulate states that the energy given off by an atom is carried away by a photon of energy hf.

It is important to understand the behavior of the hydrogen atom as an atomic system for the following reasons:

1. Much of what is learned about the hydrogen atom with its single electron can be extended to such single-electron ions as He^+ and Li^{2+}, which are hydrogen-like in their atomic structure.

2. The hydrogen atom is an ideal system for performing precise tests of theory against experiment and for improving our overall understanding of atomic structure.

3. The quantum numbers used to characterize the allowed states of hydrogen can be used to describe the allowed states of more complex atoms. This enables us to understand the periodic table of the elements, which is one of the greatest triumphs of quantum mechanics.

4. The basic ideas about atomic structure must be well understood before we attempt to deal with the complexities of molecular structures and the electronic structure of solids.

In addition to providing a theoretical derivation of the line spectrum, Bohr also explained:

(a) the limited number of lines seen in the absorption spectrum of hydrogen compared to the emission spectrum,
(b) the emission of x-rays from atoms,
(c) the chemical properties of atoms in terms of the electron shell model,
(d) how atoms associate to form molecules.

28.4 Modification of the Bohr Theory

In the three-dimensional problem of the hydrogen atom, three quantum numbers are required for each stationary state, corresponding to the three independent degrees of freedom for the electron.

The three quantum numbers which emerge from the theory are represented by the symbols n, ℓ, and m_ℓ. The quantum number n is called the **principal quantum number,** ℓ is called the **orbital quantum number,** and m_ℓ is called the **orbital magnetic quantum number.**

There are certain important relationships between these quantum numbers, as well as certain restrictions on their values. These restrictions are:

The values of n can range from 1 to ∞.
The values of ℓ can range from 0 to $n - 1$.
The values of m_ℓ can range from $-\ell$ to ℓ.

For historical reasons, **all states with the same principal quantum number are said to form a shell.** These shells are identified by the letters K, L, M, . . . , which designate the states for which $n = 1, 2, 3, \ldots$. Likewise, **the states having the same values of n and l are said to form a subshell.** The letters *s, p, d, f, g, h,* . . . are used to designate the states for which $\ell = 0, 1, 2, 3, \ldots$.

28.7 The Spin Magnetic Quantum Number

In order to completely describe a quantum state of the hydrogen atom, it is necessary to include a fourth quantum number, m_s, called the **spin magnetic quantum number.** The **spin magnetic quantum number** m_s accounts for the two closely spaced energy states corresponding to the two possible orientations of electron spin. This quantum number can have only two values, $\pm\frac{1}{2}$. In effect, this doubles the number of allowed states specified by the quantum numbers n, ℓ, and m_ℓ

28.9 The Exclusion Principle and the Periodic Table

At this point you might find it useful to review the description of the quantum numbers in Section 28.4 and summarized in Table 28.1 of the textbook.

The **exclusion principle** states that no two electrons can exist in identical quantum states. This means that no two electrons in a given atom can be characterized by the same set of quantum numbers at the same time.

X-rays are emitted by atoms when an electron undergoes a transition from an outer shell into an electron vacancy in one of the inner shells. Transitions into a vacant state in the K shell give rise to the K series of spectral lines, transitions into a vacant state in the L shell create the L series of lines, and so on. The x-ray spectrum of a metal target consists of a set of sharp characteristic lines superimposed on a broad, continuous spectrum.

28.11 Atomic Transitions

Stimulated absorption occurs when light with photons of an energy which matches the energy separation between two atomic energy levels is absorbed by an atom.

An atom in an excited state has a certain probability of returning to its original energy state. This process is called **spontaneous emission.**

When a photon with an energy equal to the excitation energy of an excited atom is incident on the atom, it can increase the probability of de-excitation. This is called **stimulated emission** and results in a second photon of energy equal to that of the incident photon.

28.12 Lasers and Holography

The following three conditions must be satisfied in order to achieve laser action:

1. The system must be in a state of **population inversion** (that is, more atoms in an excited state than in the ground state).

2. The excited state of the system must be a **metastable state,** which means its lifetime must be long compared with the usually short lifetimes of excited states. When such is the case, stimulated emission will occur before spontaneous emission.

3. The emitted photons must be confined in the system long enough to allow them to stimulate further emission from other excited atoms. This is achieved by the use of reflecting mirrors at the ends of the system. One end is made totally reflecting, and the other is slightly transparent to allow the laser beam to escape.

EQUATIONS AND CONCEPTS

The emission spectrum of hydrogen includes four prominent lines that occur at wavelengths described by a simple empirical equation.

$$\frac{1}{\lambda} = R_H\left(\frac{1}{2^2} - \frac{1}{n^2}\right) \qquad (28.1)$$

$$n = 3, 4, 5, \ldots$$

R_H is a constant called the **Rydberg constant.**

$$R_H = 1.0973732 \times 10^7 \text{ m}^{-1}$$

When the electron undergoes a transition from one allowed orbit to another, the frequency of the emitted photon is proportional to the difference in energies of the initial and final states.

$$E_i - E_f = hf \qquad (28.3)$$

The Bohr model of the atom assumed that the electron orbited the nucleus in a circular path under the influence of the Coulomb force of attraction. Also, the angular momentum of the electron about the nucleus must be quantized in units of $nh/(2\pi)$.

$$m_e v r = n\hbar \qquad (28.4)$$

$$n = 1, 2, 3, \ldots$$

While in one of the allowed orbits or stationary states (determined by quantization of the orbital angular momentum), the electron does not radiate energy.

The total energy of the hydrogen atom $(KE + PE)$ depends on the radius of the allowed orbit of the electron.

$$E = -\frac{ke^2}{2r} \qquad (28.8)$$

The electron can exist only in certain allowed orbits, the value of which can be expressed in terms of the Bohr radius, a_0.

$$r_n = \frac{n^2\hbar^2}{m_e k_e e^2} \qquad n = 1, 2, 3, \ldots \qquad (28.9)$$

This is the **Bohr radius** which corresponds to $n = 1$.

$$a_o = \frac{\hbar^2}{m_e k_e e^2} = 0.0529 \text{ nm} \qquad (28.10)$$

This is a general expression for the radius of any orbit in the hydrogen atom.

$$r_n = n^2 a_0 = n^2(0.0529 \text{ nm}) \qquad (28.11)$$

When numerical values of the constants are used, the energy level values for hydrogen can be expressed in units of electron volts (eV).

$$E_n = -\frac{13.6}{n^2} \text{ eV} \qquad (28.13)$$

The lowest energy state or ground state corresponds to the principle quantum number $n = 1$. The energy level approaches $E = 0$ as r approaches infinity. This is the ionization energy for the atom.

Comment on energy levels

A photon of frequency, f, (and wavelength λ) is emitted when an electron undergoes a transition from an initial energy level to a final lower level.

$$f = \frac{E_i - E_f}{h} \qquad (28.14)$$

$$f = \frac{m_e k_e^2 e^4}{4\pi \hbar^3}\left(\frac{1}{n_f^2} - \frac{1}{n_i^2}\right)$$

$$\frac{1}{\lambda} = R_H\left(\frac{1}{n_f^2} - \frac{1}{n_i^2}\right) \quad \text{where}$$

R_H is the theoretical value of the Rydberg constant.

$$R_H = \frac{m_e k_e^2 e^4}{4\pi c \hbar^3} \qquad (28.16)$$

The lines observed in the hydrogen spectrum can be arranged into series corresponding to assigned values of principal quantum numbers of the initial and final states.

Comment on the line spectra of hydrogen

For the Lyman series,
$n_f = 1$ and $n_i = 2, 3, 4, \ldots$

For the Balmer series,
$n_f = 2$ and $n_i = 3, 4, 5, \ldots$

For the Paschen series,
$n_f = 3$ and $n_i = 4, 5, 6, \ldots$

For the Brackett series,
$n_f = 4$ and $n_i = 5, 6, 7, \ldots$

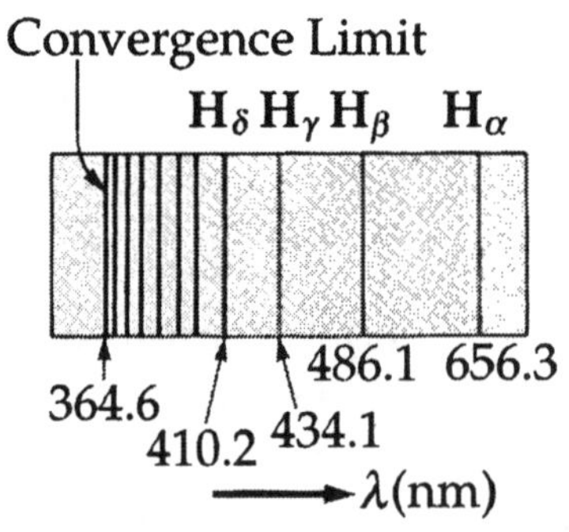

In the case of very large values of the principal quantum number, the energy differences between adjacent levels approach zero and essentially a continuous range (as opposed to a quantized set) of energy values of the emitted photon is possible. In this limit of very large quantum numbers, the classical model of the atom is reasonably accurate.

Comment on the correspondence principle

In addition to the principal quantum, n, other quantum numbers are necessary to completely and accurately specify the possible energy levels in the hydrogen atom and also in more complex atoms.

Comment on quantum numbers

All energy states with the same principal quantum number, n, form a shell. These shells are identified by the spectroscopic notation K, L, M, . . . corresponding to $n = 1, 2, 3, \ldots$.

n can range from 1 to ∞

The orbital quantum number, ℓ, determines the allowed value of orbital angular momentum. All energy states having the same values of n and ℓ form a subshell. The letter designations $s, p, d, f, \ldots$ correspond to values of $\ell = 1, 2, 3, 4, \ldots$.

ℓ can range from 0 to $(n - 1)$

The magnetic orbital quantum number determines the possible orientations of the electron's orbital angular momentum vector in the presence of external magnetic fields.

m_ℓ can range from $-\ell$ to ℓ

The spin magnetic quantum number can have only two values; these correspond to the two possible directions of the electron's intrinsic spin (either parallel or antiparallel to the direction of the orbital angular momentum vector).

$$m_s = \pm\frac{1}{2}$$

No two electrons in an atom can have the same set of quantum numbers n, ℓ, m_ℓ, and m_s.

Comment on the exclusion principle

REVIEW CHECKLIST

▷ State the basic postulates of the Bohr model of the hydrogen atom.

▷ Sketch an energy level diagram for hydrogen (include assignment of values of the principle quantum number, n), show transitions corresponding to spectral lines in the several known series, and make calculations of wavelength values.

▷ For each of the quantum numbers, n, ℓ (the orbital quantum number), m_ℓ, (the orbital magnetic quantum number), and m_s (the spin magnetic quantum number): (i) qualitatively describe what each implies concerning atomic structure, (ii) state the allowed values which may be assigned to each, and the number of allowed states which may exist in a particular atom corresponding to each quantum number.

▷ State the Pauli exclusion principle and describe its relevance to the periodic table of the elements. Show how the exclusion principle leads to the known electronic ground state configuration of the light elements.

SOLUTIONS TO SELECTED END-OF-CHAPTER PROBLEMS

5. The "size" of the atom in Rutherford's model is about 1.0×10^{-10} m. (a) Determine the speed of an electron moving about the proton using the attractive electrostatic force between an electron and a proton separated by this distance. (b) Does this speed suggest that Einsteinian relativity must be considered when studying the atom? (c) Compute the de Broglie wavelength of the electron as it moves about the proton. (d) Does this wavelength suggest that wave effects, such as diffraction and interference, must be considered when studying the atom?

Solution

(a) The force of electrostatic attraction between the electron and proton has a magnitude $F = ke^2/r^2$ and produces the needed centripetal acceleration, $a_c = v^2/r$, as the electron orbits the proton. Thus we can find the orbital speed of the electron by applying Newton's second law to the orbital motion:

$$m\frac{v^2}{r} = \frac{ke^2}{r^2} \quad \text{or, isolating the velocity,} \quad v = \sqrt{\frac{ke^2}{mr}}$$

If the electron and proton are separated by $r = 1.0 \times 10^{-10}$ m, the speed is

$$v = \sqrt{\frac{\left(8.99 \times 10^9 \text{ N}\cdot\text{m}^2/\text{C}^2\right)\left(1.60 \times 10^{-19} \text{ C}\right)^2}{\left(9.11 \times 10^{-31} \text{ kg}\right)\left(1.0 \times 10^{-10} \text{ m}\right)}} = 1.6 \times 10^6 \text{ m/s} \qquad \lozenge$$

(b) Since $v/c = \left(1.6 \times 10^6 \text{ m/s}\right)/\left(3.0 \times 10^8 \text{ m/s}\right) << 1$, the electron is not relativistic and relativity effects need not be considered. ◊

(c) The de Broglie wavelength of the electron is $\lambda = \frac{h}{p} = \frac{h}{mv}$

Therefore, $$\lambda = \frac{6.626\times 10^{-34}\ \text{J}\cdot\text{s}}{\left(9.11\times 10^{-31}\ \text{kg}\right)\left(1.6\times 10^{6}\ \text{m/s}\right)} = 4.5\times 10^{-10}\ \text{m}$$ ◊

(d) Since the de Broglie wavelength of the electron and the size of the atom are of the same order of magnitude, wave effects such as interference and diffraction should be considered when studying the atom. ◊

---

9. Show that the speed of the electron in the nth Bohr orbit in hydrogen is given by $v_n = ke^2/n\hbar$.

Solution

In the Bohr model of the atom, the force that produces centripetal acceleration and maintains the electron in orbit is the electrostatic attraction between the electron and proton. Thus, $mv^2/r = ke^2/r^2$, or

$$mv_n^{\ 2} = ke^2/r_n \qquad \textbf{[Equation 1]}$$

Here m is the mass of the electron, r_n is the radius of the nth Bohr orbit, and v_n is the speed of the electron in that orbit, h is Planck's constant, and $\hbar = h/2\pi$. According to Bohr's assumptions, the angular momentum of the electron in the nth Bohr orbit is

$$L_n = mr_nv_n = n\hbar \qquad \text{or} \qquad r_n = n\hbar/mv_n$$

Substituting this expression for the orbit radius into Equation 1 gives

$$mv_n^{\ 2} = \frac{ke^2}{n\hbar/mv_n} = \frac{mv_nke^2}{n\hbar}, \text{ which reduces to } v_n = \frac{ke^2}{n\hbar}$$ ◊

---

13. What is the energy of the photon that, when absorbed by a hydrogen atom, could cause (a) an electronic transition from the $n = 3$ state to the $n = 5$ state and (b) an electronic transition from the $n = 5$ state to the $n = 7$ state?

Solution To be absorbed, the energy of the photon must match the difference in the energy of the electron in the final and initial states. The energy of the electron in the nth orbit of a hydrogen atom is

$$E_n = -\left(13.6 / n^2\right) \text{eV}$$

(a) For an electronic transition from the $n = 3$ state to the $n = 5$ state, the required energy of the photon is

$$E_\gamma = E_5 - E_3 = -\frac{13.6}{5^2}\text{eV} - \left(-\frac{13.6}{3^2}\text{eV}\right) \quad \text{or} \quad E_\gamma = 13.6\ \text{eV}\left(\frac{1}{9} - \frac{1}{25}\right) = 0.967\ \text{eV} \quad \Diamond$$

(b) The photon energy required for a transition from the $n = 5$ state to the $n = 7$ state is

$$E_\gamma = E_7 - E_5 = -\frac{13.6}{7^2}\text{eV} - \left(-\frac{13.6}{5^2}\text{eV}\right) \quad \text{or} \quad E_\gamma = 13.6\ \text{eV}\left(\frac{1}{25} - \frac{1}{49}\right) = 0.266\ \text{eV} \quad \Diamond$$

---

17. Determine both the longest and the shortest wavelengths in (a) the Lyman series $\left(n_f = 1\right)$ and (b) the Paschen series $\left(n_f = 3\right)$ of hydrogen.

Solution The wavelength of a photon having energy E_γ may be written

$$\lambda = \frac{hc}{E_\gamma} = \frac{\left(6.626 \times 10^{-34}\ \text{J} \cdot \text{s}\right)\left(3.00 \times 10^8\ \text{m/s}\right)\left(1\ \text{eV}/1.60 \times 10^{-19}\ \text{J}\right)}{E_\gamma}$$

or

$$\lambda = \frac{1.242 \times 10^{-6}\ \text{m} \cdot \text{eV}}{E_\gamma} = \frac{1.242 \times 10^{3}\ \text{nm} \cdot \text{eV}}{E_\gamma}$$

(a) The Lyman series of spectral lines result from electronic transitions that terminate in the n = 1 state. The longest wavelength in this series is produced when the electron begins the transition from the n = 2 state, so the energy of the emitted photon (for a hydrogen atom) is

$$E_\gamma = 13.6 \text{ eV}\left(\frac{1}{1^2} - \frac{1}{2^2}\right) = 10.2 \text{ eV}$$

The longest wavelength in the Lyman series is then

$$\left(\lambda_{\text{Lyman}}\right)_{\max} = \frac{1.242 \times 10^3 \text{ nm} \cdot \text{eV}}{10.2 \text{ eV}} = 122 \text{ nm} \qquad \Diamond$$

The shortest wavelength (i.e., the series limit) occurs when an electron having zero energy ($n \to \infty$) makes the transition to the n = 1 state. Thus,

$$E_\gamma = 0 - E_1 = -\left(-\frac{13.6}{1^2}\text{eV}\right) = 13.6 \text{ eV}$$

and $$\left(\lambda_{\text{Lyman}}\right)_{\min} = \frac{1.242 \times 10^3 \text{ nm} \cdot \text{eV}}{13.6 \text{ eV}} = 91.3 \text{ nm} \qquad \Diamond$$

Note that both limits of the Lyman series are in the ultraviolet region of the spectrum.

(b) For the Paschen series, all the electronic transitions terminate in the n = 3 state. The energy of the photon emitted in such a transition is

$$E_\gamma = 13.6 \text{ eV}\left(\frac{1}{3^2} - \frac{1}{n_i^2}\right) \quad (n_i = 4,\ 5,\ 6, \ldots)$$

The longest wavelength (i.e., lowest energy) photon is produced when $n_i = 4$, so $\left(E_\gamma\right)_{\min} = 0.661$ eV:

$$\left(\lambda_{\text{Paschen}}\right)_{\max} = \frac{1.242 \times 10^3 \text{ nm} \cdot \text{eV}}{0.661 \text{ eV}} = 1.88 \times 10^3 \text{ nm} \qquad \Diamond$$

The shortest wavelength (series limit) occurs when $n_i = \infty$ and the maximum energy photon is produced. Thus, $(E_\gamma)_{max} = 1.51$ eV and

$$(\lambda_{Paschen})_{min} = \frac{1.242\times10^3 \text{ nm}\cdot\text{eV}}{1.51 \text{ eV}} = 823 \text{ nm} \qquad \Diamond$$

It is seen that both the long and short wavelength limits of the Paschen series in hydrogen are in the infrared portion of the spectrum.

---

21. A particle of charge q and mass m, moving with a constant speed, v, perpendicular to a constant magnetic field, B, follows a circular path. If the angular momentum about the center of this circle is quantized so that $mvr = n\hbar$, show that the allowed radii for the particle are

$$r_n = \sqrt{\frac{n\hbar}{qB}} \quad \text{for} \quad n = 1,2,3,\ldots.$$

Solution

The force that produces the needed centripetal acceleration to cause the particle to follow a circular path is the magnetic force $F = qBv$ exerted on the moving particle by the magnetic field. Therefore, $mv^2/r = qBv$, or $r = mv/qB$. From the quantization rule for the angular momentum ($L = mrv = n\hbar$), it is seen that

$$mv = \frac{L}{r} = \frac{n\hbar}{r}$$

Thus, the equation for the radius of the particle's path becomes

$$r = \frac{n\hbar/r}{qB} \quad \text{which reduces to } r^2 = \frac{n\hbar}{qB}$$

The allowed radii for the particle's circular path are then given by

$$r_n = \sqrt{\frac{n\hbar}{qB}} \quad \text{where} \quad n = 1, 2, 3, \ldots. \qquad \Diamond$$

---

25. Consider a hydrogen atom. (a) Calculate the frequency f of the $n=2 \to n=1$ transition and compare with the frequency f_{orb} of the electron orbital motion in the $n=2$ state. (b) Make the same calculation for the $n=10000 \to n=9999$ transition. Comment on the results.

Solution

(a) When the electron in the hydrogen atom makes a transition from the $n=n_i$ state to the $n=n_f$ state, the frequency of the emitted photon is

$$f=\frac{E\gamma}{h}=\frac{13.6\text{ eV}\left(1/n_f^2-1/n_i^2\right)\left(1.60\times10^{-19}\text{ J/1 eV}\right)}{6.626\times10^{-34}\text{ J}\cdot\text{s}}$$

and $$f_\gamma=3.28\times10^{15}\text{ Hz}\left(\frac{n_i^2-n_f^2}{n_i^2 n_f^2}\right)$$

For $n_i=2$ and $n_f=1$ $\quad f_\gamma=2.46\times10^{15}\text{ Hz}$

The orbital frequency of the electron in the original orbit is, $f_{\text{orb}}=\dfrac{1}{T}=\dfrac{v_{n_i}}{2\pi r_{n_i}}$

From the solution of Problem 28.9, the orbital speed is, $\quad v_{n_i}=\dfrac{ke^2}{n_i\hbar}=\dfrac{2\pi ke^2}{n_i h}$

Using the quantization rule for angular momentum $(L=mvr=n\hbar=nh/2\pi)$, the radius of the $n=n_i$ orbit is

$$r_{n_i}=\frac{n_i h}{2\pi m v_{n_i}}=\frac{n_i h}{2\pi m\left(2\pi ke^2/n_i\hbar\right)}=\frac{n_i^2h^2}{4\pi^2mke^2}$$

Thus, the orbital frequency becomes, $$f_{\text{orb}}=\frac{2\pi ke^2/n_i h}{2\pi\left(n_i^2h^2/4\pi^2mke^2\right)}=\frac{4\pi^2mk^2e^4}{n_i^3h^3}$$

For $n_i = 2$, this yields

$$f_{\text{orb}} = \frac{4\pi^2\left(9.11\times10^{-31}\text{ kg}\right)\left(8.99\times10^{9}\text{ N}\cdot\text{m}^2/\text{C}^2\right)^2\left(1.60\times10^{-19}\text{ C}^2\right)^4}{(2)^3\left(6.626\times10^{-34}\text{ J}\cdot\text{s}\right)^3}$$

or $$f_{\text{orb}} = 8.19\times10^{14}\text{ Hz} \quad \text{as compared to} \quad f_\gamma = 2.46\times10^{15}\text{ Hz}$$ ◊

(b) If the calculations done above are repeated with $n_i = 10\,000$ and $n_f = 9999$, the results are $f_\gamma = 6.56\times10^{3}$ Hz and $f_{\text{orb}} = 6.55\times10^{3}$ Hz. ◊

Classical theory predicts that the frequency of the emitted light should be the same as the orbital frequency of the electron. It is observed that the two frequencies differ significantly for transitions involving small quantum numbers (where the differences between the allowed energy levels are large). However, for transitions involving large quantum numbers (where very small differences exist between the allowed energies) the results of quantum theory approach those predicted by classical theory. This is an example of Bohr's correspondence principle. ◊

---

31. Determine the wavelength of an electron in the third excited orbit of the hydrogen atom.

Solution Taking $p = mv$ to be the magnitude of the linear momentum of a particle, the de Broglie wavelength of the particle is

$$\lambda = \frac{h}{p} = \frac{h}{mv}$$

From the Bohr model, an electron in one of the allowed orbits has an angular momentum of $L = mv_n r_n = n\hbar = nh/2\pi$.

Thus, the magnitude of the linear momentum is $mv_n = nh/2\pi r_n$, and the de Broglie wavelength is

$$\lambda = \frac{h}{nh/2\pi r_n} \qquad \text{or} \qquad \lambda = \frac{2\pi r_n}{n} \qquad \textbf{[Equation 1]}$$

Note that this result says that $2\pi r_n = n\lambda$, or the allowed orbits are those whose circumference is an integral multiple of the de Broglie wavelength of the electron.

The radii of the allowed electron orbits in the hydrogen atom may be written as $r_n = n^2 a_0$ where $a_0 = 0.0529 \text{ nm} = 5.29 \times 10^{-11}$ m is the Bohr radius (the radius of the innermost orbit). Hence Equation 1 becomes

$$\lambda = \frac{2\pi n^2 a_0}{n} = n(2\pi a_0) = n\left(3.32 \times 10^{-10} \text{ m}\right)$$

The de Broglie wavelength of the electron in the third excited state (i.e., for $n = 4$) in the hydrogen atom is then:

$$\lambda = 4\left(3.32 \times 10^{-10} \text{ m}\right) = 1.33 \times 10^{-9} \text{ m} = 1.33 \text{ nm}$$ ◊

33. List the possible sets of quantum numbers for electrons in the $3p$ subshell.

Solution The electron states within an atom are characterized by four quantum numbers, n, ℓ, m_ℓ, and m_s. All states in the $3p$ subshell have $n = 3$ and $\ell = 1$. For a given value of ℓ, the orbital magnetic quantum number, m_ℓ, may have $2\ell + 1$ different values, ranging from $-\ell$ to $+\ell$ in integer steps. For each value of m_ℓ, the spin magnetic quantum number may have two different values, $m_s = \pm 1/2$. Thus, there are six distinct states (each associated with a **different** combination of quantum numbers) in the $3p$ subshell. The combinations of quantum numbers associated with each of these states are summarized in the table below:

n	3	3	3	3	3	3
ℓ	1	1	1	1	1	1
m_ℓ	−1	−1	0	0	+1	+1
m_s	+1/2	−1/2	+1/2	−1/2	+1/2	−1/2

◊

37. Zirconium (Z = 40) has two electrons in an incomplete d subshell. (a) What are the values of n and ℓ for each electron? (b) What are all possible values of m_ℓ and m_s? (c) What is the electron configuration in the ground state of zirconium?

Solution (a) Zirconium, with 40 electrons, has 4 electrons outside a closed Krypton core. The Krypton core, with 36 electrons, fills all states up through the $4p$ subshell. Normally, one would expect the next 4 electrons to go into the $4d$ subshell. However, an exception to the rule occurs at this point: the $5s$ subshell fills (with 2 electrons) before the $4d$ ($n = 4$, $\ell = 2$) subshell starts filling. Thus, there are two electrons in an incomplete $4d$ subshell in Zirconium with quantum numbers of $n = 4$, and $\ell = 2$ ◊

(b) The orbital quantum number for each electron in the $4d$ subshell is $\ell = 2$. The orbital magnetic quantum number for each of these electrons may range from $-\ell$ to $+\ell$ in integer steps, or the possible values for m_ℓ for each of these electrons are $m_\ell = (0, \pm1, \pm2)$. The possible values of the spin magnetic quantum number for any electron are $m_s = \pm 1/2$. ◊

(c) The complete electron configuration for the ground state of Zirconium is:

$$1s^2 2s^2 2p^6 3s^2 3p^6 3d^{10} 4s^2 4p^6 4d^2 5s^2 = [Kr] 4d^2 5s^2$$ ◊

42. When an electron drops from the M shell ($n = 3$) to a vacancy in the K shell ($n = 1$), the measured wavelength of the emitted x-ray is found to be 0.101 nm. Identify the element.

Solution

The energy of an electron in a shell of a many electron atom may be estimated by

$$E = -Z_{eff}^2 \frac{(13.6 \text{ eV})}{n^2}$$

where $Z_{eff}e$ is the effective nuclear charge seen by that electron. An electron in the K shell $(n=1)$ is partially screened from the nuclear charge by the other electron in the K shell. Thus, $Z_{eff} = Z-1$ for this K shell electron and the estimate of its energy is $E_K = -(Z-1)^2(13.6 \text{ eV})$. When a vacancy exists in the K shell (i.e., only one electron is present in this shell), an electron in the M shell $(n=3)$ is screened from the nuclear charge by a total of 9 electrons, 1 in the K shell and 8 in the filled L shell $(n=2)$. Hence, $Z_{eff} = Z-9$ for this M shell electron and the estimate of its energy is

$$E_M = -(Z-9)^2 \frac{(13.6 \text{ eV})}{(3)^2}$$

When this electron drops from the M shell to fill the vacancy in the K shell, the energy of the emitted x-ray photon is

$$E_\gamma = E_M - E_K = \left[-\frac{(Z-9)^2}{9} + (Z-1)^2\right](13.6 \text{ eV}) \qquad \textbf{[Equation 1]}$$

If the wavelength of the emitted x-ray is observed to be $\lambda = 0.101$ nm,

$$E_\gamma = \frac{hc}{\lambda} = \frac{(6.626\times10^{-34} \text{ J}\cdot\text{s})(3.00\times10^8 \text{ m/s})}{0.101\times10^{-9} \text{ m}}\left(\frac{1 \text{ eV}}{1.60\times10^{-19} \text{ J}}\right) = 1.23\times10^4 \text{ eV}$$

Substituting this into Equation 1 and simplifying gives

$$-(Z-9)^2+9(Z-1)^2=\frac{9\left(1.23\times10^4\ \text{eV}\right)}{13.6\ \text{eV}}=8.14\times10^3 \quad\text{or}\quad 8Z^2-72=8.14\times10^3$$

This reduces to $Z=32.0$. Thus, the element must be Germanium. ◊

---

45. A laser used in a holography experiment has an average output power of 5.0 mW. The laser beam is actually a series of pulses of electromagnetic radiation at a wavelength of 632.8 nm, each having a duration of 25 ms. Calculate (a) the energy (in joules) radiated with each pulse and (b) the number of photons in each pulse.

Solution

(a) The energy emitted in a single pulse of this laser is $E=\overline{P}(\Delta t)$ where $\overline{P}$ is the average power output and Δt is the duration of the pulse.

Thus, $$E_{\text{pulse}}=\left(5.0\times10^{-3}\ \text{J/s}\right)\left(25\times10^{-3}\ \text{s}\right)=1.3\times10^{-4}\ \text{J}$$ ◊

(b) With a wavelength of $\lambda=632.8\ \text{nm}$, the energy of each photon in the beam is

$$E_\gamma=\frac{hc}{\lambda}=\frac{\left(6.626\times10^{-34}\ \text{J}\cdot\text{s}\right)\left(3.00\times10^8\ \text{m/s}\right)}{632.8\times10^{-9}\ \text{m}}=3.14\times10^{-19}\ \text{J}$$

The number of photons in each pulse is then

$$n=\frac{E_{\text{pulse}}}{E_\gamma}=\frac{1.3\times10^{-4}\ \text{J}}{3.14\times10^{-19}\ \text{J}}=4.0\times10^{14}\ \text{photons per pulse}$$ ◊

---

56. A pi meson (π^-) of charge $-e$ and mass 273 times greater than that of the electron is captured by a helium nucleus (Z = +2) as shown in Figure P28.56. (a) Draw an energy level diagram (in units of eV) for this "Bohr-type" atom up to the first six energy levels. (b) When the pi meson makes a transition between two orbits, a photon is emitted that Compton scatters off a free electron initially at rest, producing a scattered photon of wavelength $\lambda' = 0.089\,929\,3$ nm at an angle of $\theta = 50.00°$, as shown on the right-hand side of Figure P28.56. Between which two orbits did the pi meson make a transition?

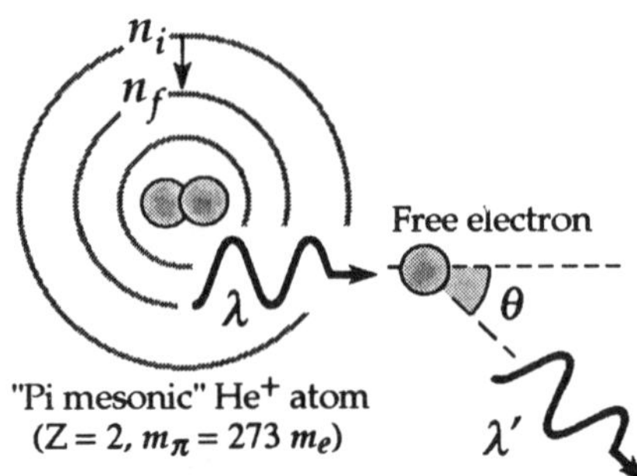

Figure P28.56

Solution (a) The energy of the allowed levels in a one-electron atom with a nuclear charge of $+Ze$ is $E_n = E_1/n^2$, where the energy of the $n = 1$ level is

$$E_1 = -\frac{m_e k^2 Z^2 e^4}{2\hbar^2} = -Z^2(13.60\ \text{eV})$$

If the orbiting electron, of charge $-e$ and mass m_e, is replaced by a pi meson having the same charge but of mass $m_\pi = 273 m_e$, the energy of the $n = 1$ level is changed to

$$E_1' = -\frac{m_\pi k^2 Z^2 e^4}{2\hbar^2} = 273\left(-\frac{m_e k^2 Z^2 e^4}{2\hbar^2}\right) = 273E_1 = -Z^2(273)(13.60\ \text{eV})$$

Simplifying, we find that $E_1' = -Z^2\left(3.713\times10^3\ \text{eV}\right)$. The energy of the levels in a pi-mesonic He$^+$ atom $(Z = 2)$ are then

$$E_n = \frac{E_1'}{n^2} = -\frac{(2)^2\left(3.713\times10^3\ \text{eV}\right)}{n^2} = -\frac{1.485\times10^4\ \text{eV}}{n^2}$$

The energies of the first six energy levels in this atom are:

$E_1 = -1.485 \times 10^4$ eV $\qquad E_2 = -3.713 \times 10^3$ eV $\qquad E_3 = -1.650 \times 10^3$ eV

$E_4 = -928.1$ eV $\qquad E_5 = -594.0$ eV $\qquad E_6 = -412.5$ eV ◊

(b) When a photon scatters from a free electron, the Compton shift in the wavelength is $\Delta\lambda = \lambda' - \lambda = \lambda_c(1-\cos\theta)$ where the Compton wavelength is $\lambda_c = 0.00243$ nm. Thus, at a scattering angle of $\theta = 50.00°$ the shift is $\Delta\lambda = (0.00243 \text{ nm})(1-\cos 50.00°) = 8.680 \times 10^{-4}$ nm. Hence, if the measured wavelength of the scattered photon is $\lambda' = 0.0899293$ nm, the wavelength of the emitted photon must have been

$$\lambda = \lambda' - \Delta\lambda = 0.08906 \text{ nm} = 8.906 \times 10^{-11} \text{ m}$$

The energy the atom gave up in this transition is $\Delta E = E_\gamma = hc/\lambda$, or

$$\Delta E = \frac{\left(6.626 \times 10^{-34} \text{ J}\cdot\text{s}\right)\left(3.000 \times 10^8 \text{ m/s}\right)}{8.906 \times 10^{-11} \text{ m}}\left(\frac{1 \text{ eV}}{1.602 \times 10^{-19} \text{ J}}\right) = 1.393 \times 10^4 \text{ eV}$$

Comparing this to the energies of the levels found in part (a), it is observed that the transition must terminate on the $n = 1$ level if the atom is to give up this much energy. Thus, $n_f = 1$ and the energy of the initial level must have been

$$E_{n_i} = E_1 + \Delta E = -1.485 \times 10^4 \text{ eV} + 1.393 \times 10^4 \text{ eV} = -920 \text{ eV}$$

This is very close to the value computed earlier for the $n = 4$ level. Thus, the transition was from the $n = 4$ level to the $n = 1$ level. ◊

CHAPTER SELF-QUIZ

1. The restriction that no more than one electron may occupy a given quantum state in an atom was first stated by which of the following scientists?
   a. Bohr
   b. de Broglie
   c. Heisenberg
   d. Pauli

2. The ionization energy for the hydrogen atom is 13.6 eV. What is the energy of a photon that is emitted, as a hydrogen atom makes a transition between the $n = 5$ and $n = 3$ states?
   a. 0.544 eV
   b. 0.967 eV
   c. 1.51 eV
   d. 10.2 eV

3. When a cool gas is placed between a glowing wire filament source and a diffraction grating, the resultant spectrum from the grating is which one of the following?
   a. line emission
   b. line absorption
   c. continuous
   d. monochromatic

4. The Balmer series of hydrogen is comprised of transitions from higher levels to the $n = 2$ level. If the first line in that series has a wavelength 653 nm, what wavelength corresponds to the transition from $n = 5$ to $n = 2$?
   a. 390 nm
   b. 432 nm
   c. 503 nm
   d. 630 nm

5. If the radius of the electron orbit in the $n = 1$ level of the hydrogen atoms is 0.053 nm, what is its radius for the $n = 4$ level? (Assume the Bohr model is valid.)
   a. 0.11 nm
   b. 0.21 nm
   c. 0.48 nm
   d. 0.85 nm

6. The quantum mechanical model of the hydrogen atom requires that if the orbital quantum number = 4, how many possible substates are permitted?
   a. 4
   b. 8
   c. 16
   d. 18

7. Atoms which absorb ultraviolet photons and in turn emit photons in the visible range are used in which type of device?
   a. incandescent lamp
   b. radar
   c. fluorescent lamp
   d. electron microscope

8. The four visible colors emitted by hydrogen atoms are produced by electrons that
   a. start in the ground state
   b. end up in the ground state
   c. start in the level with $n = 2$
   d. end up in the level with $n = 2$

9. In the Bohr model of the atom, the orbits where electrons move fastest have
   a. the least energy
   b. the most energy
   c. the biggest radius
   d. the greatest angular momentum

10. The familiar yellow light from a sodium vapor street lamp results from the $3p \rightarrow 3s$ transition in Na. Determine the wavelength of the light given off if the energy difference $E_{3p} - E_{3s} = 2.10$ eV.
    a. 560 nm
    b. 575 nm
    c. 590 nm
    d. 600 nm

11. A hydrogen atom in the ground state absorbs a 12.1-eV photon. To what level is the electron promoted? (The ionization energy of hydrogen is 13.6 eV.)
    a. $n = 2$
    b. $n = 3$
    c. $n = 4$
    d. $n = 5$

12. An energy of 13.6 eV is needed to remove the electron from the ground state of a hydrogen atom. If a single photon accomplishes this task, what wavelength must the photon have?
    a. 61 nm
    b. 71 nm
    c. 81 nm
    d. 91 nm

Chapter 29

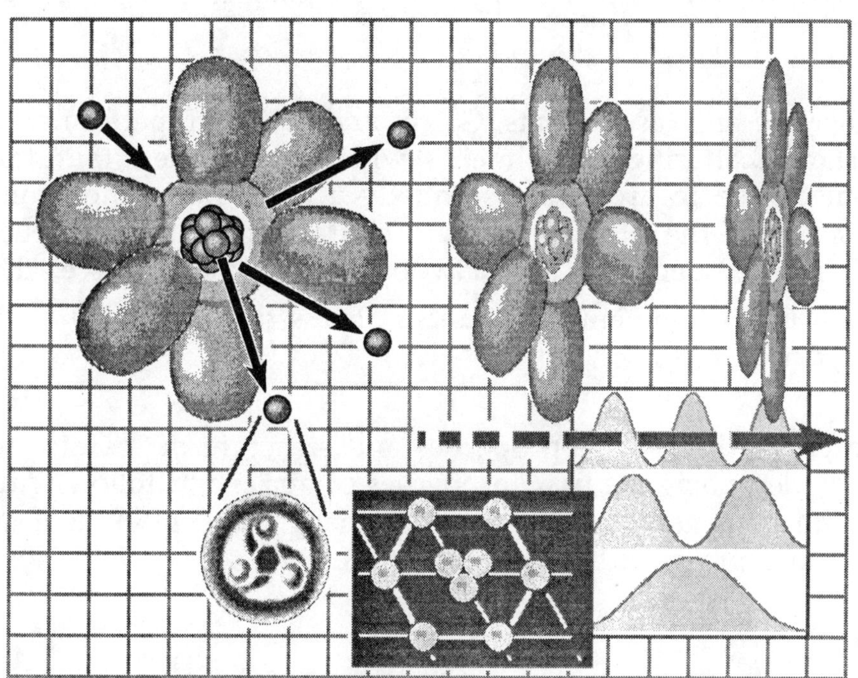

NUCLEAR PHYSICS

Chapter 29

NUCLEAR PHYSICS

In 1896, the year that marks the birth of nuclear physics, Henri Becquerel (1852-1908) discovered radioactivity in uranium compounds. Pioneering work by Rutherford showed that the radiation was of three types, which he called alpha, beta, and gamma rays. These types are classified according to the nature of their electric charge and according to their ability to penetrate matter.

In 1911 Rutherford and his students, Geiger and Marsden, performed a number of important scattering experiments which established that the nucleus of an atom can be regarded as essentially a point mass and point charge and that most of the atomic mass is contained in the nucleus. Furthermore, such studies demonstrated a wholly new type of force, the **nuclear force,** which is predominant at distances of less than about 10^{-14} m and zero at great distances.

In this chapter we discuss the properties and structure of the atomic nucleus. We start by describing the basic properties of nuclei and follow with a discussion of the phenomenon of radioactivity. Finally, we explore nuclear reactions and the various processes by which nuclei decay.

NOTES FROM SELECTED CHAPTER SECTIONS

29.1 Some Properties of Nuclei

Important quantities in the description of nuclear properties are:

1. The **atomic number,** Z, which equals the number of protons in the nucleus.

2. The **neutron number,** N, which equals the number of neutrons in the nucleus.

3. The **mass number,** A, which equals the number of nucleons (neutrons plus protons) in the nucleus.

The nuclei of all atoms of a particular element contain the same number of protons but often contain different numbers of neutrons. Nuclei that are related in this way are called **isotopes.** The isotopes of an element have the same Z value but different N and A values.

The **atomic mass unit,** u, is defined such that the mass of the isotope ^{12}C is exactly 12 u.

Experiments have shown that most nuclei are approximately spherical and all **have nearly the same density.** Nucleons combine to form a nucleus **as though** they were tightly packed spheres. The stability of nuclei is due to the **nuclear force.** This is a **short range, attractive** force which acts between all nuclear particles. Nuclei have **intrinsic angular momentum** which is quantized by the **nuclear spin quantum number** which may be integer or half integer.

29.2 Binding Energy

The total mass of a nucleus is always less than the sum of the masses of its individual nucleons. The **binding energy** of the nucleus is mass difference multiplied by c^2. Energy must be added to a nucleus in order to separate the nucleus into neutrons and protons.

29.3 Radioactivity

There are three processes by which a radioactive substance can undergo decay: alpha (α) decay, where the emitted particles are ^{4}He nuclei; beta (β) decay, in which the emitted particles are either electrons or positrons; and gamma (γ) decay, in which the emitted "rays" are high-energy photons. A positron is a particle similar to the electron in all respects except that it has a charge of $+e$ (the antimatter twin of the electron). The symbol β^- is used to designate an electron, and β^+ designates a positron.

The three types of radiation have quite different penetrating powers. Alpha particles barely penetrate a sheet of paper, beta particles can penetrate a few millimeters of aluminum, and gamma rays can penetrate several centimeters of lead.

The decay rate, or activity, R, of a sample is defined as the number of decays per second. The half life of a radioactive substance is the time it takes half of a given number of radioactive nuclei to decay. The SI unit of activity is the **becquerel** (Bq), where 1 Bq = 1 decay/s.

29.4 The Decay Processes

Alpha decay can occur because, according to quantum mechanics, some nuclei have barriers that can be penetrated by the alpha particles (the tunneling process). This process is energetically more favorable for those nuclei having a large excess of neutrons. A nucleus can undergo beta decay in two ways. It can emit either an electron (β^-) and an antineutrino ($\bar{\nu}$) or a positron (β^+) and a neutrino (ν). In the electron-capture process, the nucleus of an atom absorbs one of its own electrons (usually from the K shell) and emits a neutrino.

The neutrino has the following properties:

1. It has zero electric charge.
2. It has a rest mass smaller than that of the electron; and, in fact, its mass may be zero (although recent experiments suggest that this may not be true).
3. It has a spin of $\frac{1}{2}$, which satisfies the law of conservation of angular momentum.
4. It interacts very weakly with matter and is therefore very difficult to detect.

In gamma decay, a nucleus in an excited state decays to its ground state and emits a gamma ray. The Q value (disintegration energy) is the energy released as a result of the decay process.

29.5 Natural Radioactivity

Radioactive nuclei are generally classified into two groups: (1) unstable nuclei found in nature, which give rise to what is called **natural radioactivity,** and (2) nuclei produced in the laboratory through nuclear reactions, which exhibit **artificial radioactivity.**

29.6 Nuclear Reactions

Nuclear reactions are events in which collisions change the identity or properties of nuclei. The total energy released as a result of a nuclear reaction is called the **reaction energy,** Q.

An **endothermic reaction** is one in which Q is negative and the minimum energy for which the reaction will occur is called the **threshold energy.**

EQUATIONS AND CONCEPTS

Most nuclei are approximately spherical in shape and have an average radius which is proportional to the cube root of the mass number or total number of nucleons. This means that the volume is proportional to A and that all nuclei have nearly the same density.

$$r = r_0 A^{1/3} \quad (29.1)$$

The number of radioactive nuclei in a given sample which undergo decay during a time interval Δt depends on the number of nuclei present. The number of decays depends also on the decay constant, λ, which is characteristic of a particular isotope.

$$\Delta N = -\lambda N \Delta t \quad (29.2)$$

The decay rate or activity, R, of a sample of radioactive nuclei is defined as the number of decays per second.

$$R = \lambda N \quad (29.3)$$

Activity can be expressed in units of becquerels or curies.

$$1\ \text{Ci} \equiv 3.70 \times 10^{10}\ \text{decays/s} \quad (29.6)$$

$$1\ \text{Bq} = 1\ \text{decay/s} \quad (29.7)$$

The number of nuclei in a radioactive sample decreases exponentially with time. The plot of number of nuclei, N, versus elapsed time, t, is called a decay curve.

$$N = N_0 e^{-\lambda t} \tag{29.4}$$

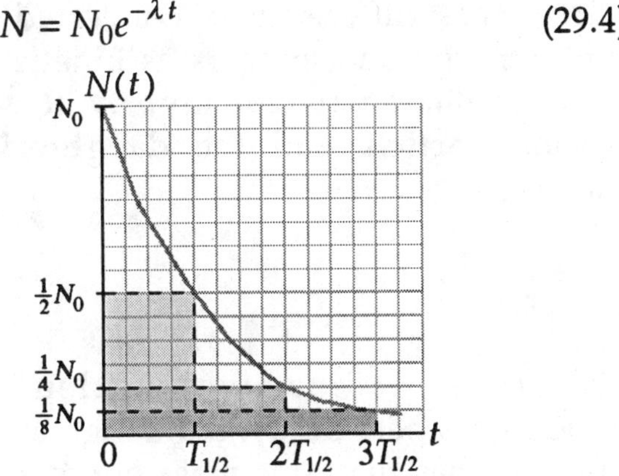

The half-life is the time required for half of a given number of radioactive nuclei to decay.

$$T_{1/2} = \frac{\ln 2}{\lambda} = \frac{0.693}{\lambda} \tag{29.5}$$

When a nucleus decays (a process called transmutation) by alpha emission, the parent nucleus loses two neutrons and two protons. Alpha decay can be represented symbolically in terms of the parent nucleus, X, and the daughter nucleus, Y. An example is the decay of $^{238}_{92}\text{U}$.

$$^{A}_{Z}\text{X} \rightarrow \,^{A-4}_{Z-2}\text{Y} + \,^{4}_{2}\text{He} \tag{29.8}$$

$$^{238}_{92}\text{U} \rightarrow \,^{234}_{90}\text{Th} + \,^{4}_{2}\text{He} \tag{29.9}$$

In order for alpha emission to occur, the mass of the parent nucleus must be greater than the combined mass of the daughter nucleus and the emitted alpha particle.

Comment on alpha decay

The mass difference is converted into energy and appears as kinetic energy shared (unequally) by the alpha particle and the daughter nucleus.

When a radioactive nucleus undergoes beta decay, the daughter nucleus has the same mass number as the parent nucleus but the charge number (or atomic number) increases by one. Beta decay can be shown symbolically and by example with an isotope of carbon.

$$ {}^{A}_{Z}X \rightarrow {}^{A}_{Z+1}Y + e^{-1} \quad (29.11)$$

$$ {}^{14}_{6}C \rightarrow {}^{14}_{7}N + e^{-1} \quad (29.12)$$

The electron that is emitted is created within the parent nucleus by a process which can be represented by a neutron transformed into a proton and an electron.

$$ {}^{1}_{0}n \rightarrow {}^{1}_{1}p + e^{-1} \quad (29.14)$$

The total energy released in beta decay is greater than the combined kinetic energies of the electron and the daughter nucleus. This difference in energy is associated with a third particle called a neutrino.

Comment on beta decay

Nuclei which undergo alpha or beta decay are often left in an excited energy state. The nucleus returns to the ground state by emission of one or more photons. Gamma ray emission results in no change in mass number or atomic number.

$$^{12}_{5}B \rightarrow {}^{12}_{6}C^{*} + e^{-1} \qquad (29.17)$$

$$^{12}_{6}C^{*} \rightarrow {}^{12}_{6}C + \gamma \qquad (29.18)$$

Artificial transmutations (nuclear reactions) can be induced by bombarding stable target nuclei with energetic particles.

$$^{4}_{2}He + {}^{14}_{7}N \rightarrow {}^{17}_{8}O + {}^{1}_{1}H \qquad (29.20)$$

Comment on energy conservation in nuclear reactions: Q values

The quantity of energy required to balance the equation representing a nuclear reaction (e.g. Eq. 29.20) is called the Q value of the reaction. The Q value can be calculated in terms of the total mass of the reactants minus the total mass of the products or as the kinetic energy of the products minus the kinetic energy of the reactants. Q is positive for exothermic reactions and negative for endothermic reactions.

Endothermic reactions are characterized by a threshold energy (or minimum kinetic energy of the incoming particle) which is required in order for energy and momentum to be conserved. Note that m is the mass of the incident particle and M is the mass of the target particle.

$$KE_{\text{min}} = \left(1 + \frac{m}{M}\right)|Q| \qquad (29.23)$$

SUGGESTIONS, SKILLS, AND STRATEGIES

The rest energy of a particle is given by $E = mc^2$. It is therefore often convenient to express the unified mass unit in terms of its energy equivalent, $1\ \text{u} = 1.660\,559 \times 10^{-27}$ kg or $1\ \text{u} = 931.50\ \text{MeV}/c^2$. When masses are expressed in units of u, energy values are then $E = m(931.50\ \text{MeV}/\text{u})$.

Equation 29.4 can be solved for the particular time t after which the number of remaining nuclei will be some specified fraction of the original number N_0. This can be done by taking the natural log of each side of Equation 29.4 to find

$$t = \left(\frac{1}{\lambda}\right)\ln\left(\frac{N_0}{N}\right)$$

REVIEW CHECKLIST

▷ Account for nuclear binding energy in terms of the Einstein mass-energy relationship. Describe the basis for energy release by fission and fusion in terms of the shape of the curve of binding energy per nucleon vs. mass number.

▷ Identify each of the components of radiation that are emitted by the nucleus through natural radioactive decay and describe the basic properties of each. Write out typical equations to illustrate the processes of transmutation by alpha and beta decay.

▷ State, and apply to the solution of related problems, the formula which expresses decay rate as a function of decay constant and number of radioactive nuclei. Also apply the exponential formula which expresses the number of remaining radioactive nuclei left as a function of elapsed time, decay constant, or half-life, and the initial number of nuclei.

▷ Calculate the Q value of given nuclear reactions and determine the threshold energy of endothermic reactions.

SOLUTIONS TO SELECTED END-OF-CHAPTER PROBLEMS

5. (a) Find the speed an alpha particle requires to come within 3.2×10^{-14} m of a stationary gold nucleus. (b) Find the energy of the alpha particle in MeV.

Solution (a) Assuming a head-on collision, the electrostatic potential energy of the alpha particle at the distance of closest approach is equal to its kinetic energy when very far from the nucleus (i.e., at $r \approx \infty$). Thus, we can find the speed of the alpha particle when it is far from the nucleus:

$$\frac{1}{2} m_\alpha v^2 = \frac{k(2e)(Ze)}{r_{\min}} \qquad \text{and} \qquad v = \sqrt{\frac{4kZe^2}{m_\alpha r_{\min}}}$$

For a gold nucleus, $Z = 79$. The mass of an alpha particle is

$$m_\alpha \approx 4 \text{ u} = 4\left(1.66 \times 10^{-27} \text{ kg}\right) = 6.64 \times 10^{-27} \text{ kg}$$

If $r_{\min} = 3.2 \times 10^{-14}$ m, the required speed is then

$$v = \sqrt{\frac{4\left(8.99 \times 10^9 \text{ N} \cdot \text{m}^2/\text{C}^2\right)(79)\left(1.60 \times 10^{-19} \text{ C}\right)^2}{\left(6.64 \times 10^{-27} \text{ kg}\right)\left(3.2 \times 10^{-14} \text{ m}\right)}} = 1.9 \times 10^7 \text{ m/s} \qquad \Diamond$$

(b) The final potential energy, and hence initial kinetic energy, of the alpha particle is

$$KE = \frac{k(2e)(79e)}{r_{\min}} = \frac{158\left(8.99 \times 10^9 \text{ N} \cdot \text{m}^2/\text{C}^2\right)\left(1.6 \times 10^{-19} \text{ C}\right)^2}{3.2 \times 10^{-14} \text{ m}}\left(\frac{1 \text{ MeV}}{1.6 \times 10^{-13} \text{ J}}\right)$$

Solving, we find the initial kinetic energy to be $KE = 7.1$ MeV $\Diamond$

11. A pair of nuclei for which $Z_1 = N_2$ and $Z_2 = N_1$ are called **mirror isobars** (the atomic and neutron numbers are interchanged). Binding energy measurements on such pairs can be used to obtain evidence of the charge independence of nuclear forces. Charge independence means that the proton-proton, proton-neutron, and neutron-neutron forces are approximately equal. Calculate the difference in binding energy for the two mirror nuclei, $^{15}_{8}\text{O}$ and $^{15}_{7}\text{N}$.

Solution The binding energy of a nucleus, $^{A}_{Z}\text{X}$, is given by $E_b = (\Delta m)c^2$ where the mass deficit is

$$\Delta m = \left[Zm_{^1_1\text{H}} + (A-Z)m_n\right] - m_{^A_Z\text{X}}$$

Here, Z is the atomic number (number of protons in the nucleus), $(A-Z)$ is the number of neutrons, m_n is the mass of a neutron, $m_{^1_1\text{H}}$ is the atomic mass of hydrogen, and $m_{^A_Z\text{X}}$ is the atomic mass of this isotope. Atomic masses from Appendix B may be used in this calculation since the masses of the electrons cancel.

For $^{15}_{8}\text{O}$, the mass deficit is

$$\Delta m = 8(1.007\,825\text{ u}) + 7(1.008\,665\text{ u}) - 15.003\,065\text{ u} = 0.120\,190\text{ u}$$

Thus, the binding energy of $^{15}_{8}\text{O}$ is $E_b = 0.120\,190\text{ u}\cdot c^2$. The energy equivalent of the atomic mass unit is $1\text{ u}\cdot c^2 = 931.5\text{ MeV}$, so the binding energy of $^{15}_{8}\text{O}$ becomes

$$E_b = (0.120\,190\text{ u})(931.5\text{ MeV/u}) = 112.0\text{ MeV}$$

For $^{15}_{7}\text{N}$, $\Delta m = 7(1.007\,825\text{ u}) + 8(1.008\,665\text{ u}) - 15.000\,108\text{ u} = 0.123\,987\text{ u}$

In this case, the binding energy is $E_b = (0.123\,987\text{ u})(931.5\text{ MeV/u}) = 115.5\text{ MeV}$

The difference in the binding energies of these mirror isobars is 3.5 MeV ◊

15. The half-life of ^{131}I is 8.04 days. (a) Calculate the decay constant for this isotope. (b) Find the number of ^{131}I nuclei necessary to produce a sample with an activity of 0.50 μCi.

Solution

(a) If there are N_0 radioactive nuclei present at $t = 0$, the number present at time t later is $N = N_0 e^{-\lambda t}$ where λ is the decay constant for this type nucleus. When $N = N_0/2$, the time is $t = T_{1/2}$ (the half-life).

Thus, $1/2 = e^{-\lambda T_{1/2}}$ or $e^{+\lambda T_{1/2}} = 2$. Solving this for the decay constant,

$$\lambda = \frac{\ln 2}{T_{1/2}}$$

Since the half-life of ^{131}I is $T_{1/2} = 8.04$ days, its decay constant is

$$\lambda = \frac{\ln 2}{(8.04 \text{ day})(8.64 \times 10^4 \text{ s}/1 \text{ day})} = 9.98 \times 10^{-7} \text{ s}^{-1} \qquad \Diamond$$

(b) The curie is a unit of activity given by $1 \text{ Ci} = 3.7 \times 10^{10}$ decays/s. If there are N radioactive nuclei present, the decay rate is $R = |\Delta N / \Delta t| = \lambda N$. Thus, if a sample of ^{131}I has a decay rate of $0.50\ \mu\text{Ci}$, the number of nuclei present is

$$N = \frac{R}{\lambda} = \frac{0.50 \times 10^{-6} \text{ Ci}}{9.98 \times 10^{-7} \text{ s}^{-1}} \left(\frac{3.7 \times 10^{10} \text{ decays/s}}{1 \text{ Ci}} \right) = 1.9 \times 10^{10} \qquad \Diamond$$

23. A freshly prepared sample of a certain radioactive isotope has an activity of 10.0 mCi. After 4.00 h, the activity is 8.00 mCi. (a) Find the decay constant and half-life of the isotope. (b) How many atoms of the isotope were contained in the freshly prepared sample? (c) What is the sample's activity 30 h after it is prepared?

Solution

(a) The variation in the activity of a radioactive sample with time is given by $R = R_0 e^{-\lambda t}$ where R_0 is the activity at $t = 0$. Hence, $e^{+\lambda t} = R_0/R$ and the decay constant is

$$\lambda = \frac{\ln(R_0/R)}{t}$$

If $R_0 = 10.0$ mCi and the activity at $t = 4.00$ h is $R = 8.00$ mCi, then the decay consant and the half-life are

$$\lambda = \frac{\ln(10/8)}{4.00\text{ h}} = 5.58\times10^{-2}\text{ h}^{-1} \qquad T_{1/2} = \frac{\ln 2}{\lambda} = \frac{\ln 2}{5.58\times10^{-2}\text{ h}^{-1}} = 12.4\text{ h} \qquad \lozenge$$

(b) The initial activity is related to the number of radioactive atoms in the newly prepared sample by $R_0 = \lambda N_0$. Hence,

$$N_0 = \frac{R_0}{\lambda} = \frac{10.0\times10^{-3}\text{ Ci}}{5.58\times10^{-2}\text{ h}^{-1}}\left(\frac{3.70\times10^{10}\text{ s}^{-1}}{1\text{ Ci}}\right)\left(\frac{3600\text{ s}}{1\text{ h}}\right) = 2.39\times10^{13} \qquad \lozenge$$

(c) At $t = 30$ h, the activity of this sample will be $R = R_0 e^{-\lambda t}$, or

$$R = (10.0\text{ mCi})e^{-(5.58\times10^{-2}\text{ h}^{-1})(30\text{ h})} = 1.9\text{ mCi} \qquad \lozenge$$

27. The mass of ^{56}Fe is 55.9349 u and the mass of ^{56}Co is 55.9399 u. Which isotope decays into the other, and by what process?

Solution Since the parent nucleus is giving up energy, and energy is equivalent to mass, the daughter nucleus must be less massive than the parent. Thus, the cobalt must decay into iron. ◊

To determine the decay process involved, realize that both charge and the total mass number must be conserved in the decay. Since both the parent and daughter nuclei in this decay have same mass number (56), the emitted particle must have a mass number of zero. The parent nucleus, $^{56}_{27}Co$, has 27 protons while the daughter, $^{56}_{26}Fe$, has 26 protons. Hence, the emitted particle must have a charge of $+e$. The particle that fits both of these criteria is the positron e^+. This must be positron decay (also called β^+ or e^+ emission). ◊

The decay equation is $^{56}_{27}Co \rightarrow ^{56}_{26}Fe + e^+ + \nu$. Note that the neutrino, ν, must be produced in this decay to conserve electron-lepton number which is discussed in the next chapter.

---

31. An ^3H nucleus beta decays into ^3He by creating an electron and an antineutrino according to the reaction $^3_1H \rightarrow ^3_2He + e^- + \bar{\nu}$. Use Appendix B to determine the total energy released in this reaction.

Solution The beta decay of a parent nucleus P into a daughter nucleus D is represented by the general decay equation: $^A_ZP \rightarrow {}_{Z+1}^{A}D + {}_{-1}^{0}e + \bar{\nu}$. The Q value (total energy released) in such a decay is given by

$$Q = (\Delta m)c^2 = \left[(\text{mass of parent nucleus}) - (\text{mass of daughter nucleus}) - m_e\right]c^2$$

Here, m_e is the mass of an electron. Note that mass of the anti-neutrino is zero. These **nuclear** masses may be written in terms of neutral atomic masses, m_P and m_D of the parent and daughter respectively as:

$$(\text{mass of parent nucleus}) = m_P - Zm_e$$

and $$(\text{mass of daughter nucleus}) = m_D - (Z+1)m_e$$

The Q value then becomes $Q = [m_P - Zm_e - m_D + (Z+1)m_e - m_e]c^2$ or $Q = [m_P - m_D]c^2$. Observe that the masses of the electrons have all canceled.

For the given decay, $Q = \left[m_{^3_1\text{H}} - m_{^3_2\text{He}}\right]c^2$. From Appendix B, we can then look up the appropriate values, and calculate the total released energy:

$$Q = (3.016\,049\text{ u} - 3.016\,029\text{ u})\left(931.5\ \frac{\text{MeV}}{\text{u}}\right) = 1.863\times10^{-2}\text{ MeV} = 18.63\text{ keV} \quad \Diamond$$

---

39. The first known reaction in which the product nucleus was radioactive (achieved in 1934) was one in which $^{27}_{13}\text{Al}$ was bombarded with alpha particles. Produced in the reaction were a neutron and a product nucleus. (a) What was the product nucleus? (b) Find the Q value of the reaction.

Solution (a) The reaction was $^{27}_{13}\text{Al} + ^4_2\text{He} \rightarrow ^A_Z\text{X} + ^1_0\text{n}$ where ^A_ZX is the product nucleus to be identified. Since charge must be conserved in the reaction, it is necessary that $13+2 = Z+0$, or $Z = 15$. Thus, the product nucleus is an isotope of phosphorus. Requiring that the total mass number be the same before and after the reaction, it is seen that $27+4 = A+1$ or $A = 30$, and the product nucleus is $^{30}_{15}\text{P}$. ◊

(b) The Q value (total energy released) of the reaction is $Q = (\Delta m)c^2$ where $\Delta m = m_{^{27}\text{Al}} + m_{^4\text{He}} - m_{^{30}\text{P}} - m_{\text{n}}$. Hence, using Table 29.4 and Appendix B,

$$Q = (26.981\,538 + 4.002\,602 - 29.978\,310 - 1.008\,665)\text{u}\cdot c^2$$

or $$Q = \left(-2.8368\times10^{-3}\text{ u}\right)(931.5\text{ MeV/u}) = -2.642\text{ MeV} \quad \Diamond$$

---

43. When ^{18}O is struck by a proton, ^{18}F and another particle are produced. (a) What is the other particle? (b) This reaction has a Q value of –2.453 MeV, and the atomic mass of ^{18}O is 17.999 160 u. What is the atomic mass of ^{18}F?

Solution

(a) The reaction equation is $^{18}_{8}\text{O} + ^{1}_{1}\text{H} \rightarrow ^{18}_{9}\text{F} + ^{A}_{Z}\text{X}$ where $^{A}_{Z}\text{X}$ is the unknown particle. Requiring that charge be conserved gives $8+1=9+Z$, or $Z=0$. Thus, the unknown must be neutral. Balancing the mass number on the left and right sides of the reaction equation gives $18+1=18+A$, so $A=1$. Therefore, the unknown particle is neutral and has a mass number of 1. This is a neutron, $^{1}_{0}\text{n}$. ◊

(b) The Q value of a reaction is $Q=(\Delta m)c^2=(\Delta m)(931.5\text{ MeV/u})$. If $Q=-2.453\text{ MeV}$ for this reaction, the mass deficit must be

$$\Delta m = m_{^{18}_{8}\text{O}} + m_{^{1}_{1}\text{H}} - m_{^{18}_{9}\text{F}} - m_{^{1}_{0}\text{n}} = \frac{Q}{931.5\text{ MeV/u}} = \frac{-2.453\text{ MeV}}{931.5\text{ MeV/u}}$$

Solving, $\Delta m = -0.002\,634\text{ u}$.

Using atomic masses from Appendix B,

$$m_{^{18}_{9}\text{F}} = m_{^{18}_{8}\text{O}} + m_{^{1}_{1}\text{H}} - m_{^{1}_{0}\text{n}} + 0.002\,634\text{ u}$$

$$= 17.999\,160\text{ u} + 1.007\,825\text{ u} - 1.008\,665\text{ u} + 0.002\,634\text{ u}$$

This gives the mass of ^{18}F as

$$m_{^{18}_{9}\text{F}} = 18.000\,953\text{ u}$$ ◊

49. A patient swallows a radiopharmaceutical tagged with phosphorus-32 $\left({}^{32}_{15}\text{P}\right)$, an e^- emitter with a half-life of 14.3 days. The average kinetic energy of the emitted electrons is 700 keV. If the initial activity of the sample is 1.31 MBq, determine: (a) the number of electrons emitted in a 10-day period, (b) the total energy deposited in the body during the 10 days, and (c) the absorbed dose if the electrons are completely absorbed in 100 g of tissue.

Solution (a) The decay constant for ${}^{32}\text{P}$ is:

$$\lambda = \frac{\ln 2}{T_{1/2}} = \frac{\ln 2}{14.3 \text{ days}} = 4.85\times 10^{-2} \text{ days}^{-1}$$

Thus, the initial number of ${}^{32}\text{P}$ nuclei present is

$$N_0 = \frac{R_0}{\lambda} = \frac{1.31\times 10^6 \text{ Bq}}{4.85\times 10^{-2} \text{ days}^{-1}}\left(\frac{1 \text{ decay / s}}{1 \text{ Bq}}\right)\left(\frac{86\,400 \text{ s}}{1 \text{ day}}\right) = 2.33\times 10^{12}$$

At time t, the number of ${}^{32}\text{P}$ nuclei remaining in the sample is $N = N_0 e^{-\lambda t}$. Hence, the number of decays that have occurred in the elapsed time is $\Delta N = N_0 - N = N_0\left(1 - e^{-\lambda t}\right)$. The number of decays occurring in this sample during the first 10.0 days is

$$\Delta N = N_0\left[1 - e^{-\lambda(10.0 \text{ days})}\right] = 2.33\times 10^{12}\left[1 - e^{-\left(4.85\times 10^{-2} \text{ days}^{-1}\right)(10.0 \text{ days})}\right]$$

or $\Delta N = 8.96\times 10^{11}$

Since one β^- particle (electron) is emitted per decay, the number of electrons emitted in 10.0 days is 8.96×10^{11}. ◊

(b) Each electron deposits its kinetic energy (an average of 700 keV per electron) in the body. The total energy deposited in the first 10.0 days is

$$E = \Delta N(700 \text{ keV}) = \left(8.96\times 10^{11}\right)(700 \text{ keV})\left(\frac{1.60\times 10^{-16} \text{ J}}{1 \text{ keV}}\right) = 1.00\times 10^{-1} \text{ J} \quad ◊$$

(c) If the electrons are absorbed by 100 g of tissue, the energy deposited per unit mass is:

$$\text{dose} = \frac{E}{m} = \frac{1.00 \times 10^{-1}\ \text{J}}{0.100\ \text{kg}} = 1.00\ \text{J/kg}$$

Since the rad is a dosage unit equal to 10^{-2} J/kg, the absorbed dose in rad units is

$$\text{absorbed dose} = \left(1.00\ \frac{\text{J}}{\text{kg}}\right)\left(\frac{1\ \text{rad}}{10^{-2}\ \text{J/kg}}\right) = 100\ \text{rad} \qquad \Diamond$$

---

55. Deuterons that have been accelerated are used to bombard other deuterium nuclei, resulting in the reaction ${}_1^2\text{H} + {}_1^2\text{H} \rightarrow {}_2^3\text{He} + {}_0^1\text{n}$. Does this reaction require a threshold energy? If so, what is its value?

Solution For an endothermic reaction $(Q < 0)$, the bombarding particle must have some minimum kinetic energy to trigger the reaction. This threshold energy is given by $KE_{\text{min}} = (1 + m/M)|Q|$. Thus, to see whether the reaction requires a threshold energy (i.e., whether it is endothermic), and if so, what that energy is, one must compute the Q value of the reaction. For the given reaction, the mass deficit is

$$\Delta m = m_{{}_1^2\text{H}} + m_{{}_1^2\text{H}} - m_{{}_2^3\text{He}} - m_{{}_0^1\text{n}}$$

Using Appendix B, the Q value is found to be

$$Q = (\Delta m)c^2 = [2(2.014\,102\ \text{u}) - 3.016\,029\ \text{u} - 1.008\,665\ \text{u}](931.5\ \text{MeV/u})$$

or $\quad Q = +3.270\ \text{MeV}$

It is seen that the reaction is exothermic $(Q > 0)$.
Therefore, no threshold energy is required. $\Diamond$

---

59. The theory of nuclear astrophysics proposes that all the heavy elements such as uranium are formed in explosions ending the lives of massive stars. These supernovas release the elements into space. If we assume that at the time of explosion there were equal amounts of ^{235}U and ^{238}U, how long ago were the elements that formed our Earth released, given that the present $^{235}U/^{238}U$ ratio is 0.007? (The half-lives of ^{235}U and ^{238}U are 0.70×10^9 y and 4.47×10^9 y, respectively.)

Solution For simplicity, call the ^{235}U, isotope #1 and the ^{238}U, isotope #2. The assumption is that at the time of the explosion ($t = 0$), the ratio of the number of nuclei of each species was $N_{01}/N_{02} = 1$. At time t, the ratio of numbers of nuclei remaining will be

$$\frac{N_1}{N_2} = \frac{N_{01}e^{-\lambda_1 t}}{N_{02}e^{-\lambda_2 t}} = e^{(\lambda_2 - \lambda_1)t}$$

If the present value of this ratio is $N_1/N_2 = 0.007$, then $e^{(\lambda_2 - \lambda_1)t} = 0.007$ and the elapsed time must be:

$$t = \frac{\ln(0.007)}{(\lambda_2 - \lambda_1)} \qquad \textbf{[Equation 1]}$$

The decay constant of each isotope may be found from $\lambda = \ln 2 / T_{1/2}$ (see the solution of Problem 29.15). Thus,

$$\lambda_1 = \frac{\ln 2}{0.70 \times 10^9 \text{ y}} \qquad \text{and} \qquad \lambda_2 = \frac{\ln 2}{4.47 \times 10^9 \text{ y}}$$

Equation 1 then becomes

$$t = \frac{\ln(0.007)}{\dfrac{\ln 2}{4.47 \times 10^9 \text{ y}} - \dfrac{\ln 2}{0.70 \times 10^9 \text{ y}}} = \left[\frac{(4.47 \times 10^9 \text{ y})(0.70 \times 10^9 \text{ y})}{0.70 \times 10^9 \text{ y} - 4.47 \times 10^9 \text{ y}}\right]\frac{\ln(0.007)}{\ln 2}$$

This gives the estimated age of the Earth as: $t = 5.9 \times 10^9$ y ◊

65. A fission reactor is hit by a nuclear weapon, causing 5.0×10^6 Ci of ^{90}Sr ($T_{1/2} = 28.7$ years) to evaporate into the air. The ^{90}Sr falls out over an area of 10^4 km^2. How long will it take the activity of the ^{90}Sr to reach the agriculturally "safe" level of 2.0 μCi/m^2?

Solution

The original activity per unit area is

$$R_0 = \frac{5.0 \times 10^6 \text{ Ci}}{10^4 \text{ km}^2}\left(\frac{1 \text{ km}}{10^3 \text{ m}}\right)^2 = 5.0 \times 10^{-4} \text{ Ci/m}^2$$

After time t has elapsed, the remaining activity will be $R = R_0 e^{-\lambda t}$. The decay constant for ^{90}Sr is found from $\lambda = \ln 2 / T_{1/2} = \ln 2 / 28.7$ y. The time when the activity reaches the "safe" level of $R = 2.0\ \mu\text{Ci/m}^2$ may then be found from $e^{\lambda t} = R_0/R$ or

$$t = \frac{\ln(R_0/R)}{\lambda} = \frac{\ln\left[\left(5.0 \times 10^{-4} \text{ Ci/m}^2\right)/\left(2.0 \times 10^{-6} \text{ Ci/m}^2\right)\right]}{(\ln 2)/28.7 \text{ y}}$$

This reduces to $t = (28.7 \text{ y})\dfrac{\ln\left(2.5 \times 10^2\right)}{\ln 2} = 2.3 \times 10^2 \text{ y}$ ◊

CHAPTER SELF-QUIZ

1. A radioactive material initially is observed to have an activity of 800 counts/sec. If four hours later it is observed to have an activity of 200 counts/sec, what is its half-life?
   a. 1 hour
   b. 2 hours
   c. 4 hours
   d. 8 hours

2. Radium-226 decays to Radon-222 by emitting which of the following?
   a. beta
   b. alpha
   c. gamma
   d. positron

3. An endothermic nuclear reaction occurs as a result of the collision of two reactant nuclei. If the Q value of this reaction is –2.17 MeV, which of the following describes the minimum kinetic energy needed in the reactant nuclei if the reaction is to occur?
   a. equal to 2.17 MeV
   b. greater than 2.17 MeV
   c. less than 2.17 MeV
   d. exactly half of 2.17 MeV

4. The ratio of the numbers of neutrons to protons in the nucleus of naturally occurring isotopes tends to vary with atomic number in what manner?
   a. increases with greater atomic number
   b. decreases with greater atomic number
   c. is maximum for atomic number = 60
   d. remains constant for entire range of atomic numbers

5. Electron emission in the beta decay process results in the daughter nucleus differing in what manner from the parent?
   a. atomic mass increases by one
   b. atomic number decreases by two
   c. atomic number increases by one
   d. atomic mass decreases by two

6. What energy must be added or given off in a reaction where one hydrogen atom and two neutrons are combined to form a tritium atom? (atomic masses for each: hydrogen, 1.007 825 u; neutron, 1.008 665 u; tritium, 3.016 049 u; also, $1\ \text{u} \times c^2 = 931.5\ \text{MeV}$)
   a. 8.48 MeV added
   b. 8.48 MeV given off
   c. 10.3 MeV given off
   d. 10.3 MeV added

7. Tritium has a half-life of 12.3 years. How many years will elapse when the radioactivity of a tritium sample diminishes to 10% of its original value?
   a. 31 years
   b. 41 years
   c. 84 years
   d. 123 years

8. If there are 128 neutrons in Pb-210, how many neutrons are found in the nucleus of Pb-206?
   a. 122
   b. 124
   c. 126
   d. 130

9. What particle is emitted when P-32 decays to S-32? (Atomic numbers of P and S are, respectively, 15 and 16.)
   a. alpha
   b. electron
   c. positron
   d. gamma quantum

10. The binding energy of a nucleus is equal to
    a. the energy needed to remove one of the nucleons
    b. the average energy with which any nucleon is bound in the nucleus
    c. the energy needed to separate all the nucleons from each other
    d. the mass of the nucleus times c^2.

11. The half-life of radioactive iodine-137 is 8 days. Find the number of ^{131}I nuclei necessary to produce a sample of activity 1.0 μCi.
    (1 Curie = 3.70×10^{10} decays/second)
    a. 4.6×10^9
    b. 3.7×10^{10}
    c. 7.6×10^{12}
    d. 8.1×10^{13}

12. What is the Q-value for the reaction $^9\text{Be} + \alpha \rightarrow {}^{12}\text{C} + \text{n}$?
    ($m_\alpha = 4.0026$ u, $m_{\text{Be}} = 9.01218$ u, $m_\text{C} = 12.0000$ u, $m_n = 1.008\,665$ u, and 1 u = 931.5 MeV/c^2)
    a. 8.40 MeV
    b. 7.30 MeV
    c. 6.20 MeV
    d. 5.70 MeV

Chapter 30

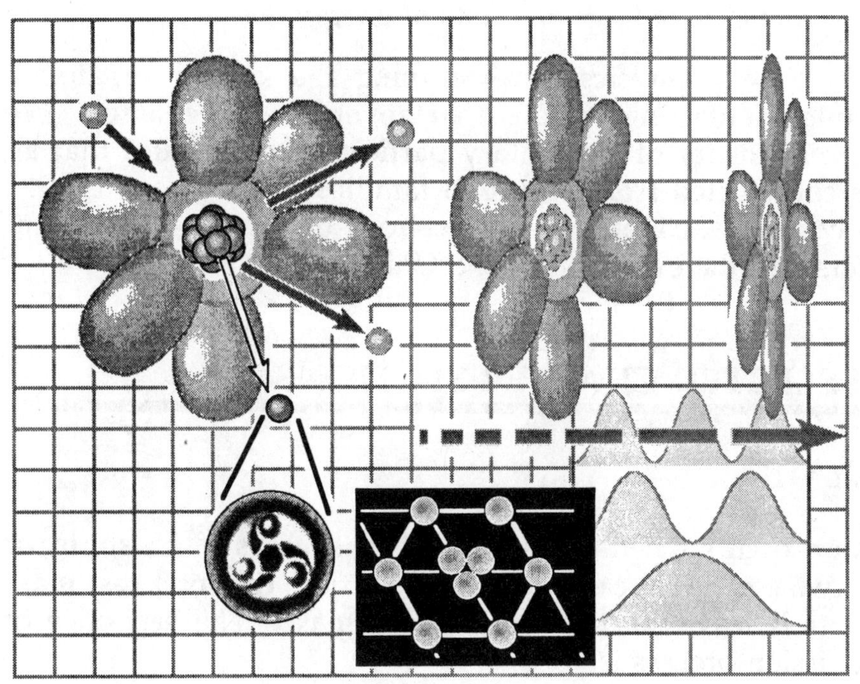

NUCLEAR ENERGY AND ELEMENTARY PARTICLES

Chapter 30

NUCLEAR ENERGY AND ELEMENTARY PARTICLES

In this concluding chapter we discuss the two means by which energy can be derived from nuclear reactions. These two techniques are fission, in which a nucleus of large mass number splits, or fissions, into two smaller nuclei; and fusion, in which two light nuclei fuse to form a heavier nucleus. In either case, there is a release of large amounts of energy.

We end our study of physics by examining the known subatomic particles and the fundamental interactions that govern their behavior. We also discuss the current theory of elementary particles, which states that all matter in nature is constructed from only two families of particles: quarks and leptons. Finally, we describe how clarifications of such models might help scientists understand the evolution of the Universe.

NOTES FROM SELECTED CHAPTER SECTIONS

30.1 Nuclear Fission

Nuclear fission occurs when a heavy nucleus, such as ^{235}U, splits, or fissions, into two smaller nuclei. In such a reaction, **the total rest mass of the products is less than the original rest mass.** The sequence of events in the fission process is:

1. The ^{235}U nucleus captures a thermal (slow-moving) neutron.
2. This capture results in the formation of $^{236}U^*$, and the excess energy of this nucleus causes it to undergo violent oscillations.
3. The $^{236}U^*$ nucleus becomes highly distorted, and the force of repulsion between protons in the two halves of the dumbbell shape tends to increase the distortion.
4. The nucleus splits into two fragments, emitting several neutrons in the process.

Chapter 30

30.2 Nuclear Reactors

A nuclear reactor is a system designed to maintain a **self-sustained chain reaction.**

The **reproduction constant** K is defined as the average number of neutrons released from each fission event that will cause another event. In a power reactor, it is necessary to maintain a value of K close to 1. Under this condition, the reactor is said to be **critical.**

Important factors and processes relative to the design and operation of a nuclear reactor include: neutron leakage, regulating neutron energies, neutron capture by nonfissioning nuclei, control of the power level, and provisions to ensure reactor safety.

30.3 Nuclear Fusion

Nuclear fusion is a process in which two light nuclei combine to form a heavier nucleus. A great deal of energy is released in such a process. The major obstacle in obtaining useful energy from fusion is the large Coulomb repulsive force between the charged nuclei at close separations. Sufficient energy must be supplied to the particles to overcome this Coulomb barrier and thereby enable the nuclear attractive force to take over.

The temperature at which the power generation exceeds the loss rate is called the **critical ignition temperature.** The **confinement time** is the time the interacting ions are maintained at a temperature equal to or greater than the temperature required for the reaction to proceed successfully.

30.5 The Fundamental Forces in Nature

There are four fundamental forces in nature: **strong** (hadronic), **electromagnetic, weak,** and **gravitational.**

- The **strong force** is responsible for the binding of neutrons and protons into nuclei. It is very short-ranged and is negligible for separations greater than the approximate size of the nucleus.

- The **electromagnetic force** is responsible for the binding of atoms and molecules. It is a long-range force that decreases in strength as the inverse square of the separation between interacting particles.

- The **weak force** is a short-range nuclear force that tends to produce instability in certain nuclei.

- The **gravitational force** is the weakest of all the fundamental forces. It is a long-range force that holds the planets, stars, and galaxies together and its effect on the elementary particles is negligible.

The electromagnetic and weak forces are now considered to be manifestations of a single force called the **electroweak** force.

The fundamental forces are described in terms of particle or quanta exchanges which **mediate** the forces. The electromagnetic force is mediated by photons, which are the quanta of the electromagnetic field. Likewise, the strong force is mediated by field particles called **gluons,** the weak force is mediated by particles called the W and Z **bosons,** and the gravitational force is mediated by quanta of the gravitational field called **gravitons.**

30.6 Positrons and Other Antiparticles

An antiparticle and a particle have the same mass, but opposite charge. Furthermore, other properties may have opposite values such as lepton number and baryon number. It is possible to produce particle-antiparticle pairs in nuclear reactions if the available energy is greater than $2mc^2$, where m is the mass of the particle (or antiparticle).

Pair production is a process in which a gamma ray with an energy of at least 1.02 MeV interacts with a nucleus and an electron-positron pair is created.

Pair annihilation is an event in which an electron and a positron can annihilate to produce two gamma rays, each with an energy of at least 0.511 MeV.

30.7 Mesons and the Beginning of Particle Physics

The interaction between two particles can be represented in a diagram called a **Feynman diagram** (See Figure 30.7 in the textbook.) The strong force is mediated by pions, while the electromagnetic force is mediated by photons. The graviton, which is the mediator of the gravitational force, has yet to be observed. The W and Z particles mediate the weak force.

30.8 Classification of Particles

All particles (other than photons) can be classified into two categories: **hadrons,** which interact through the strong force, and **leptons,** which participate in the weak interaction.

There are two classes of hadrons: **mesons** and **baryons,** which are grouped according to their masses and spins. It is believed that hadrons are composed of units called **quarks** which are more fundamental in nature.

Leptons have no structure or size and are therefore considered to be truly elementary particles.

30.9 Conservation Laws

In all reactions and decays, quantities such as energy, linear momentum, angular momentum, electric charge, baryon number, and lepton number are strictly conserved. Certain particles have properties called **strangeness** and **charm**. These unusual properties are conserved only in those reactions and decays that occur via the strong force.

Whenever a nuclear reaction or decay occurs, the sum of the baryon numbers before the process must equal the sum of the baryon numbers after the process.

The sum of the electron-lepton numbers before a reaction or decay must equal the sum of the electron-lepton numbers after the reaction or decay. Similar laws apply to muon-lepton and tau-lepton numbers.

30.10 Strange Particles and Strangeness

Strange particles are always produced in pairs by the strong interaction, and decay very slowly — a characteristic of the weak interaction.

Whenever a nuclear reaction or decay occurs, the sum of the strangeness numbers before the process must equal the sum of the strangeness numbers after the process.

30.12 Quarks

Recent theories in elementary particle physics have postulated that all hadrons are composed of smaller units known as **quarks.** Quarks have a fractional electric charge and a baryon number of 1/3. There are six flavors of quarks, up (u), down (d), strange (s), charmed (c), top (t), and bottom (b). All baryons contain three quarks, while all mesons contain one quark and one antiquark.

According to the theory of **quantum chromadynamics,** quarks have a property called **color,** and the strong force between quarks is referred to as the **color force.**

EQUATIONS AND CONCEPTS

The fission of an uranium nucleus by bombardment with a low energy neutron results in the production of fission fragments and typically two or three neutrons. The energy released in the fission event appears in the form of kinetic energy of the fission fragments and the neutrons.

$$ {}_{0}^{1}n + {}_{92}^{235}U \rightarrow {}_{92}^{236}U^{*} \rightarrow X + Y + \text{neutrons} \qquad (30.1)$$

In nuclear fission,

1. The ^{235}U nucleus captures a thermal (slow-moving) neutron.
2. This capture results in the formation of $^{236}U^*$, and the excess energy of this nucleus causes it to violently oscillate.
3. The nucleus becomes highly distorted, and the force of repulsion between protons in the two halves of the dumbell shape tends to increase the distortion.
4. The nucleus splits into two fragments, emitting several neutrons in the process.

Comment on the sequence of events in the fission process.

These fusion reactions seem to be most likely to be used as the basis of the design and operation of a fusion power reactor. The Q values refer to the energy released from each reaction.

$$ {}^{2}_{1}\text{H} + {}^{2}_{1}\text{H} \rightarrow {}^{3}_{2}\text{He} + {}^{1}_{0}\text{n} \quad (Q = 3.27\text{ MeV}) \tag{30.4}$$

$$ {}^{2}_{1}\text{H} + {}^{2}_{1}\text{H} \rightarrow {}^{3}_{1}\text{H} + {}^{1}_{1}\text{H} \quad (Q = 4.03\text{ MeV})$$

$$ {}^{2}_{1}\text{H} + {}^{3}_{1}\text{H} \rightarrow {}^{4}_{2}\text{He} + {}^{1}_{0}\text{n} \quad (Q = 17.59\text{ MeV})$$

Lawson's criterion states the conditions under which a net power output of a fusion reactor is possible. In these expressions, n is the plasma density (number of ions per cubic cm) and τ is the plasma confinement time (the time during which the interacting ions are maintained at a temperature equal to or greater than that required for the reaction to proceed).

$$n\tau \geq 10^{14}\text{ s/cm}^3 \quad (\text{D} - \text{T interaction}) \tag{30.5}$$

$$n\tau \geq 10^{16}\text{ s/cm}^3 \quad (\text{D - D interaction})$$

Pions and muons are very unstable particles. Shown is the sequence of decays for π^-.

$$\pi^- \rightarrow \mu^- + \bar{\nu} \tag{30.6}$$

$$\mu^- \rightarrow \text{e}^- + \nu + \bar{\nu}$$

REVIEW CHECKLIST

- ▷ Write an equation which represents a typical fission event and describe the sequence of events which occurs during the fission process. Use data obtained from the binding energy curve to estimate the disintegration energy of a typical fission event.
- ▷ Describe the basis of energy release in fusion and write out several nuclear reactions which might be used in a fusion-powered reactor.
- ▷ Outline the classification of elementary particles, and mention several characteristics of each group.
- ▷ Know the broad classification of particles and the characteristic properties of the several classes (relative mass value, spin, decay mode).
- ▷ Determine whether or not a suggested decay can occur based on the conservation of baryon number and the conservation of lepton number. Determine whether or not a predicted reaction/decay will occur based on the conservation of strangeness for the strong and electromagnetic interactions.

SOLUTIONS TO SELECTED END-OF-CHAPTER PROBLEMS

3. Find the energy released in the following fission reaction:

$$ {}_0^1\text{n} + {}_{92}^{235}\text{U} \rightarrow {}_{38}^{88}\text{Sr} + {}_{54}^{136}\text{Xe} + 12\,{}_0^1\text{n} $$

Solution The disintegration energy, Q, released in this fission of a ^{235}U nucleus is $Q = \left[\left(m_{^{235}\text{U}} + m_{\text{n}}\right) - \left(m_{^{88}\text{Sr}} + m_{^{136}\text{Xe}} + 12m_{\text{n}}\right)\right]c^2$. Using the atomic masses given in Appendix B, this becomes

$$ Q = \left[235.043\,924\text{ u} - 87.905\,618\text{ u} - 135.907\,215\text{ u} - 11(1.008\,665\text{ u})\right]c^2 $$

or $$ Q = (0.135\,776\text{ u})(931.5\text{ Mev/u}) $$

The energy released in this fission reaction is: $Q = 126.5\text{ MeV}$ ◊

7. An all-electric home uses approximately 2000 kWh of electric energy per month. How much ^{235}U would be required to provide this house with its energy needs for one year? (Assume 100% conversion efficiency and 208 MeV released per fission.)

Solution

The total energy required to meet the energy needs of this house for one year is

$$E = \left(2000\ \frac{\text{kWh}}{\text{month}}\right)\left(3.60\times10^{6}\ \frac{\text{J}}{\text{kWh}}\right)\left(12\ \frac{\text{months}}{\text{y}}\right) = 8.64\times10^{10}\ \text{J/y}$$

Thus, assuming 100% conversion efficiency and 208 Mev released per fission, the number of fission events required per year is

$$N = \frac{E}{208\ \text{Mev}} = \frac{8.64\times10^{10}\ \text{J/y}}{208\ \text{Mev}}\left(\frac{1\ \text{MeV}}{1.60\times10^{-13}\ \text{J}}\right) = 2.60\times10^{21}\ \text{y}^{-1}$$

The number of moles of ^{235}U that will be required each year is then

$$n = \frac{N}{\text{Avogadro's number}} = \frac{2.60\times10^{21}\ \text{atoms/y}}{6.02\times10^{23}\ \text{atoms/mol}} = 4.31\times10^{-3}\ \text{mol/y}$$

Since 1 mole of ^{235}U has a mass of 235 g, the mass of ^{235}U required to supply the energy needs of this home for one year is

$$m = nM = \left(4.31\times10^{-3}\ \text{mol/y}\right)\left(235\ \text{g/mol}\right) = 1.01\ \text{g}$$ ◊

9. Suppose that the water exerts an average frictional drag of 1.0×10^5 N on a nuclear-powered ship. How far can the ship travel per kilogram of fuel if the fuel consists of enriched uranium containing 1.7% of the fissionable isotope ^{235}U, and the ship's engine has a efficiency of 20%? (Assume 208 MeV released per fission event.)

Solution

The mass of fissionable material in 1 kilogram (or 1000 grams) of the enriched fuel is

$$m = 0.017(1000 \text{ g}) = 17 \text{ g}$$

The number of atoms (and hence nuclei) in this quantity of ^{235}U is $N = N_A(m/M)$, where N_A is Avogadro's number and M is the molecular weight of ^{235}U. Thus,

$$N = \left(6.02 \times 10^{23} \text{ gmol}^{-1}\right)\left(\frac{17 \text{ g}}{235 \text{ g/gmol}}\right) = 4.4 \times 10^{22}$$

The energy released when this number of ^{235}U nuclei undergo fission, with a release of an average of 208 MeV per fission event, is

$$E = N(208 \text{ MeV}) = \left(4.4 \times 10^{22}\right)(208 \text{ MeV}) = 9.1 \times 10^{24} \text{ MeV}$$

or $$E = \left(9.1 \times 10^{24} \text{ MeV}\right)\left(\frac{1.6 \times 10^{-13} \text{ J}}{1 \text{ MeV}}\right) = 1.5 \times 10^{12} \text{ J}$$

If the engine has an efficiency of 20%, the useful work output is $W = (0.20)E$, or

$$W = (0.20)\left(1.5 \times 10^{12} \text{ J}\right) = 3.0 \times 10^{11} \text{ J}$$

The applied force required to overcome the frictional drag is $F = 1.0 \times 10^5$ N. From the definition of work ($W = FS\cos\theta$ with $\theta = 0°$ in this case), the distance this quantity of work output from the engine can move the ship is seen to be

$$S = \frac{W}{F} = \frac{3.0 \times 10^{11}\ \text{J}}{1.0 \times 10^{5}\ \text{N}} = 3.0 \times 10^{6}\ \text{m} = 3.0 \times 10^{3}\ \text{km}$$

or
$$S = 3.0 \times 10^{3}\ \text{km}\left(\frac{1\ \text{mi}}{1.609\ \text{km}}\right) = 1.9 \times 10^{3}\ \text{mi}$$ ◊

13. If an all-electric home uses approximately 2000 kWh of electric energy per month, how many fusion events described by the reaction ${}^{2}_{1}\text{H} + {}^{3}_{1}\text{H} \rightarrow {}^{4}_{2}\text{He} + {}^{1}_{0}\text{n}$ would be required to keep this house running for one year?

Solution In each fusion event, the energy released is

$$Q = \left[m_{{}^{2}_{1}\text{H}} + m_{{}^{3}_{1}\text{H}} - m_{{}^{4}_{2}\text{He}} - m_{{}^{1}_{0}\text{n}}\right]c^2$$

$$Q = [2.014\,102\ \text{u} + 3.016\,049\ \text{u} - 4.002\,602\ \text{u} - 1.008\,665\ \text{u}]c^2$$

or
$$Q = (0.018\,884\ \text{u})\left(\frac{931.5\ \text{MeV}}{1\ \text{u}}\right) = 17.59\ \text{MeV}\left(\frac{1.60 \times 10^{-13}\ \text{J}}{1\ \text{MeV}}\right) = 2.81 \times 10^{-12}\ \text{J}$$

The total energy required to keep the house running for one year is

$$E = \left(2000\ \frac{\text{kWh}}{\text{month}}\right)\left(3.60 \times 10^{6}\ \frac{\text{J}}{\text{kWh}}\right)\left(12\ \frac{\text{months}}{\text{y}}\right) = 8.64 \times 10^{10}\ \text{J/y}$$

Thus, the total number of fusion events required is

$$n = \frac{E}{Q} = \frac{8.64 \times 10^{10}\ \text{J/y}}{2.81 \times 10^{-12}\ \text{J/event}} = 3.07 \times 10^{22}\ \text{events/y}$$ ◊

17. A photon with an energy of 2.09 GeV creates a proton-antiproton pair in which the proton has a kinetic energy of 95.0 MeV. What is the kinetic energy of the antiproton?

Solution

The total energy of each particle created is the sum of its rest energy and its kinetic energy. Requiring the total energy after this pair production event equal the total energy before,

$$\left(E_{Rp} + KE_p\right) + \left(E_{R\bar{p}} + KE_{\bar{p}}\right) = E_\gamma$$

or the kinetic energy of the antiproton is

$$KE_{\bar{p}} = E_\gamma - E_{R\bar{p}} - E_{Rp} - KE_p$$

The energy of the photon is given as $E_\gamma = 2.09 \text{ GeV} = 2.09 \times 10^3 \text{ MeV}$. Looking at Table 30.2, we see that the rest energy of both the proton and the antiproton is

$$E_{Rp} = E_{R\bar{p}} = m_p c^2 = 938.3 \text{ MeV}$$

If the kinetic energy of the proton is observed to be 95.0 MeV, the kinetic energy of the antiproton is

$$KE_{\bar{p}} = 2.09 \times 10^3 \text{ MeV} - 2(938.3 \text{ MeV}) - 95.0 \text{ MeV}$$

or $$KE_{\bar{p}} = 118 \text{ MeV}$$ ◊

19. One of the mediators of the weak interaction is the Z^0 boson, whose mass is 96 GeV/c^2. Use this information to find an approximate value for the range of the weak interaction.

Solution

If the Z^0 boson mediates the weak interaction (i.e., the weak interaction between particles consists of the exchange of a virtual Z^0 boson), the uncertainty in the energy of the exchange particle cannot be less than the rest energy of the Z^0 boson. Otherwise, the weak interaction would be found to violate the principle of conservation of energy.

The range of the weak interaction can then be approximated by determining the distance the exchange particle (a virtual Z^0 boson) could travel in its maximum lifetime. From the uncertainty relation $\Delta E \Delta t \approx \hbar$, this maximum lifetime is

$$\Delta t \approx \frac{\hbar}{\Delta E} = \frac{\hbar}{E_R}$$

where the rest energy of the Z^0 boson is $E_R = mc^2 = 96\ \text{GeV}$. Thus,

$$\Delta t \approx \frac{1.05\times10^{-34}\ \text{J}\cdot\text{s}}{96\ \text{GeV}}\left[\frac{1\ \text{GeV}}{1.60\times10^{-10}\ \text{J}}\right] \quad \text{or} \quad \Delta t \approx 6.8\times10^{-27}\ \text{s}$$

The maximum distance the exchange particle could travel in this time (i.e., the approximate value for the range of the interaction) is

$$d \approx c(\Delta t) = \left(3.0\times10^{8}\ \text{m/s}\right)\left(6.8\times10^{-27}\ \text{s}\right) = 2.0\times10^{-18}\ \text{m} \qquad \lozenge$$

---

23. Identify the unknown particle on the left side of the reaction

$$? + p \rightarrow n + \mu^+.$$

Solution The proton, neutron and μ^+ are hadrons (i.e., particles which interact through the strong interaction). The strong interaction always conserves several quantities, namely charge, Q; baryon number, B; electron-lepton number, L_e; muon-lepton number, L_μ; tau-lepton number, L_τ; and strangeness, S. These conservation laws may be used to identify the unknown particle, X, in the reaction $X + p \rightarrow n + \mu^+$. The table below summarizes the properties of the particles present before and after the reaction (see Table 30.2 in the textbook). Note that the property values of an antiparticle are the negative of those for the corresponding particle.

	Particles before reaction			Particles after reaction		
Property	Particle X	Proton	Total	Neutron	Antimuon	Total
Q	Q_X	+e	Q_X+e	0	+e	+e
B	B_X	+1	B_X+1	+1	0	+1
L_e	L_{eX}	0	L_{eX}	0	0	0
L_μ	$L_{\mu X}$	0	$L_{\mu X}$	0	−1	−1
L_τ	$L_{\tau X}$	0	$L_{\tau X}$	0	0	0
S	S_X	0	S_X	0	0	0

If the reaction is to conserve each of the properties listed, the entries in the total column before reaction must equal the entries in the total column after reaction. Thus, $Q_X + e = +e$, $B_X + 1 = +1$, $L_{eX} = 0$, $L_{\mu X} = -1$, $L_{\tau X} = 0$, and $S_X = 0$. The property values for the unknown particle are therefore,

$$Q_X = 0, \quad B_X = 0, \quad L_{eX} = 0, \quad L_{\mu X} = -1, \quad L_{\tau X} = 0, \quad \text{and} \quad S_X = 0.$$

The particle with these properties is $\bar{\nu}_\mu$, the anti-muon-neutrino. ◊

32. Fill in the missing particle. Assume that (a) occurs via the strong interaction while (b) and (c) involve the weak interaction.

(a) $K^+ + p \rightarrow\ ?\ + p$ (b) $\Omega^- \rightarrow\ ?\ + \pi^-$ (c) $K^+ \rightarrow\ ?\ + \mu^+ + \nu_\mu$

Solution To identify the missing particle in each reaction or decay, one must consider several conservation laws: total charge, Q; total baryon number, B; total lepton numbers (electron-lepton number, L_e, muon-lepton number, L_μ, and tau-lepton number, L_τ); and strangeness, S. The strong interaction conserves all of these properties, including strangeness $(\Delta S = 0)$. The weak interaction may violate conservation of strangeness, but never by more than one unit. Thus, $\Delta S = 0, \pm 1$ for weak interactions.

(a) Consider the reaction $K^+ + p \rightarrow X + p$. If it occurs via the strong interaction, each property listed above is conserved. This is summarized in the following table:

Q	$+e + e \rightarrow Q_X + e$	$\Delta Q = 0 \quad \Rightarrow \quad Q_X = +e$
B	$0 + 1 \rightarrow B_X + 1$	$\Delta B = 0 \quad \Rightarrow \quad B_X = 0$
L_e	$0 + 0 \rightarrow L_{eX} + 0$	$\Delta L_e = 0 \quad \Rightarrow \quad L_{eX} = 0$
L_μ	$0 + 0 \rightarrow L_{\mu X} + 0$	$\Delta L_\mu = 0 \quad \Rightarrow \quad L_{\mu X} = 0$
L_τ	$0 + 0 \rightarrow L_{\tau X} + 0$	$\Delta L_\tau = 0 \quad \Rightarrow \quad L_{\tau X} = 0$
S	$+1 + 0 \rightarrow S_X + 0$	$\Delta S = 0 \quad \Rightarrow \quad S_X = +1$

Thus, the unknown particle in this reaction must be a non-baryon $(B = 0)$, a non-lepton $\left(L_e = L_\mu = L_\tau = 0\right)$, with charge $Q = +e$, and strangeness $S = +1$. From Table 30.2, the particle with these properties is the Kaon, K^+. This must be an elastic scattering process. ◊

(b) The application of the various conservation laws to the decay $\Omega^- \to X + \pi^-$ (via the weak interaction) is summarized as:

Q	$-e \to Q_X - e$	$\Delta Q = 0 \Rightarrow Q_X = 0$
B	$+1 \to B_X + 0$	$\Delta B = 0 \Rightarrow B_X = +1$
L_e	$0 \to L_{eX} + 0$	$\Delta L_e = 0 \Rightarrow L_{eX} = 0$
L_μ	$0 \to L_{\mu X} + 0$	$\Delta L_\mu = 0 \Rightarrow L_{\mu X} = 0$
L_τ	$0 \to L_{\tau X} + 0$	$\Delta L_\tau = 0 \Rightarrow L_{\tau X} = 0$
S	$-3 \to S_X + 0$	$\Delta S = 0, \pm 1 \Rightarrow S_X = -4, -3, -2$

Therefore, the unknown particle must be a neutral baryon having strangeness of -4, -3, or -2. From Table 30.2, the only known particle meeting these criteria is the neutral xi (or cascade) particle, Ξ^0. ◊

(c) Assuming that the kaon decay, $K^+ \to X + \mu^+ + \nu_\mu$, involves the weak interaction, the observance of the conservation laws is summarized as:

Q	$+e \to Q_X + e + 0$	$\Delta Q = 0 \Rightarrow Q_X = 0$
B	$0 \to B_X + 0 + 0$	$\Delta B = 0 \Rightarrow B_X = 0$
L_e	$0 \to L_{eX} + 0 + 0$	$\Delta L_e = 0 \Rightarrow L_{eX} = 0$
L_μ	$0 \to L_{\mu X} - 1 + 1$	$\Delta L_\mu = 0 \Rightarrow L_{\mu X} = 0$
L_τ	$0 \to L_{\tau X} + 0 + 0$	$\Delta L_\tau = 0 \Rightarrow L_{\tau X} = 0$
S	$+1 \to S_X + 0 + 0$	$\Delta S = 0, \pm 1 \Rightarrow S_X = 0, +1, +2$

Thus, the unknown particle must be a neutral meson (i.e., non-baryon and non-lepton) with a strangeness of $S = 0, +1$, or $+2$. The only particle in Table 30.2 which has these properties and also has a mass less than that of the K^+ particle is the neutral pion, π^0. ◊

38. Neglect binding energies and estimate the mass of the *u* and *d* quarks from the mass of the proton and neutron.

Solution

The quark composition of the proton is *uud* while that of the neutron is *udd* (see Table 30.4 from the textbook). Therefore, if binding energies are neglected, the masses of the proton and the neutron should be given in terms of the quark masses as:

$$m_p = 2m_u + m_d \qquad \textbf{[Equation 1]}$$

and

$$m_n = m_u + 2m_d \qquad \textbf{[Equation 2]}$$

Combining Equations 1 and 2, it is observed that

$$2m_p - m_n = 4m_u + 2m_d - m_u - 2m_d = 3m_u$$

Thus, the estimate for the mass of the *up* quark is

$$m_u = \frac{2m_p - m_n}{3} = \frac{2\left(938.3 \text{ MeV}/c^2\right) - 939.6 \text{ MeV}/c^2}{3} = 312.3 \text{ MeV}/c^2 \ \lozenge$$

Then, from Equation 1, the estimate of the mass of the *down* quark is

$$m_d = m_p - 2m_u = 938.3 \text{ MeV}/c^2 - 2\left(312.3 \text{ MeV}/c^2\right) = 313.7 \text{ MeV}/c^2 \ \lozenge$$

The difference between these estimates and the values given in Table 30.5 of the textbook is due, at least in part, to neglect of the binding energies of the proton and neutron.

---

41. A K^0 particle at rest decays into a π^+ and a π^-. What will be the speed of each of the pions? The mass of the K^0 is 497.7 MeV/c^2 and the mass of each pion is 139.6 MeV/c^2.

Solution

Since the K^0 particle was at rest before decay, the total linear momentum both before and after the decay is zero. Thus, the two pions must travel in opposite directions with equal magnitude momenta. Since the two pions have the same momentum and the same mass, their total energies must also be equal. Hence, conservation of energy gives $(E_R)_{K^0} = E_{\pi^+} + E_{\pi^-} = 2E_{\pi^\pm}$, and the total energy of either pion must be

$$E_{\pi^\pm} = \frac{(E_R)_{K^0}}{2} = \frac{497.7 \text{ MeV}}{2} = 248.9 \text{ MeV}$$

Observe that this result is much larger than the rest energy of a pion. Hence, relativistic effects must be considered. The total energy is

$$E = \gamma mc^2 = \gamma E_R \qquad \text{or} \qquad \frac{E_R}{E} = \frac{1}{\gamma}$$

Thus,

$$\frac{1}{\gamma} = \sqrt{1-(v/c)^2} = \frac{(E_R)_{\pi^\pm}}{E_{\pi^\pm}} = \frac{139.6 \text{ MeV}}{248.9 \text{ MeV}} = 0.5609$$

and

$$(v/c)^2 = 1 - (0.5609)^2 = 0.6854$$

This yields $v = 0.8279\,c$ as the speed of each pion after the decay. ◊

45. Each of the following decays is forbidden. For each process, determine a conservation law that is violated. (a) $\mu^- \to e^- + \gamma$ (b) $n \to p + e^- + \nu_e$ (c) $\Lambda^0 \to p + \pi^0$ (d) $p \to e^+ + \pi^0$ (e) $\Xi^0 \to n + \pi^0$

Solution (a) The first decay, $\mu^- \to e^- + \gamma$, violates conservation of both electron-lepton number and muon-lepton number as seen below:

L_e: $0 \to +1 + 0 \neq 0$ (thus, $\Delta L_e \neq 0$)
L_μ: $+1 \to 0 + 0 \neq +1$ (also, $\Delta L_\mu \neq 0$)

Since all interactions must conserve both of these quantities, this decay is forbidden. ◊

(b) The decay $n \to p + e^- + \nu_e$ violates conservation of electron-lepton number as seen below, and hence is forbidden.

L_e: $0 \to 0 + 1 + 1 \neq 0$ (Therefore, $\Delta L_e \neq 0$) ◊

(c) In the decay $\Lambda^0 \to p + \pi^0$, the total charge is zero before decay and +e after decay. Since charge must be conserved, this decay is forbidden. ◊

(d) As seen below, the decay $p \to e^+ + \pi^0$ does not conserve baryon number.

B: $+1 \to 0 + 0 \neq +1$ ($\Delta B \neq 0$, so this decay is forbidden) ◊

(e) The decay $\Xi^0 \to n + \pi^0$ violates the conservation of strangeness as shown below:

S: $-2 \to 0 + 0 \neq -2$ ($\Delta S = -2$)

While the weak interaction can violate the conservation of strangeness, it never violates it by two units (i.e., $\Delta S = 0, \pm 1$ for the weak interaction). All other interactions conserve strangeness. Thus, this decay is forbidden. ◊

49. If baryon number is not conserved, then one possible mechanism by which a proton can decay is

$$p \rightarrow e^{+} + \gamma$$

(a) Show that this reaction violates conservation of baryon number. (b) Assuming that this reaction occurs, and that the proton is initially at rest, determine the energy and momentum of the photon after the reaction. (**Hint:** Recall that energy and momentum must be conserved in the reaction.) (c) Determine the speed of the positron after the reaction.

Solution

(a) The proton is a baryon, but neither the positron nor the photon are baryons. In this proposed decay, there is one baryon before decay and no baryons afterward. Thus, the total baryon number is not conserved. ◊

(b) If this decay did occur, with the proton initially at rest, the total linear momentum before decay would be zero. Thus, the total momentum after decay must also be zero. This means that the positron and the photon must travel in opposite directions with momenta of equal magnitudes (i.e., $p_e = p_\gamma$). Photons have zero rest energy, so the total energy of a photon is $E_\gamma = p_\gamma c$. The total energy of the positron is $E_e^2 = (p_e c)^2 + (E_{Re})^2$. Since the momenta of the positron and the photon have equal magnitudes, $(p_e c)^2 = (p_\gamma c)^2 = E_\gamma{}^2$, and the equation for the total energy of the positron becomes:

$$E_e{}^2 = E_\gamma{}^2 + (E_{Re})^2 \qquad \textbf{[Equation 1]}$$

From conservation of energy, $E_{\text{total before}} = E_{\text{total after}}$, or $E_{Rp} = E_e + E_\gamma$. Rewriting this as $E_e = E_{Rp} - E_\gamma$ and squaring gives:

$$E_e^2 = (E_{Rp})^2 - 2E_{Rp}E_\gamma + E_\gamma{}^2 \qquad \textbf{[Equation 2]}$$

Equating Equation 2 to Equation 1 gives $E_\gamma^{\,2} + (E_{Re})^2 = (E_{Rp})^2 - 2E_{Rp}E_\gamma + E_\gamma^{\,2}$, which reduces to

$$E_\gamma = \frac{(E_{Rp})^2 - (E_{Re})^2}{2E_{Rp}}$$

Since the rest energies of the proton and the positron are 938.3 MeV and 0.511 MeV respectively, the energy of the photon is

$$E_\gamma = \frac{(938.3\text{ MeV})^2 - (0.511\text{ MeV})^2}{2(938.3\text{ MeV})} = 469\text{ MeV} \qquad \lozenge$$

Therefore, the momentum of the photon is $p_\gamma = E_\gamma / c = 469\text{ MeV} / \text{c}$ ◊

(c) From conservation of energy, the total energy of the positron is

$$E_e = E_{Rp} - E_\gamma = 938.3\text{ MeV} - 469\text{ MeV} = 469\text{ MeV}$$

But, $E_e = \gamma E_{Re}$. Thus, $\frac{1}{\gamma^2} = 1 - (v/c)^2 = \left(\frac{E_{Re}}{E_e}\right)^2$

Hence, the speed of the positron after the decay is

$$v = c\sqrt{1 - (E_{Re}/E_e)^2} = c\sqrt{1 - (0.511\text{ MeV}/469\text{ MeV})^2} = 0.9999994c \qquad \lozenge$$

CHAPTER SELF-QUIZ

1. If the controlled fusion process were to be proven feasible, which of the following would be the main source of fuel?
   a. air
   b. water
   c. coal
   d. petroleum

2. When several small pieces of fissionable material are combined into one large spherical body, what happens to the overall surface area?
   a. increases
   b. decreases
   c. remains the same
   d. becomes larger or smaller depending on the element

3. Nuclear fusion involves combining nuclei of elements that fall into what category?
   a. metals
   b. non-metals
   c. high atomic numbers
   d. low atomic numbers

4. The moderator material in a nuclear fission reactor is intended to fulfill which of the following purposes?
   a. absorb neutrons
   b. create new neutrons
   c. accelerate neutrons
   d. decelerate neutrons

5. Assume that (i) the energy released per fission event of U-235 is 208 MeV and (ii) 30% of the nuclear energy released in a power plant is ultimately converted to usable electrical energy. Approximately how many fission events will occur in one second in order to provide the 2.00 kW electrical power needs of a typical home? (1 eV = 1.60×10^{-19} J)
   a. 3.00×10^{10} fissions per sec
   b. 2.00×10^{14} fissions per sec
   c. 5.00×10^{18} fissions per sec
   d. 3.00×10^{21} fissions per sec

6. Cadmium control rods used in a nuclear fission reactor are intended to serve what purpose?
   a. absorb neutrons
   b. create neutrons
   c. accelerate neutrons
   d. decelerate neutrons

7. A proton and a neutron combine in a fusion process to form a stable deuterium nucleus. Which of the following statements best applies in describing the mass of the deuterium nucleus?
   a. less than sum of proton and neutron masses
   b. equal to sum of proton and neutron masses
   c. greater than sum of proton and neutron masses
   d. exactly equal to twice proton mass

8. A fast neutron will lose more kinetic energy if it collides with
   a. a lead atom
   b. a carbon atom
   c. a hydrogen atom
   d. an electron

9. Which type of particle is most massive?
   a. leptons
   b. baryons
   c. mesons
   d. photons

10. The spin of all quarks is
    a. 0
    b. 1/2
    c. 1
    d. 1/3 or 2/3

11. Which of the following particle reactions cannot occur?
    a. $\bar{n} \rightarrow \bar{p} + e^+ + \nu$
    b. $p + p \rightarrow p + p + p + \bar{p}$
    c. $\pi^0 \rightarrow \gamma + \gamma$
    d. $p + p \rightarrow p + \pi^+$

12. Which of the following reactions violates conservation of strangeness?
    a. $\Omega^- \rightarrow 3\gamma$
    b. $\bar{p} + p \rightarrow \overline{\Lambda^0} + \Lambda^0$
    c. $\bar{n} \rightarrow e^+ + \bar{p} + \nu_e$
    d. $e^+ + e^- \rightarrow 2\gamma$

ANSWERS TO CHAPTER SELF-QUIZ

Chapter 1	1. **b**	2. **d**	3. **d**	4. **c**	5. **a**	6. **c**
	7. **b**	8. **c**	9. **a**	10. **d**	11. **a**	12. **b**
Chapter 2	1. **a**	2. **b**	3. **a**	4. **b**	5. **d**	6. **d**
	7. **d**	8. **b**	9. **a**	10. **a**	11. **a**	12. **c**
Chapter 3	1. **b**	2. **c**	3. **d**	4. **d**	5. **b**	6. **d**
	7. **d**	8. **a**	9. **d**	10. **c**	11. **a**	12. **d**
Chapter 4	1. **b**	2. **a**	3. **b**	4. **b**	5. **a**	6. **d**
	7. **a**	8. **c**	9. **b**	10. **a**	11. **c**	12. **d**
Chapter 5	1. **c**	2. **d**	3. **d**	4. **a**	5. **a**	6. **c**
	7. **c**	8. **d**	9. **c**	10. **a**	11. **b**	12. **a**
Chapter 6	1. **b**	2. **d**	3. **d**	4. **d**	5. **a**	6. **b**
	7. **a**	8. **a**	9. **b**	10. **b**	11. **d**	12. **b**
Chapter 7	1. **b**	2. **a**	3. **c**	4. **d**	5. **d**	6. **d**
	7. **b**	8. **b**	9. **a**	10. **a**	11. **b**	12. **a**
Chapter 8	1. **c**	2. **b**	3. **c**	4. **b**	5. **d**	6. **b**
	7. **b**	8. **b**	9. **d**	10. **d**	11. **c**	12. **a**
Chapter 9	1. **a.**	2. **a**	3. **d**	4. **c**	5. **d**	6. **c**
	7. **a**	8. **b**	9. **d**	10. **d**	11. **a**	12. **a**
Chapter 10	1. **c**	2. **b**	3. **d**	4. **a**	5. **a**	6. **d**
	7. **a**	8. **c**	9. **c**	10. **b**	11. **b**	12. **c**
Chapter 11	1. **a**	2. **d**	3. **c**	4. **d**	5. **d**	6. **b**
	7. **a**	8. **d**	9. **d**	10. **b**	11. **b**	12. **b**
Chapter 12	1. **d**	2. **a**	3. **c**	4. **a**	5. **d**	6. **b**
	7. **b**	8. **c**	9. **c**	10. **d**	11. **b**	12. **a**
Chapter 13	1. **c**	2. **b**	3. **c**	4. **a**	5. **b**	6. **d**
	7. **d**	8. **d**	9. **b**	10. **b**	11. **c**	12. **c**
Chapter 14	1. **d**	2. **a**	3. **b**	4. **c**	5. **a**	6. **c**
	7. **d**	8. **a**	9. **c**	10. **a**	11. **a**	12. **d**
Chapter 15	1. **d**	2. **c**	3. **d**	4. **a**	5. **b**	6. **b**
	7. **c**	8. **c**	9. **b**	10. **a**	11. **a**	12. **a**

Chapter 16	1. **d**	2. **b**	3. **b**	4. **a**	5. **d**	6. **d**
	7. **a**	8. **a**	9. **a**	10. **c**	11. **c**	12. **d**
Chapter 17	1. **c**	2. **c**	3. **b**	4. **d**	5. **c**	6. **a**
	7. **b**	8. **d**	9. **b**	10. **c**	11. **b**	12. **c**
Chapter 18	1. **d**	2. **a**	3. **a**	4. **a**	5. **b**	6. **a**
	7. **b**	8. **d**	9. **a**	10. **b**	11. **b**	12. **c**
Chapter 19	1. **b**	2. **b**	3. **c**	4. **b**	5. **b**	6. **b**
	7. **a**	8. **a**	9. **c**	10. **a**	11. **c**	12. **c**
Chapter 20	1. **a**	2. **d**	3. **b**	4. **d**	5. **d**	6. **a**
	7. **d**	8. **b**	9. **b**	10. **b**	11. **b**	12. **a**
Chapter 21	1. **c**	2. **b**	3. **a**	4. **c**	5. **d**	6. **b**
	7. **b**	8. **a**	9. **b**	10. **c**	11. **d**	12. **a**
Chapter 22	1. **a**	2. **a**	3. **c**	4. **c**	5. **b**	6. **b**
	7. **a**	8. **c**	9. **c**	10. **a**	11. **d**	12. **d**
Chapter 23	1. **b**	2. **d**	3. **d**	4. **b**	5. **c**	6. **a**
	7. **b**	8. **c**	9. **b**	10. **b**	11. **c**	12. **c**
Chapter 24	1. **c**	2. **b**	3. **c**	4. **b**	5. **d**	6. **c**
	7. **b**	8. **b**	9. **b**	10. **c**	11. **a**	12. **c**
Chapter 25	1. **c**	2. **c**	3. **b**	4. **b**	5. **a**	6. **d**
	7. **b**	8. **d**	9. **b**	10. **a**	11. **b**	12. **c**
Chapter 26	1. **b**	2. **d**	3. **b**	4. **c**	5. **d**	6. **b**
	7. **a**	8. **c**	9. **c**	10. **b**	11. **c**	12. **b**
Chapter 27	1. **d**	2. **b**	3. **a**	4. **b**	5. **d**	6. **c**
	7. **b**	8. **c**	9. **a**	10. **a**	11. **c**	12. **c**
Chapter 28	1. **d**	2. **b**	3. **b**	4. **b**	5. **d**	6. **d**
	7. **c**	8. **d**	9. **a**	10. **c**	11. **b**	12. **d**
Chapter 29	1. **b**	2. **b**	3. **b**	4. **a**	5. **c**	6. **b**
	7. **b**	8. **b**	9. **b**	10. **c**	11. **b**	12. **d**
Chapter 30	1. **b**	2. **b**	3. **d**	4. **d**	5. **b**	6. **a**
	7. **a**	8. **c**	9. **b**	10. **b**	11. **d**	12. **a**